T0178226

LONDON MATHEMATICAL SOCIETY LECTURE NOTE SERIES

Managing Editor: Professor N.J. Hitchin, Mathematical Institute,
University of Oxford, 24–29 St Giles, Oxford OX1 3LB, United Kingdom

The titles below are available from booksellers, or from Cambridge University Press at www.cambridge.org/mathematics.

London Mathematical Society Lecture Note Series: 352

Number Theory and Polynomials

JAMES McKEE
Royal Holloway, University of London

CHRIS SMYTH
University of Edinburgh

CAMBRIDGE
UNIVERSITY PRESS

CAMBRIDGE
UNIVERSITY PRESS

University Printing House, Cambridge CB2 8BS, United Kingdom

One Liberty Plaza, 20th Floor, New York, NY 10006, USA

477 Williamstown Road, Port Melbourne, VIC 3207, Australia

314-321, 3rd Floor, Plot 3, Splendor Forum, Jasola District Centre, New Delhi - 110025, India

103 Penang Road, #05-06/07, Visioncrest Commercial, Singapore 238467

Cambridge University Press is part of the University of Cambridge.

It furthers the University's mission by disseminating knowledge in the pursuit of education, learning and research at the highest international levels of excellence.

www.cambridge.org
Information on this title: www.cambridge.org/9780521714679

© Cambridge University Press 2008

First published 2008

A catalogue record for this publication is available from the British Library

ISBN 978-0-521-71467-9 Hardback

Contents

The trace problem for totally positive algebraic integers
Julián Aguirre and Juan Carlos Peral,

Mahler's measure: from Number Theory to Geometry

Explicit calculation of elliptic fibrations of $K3$-surfaces and their Belyi-maps

The merit factor problem

Barker sequences and flat polynomials

The Hansen-Mullen primitivity conjecture: completion of proof

An inequality for the multiplicity of the roots of a polynomial

Newman's inequality for increasing exponential sums

On primitive divisors of $n^2 + b$

Irreducibility and greatest common divisor algorithms for sparse polynomials

Consequences of the continuity of the monic integer transfinite diameter

Nonlinear recurrence sequences and Laurent polynomials

Conjugate algebraic numbers on conics: a survey

On polynomial ergodic averages and square functions

Preface

This volume is the proceedings of a workshop on 'Number Theory and Polynomials' held at Bristol University, 3–7 April 2006, with about fifty participants. The workshop was the first in a series of workshops sponsored by the Heilbronn Institute for Mathematical Research. During the meeting, the participants exchanged lectures, had informal discussions, and posed problems in the broad subject area defined by the theme of the workshop. Some of the articles in these proceedings are, in whole or in part, the direct outcome of questions posed and ideas raised during the workshop.

The meeting and the proceedings shared the aim of bringing together number-theorists with varied backgrounds having a common interest in problems concerning polynomials. Many of the overseas participants were supported by the Heilbronn Institute for Mathematical Research. The articles in the proceedings are not intended to be a record of the lectures at the meeting: some of the papers are more extensive than the corresponding talks, some of the talks are not represented by papers, and non-speakers were also invited to submit papers on the theme of the workshop. Expository papers and surveys were encouraged, and many of the submissions are of this form. It is hoped that this collection of papers will form a useful resource for new and old researchers in the field.

The papers in the proceedings were refereed individually to a high standard, and not all submissions were accepted. We take this opportunity to thank the small army of referees who gave of their time and expertise so willingly. We are grateful to all the participants, to the speakers, to the authors of the papers, to the London Mathematical Society, to the staff at Bristol University, and to the Heilbronn Coordinator, Cathy Badley.

James McKee
Chris Smyth
Royal Holloway and Edinburgh, April 2007.

Index of authors

Mateja Prešern 89
  Department of Mathematics, University of Glasgow, Glasgow G12 8QW,
  UK.
  mp@maths.gla.ac.uk

Igor E. Pritsker 255
  Department of Mathematics, Oklahoma State University,
  Stillwater, OK 74078, USA.
  igor@math.okstate.edu

Georges Rhin 277
  UMR CNRS 7122, Département de Mathématiques, UFR MIM,
  Université de Metz, Ile du Saulcy, 57045 Metz Cedex 01, France.
  rhin@math.univ-metz.fr

Andrzej Schinzel 155
  Institute of Mathematics, Polish Academy of Sciences, ul. Śniadeckich 8,
  00-956, Warsaw, Poland.
  A.Schinzel@impan.gov.pl

Eira Scourfield 286
  Department of Mathematics, Royal Holloway, University of London, Egham,
  Surrey TW20 0EX, UK.
  e.scourfield@rhul.ac.uk

Jean-Pierre Serre 1
  College de France, 75231 Paris, Cedex 05, France.
  serre@dmi.ens.fr

Christopher D. Sinclair 312
  Department of Mathematics, University of Colorado at Boulder Campus,
  Box 395, Boulder, Colorado 80309-0395, USA.
  sinclair@math.ubc.ca

Chris Smyth 322
  School of Mathematics and Maxwell Institute for Mathematical Sciences,
  University of Edinburgh, James Clerk Maxwell Building, King's Buildings,
  Mayfield Road, Edinburgh EH9 3JZ, UK.
  c.smyth@ed.ac.uk

Jeffrey Vaaler 312
  Department of Mathematics, The University of Texas at Austin,
  1 University Station, C1200 Austin, Texas 78712, USA.
  vaaler@math.utexas.edu

Qiang Wu 277
  Department of Mathematics, Southwest University of China,
  2 Tiansheng Road Beibei, 400715 Chongqing, China.
  qiangwu@swu.edu.cn

Workshop Participants

J. Aguirre	G. Everest	R. Pinch
I. Aliev	R. Ferguson	M. Presern
F. Amoroso	M. Filaseta	I. Pritsker
R. Baker	L. Gallardo	E. Rees
M.J. Bertin	A. Glass	G. Rhin
F. Beukers	G. Greaves	A. Schinzel
B. Birch	K. Hare	E. Scourfield
P. Borwein	J. Hilmar	P. Shiu
R. Chapman	A. Hone	C. Sinclair
C. Christopoulos	M. Huxley	M. Singh
F. Clarke	L. Kilford	C. Smyth
C. Cocks	S. Kristensen	J. Spring
S. Cohen	A. Lauder	Y. Tourigny
E. Crane	J. McKee	P. Walker
J.L. Davison	M. Mossinghoff	S. Wilson
A. Dubickas	R. Nair	Q. Wu
T. Erdélyi	N. Peatfield	M. Zieve

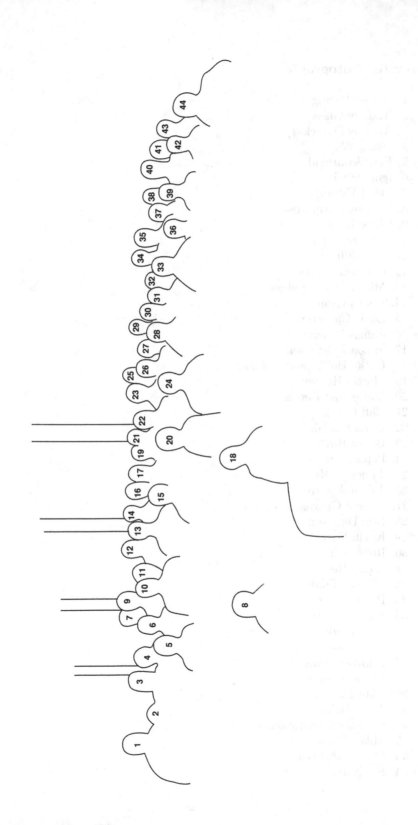

Key to photograph:

1. Joseph Spring
2. Andrew Glass
3. Arturas Dubickas
4. Qiang Wu
5. Eira Scourfield
6. Igor Pritsker
7. Lloyd Kilford
8. Francesco Amoroso
9. Chris Smyth
10. Georges Rhin
11. Jan Hilmar
12. Graham Everest
13. Michael Mossinghoff
14. Les Davison
15. Louis Gallardo
16. Tamás Erdélyi
17. Simon Kristensen
18. Cathy Hobbs and son James
19. Martin Huxley
20. Marie José Bertin
21. Cliff Cocks
22. Chris Sinclair
23. Bryan Birch
24. Peter Shiu
25. James McKee
26. Julián Aguirre
27. George Greaves
28. Ron Ferguson
29. Kevin Hare
30. Robin Chapman
31. Elmer Rees
32. Michael Filaseta
33. Peter Borwein
34. Francis Clarke
35. Frits Beukers
36. Steve Cohen
37. Andrzej Schinzel
38. Iskander Aliev
39. Alan Lauder
40. Andy Hone
41. Christos Christopoulos
42. Mike Zieve
43. Mateja Presern
44. Kit Nair

THE TRACE PROBLEM FOR TOTALLY POSITIVE ALGEBRAIC INTEGERS

JULIÁN AGUIRRE AND JUAN CARLOS PERAL,
WITH AN APPENDIX BY JEAN-PIERRE SERRE

ABSTRACT. Suppose that $P(x) = x^d + a_1 x^{d-1} + \cdots + a_d$ is a polynomial with integer coefficients, irreducible, and with all roots real and positive. In a remarkable paper of 1918, I. Schur proved that if $c < \sqrt{e}$, then there are only finitely many such polynomials for which the average of the roots, equal to $-a_1/d$, is less than c. The Schur-Siegel-Smyth trace problem asks for the largest value of c for which the same conclusion holds. In this paper we give an account of the history of the problem, the latest results, and its relations with other problems in number theory.

1. INTRODUCTION

An *algebraic number* is a complex number α that satisfies a polynomial equation

$$a_0 x^n + a_1 x^{n-1} + \cdots + a_{n-1} x + a_n = 0,$$

where the coefficients $a_k \in \mathbb{Z}$, the ring of integers. If the leading coefficient a_0 equals 1, then α is said to be an *algebraic integer*. The set of all algebraic numbers is a field, while the set of all algebraic integers, that we shall denote by $\mathbb{A}$, is a ring. Given $\alpha \in \mathbb{A}$ there is a unique monic polynomial $P \in \mathbb{Z}[x]$, the ring of all polynomials in one indeterminate with integer coefficients, such that both $P(\alpha) = 0$, and also if $Q \in \mathbb{Z}[x]$ is such that $Q(\alpha) = 0$, then P divides Q in $\mathbb{Z}[x]$. This polynomial P is irreducible, and is called the *minimal polynomial* of α; its degree is called the *degree* of α.

Let α be an algebraic integer of degree d and let $P(x) = x^d + a_1 x^{d-1} + \cdots + a_d$ be its minimal polynomial. Then P has d different roots $\alpha_1, \ldots, \alpha_d$, which are called the *conjugates* of α. We have

$$P(x) = (x - \alpha_1) \ldots (x - \alpha_d).$$

If all the conjugates of $\alpha \in \mathbb{A}$ are real, then α is said to be *totally real*; if they are all positive, then α is said to be *totally positive*. The set of all totally positive algebraic integers will be denoted by $\mathbb{A}_+$.

2000 *Mathematics Subject Classification.* 11R06.

Key words and phrases. Totally positive algebraic integers, Schur-Siegel-Smyth trace problem.

J. Aguirre supported by grant 9/UPV127.310-15969/2004 of the Universidad del País Vasco.

1

Associated with $\alpha \in \mathbb{A}$ there are several quantities of interest in algebraic number theory, among them the *trace*

$$\mathrm{Trace}(\alpha) = \sum_{k=1}^{d} \alpha_k = -a_1,$$

the *norm*

$$\mathrm{Norm}(\alpha) = \prod_{k=1}^{d} \alpha_k = (-1)^d a_d,$$

and the *discriminant*

$$\mathrm{Disc}(\alpha) = \Delta(\alpha_1, \ldots, \alpha_d),$$

where Δ is the function defined by

$$\Delta(x_1, \ldots, x_d) = \prod_{1 \le i < j \le d} (x_i - x_j)^2. \tag{1}$$

For a monic polynomial $P \in \mathbb{Z}[x]$, $\mathrm{Trace}(P)$, $\mathrm{Norm}(P)$ and $\mathrm{Disc}(P)$ are defined as $\mathrm{Trace}(\alpha)$, $\mathrm{Norm}(\alpha)$ and $\mathrm{Disc}(\alpha)$, where α is any root of P. It is clear that the trace and the norm are integers, and it turns out that so is the discriminant. The *resultant* of two polynomials $P(x) = a_0 x^n + \cdots + a_n$ of degree n and $Q(x) = b_0 x^m + \cdots + b_m$ of degree m is defined as

$$\mathrm{Resultant}(P, Q) = a_0^m \prod_{P(x)=0} Q(x),$$

that is, a_0^m times the product of the values of Q on the roots of P. If P and Q have integer coefficients, then $\mathrm{Resultant}(P, Q)$ is also an integer. Moreover, $\mathrm{Resultant}(P, Q) = 0$ if and only if P and Q have a common root. In particular, if $P, Q \in \mathbb{Z}[x]$ are coprime, then $\left| \mathrm{Resultant}(P, Q) \right| \ge 1$. All the above facts about algebraic integers can be found in any text on algebraic number theory, for instance [3].

We shall also use the family of measures defined by

$$M_p(\alpha) = \left(\frac{1}{d} \sum_{k=1}^{d} |\alpha_k|^p \right)^{1/p}, \quad p > 0.$$

If $\alpha \in \mathbb{A}_+$, then $\mathrm{Trace}(\alpha) = d \cdot M_1(\alpha)$. It follows from the inequality between the arithmetic and the geometric means that

$$M_p(\alpha) \ge \left| \mathrm{Norm}(\alpha) \right|^{1/d},$$

and thus that $M_p(\alpha) > 1$ unless $\alpha = 0$ or $\alpha = \pm 1$. The spectrum of the measure M_p is defined as the set

$$\mathcal{T}_p = \{ M_p(\alpha) : \alpha \in \mathbb{A}_+, \alpha \ne 1 \}.$$

For each positive integer n, $\theta_n = 4\cos^2(\pi/(2\,n)) \in \mathbb{A}_+$. Its minimal polynomial is a factor of

$$P_n(x) = x^{[n/2]} + \sum_{k=1}^{[n/2]}(-1)^k \frac{n}{k}\binom{n-k-1}{k-1}x^{[n/2]-k}$$

$$= x^{[n/2]} - n\,x^{[n/2]-1} + \frac{n(n-3)}{2}\,x^{[n/2]-2} - \cdots \pm a_{[n/2]},$$

where $[\,\cdot\,]$ is the integer part function, $a_{[n/2]} = \pm 2$ if n is even, and $a_{[n/2]} = \pm n$ if n is odd. Eisenstein's irreducibility criterion implies that if n is an odd prime or a power of two, then P_n is irreducible. It follows that $M_1(\theta_p) = 2\,p/(p-1)$ if p is an odd prime, and $M_1(\theta_{2^n}) = 2$ for all positive integers n. Thus 2 is a limit point of $\mathcal{T}_1$, and there is an infinite number of totally positive algebraic integers, of different degree, for which the value of M_1 is 2. The Schur-Siegel-Smyth trace problem, as stated by Peter Borwein in [4], is the following.

Schur-Siegel-Smyth Trace Problem. *Given any $\epsilon > 0$, prove that the set*

$$\{\,\alpha \in \mathbb{A}_+ : M_1(\alpha) < 2 - \epsilon\,\}$$

is finite, and if possible, find all its elements.

In other words, the problem asks whether 2 is in fact the smallest limit point of $\mathcal{T}_1$. A more general form of the problem is to find the structure of $\mathcal{T}_1$. Of course the same problem can be posed for each of the sets $\mathcal{T}_p$, $p > 0$, but our main concern will be with the case $p = 1$.

Sometimes the problem is stated for the class of totally real algebraic integers instead of for the class of totally positive algebraic integers. However both problems are equivalent, since if α is totally real, then $\alpha^2 \in \mathbb{A}_+$ and $M_p(\alpha^2) = (M_{2p}(\alpha))^2$.

The rest of the paper is divided into four sections and two appendices. Section 2 is devoted to the work of I. Schur, C.L. Siegel and C.J. Smyth on the trace problem. In Section 3 we explain the method of auxiliary functions and give the best results known. Section 4 deals with the relation between the trace problem and the integer Chebyshev problem, and Section 5 is dedicated to the special case of cyclotomic algebraic integers. Appendix A gives the best result, as far as we know, for the trace problem. Appendix B contains a letter from J.-P. Serre to C. Smyth. Appendix B contains a letter from J.-P. Serre to C. Smyth.

ACKNOWLEDGEMENTS

We wish to thank C.J. Smyth for providing us with a copy of J.-P. Serre's letters, and for several suggestions that have resulted in an improvement of the paper. We also wish to thank J.-P. Serre for kindly giving permission to publish his letter to Smyth as Appendix B, as well as for his suggestions for putting it into context.

2. Earlier results

In this section we describe the results obtained by I. Schur, C.L. Siegel and C.J. Smyth.

The work of I. Schur. The first result on the trace problem appears in I. Schur's 1918 paper [15], and is based on the following inequality for the function Δ defined in (1):

Theorem (Schur [15, Satz II]). *The maximum of $\Delta(x_1, \ldots, x_d)$ over the set of real n-tuples $(x_1, \ldots, x_d)$ such that $x_1^2 + \cdots + x_d^2 \leq 1$ is*

$$\mu_d = (d^2 - d)^{-\frac{1}{2}(d^2 - d)} \prod_{k=2}^{d} k^k.$$

It follows from Euler's summation formula that

$$\prod_{k=2}^{d} k^k = e^{\sum_{k=2}^{d} k \log k} = O\left(d^{\frac{1}{2}(d^2 + d) + \frac{1}{12}} e^{-\frac{d^2}{4}}\right),$$

and then

$$\mu_d = O\left(d^{\frac{1}{2}(3d - d^2) + \frac{1}{12}} e^{-\frac{1}{4}(2d - d^2)}\right). \tag{2}$$

Schur considers next totally real algebraic integers α of degree d, with minimal polynomial $x^d + a_1 x^{d-1} + \cdots + a_d$, and such that $\alpha_1^2 + \cdots + \alpha_d^2 \leq \gamma d$ for some $\gamma > 0$. The definition of the discriminant implies that $\mathrm{Disc}(\alpha) > 0$, and since the discriminant is an integer, we have

$$1 \leq \mathrm{Disc}(\alpha) \leq (\gamma d)^{\frac{1}{2}(d^2 - d)} \mu_d = O\left(e^{\frac{d}{4}} d^{d + \frac{1}{12}} \left(e^{-\frac{1}{2}} \gamma\right)^{\frac{1}{2}(d^2 - d)}\right). \tag{3}$$

If $\gamma < \sqrt{e} = 1.648721\ldots$, then the right hand side of (3) converges to zero as d goes to infinity. Since on the other hand $\mathrm{Disc}(\alpha) \geq 1$, there exists a positive integer d_0 such that $d \leq d_0$. Moreover, $|\alpha_k| \leq \sqrt{\gamma d}$ for $1 \leq k \leq d$. Thus

$$|a_k| = \sum_{1 \leq i_1 < i_2 < \cdots < i_k \leq d} \alpha_{i_1} \alpha_{i_2} \ldots \alpha_{i_k} \leq \binom{d}{k} (\gamma d)^{\frac{k}{2}}.$$

This concludes the proof of the following:

Theorem (Schur [15, Satz VIII]). *Let γ be a positive constant such that $\gamma < \sqrt{e}$. Then the number of totally real algebraic integers α such that*

$$a_1^2 - a_2 = \frac{\alpha_1^2 + \cdots + \alpha_d^2}{d} \leq \gamma$$

is finite.

Finally, using the observation made in the introduction, this theorem is restated as:

Theorem (Schur [15, Satz XI]). *Let γ be a positive constant such that $\gamma < \sqrt{e}$. Then the number of totally positive algebraic integers α such that*

$$\frac{\alpha_1 + \cdots + \alpha_d}{d} \leq \gamma$$

is finite.

The work of C.L. Siegel. The next advance is due to C.L. Siegel in his 1945 paper [16]. The first result in the paper is an improvement of the classical inequality between the arithmetic and the geometric means involving the function Δ. Given an integer $d \geq 2$ define a polynomial P and a rational function Q by

$$P(t) = \frac{1}{d!} \prod_{k=0}^{d-2} \left(\frac{t+k}{d-k}\right)^{d-k-1}, \qquad Q(t) = \prod_{k=1}^{d-1} \left(1 + \frac{d-k}{t+k-1}\right).$$

Theorem (Siegel [16, Theorem I]). *Let $x_1, \ldots, x_d$ be positive real numbers such that $\Delta(x_1, \ldots, x_d) \neq 0$, and let μ be the unique positive solution of the algebraic equation*

$$P(\mu) = \frac{(x_1 \ldots x_d)^{d-1}}{\Delta(x_1, \ldots, x_d)};$$

then

$$\left(\frac{x_1 + \cdots + x_d}{d}\right)^d \geq Q(\mu) \, x_1 \ldots x_d. \tag{4}$$

The polynomial P has positive coefficients and $P(0) = 0$, so that μ is well defined. Since moreover $Q(\mu) > 1$, (4) is in fact an improvement of the arithmetic-geometric inequality. If $\alpha \in \mathbb{A}_+$, (4) can be rewritten as

$$\left(M_1(\alpha)\right)^d \geq Q(\mu) \operatorname{Norm}(\alpha),$$

where μ is the unique positive solution of $P(\mu) = \operatorname{Norm}(\alpha)^{d-1} / \operatorname{Disc}(\alpha)$. Since $\operatorname{Norm}(\alpha)$ is positive, it follows that

$$\left(M_1(\alpha)\right)^{d(d-1)} \geq \operatorname{Disc}(\alpha) P(\mu) Q^{d-1}(\mu). \tag{5}$$

This inequality is the starting point for the proof of the following two theorems dealing with the trace problem.

Theorem (Siegel [16, Theorem II]). *Let ϑ be the positive root of the transcendental equation*[†]

$$(1 + \vartheta) \log(1 + \vartheta^{-1}) + \frac{\log \vartheta}{1 + \vartheta} = 1,$$

and let $\lambda_0 = e(1 + \vartheta^{-1})^{-\vartheta} = 1.7336\ldots$. Then for any $\lambda \in (1, \lambda_0)$ the set

$$\{\alpha \in \mathbb{A}_+ : M_1(\alpha) < \lambda\}$$

is finite.

[†]There is a misprint in the paper. The '$-$' on the left hand side should be a '$+$'.

Theorem (Siegel [16, Theorem III]). *The only $\alpha \in \mathbb{A}_+$ with $M_1(\alpha) \le 3/2$ are $\alpha = 1$ and $\alpha = (3 \pm \sqrt{5})/2$, the roots of the polynomial $x^2 - 3x + 1$.*

These theorems imply in particular that $3/2$ is the smallest point in $\mathcal{T}_1$ and that it is isolated. Siegel finds remarkable that they imply a refinement of Minkowski's inequality between the discriminant and the degree of totally real algebraic fields of sufficiently large degree.

The work of C.J. Smyth. Stimulated by McAuley's Master Thesis [12], C.J. Smyth carries out in his 1984 paper [19] a detailed analysis, both theoretical and numerical, of the structure of the sets $\mathcal{T}_p$ for $p > 0$ (defined in terms of totally real instead of totally positive algebraic integers). His main result for the case $p = 1$, translated to the language we have been using, is as follows.

Theorem (Smyth [19, Theorem 1]).

(1) *The smallest three elements of $\mathcal{T}_1$ are isolated, and are the only elements of $\mathcal{T}_1$ in the interval $(1, 1.7719)$:*

$$(1, 1.7719) \cap \mathcal{T}_1 = \left\{ \frac{3}{2}, \frac{5}{3}, \frac{7}{4} \right\}.$$

These values are $M_1(\alpha)$, where $\alpha \in \mathbb{A}_+$ is a root of one of the polynomials $x^2 - 3x + 1$, $x^3 - 5x^2 + 6x - 1$, $x^4 - 7x^3 + 13x^3 - 7x + 1$, $x^4 - 7x^3 + 14x^3 - 8x + 1$.
(2) *The set $\mathcal{T}_1$ is dense in $[2, +\infty)$.*

From this theorem we see that the structure of $\mathcal{T}_1$ is undetermined only in the interval $(1.7719, 2)$. Let us remark again that Smyth proves similar results for all $p > 0$.

The ideas for proving the above theorem had already been developed by Smyth in [17, 18] to treat the corresponding problem for the measure

$$\Omega(\alpha) = \left(\prod_{k=1}^{d} \max(1, |\alpha_k|) \right)^{1/d}.$$

The methods for proving (1) and (2) are quite different. We will explain with some detail in the next section the method used to prove (1), known as the method of auxiliary functions, which can be applied to a large class of problems in the theory of polynomials with integer coefficients. Whereas Schur's and Siegel's results were based on inequalities for the discriminant of an algebraic integer, the method of auxiliary functions exploits an inequality for the resultant of two polynomials, one of them being the minimal polynomial of an algebraic integer.

3. THE METHOD OF AUXILIARY FUNCTIONS

Suppose that somehow we are able to find a polynomial $Q \in \mathbb{Z}[x]$ and real constants $y > 0$, c such that

$$x - y \log |Q(x)| \geq c \quad \text{for all } x > 0. \tag{6}$$

If $\alpha \in \mathbb{A}_+$ has conjugates $\alpha_1, \ldots, \alpha_d$, then

$$\alpha_k - y \log |Q(\alpha_k)| \geq c, \quad 1 \leq k \leq d.$$

Adding these inequalities and dividing by d we get

$$M_1(\alpha) \geq c + y \log \left| \prod_{k=1}^{d} Q(\alpha_k) \right| = c + y \log |\operatorname{Resultant}(P, Q)|,$$

where P is the minimal polynomial of α. If $Q(\alpha) \neq 0$, then

$$|\operatorname{Resultant}(P, Q)| \geq 1$$

and $M_1(\alpha) \geq c$. Thus inequality (6) implies that

$$(1, c) \cap \mathcal{T}_1 \subset \{ M_1(\alpha) : Q(\alpha) = 0 \},$$

and in particular that $(1, c) \cap \mathcal{T}_1$ is finite. Define the constant $\mathcal{K}$ as

$$\mathcal{K} = \sup_{Q \in \mathbb{Z}[x], Q \neq 0, y > 0} \left\{ \inf_{x > 0} \left(x - y \log |Q(x)| \right) \right\}. \tag{7}$$

Reasoning as above, it is easy to see that

$$(1, c) \cap \mathcal{T}_1 \text{ is finite for all } c < \mathcal{K}.$$

What Smyth did to prove the first part of his theorem is to compute explicitly a polynomial $Q \in \mathbb{Z}[x]$ and a constant $y > 0$ such that

$$x - y \log |Q(x)| \geq 1.7719$$

for all $x > 0$, proving that $\mathcal{K} > 1.7719$.

How does one find such Q and a? In practice, one chooses N irreducible polynomials $Q_i \in \mathbb{Z}[x]$ and solves the optimization problem

$$\sup \left\{ \min_{x > 0} \left(x - \sum_{k=1}^{N} c_k \log |Q_k(x)| \right) \right\}, \tag{8}$$

where the supremum is taken over all N-tuples $(c_1, \ldots, c_N) \in \mathbb{R}^N$ with $c_k > 0$ for $1 \leq k \leq N$. The function $x - \sum_{k=1}^{N} c_k \log |Q_k(x)|$ is called an *auxiliary function*. The method can be adapted to study other measures. For instance, if we change x to x^p, $p > 0$, then we obtain results about the sets $\mathcal{T}_p$; changing x to $\max(0, \log x)$ will provide information on the spectrum of the Mahler measure. It should be noted that in Smyth's original approach, (8) appears as the dual of another optimization problem on the set of all probability measures on $(0, +\infty)$.

When we apply the method of the auxiliary functions we are confronted with two different problems:

(1) Find appropriate polynomials Q_k.
(2) Once the polynomials have been chosen, find the values of the coefficients c_k that maximize (8).

The polynomials. To apply the method of auxiliary functions, one needs to choose the polynomials Q_k in (8). There are heuristic rules to select them:

- They should have positive roots.
- They should have *small* coefficients.
- They should have *small* trace. The reason for this is the following: to prove that $M_1(\alpha) < c$, all polynomials of degree d whose trace is smaller than $c \cdot d$ must appear in (8).

For small positive integers d and T, it is possible to give a complete list of all monic irreducible polynomials with integer coefficients, positive roots, degree d and trace T. In [20] a complete list of such $Q \in \mathbb{Z}[x]$ with

$$\mathrm{Trace}(Q) - \deg(Q) \le 6$$

is given, and all $Q \in \mathbb{Z}[x]$ with $\deg(Q) = 10$ and $\mathrm{Trace}(Q) = 18$ are listed in [13]. Table 1 gives for each $1 \le d \le 10$ the smallest possible trace T of a totally positive algebraic integer of degree d, the corresponding value of M_1, and the number N_d of monic irreducible polynomials with positive roots having such degree and trace. Some of them appear in Table 2.

TABLE 1. Number of polynomials of a given degree and trace as small as possible

d	T	M_1	N_d
1	1	1.000	1
2	3	1.500	1
3	5	1.660	1
4	7	1.750	2
5	9	1.800	4
6	11	1.833	11
7	13	1.857	40
8	15	1.875	151
9	17	1.889	686
10	18	1.800	3

Smyth's theorem is used by O. Debarre in [7] to prove a result on curves on simple abelian varieties. He also conjectures that if $\alpha \in \mathbb{A}_+$ is of degree d, then $\mathrm{Trace}(\alpha) \ge 2d - 1$, and if equality holds, then $\mathrm{Norm}(\alpha) = 1$. The first part of the conjecture is false, since by Corollary 3 in [14], for infinitely many d there exists $\alpha \in \mathbb{A}_+$ with

$$\deg(\alpha) = d \quad \text{and} \quad \text{Trace}(d) \le 2\,d - \frac{1}{4}\,\frac{\log\log d}{\log\log\log d}\,.$$

The smallest d such that there exists $\alpha \in \mathbb{A}_+$ of degree d and

$$\text{Trace}(\alpha) < 2\,d - 1$$

is $d = 10$. The second part of the conjecture does not hold either, as is shown by the polynomial

$$x^8 - 15\,x^7 + 89\,x^6 - 268\,x^5 + 438\,x^4 - 385\,x^3 + 169\,x^2 - 32\,x + 2.$$

The optimization algorithm. Once the polynomials Q_k have been chosen, it remains to find the coefficients c_k that maximize (8). This can be done by semi-infinite linear programing, as in [19], or by a variant of the second Remes algorithm, as in [1].

Latest results. New polynomials, better optimization algorithms and more powerful computers have produced a series of improvements in the trace problem:

- $\mathcal{K} > 1.7735$ (1997, Flammang, Grandcolas & Rhin [8]),
- $\mathcal{K} > 1.7783$ (2004, McKee & Smyth [13]),
- $\mathcal{K} > 1.7800$ (2006, Aguirre, Bilbao & Peral [1]),
- $\mathcal{K} > 1.7822$ (2006, Flammang, personal communication to C. Smyth),
- $\mathcal{K} > 1.7836$ (2006, Aguirre & Peral [2]).

The best current result as far as we know* is $\mathcal{K} > 1.784109$ and is due to the authors. It is included in Appendix A.

Considering for $\xi > 0$ the optimization problem

$$\sup\left\{\min_{x>\xi}\left(x - \sum_{k=1}^{N} c_k \log |Q_k(x)|\right)\right\} \tag{9}$$

instead of (8), it is possible to obtain a different type of inequality for the trace of $\alpha \in \mathbb{A}_+$ with conjugates $\alpha_1 < \alpha_2 < \cdots < \alpha_d$:

- $M_1(\alpha) > 1.60 + \alpha_1$ (1997, Flammang, Rhin & Smyth [9]),
- $M_1(\alpha) > 1.66 + \alpha_1$ (2006, Aguirre, Bilbao & Peral [1]).

These inequalities hold for all $\alpha \in \mathbb{A}_+$ except for 26 explicit exceptions and their integer translates.

---

*Added in proof: V. Flammang has proved recently that $\mathcal{K} > 1.78702$.

The limits of the method. How far is it possible to go with the method of auxiliary functions? Is it possible to solve the trace problem? Unfortunately the answer is no.

C. Smyth proved in [21] that $\mathcal{K} < 2 - 10^{-41}$. This was then improved by J.-P. Serre in a private letter (24 February 1998) to Smyth, whose contents appear as Appendix B. Serre proves that if Q is a polynomial with *real* coefficients with leading and constant coefficients of modulus at least 1, and if $y > 0$, $z > 0$, $c \in \mathbb{R}$ are constants such that

$$x - z \log x - y \log |Q(x)| \geq c \qquad \text{for all } x > 0, \tag{10}$$

then $c \leq 1.898302\ldots$. It follows that

$$\mathcal{K} \leq 1.898302\ldots . \tag{11}$$

Since Serre's result is for polynomials with real coefficients, it is possible that the inequality is in fact strict. We see from (11) that it is impossible to show using the method of auxiliary functions that $(1, c) \cap \mathcal{T}_1$ is finite for any $c \geq 1.898302\ldots$. Moreover, to prove for instance that $(1, 1.89) \cap \mathcal{T}_1$ is finite, the sum in (8) should include all the polynomials referred to in Table 1. We believe that the computational problem is intractable, and will remain so for a long time.

In a subsequent letter (31 March 1998) Serre proved that the upper bound for c is optimal. For any $c < 1.898302\ldots$ there exist constants $y > 0$, $z > 0$, and a polynomial

$$Q(x) = \prod_{k=1}^{n} (x - \lambda_k), \quad \lambda_k \in [a, b], \quad \prod_{k=1}^{n} \lambda_k \geq 1,$$

where $a = 0.08735\ldots$, $b = 4.41107\ldots$, such that (10) holds.

For optimal c, the corresponding values of y and z are $y = 1.628472\ldots$ and $z = 0.620741\ldots$. The extremal situation is described by a measure, giving the limiting density function of the zeroes of a sequence of real polynomials Q_n whose degrees go to infinity with n. This measure has support on the interval $[a, b]$, and is obtained by projecting the uniform probability measure on the unit circle to $[a, b]$. The result is that, with this optimal choice,

$$f_{\text{opt}}(x) = x - z \log x - \int_{a}^{b} \log |x - y| \frac{\sqrt{(y - a)(b - y)}}{\pi y} \, dy.$$

This function is constant on $[a, b]$, equal to $1.898302\ldots$, and increases to infinity both as $x \to 0$ and $x \to \infty$. See Figure 1, where f_{opt} is compared with the auxiliary function of Appendix A.

Serre's interest in the problem comes from its connection with the counting of points on curves over finite fields. For a curve $\mathcal{C}$ of genus g over a finite field $\mathbb{F}_q$, the number of points of $\mathcal{C}$ in $\mathbb{F}_q$ is given by Weil as $q + 1 - \text{trace}(P)$, where P is a monic integer polynomial of degree g having all its roots in the interval $[-2\sqrt{q}, 2\sqrt{q}]$. The roots of P are of the form $\pi + \bar{\pi}$, where the π, of

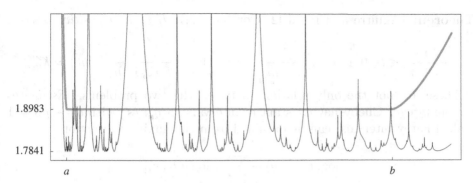

FIGURE 1. The optimal function f_{opt} and the auxiliary function of Appendix A.

modulus $\sqrt{q}$, are the eigenvalues of the Frobenius map on the Jacobian of $\mathcal{C}$. Thus by a suitable integer translation, the number of points can be expressed in terms of a sum of traces of totally positive algebraic integers whose degrees sum up to g. See [11] for details.

4. THE TRACE PROBLEM AND THE INTEGER CHEBYSHEV PROBLEM

Given $N \in \mathbb{N}$ and a compact interval $[a, b] \subset \mathbb{R}$ (or in general, a compact subset of the complex numbers), the integer Chebyshev problem asks for the polynomial of degree N with integer coefficients of minimal uniform norm on $[a, b]$. Let

$$t_{\mathbb{Z}}(a, b) = \inf_{N \in \mathbb{N}} \left(\min \left\{ \sup_{a \leq x \leq b} |P(x)|^{\frac{1}{\deg(P)}} : P \in \mathbb{Z}[x], P \neq 0, \deg(P) \leq N \right\} \right).$$

The constant $t_{\mathbb{Z}}(a, b)$ is known as the *integer Chebyshev constant* of the interval $[a, b]$. If $b - a \geq 4$, then $t_{\mathbb{Z}}(a, b) = (b - a)/4$, but no exact value of the integer Chebyshev constant is known for any interval of length less than 4. The connection between the integer Chebyshev problem and the trace problem for totally positive algebraic integers is explained in [1, 2, 5, 9]. More precisely, we have the following result.

Theorem (Borwein and Erdélyi [5, Proposition 4.1]). *Suppose that m is a positive integer and that*

$$1 < \delta < \frac{1}{t_{\mathbb{Z}}(0, 1/m)} - m.$$

Then the set $(1, \delta) \cap \mathcal{T}_1$ is finite.

We see that it is possible to obtain results about the trace problem from estimates of the integer Chebyshev constant of intervals $[0, 1/m]$. As the following theorem shows, the constant $\mathcal{K}$ is also related to the integer Chebyshev problem.

Theorem (Aguirre and Peral [2, Corollary 2]). *If m is a positive integer, then*

$$\frac{1}{\mathcal{K}+m} < t_{\mathbb{Z}}(0,1/m) < \frac{1}{1+m} \quad \text{and} \quad \lim_{m \to \infty} \left(\frac{1}{t_{\mathbb{Z}}(0,1/m)} - m \right) = \mathcal{K}.$$

These are not the only relations between the two problems. Estimates for the integer Chebyshev constant of intervals $[p/q, r/s]$ with $qr - ps = 1$ (called Farey intervals) can be obtained from the auxiliary function

$$\log(x + s/q) - \sum_{k=1}^{N} c_k \log |Q_k(x)|$$

(see [2]). The same algorithms and computer code developed for the trace problem can then be used for the integer Chebyshev problem.

5. CYCLOTOMIC INTEGERS

A *cyclotomic integer* is an algebraic integer α in a cyclotomic field, that is, there exists a root of unity ζ such that $\alpha \in \mathbb{Q}(\zeta)$, the smallest field containing the rationals and ζ. If α is a cyclotomic integer, then $|\alpha|^2 \in \mathbb{A}_+$, and since the Galois group $\mathcal{G}$ of $\mathbb{Q}(\zeta)$ is abelian, $|\alpha^\sigma|^2 = (|\alpha|^2)^\sigma$ for all $\sigma \in \mathcal{G}$, where α^σ represents the action of σ on α. It follows that

$$M_1(|\alpha|^2) = \frac{1}{|\mathcal{G}|} \sum_{\sigma \in \mathcal{G}} |\alpha^\sigma|^2 = \frac{1}{|\mathcal{G}|} \sum_{\sigma \in \mathcal{G}} (|\alpha|^2)^\sigma, \tag{12}$$

where $|\mathcal{G}|$ is the order of $\mathcal{G}$.

This property has been used by J.W.S. Cassels in [6] to settle a conjecture of R.M. Robinson about sums of roots of unity. As part of the proof, Cassels proves the following result.

Theorem (Cassels [6, Lemma 3]). *Suppose that α is a cyclotomic integer which is neither a root of unity nor representable as the sum of two roots of unity. Then*

$$M_1(|\alpha|^2) \geq 2.$$

Using this theorem it is possible to solve the trace problem for cyclotomic integers, proving that 2 is the smallest limit point of

$$\{ M_1(|\alpha|^2) : \alpha \text{ is a cyclotomic integer} \}.$$

Identity (12) has been used jointly with lower estimates on the trace of totally real positive algebraic integers to obtain results on the values of irreducible characters of finite groups. If χ is an irreducible character of a finite group G, then $\chi(g)$ is a cyclotomic integer for all $g \in G$. Problem (3.15) in the book [10] asks for the proof of the following result, atributed to Thompson: if χ is an irreducible character of the a finite group G, then $\chi(g)$ is either zero or a root of unity for more than a third of the elements $g \in G$.

TABLE 2. Values of c_k and Q_k in (14)

k	c_k	Q_k
1	0.5454395499	x
2	0.5063717310	$1-x$
3	0.0794899337	$2-x$
4	0.1888620827	$1-3x+x^2$
5	0.0214329965	$1-4x+x^2$
6	0.0112634625	$2-4x+x^2$
7	0.0809965525	$1-6x+5x^2-x^3$
8	0.0061468586	$1-8x+6x^2-x^3$
9	0.0052343340	$1-9x+6x^2-x^3$
10	0.0329631947	$1-7x+13x^2-7x^3+x^4$
11	0.0293633546	$1-8x+14x^2-7x^3+x^4$
12	0.0123766117	$1-11x+29x^2-26x^3+9x^4-x^5$
13	0.0127623387	$1-12x+31x^2-27x^3+9x^4-x^5$
14	0.0064234181	$1-13x+32x^2-27x^3+9x^4-x^5$
15	0.0131772288	$1-15x+35x^2-28x^3+9x^4-x^5$
16	0.0017897254	$1-14x+51x^2-72x^3+43x^4-11x^5+x^6$
17	0.0026454115	$1-15x+59x^2-78x^3+44x^4-11x^5+x^6$
18	0.0005943490	$1-18x+63x^2-79x^3+44x^4-11x^5+x^6$
19	0.0024309470	$1-15x+71x^2-144x^3+136x^4-62x^5+13x^6-x^7$
20	0.0043077189	$1-16x+78x^2-157x^3+143x^4-63x^5+13x^6-x^7$
21	0.0002265137	$1-17x+81x^2-158x^3+143x^4-63x^5+13x^6-x^7$
22	0.0012139399	$1-17x+82x^2-159x^3+143x^4-63x^5+13x^6-x^7$
23	0.0015863832	$1-18x+89x^2-172x^3+150x^4-64x^5+13x^6-x^7$
24	0.0014865771	$1-15x+83x^2-220x^3+303x^4-220x^5+83x^6-15x^7+x^8$
25	0.0026334873	$1-24x+194x^2-743x^3+1526x^4-1798x^5+1265x^6$ $-537x^7+134x^8-18x^9+x^{10}$
26	0.0048437178	$1-24x+200x^2-766x^3+1560x^4-1822x^5+1273x^6$ $-538x^7+134x^8-18x^9+x^{10}$
27	0.0032411370	$1-24x+206x^2-813x^3+1662x^4-1920x^5+1320x^6$ $-549x^7+135x^8-18x^9+x^{10}$
28	0.0010061264	$1-26x+265x^2-1388x^3+4177x^4-7677x^5+8944x^6$ $-6752x^7+3322x^8-1050x^9+204x^{10}-22x^{11}+x^{12}$
29	0.0032149270	$1-29x+314x^2-1676x^3+5007x^4-9012x^5+10213x^6$ $-7474x^7+3561x^8-1092x^9+207x^{10}-22x^{11}+x^{12}$
30	0.0012158245	$1-32x+361x^2-1941x^3+5726x^4-10061x^5+11086x^6$ $-7897x^7+3678x^8-1109x^9+208x^{10}-22x^{11}+x^{12}$
31	0.0011466823	$1-33x+377x^2-2009x^3+5846x^4-10166x^5+11134x^6$ $-7908x^7+3679x^8-1109x^9+208x^{10}-22x^{11}+x^{12}$

Appendix A. The best lower estimate of the constant $\mathcal{K}$

Theorem (Aguirre & Peral, 2007). *We have*

$$\mathcal{K} > 1.784109 . \tag{13}$$

Proof. It is enough to check the inequality

$$x - \sum_{k=1}^{31} c_k \log |Q_k(x)| > 1.784109 \quad \text{for all } x > 0, \tag{14}$$

where the coefficients c_k and the polynomials Q_k are given in Table 2. $\square$

Appendix B. The letter from J.-P. Serre to C.J. Smyth

Notation. P is a polynomial with real coefficients, satisfying conditions:
  a) the highest coefficient of P has modulus ≥ 1;
  b) the lowest coefficient of P has modulus ≥ 1.
p is the degree of P; $\gamma = v/p$, where x^v is the highest power of x dividing $P(x)$. $R(x) = x^{-v} P(x)$ and $Q(x) = x^p P(1/x)$.

We are interested in
$$x \geq c + y \log |P(x)| \quad \text{for all} \quad x > 0,$$

and we write y as t/p. (The point of these conventions is that what is really important is $|P|^{1/\deg(P)}$.)
  The basic inequality is

$$(1) \quad x > c + \frac{t}{p} \log |P(x)| \quad \text{for all} \quad x > 0,$$

and one wants to show that this implies $c < 1.9$ (more precisely: $c < 1.8983021$).
  Of course, one can rewrite (1) as:

$$(2) \quad x \geq c + t\gamma \log(x) + (1 - \gamma)t \frac{1}{q} \log |R(x)|,$$

with $q = \deg(R) = p(1 - \gamma)$.
  The strategy will be to prove two inequalities for c, namely:

$$(3) \quad c/t \leq 1 - \log(t/2) - \frac{1+\gamma}{2} \log(1 + \gamma) - \frac{1-\gamma}{2} \log(1 - \gamma)$$

and

$$(4) \quad c/t \leq \gamma + \frac{1+\gamma}{2} \log(1 + \gamma) - \frac{1-\gamma}{2} \log(1 - \gamma) - \gamma \log(2t\gamma^2).$$

For any γ, t call $a(\gamma, t) = $ right side of (3), and $b(\gamma, t) = $ right side of (4). Define $c(\gamma, t)$ by

$$(5) \quad c(\gamma, t) = t \cdot \inf(a(\gamma, t), b(\gamma, t)),$$

so that we may sum up (3) and (4) as:

(6) $c \le c(\gamma, t)$.

The last step will be to prove (with computer help):

(7)   The upper bound of $c(\gamma, t)$, for $t > 0$, $0 < \gamma < 1$, is < 1.8983021.

This will complete the proof.
Let me give some details.

1. Integration formulae

I shall consider intervals (a, b), with $0 < a < b$. I call L the capacity of such an interval, i.e. $(b - a)/4$. On (a, b) I put the measure called 'equilibrium distribution' in capacity theory. If one views the interval as the projection of a circle in the plane, it is the image of the rotation invariant measure on the circle, normalized so that its total mass is equal to 1. If $f(x)$ is an integrable function on the interval, I shall write $\int_a^b f(x)$, or just $\int f(x)$, for its integral with respect to that measure. The few integration formulae I shall need are:

(8) $\displaystyle\int_a^b \log|x - \lambda| \ge \log L$ for every $\lambda \in \mathbb{C}$.

(This follows from the very definition of the equilibrium measure.)

As a consequence, if K is a polynomial of degree k, with highest coefficient of modulus ≥ 1, we have:

(9) $\displaystyle\int_a^b \frac{1}{k} \log|K(x)| \ge \log L$.

The inequality (8) is in fact an equality when λ belongs to the interval. I shall need an extra case, namely $\lambda = 0$:

(10) $\displaystyle\int_a^b \log x = \log L + \log\left(1 + u/2 + \sqrt{u + u^2/4}\right)$,

where $u = a/L$. (Remember that I am assuming $a > 0$.)
Moreover:

(11) $\displaystyle\int_a^b x = (a + b)/2$.

(Clear by using the symmetry around the middle of the interval.)

(12) $\displaystyle\int_a^b \frac{1}{x} = 1/\sqrt{ab}$.

(This follows by a change of variables from $\frac{1}{2\pi}\int_0^{2\pi} \frac{d\varphi}{C + \cos\varphi} = \frac{1}{\sqrt{C^2 - 1}}$, $C > 1$.)

2. Proof of the first inequality

One integrates (2) over any interval (a, b) with $0 < a < b$. If one defines u by $u = a/L$ and z by $z = 1 + u/2 + \sqrt{(u + u^2/4)}$, one gets:

(13) $(a + b)/2 \geq c + t\gamma(\log L + \log z) + t(1 - \gamma)\log L$,

i.e.

(14) $c/t \leq (a + b)/2t - \log L - \gamma \log z$.

It is convenient to write everything in terms of $L > 0$ and $z > 1$: we have $u + 2 = z + z^{-1}$, $a = L(z + z^{-1} - 2)$, $(a + b)/2 = (z + z^{-1})L$, and (14) becomes:

(15) $c \leq L(z + z^{-1}) - t \log L - t\gamma \log z$ (for every $L > 0$, $z > 1$).

One now optimizes this inequality. It is not hard to see that the optimal choice $\big($for a given pair $(\gamma, t)\big)$ is $z = \sqrt{(1 + \gamma)/(1 - \gamma)}$, $L = \frac{t}{2}\sqrt{1 - \gamma^2}$, which corresponds to:

$$a = t\left(1 - \sqrt{1 - \gamma^2}\right), \quad b = t\left(1 + \sqrt{1 - \gamma^2}\right).$$

Then (15) gives:

(16) $c \leq t - t \cdot \log\left(\dfrac{t}{2}\sqrt{1 - \gamma^2}\right) - t\gamma \cdot \log\sqrt{(1 + \gamma)/(1 - \gamma)}$,

which is equivalent to (3).

2. Proof of the second inequality

Recall that $P(x) = x^{\gamma p}R(x)$, where R is a polynomial with non-zero constant term. Call Q the reciprocal polynomial. We have

(17) $P(1/x) = x^{-p}Q(x)$,

and the highest coefficient of Q has modulus ≥ 1 (this is where hypothesis (b) is used). By writing $1/x$ instead of x in (1), we get:

(18) $1/x \geq c - t \log x + \dfrac{t}{p} \log|Q(x)|$,

which I prefer to write as:

(19) $1/x \geq c - t \log x + t(1 - \gamma)\dfrac{1}{q} \log|Q(x)|$,

where $q = (1 - \gamma)p = \deg R = \deg Q$.

As before, we integrate (19) over an interval (a, b), and we make the same change of variables: $u = a/L$, $z = 1 + u/2 + \sqrt{(u + u^2/4)}$. The integral of $1/x$ is $1/\sqrt{ab} = 1/L(z - z^{-1})$, and we obtain:

(20) $c \leq 1/L(z - z^{-1}) + t\gamma \log L + t \log z$.

If we make the optimal choice of (z, L), which is:

$$z = \sqrt{(1+\gamma)/(1-\gamma)}, \quad L = 1/t\gamma(z - z^{-1}) = \sqrt{1 - \gamma^2}/2t\gamma^2,$$

we get:

(21) $\quad c/t \leq \gamma + \gamma \log\left(\sqrt{1 - \gamma^2}/2t\gamma^2\right) + \log\left(\sqrt{(1+\gamma)/(1-\gamma)}\right),$

which is equivalent to (4) .

4. Upper bound of $c(\gamma, t)$

It remains to determine the maximum of the function $c(\gamma, t)$ defined by formula (5):

$$c(\gamma, t) = t \cdot \inf\big(a(\gamma, t), b(\gamma, t)\big),$$

where

$$a(\gamma, t) = 1 - \log(t/2) - \frac{1+\gamma}{2}\log(1+\gamma) - \frac{1-\gamma}{2}\log(1-\gamma)$$

and

$$b(\gamma, t) = \gamma + \frac{1+\gamma}{2}\log(1+\gamma) - \frac{1-\gamma}{2}\log(1-\gamma) - \gamma\log(2t\gamma^2).$$

Now t and γ are free variables, subject only to $t > 0$ and also $0 < \gamma < 1$ (one could also accept the limiting cases $\gamma = 0, 1$: they give a poor value of c, as may be expected). Since we can check that $1.898\ldots$ is a value of c, this already gives a small variation range for t e.g. $1 \leq t \leq 3$. (Indeed, one can check that $a(\gamma, t) < 1 - \log(t/2)$ hence $c(\gamma, t) \leq t - t\log(t/2)$. One is led to check that $t - t\log(t/2) > 1.89$ implies $1 \leq t \leq 3$, which is true, with a lot of room to spare.) This shows that the maximum of $c(\gamma, t)$ occurs at some *interior point* of the range.

The next remark is that neither $t\,a(\gamma, t)$, nor $t\,b(\gamma, t)$, have a maximum *inside* the range. For $t\,a$, this is very simple, since the partial derivative relative to γ is easily shown to be everywhere < 0. For $t\,b(\gamma, t)$, one needs to compute both partial derivatives, and check that they cannot be both zero.

When this is done, it is obvious that the maximal value of $c(\gamma, t)$ can be attained *only at a point* (γ, t) *where* $a(\gamma, t) = b(\gamma, t)$.

If one writes down the equation $a(\gamma, t) = b(\gamma, t)$, one finds a relation which gives t as a function of γ. More precisely, one finds:

(22) $\quad (1 - \gamma)\log(t/2) = e(\gamma),$

where:

(23) $\quad e(\gamma) = 1 - \gamma + \gamma\log(4\gamma^2) - (1+\gamma)\log(1+\gamma).$

Hence:

(24) $\quad t = 2 \cdot \exp(e(\gamma)/(1-\gamma)).$

Plugging this value of t inside the definition of $c(\gamma, t)$ gives a function:

$$(25) \quad C(\gamma) = c(\gamma, t) = c\big(\gamma, 2 \cdot \exp\big(2\exp\big(e(\gamma)/(1-\gamma)\big)\big)\big).$$

The last step is to compute the maximum of $C(\gamma)$ and show that it is approximately $1.89830200\dots$. I do not have an actual proof for that. What I did was first to program $C(\gamma)$ on my SHARP pocket calculator, and made a table of values, for which the approximate value $1.898\dots$ was pretty clear. Then I used the MAPLE system on my portable computer to draw up graphs of C and its derivative C' (and also C''', for good measure). The graph of C' (and the values of C'' which seem to be everywhere < -1) gave me the optimal value of γ as being around* 0.27598, with a corresponding t close to $2.2492\dots$ and C close to $1.898302009\dots$. I could easily get more precise estimates* by using the PARI program, but there is not much point in doing so. A more serious problem would be to transform this last part into an actual proof. You must know what kind of methods are best suited for such things. The trouble with $C(\gamma)$ is that, although it is an 'elementary' function, it is rather complicated to write down, and its derivative is even worse! I have tried changing variables for γ by putting $\gamma = \sin\varphi$, which allows one to write $1 + \gamma$ and $1 - \gamma$ in a nice trigonometric form, but that does not seem to simplify the computations much.

REFERENCES

[1] J. Aguirre, M. Bilbao, J.C Peral, The trace of totally positive algebraic integers, *Math. Comp.* **75** (2006), no. 253, 385–393.

[2] J. Aguirre, J.C. Peral, The integer Chebyshev constant of Farey intervals, *Publicacions Matemàtiques de la UAB* (to appear).

[3] Ş. Alaca, K.S. Williams, *Introductory algebraic number theory*, Cambridge University Press, Cambridge 2004.

[4] P. Borwein, *Computational excursions in analysis and number theory*, CMS Books in Mathematics/Ouvrages de Mathématiques de la SMC, 10, Springer-Verlag, New York 2002.

[5] P. Borwein, T. Erdélyi, The integer Chebyshev problem, *Math. Comp.* **65** (1996), no. 214, 661–681.

[6] J.W.S. Cassels, On a conjecture of R.M. Robinson about sums of roots of unity, *J. Reine Angew. Math.* **238** (1969), 112–131.

[7] O. Debarre, Degrees of curves in abelian varieties, *Bull. Soc. Math. France* **122**, (1994), no. 3, 343–361.

[8] V. Flammang, M. Grandcolas, G. Rhin, Small Salem numbers, *Number theory in progress*, Vol. 1 (Zakopane-Kościelisko 1997), 165–168, de Gruyter, Berlin, 1999.

[9] V. Flammang, G. Rhin, C.J. Smyth, The integer transfinite diameter of intervals and totally real algebraic integers, *J. Théor. Nombres Bordeaux* **9** (1997), no. 1, 137–168.

[10] I.M. Isaacs, *Character theory of finite groups*, (corrected reprint of the 1976 original [Academic Press, New York]), Dover Publishers, New York, 1994.

[11] K. Lauter, Geometric methods for improving the upper bounds on the number of rational points on algebraic curves over finite fields, with an appendix in French by J.-P. Serre, *J. Algebraic Geom.* **10** (2001), no. 1, 19–36.

---

*0.27598154

[12] M.J. McAuley, *Topics in J-fields and a diameter problem*, MSc thesis, Univ. of Adelaide 1981.

[13] J. McKee, C.J. Smyth, Salem numbers of trace −2 and traces of totally positive algebraic integers, *Algorithmic number theory*, Lecture Notes in Computer Science **3076**, 327–337, Springer Verlag, Berlin, 2004.

[14] _____, _____, There are Salem numbers of every trace, *Bull. London Math. Soc.* **37** (2005), no. 1, 25–36.

[15] I. Schur, Über die Verteilung der Wurzeln bei gewissen algebraischen Gleichungen mit ganzzahligen Koeffizienten, *Math. Z.* **1** (1918), no. 4, 377–402.

[16] C.L. Siegel, The trace of totally positive and real algebraic integers, *Ann. of Math.* (2) **46** (1945), 302–312.

[17] C.J. Smyth, On the measure of totally real algebraic integers, *J. Austral. Math. Soc. Ser. A* **30** (1980/81), no. 2, 137–149.

[18] _____, On the measure of totally real algebraic integers. II, *Math. Comp.* **37** (1981), no. 155, 205–208.

[19] _____, The mean values of totally real algebraic integers, *Math. Comp.* **42** (1984), no. 166, 663–681.

[20] _____, Totally positive algebraic integers of small trace, *Ann. Inst. Fourier (Grenoble)* **34** (1984), no. 3, 1–28.

[21] _____, An inequality for polynomials, *Number theory* (Ottawa, ON, 1996), CRM Proc. Lecture Notes **19**, 315–321, Amer. Math. Soc., Providence, RI, 1999.

MAHLER'S MEASURE: FROM NUMBER THEORY TO GEOMETRY

MARIE JOSÉ BERTIN

ABSTRACT. We survey the latest developments concerning explicit logarithmic Mahler measures for polynomials defining rational curves, elliptic curves or $K3$ hypersurfaces.

1. INTRODUCTION

The first person interested in Mahler's measure was D.H. Lehmer. In his famous paper [20] he asked the following question (still unsolved): does there exist a monic irreducible polynomial P, not cyclotomic, with integer coefficients, such that

$$\Omega(P) := \prod_{P(\alpha)=0} \max(|\alpha|, 1) < \Omega(P_0) \simeq 1.17628 \,?$$

Here P_0 is the Lehmer polynomial

$$x^{10} + x^9 - x^7 - x^6 - x^5 - x^4 - x^3 + x + 1.$$

In fact

$$\Omega(P) = M(P),$$

where $M(P)$ denotes the measure of the polynomial P introduced by Mahler in 1962. The *logarithmic Mahler measure* of a multivariate polynomial P is defined by

$$m(P) := \frac{1}{(2\pi i)^n} \int_{\mathbb{T}^n} \log|P(x_1, \cdots, x_n)| \frac{dx_1}{x_1} \cdots \frac{dx_n}{x_n}$$

and the *Mahler measure* by

$$M(P) = \exp\big(m(P)\big).$$

By Jensen's formula, if $P \in \mathbb{Z}[X]$ is monic, then

$$M(P) = \prod_{P(\alpha)=0} \max(|\alpha|, 1).$$

2000 *Mathematics Subject Classification*. Primary: 11G05, 14H50, 14H52, 14J28; secondary: 11G15, 14H10.

Key words and phrases. Mahler's Measure, Bloch groups, Elliptic dilogarithm, Exotic relations, Eisenstein-Kronecker Series, L-series of K3 hypersurfaces.

The first partial answer to Lehmer's question is due to Breusch (1951)[14]:

$$M(P) \geq 1.1796\ldots$$

if P is nonreciprocal and $P(x) \neq \pm x$. In 1971 Smyth [27] independently sharpened this, for the same P, to

$$M(P) \geq M(x^3 - x - 1) \simeq 1.3247\ldots.$$

The obstruction to answering Lehmer's question is therefore the reciprocal polynomials, in particuliar Salem polynomials [5]. A reciprocal polynomial is a palindromic polynomial and a Salem polynomial is an irreducible reciprocal polynomial with a unique root outside the unit circle, hence a unique root inside the unit circle and the other roots on the unit circle.

Boyd's limit formula (1981) [8]

$$\lim_{N \to +\infty} m\big(P(x, x^N)\big) = m\big(P(x, y)\big)$$

whenever the left hand term contains an infinity of different measures, was a hope for getting small measures in one variable from small measures in two variables. At that time, Boyd computed numerically

$$M\big((x + 1)y^2 + (x^2 + x + 1)y + x(x + 1)\big) = 1.25542\ldots,$$

$$M\big(y^2 + (x^2 + x + 1)y + x^2\big) = 1.28573\ldots,$$

and these are the smallest known measures in two variables. Notice that these polynomials define elliptic curves. In the same year, Smyth obtained the first explicit Mahler measures [8]:

$$m(x + y + 1) = L'(\chi_{-3}, -1),$$

$$m(x + y + z + 1) = \frac{7}{2\pi^2}\zeta(3),$$

where χ_{-3} denotes the odd quadratic character of conductor 3 and

$$L'(\chi_{-3}, -1) = \frac{3\sqrt{3}}{4\pi}L(\chi_{-3}, 2)$$

is derived from the functional equation of the Dirichlet L-series.

Then we must await Deninger's guess of the formula [15] (1996)

$$m(x + \frac{1}{x} + y + \frac{1}{y} + 1) \stackrel{?}{=} \frac{15}{4\pi^2}L(E, 2) = L'(E, 0),$$

where the Laurent polynomial defines an elliptic curve E of conductor 15 and $L(E, 2)$ its L-series at $s = 2$. The last equality comes from the functional equation. (A question-mark over an equals sign means that the relation is verified numerically up to fifty decimals.) Since then, there has been an abundant literature in this area, three conferences on Mahler measure and developments in many mathematics domains; see for example [10].

I want to focus on two questions:

- There are experimental relations between the Mahler measure of different polynomials. Can we prove these relations? What do they encode?
- How is the geometry of the curve or the surface involved in the explicit expressions?

2. Curves of genus 0

Let me take an example. For the polynomial

$$P = y^2(x+1)^2 + 2y(x^2 - 6x + 1) + (x+1)^2,$$

Boyd guessed (1998) [9] that

$$m(P) \overset{?}{=} 4L'(\chi_{-4}, -1) = \frac{8}{\pi} L(\chi_{-4}, 2),$$

where

$$L(\chi_{-4}, 2) = 1 - \frac{1}{3^2} + \frac{1}{5^2} + \cdots = G,$$

G being Catalan's constant. Here P defines a cubic curve C with $(1,1)$ as double point. Putting $x = 1 + X$ and $y = 1 + Y$ and completing the square, we find

$$\left(Y(X+2)^2 + 2X^2\right)^2 = -16X^2(X+1).$$

Hence we get the parametrisation of the two branches γ_1 and γ_2 of the curve C:

$$x_1 = -t^2, \qquad y_1 = -\left(\frac{1+t}{1-t}\right)^2;$$

$$x_2 = -t^2, \qquad y_2 = -\left(\frac{1-t}{1+t}\right)^2.$$

But

$$m(P) = \frac{1}{(2\pi i)^2} \int_{|x|=1} \int_{|y|=1} \log|P(x,y)| \frac{dx}{x}\frac{dy}{y}$$

$$= \frac{1}{2\pi i} \int_{|x|=1} \log\left(\max(|y_1|, |y_2|)\right) \frac{dx}{x} \quad \text{(by Jensen's formula)}$$

$$= \frac{1}{2\pi i} \int_{\Gamma} \eta_2(2)(x,y),$$

with

$$\eta_2(2)(x,y) = i \log|y| \, d\arg x - i \log|x| \, d\arg y$$

being a differential form on the variety Γ (Maillot's trick [22]), where

$$\Gamma = \left\{(x,y) \in \mathbb{C}^2 / \ (x,y) \in C, \ |x| = 1, \ |y| \geq 1\right\}.$$

Using this parametrisation,

$$2m(P) = -\frac{1}{2\pi i} \int_{\gamma_1} \eta_2(2)(x_1(t), y_1(t)) - \frac{1}{2\pi i} \int_{\gamma_2} \eta_2(2)(x_2(t), y_2(t)).$$

Now $\eta_2(2)$ is related to the Bloch-Wigner dilogarithm D:

$$D(x) := \Im \operatorname{Li}_2(x) + \log |x| \arg(1 - x),$$

which is an univalued function, real analytic on $\mathbb{P}^1(\mathbb{C}) \setminus \{0, 1, \infty\}$ and continuous on $\mathbb{P}^1(\mathbb{C})$. The dilogarithm Li_2 is defined as

$$\operatorname{Li}_2(x) := \sum_{n=1}^{\infty} \frac{x^n}{n^2}.$$

If we set

$$\hat{D}(x) = iD(x),$$

then the relation is the following:

$$d\hat{D}(x) = \eta_2(2)(x, 1 - x).$$

The differential $\eta_2(2)$ satisfies

- multiplicativity in each variable,
- antisymmetry,
- if $\alpha \neq \beta$ then

$$\eta_2(2)(t - \alpha, t - \beta) = \eta_2(2)\left(\frac{t - \alpha}{t - \beta}, 1 - \frac{t - \alpha}{t - \beta}\right)$$
$$+ \eta_2(2)(t - \alpha, \alpha - \beta) + \eta_2(2)(\beta - \alpha, t - \beta).$$

Thus the Mahler measure can be expressed as

$$2m(P) = -\frac{1}{2\pi i} \int_{\gamma_1} 4d\hat{D}(-t) - 4d\hat{D}(t) - \frac{1}{2\pi i} \int_{\gamma_2} -4d\hat{D}(-t) + 4d\hat{D}(t)$$
$$= \frac{2}{\pi}\left[D(t) - D(-t)\right]_{-i}^{i} + \frac{2}{\pi}\left[D(-t) - D(t)\right]_{i}^{-i}$$
$$= \frac{16}{\pi}D(i)$$
$$= 8d_4.$$

Thus Boyd's guess is proved. So, for some genus 0 curves, the Mahler measure encodes the Bloch-Wigner dilogarithm, and hence the Bloch groups.

2.1. The Bloch groups.

Let F be a field. Let us define the group of relations

$$R_2(F) \subset \mathbb{Z}[\mathbb{P}^1_F],$$

$$R_2(F) := [x] + [y] + [1 - xy] + \left[\frac{1 - x}{1 - xy}\right] + \left[\frac{1 - y}{1 - xy}\right].$$

The second Bloch group is the quotient group

$$B_2(F) := \mathbb{Z}[\mathbb{P}^1_F]/R_2(F).$$

There is a homomorphism δ_1^2:

$$B_2(F) \xrightarrow{\delta_1^2} \Lambda^2 F^*$$

defined by
$$\delta_1^2([x]_2) = x \wedge (1-x).$$
The class of x in $B_2(F)$, $[x]_2$, behaves like a Bloch-Wigner dilogarithm.
The cohomology of the complex
$$B_F(2) \otimes \mathbb{Q} : B_2(F)_\mathbb{Q} \xrightarrow{\delta_1^2} (\Lambda^2 F^*)_\mathbb{Q}$$
is related to K-theory by Matsumoto's theorem
$$H^2\big(B_F(2)\big) \simeq K_2(F).$$

Lalin [18] has proved that Smyth's first result can be treated in that context. Vandervelde [31] has given a class of polynomials defining rational curves to which this applies. For this class of polynomials, the Mahler measure can be expressed in terms of dilogarithms of algebraic numbers up to possibly a term in $\zeta_F(2)$. One of the conditions for such polynomials is to be *tempered*.

Definition 1. *A polynomial in two variables is tempered if the polynomials corresponding to the faces of its Newton polygon have roots of unity as their only zeros.*

When drawing the convex hull of points (i,j) in $\mathbb{Z}^2$ corresponding to the monomials $a_{i,j}x^i y^j$, $a_{i,j} \neq 0$, you also draw points of $\mathbb{Z}^2$ located on the faces. The polynomial of the face is a polynomial in one variable t which is a combination of the monomials 1, t, t^2, ... The coefficients of the combination are given when going along the face, that is $a_{i,j}$ if the lattice point of the face belongs to the convex hull and 0 otherwise. For example, the polynomial

$$y^2 + y + x^2 + x + 1$$

is tempered, since its Newton polygon corresponds to

$$\begin{array}{ccc} 1 & & \\ 1 & 0 & \\ 1 & 1 & 1 \end{array}$$

and the polynomials of the faces are
$$1+t^2, \qquad 1+t+t^2, \qquad 1+t+t^2.$$

The polynomial
$$P := (x^2 + x - 1)y^2 + (x^2 + 5x + 1)y - x^2 + x + 1$$
is not tempered, since its Newton polygon corresponds to

$$\begin{array}{ccc} -1 & 1 & 1 \\ 1 & 5 & 1 \\ 1 & 1 & -1 \end{array}$$

and the polynomials of the faces are all equal to $\pm(t^2 + t - 1)$. However, even in that case, the Mahler measure may have an expression of the same shape.

Indeed, Boyd guessed for $m(P)$ the formula [9]

$$A := m(P) \overset{?}{=} \frac{2}{3} \log \phi + \frac{1}{6} d_{15};$$

$$d_{15} = L'(\chi_{-15}, -1),$$

where χ_{-15} is the odd primitive character of conductor 15.

Bloch's formula [9] gives $L'(\chi_{-f}, -1)$ for odd primitive χ_f as a combination of Bloch-Wigner dilogarithms,

$$L'(\chi_{-f}, -1) = \frac{f}{4\pi} \sum_{m=1}^{f} \chi_{-f}(m) D(\zeta_f^m),$$

where ζ_f denotes an f-th root of unity. Applying this formula, we find

$$\frac{1}{6} L'(\chi_{-15}, -1) = \frac{5}{4\pi} \left[D(\zeta_{15}) + D(\zeta_{15}^2) + D(\zeta_{15}^4) + D(\zeta_{15}^8) \right].$$

Now, using a parametrisation of the two branches of the curve

$$x = j \pm \frac{t}{\sqrt{5}},$$

$$y = \left(\frac{\pm\sqrt{5} - 1}{2} \right) \left(\frac{2t \mp 2\sqrt{5} - 5 \pm i\sqrt{15}}{2t + 5 \pm i\sqrt{15}} \right),$$

we get

$$A = \frac{2}{3} \log \left(\frac{1 + \sqrt{5}}{2} \right) - D \left(-j^2 \frac{1 - \sqrt{5}}{2} \right) - D \left(j \frac{1 - \sqrt{5}}{2} \right)$$

$$- D \left(-j \frac{1 + \sqrt{5}}{2} \right) - D \left(j^2 \frac{1 + \sqrt{5}}{2} \right).$$

Numerically, correct to fifty decimal places, we check that

$$C := -D \left(-j^2 \frac{1 - \sqrt{5}}{2} \right) - D \left(j \frac{1 - \sqrt{5}}{2} \right)$$

$$- D \left(-j \frac{1 + \sqrt{5}}{2} \right) - D \left(j^2 \frac{1 + \sqrt{5}}{2} \right)$$

$$\overset{?}{=} E := \frac{5}{4} \left[D(\zeta_{15}) + D(\zeta_{15}^2) + D(\zeta_{15}^4) + D(\zeta_{15}^8) \right],$$

where ζ_{15} is a 15-th root of unity; hence the formula guessed by Boyd.

So, even if the polynomial is not tempered, there is an underlying relation between dilogarithms. Hence the objects are living in a Bloch group.

By Galois descent in the Bloch group, C and E live in $B_2\big(\mathbb{Q}(\sqrt{-15})\big)_{\mathbb{Q}}$. I have not yet guessed to which element in $B_2\big(\mathbb{Q}(\sqrt{-15})\big)_{\mathbb{Q}}$ the numbers C and E correspond.

3. CURVES OF GENUS 1

Let us consider the polynomials

$$P = (x+1)(y+1)(x+y+1) + xy$$

and

$$Q = y'^2 + (x'^2 + 2x' - 1)y' + x'^3.$$

Boyd [9] suspected and I proved [2] in 2004 the following relation between the Mahler measure of the two polynomials.

Theorem 1. [2] *We have*

$$7m(Q) = 5m(P).$$

What does this mean? If P defines a 'good' curve of genus 1, that is, if P is tempered, such an equality encodes the K-theory of an elliptic curve.

3.1. **The elliptic regulator.** Let F be a field. By Matsumoto's theorem, the second group of K-theory $K_2(F)$ can be described in terms of symbols $\{f, g\}$, for f and $g \in F^*$ and relations between them.

The relations are

- $\{f_1 f_2, g\} = \{f_1, g\} + \{f_2, g\}$,
- $\{f, g_1 g_2\} = \{f, g_1\} + \{f, g_2\}$,
- $\{1 - f, f\} = 0$.

For example, if v is a discrete valuation on F with maximal ideal $\mathcal{M}$ and residual field k, the Tate tame symbol

$$(x, y)_v \equiv (-1)^{v(x)v(y)} \frac{x^{v(y)}}{y^{v(x)}} \qquad (\mathrm{mod}\ \mathcal{M})$$

defines a homomorphism

$$\lambda_v : K_2(F) \to k^*.$$

Let E be an elliptic curve on $\mathbb{Q}$ and $\mathbb{Q}(E)$ its rational function field. To any $P \in E(\overline{\mathbb{Q}})$ is associated a valuation on $\mathbb{Q}(E)$ that gives the homomorphism

$$\lambda_P : K_2\big(\mathbb{Q}(E)\big) \to \mathbb{Q}(P)^*$$

and the exact sequence

$$0 \to K_2(E) \otimes \mathbb{Q} \to K_2\big(\mathbb{Q}(E)\big) \otimes \mathbb{Q} \xrightarrow{\lambda} \bigsqcup_{P \in E(\overline{\mathbb{Q}})} \mathbb{Q}(P)^* \otimes \mathbb{Q} \to \cdots.$$

By definition $K_2(E)$ is defined modulo torsion by

$$K_2(E) \simeq \ker \lambda = \cap_P \ker \lambda_P \subset K_2\big(\mathbb{Q}(E)\big).$$

By a theorem due to Rodriguez Villegas [25], if a tempered polynomial, $P \in \mathbb{Q}[x^\pm, y^\pm]$ defines a smooth curve C, we obtain

$$\{x, y\}^N \in K_2(C).$$

In particuliar, if

$$P(x, y) = (x + y + 1)(x + 1)(y + 1) + xy,$$

then

$$\{x, y\} \in K_2(E).$$

Let f and g be in $\mathbb{Q}(E)^*$. Define

$$\eta(f, g) = \log|f| \, \mathrm{d}\arg g - \log|g| \, \mathrm{d}\arg f.$$

Definition 2. *The elliptic regulator r of E is given by*

$$r : K_2(E) \rightarrow \mathbb{R}$$
$$\{f, g\} \mapsto \tfrac{1}{2\pi} \int_\gamma \eta(f, g)$$

for a suitable loop γ generating the subgroup $H_1(E, \mathbb{Z})^- \subset H_1(E, \mathbb{Z})$, where the complex conjugation acts by -1.

3.2. The elliptic dilogarithm $D^E(P)$.

There are two representations of the complex structure of an elliptic curve E: the first as a lattice and the second as a Tate curve.

$$E(\mathbb{C}) \quad \simeq \quad \mathbb{C}/(\mathbb{Z} + \tau\mathbb{Z}) \quad \rightarrow \quad \mathbb{C}^*/q^{\mathbb{Z}}$$
$$\big(\mathcal{P}(u), \mathcal{P}'(u)\big) \quad \mapsto \quad u(\mathrm{mod}\,\Lambda) \quad \mapsto \quad e^{2\pi i u} = z.$$

Now define

$$D^E(P) = \sum_{n=-\infty}^{+\infty} D(q^n z)$$

for $P \in E(\mathbb{C})$ and D the Bloch-Wigner dilogarithm. The elliptic dilogarithm D^E can be extended to divisors on $E(\mathbb{C})$ and is also related to the elliptic regulator r.

3.3. Sketch of the proof of the theorem.

The equality $P = 0$ defines two roots y_1 and y_2 which are functions of x. Denote by y_1 the root satisfying $|y_1(x)| < 1$ if $|x| = 1$. By Jensen's formula, we get

$$m(P) = -\frac{1}{2\pi i} \int_{|x|=1} \log|y_1| \frac{dx}{x}$$
$$= \frac{1}{2\pi} \int_\sigma \eta(x, y).$$

But σ is a path on the variety $\{(x, y) \in E, \quad |x| = 1, \quad |y| \geq 1\}$ and σ generates $H_1(E, \mathbb{Z})^-$, so

$$m(P) = \pm r(\{x, y\}).$$

For the same reasons,

$$m(Q) = \pm r(\{x', y'\}).$$

Comparing these regulators with the regulator of the isomorphic elliptic curve $X_1(11)$:

$$Y^2 + Y - X^3 + X^2 = 0,$$

one gets

$$7r(\{X, Y\}) + r(\{x, y\}) = 0,$$
$$-5r(\{X, Y\}) + r(\{x', y'\}) = 0,$$

that is

$$5m(P) = 7m(Q).$$

We obtain even more: the proof of an exotic relation suspected by Bloch and Grayson [6].

If D denotes the Bloch-Wigner dilogarithm, Milnor considers $D(\zeta)$ for $\zeta^d = 1$ and conjectures, for geometric reasons, that the only relations

$$\sum_{r=1}^{d-1} a_r D(\zeta^r) = 0, \qquad a_r \in \mathbb{Z}$$

are those arising from the distribution relations

$$D(x^s) = s \sum_{\tau^s = 1} D(\tau x)$$

together with

$$D(\bar{x}) = -D(x).$$

The corresponding situation for the elliptic dilogarithm is not quite the same.

Let E be an elliptic curve over $\mathbb{Q}$ and suppose that the torsion group $E(\mathbb{Q})_{\text{tors}}$ is cyclic and put $d = \#E(\mathbb{Q})_{\text{tors}}$. Write Σ for the number of fibers of type I_n with $n \geq 3$ in the Néron model [23] and suppose $[\frac{d-1}{2}] - \Sigma > 1$. Then there should be at least $[\frac{d-1}{2}] - \Sigma - 1$ relations

$$\sum_{r=1}^{d-1} a_r D^E(\zeta^r) = 0, \qquad a_r \in \mathbb{Z}$$

where ζ is a primitive d-th root of unity and

$$D^E(\zeta^r) = \sum_{-\infty}^{+\infty} D(q^n \zeta^r).$$

Definition 3. *The previous relations are called* exotic.

If E denotes the modular elliptic curve $X_1(11)$, the earlier proof yields the exotic relation

$$3D^E(P) = 2D^E(2P),$$

if $P = (0, 0)$ is a 5-torsion point of $X_1(11)$.

4. SURFACES

Consider the family of Laurent polynomials

$$Q_k = x + \frac{1}{x} + y + \frac{1}{y} + z + \frac{1}{z}$$

$$+ xy + \frac{1}{xy} + zy + \frac{1}{zy} + xyz + \frac{1}{xyz} - k,$$

and the relation guessed by Boyd [11]

$$2m(Q_{-36}) \overset{?}{=} 4m(Q_{-6}) + m(Q_0).$$

What does this relation mean?

Computations up to high accuracy are possible thanks to the following result.

Theorem 2 ((2005)[3]). *Let* $k = -(t + 1/t) - 2$ *and*

$$t = \frac{\eta(3\tau)^4 \eta(12\tau)^8 \eta(2\tau)^{12}}{\eta(\tau)^4 \eta(4\tau)^8 \eta(6\tau)^{12}},$$

where η *is the Dedekind eta function and* $q = e^{2\pi i \tau}$. *Then the Mahler measure of the polynomial* Q_k *can be expressed in terms of the following Eisenstein-Kronecker series:*

$$m(Q_k) = \frac{\Im\tau}{8\pi^3} \sum_{(m,\kappa)\neq(0,0)} 2\left(2\Re\frac{1}{(m\tau+\kappa)^3(m\bar{\tau}+\kappa)} + \frac{1}{(m\tau+\kappa)^2(m\bar{\tau}+\kappa)^2}\right)$$

$$- 32\left(2\Re\frac{1}{(2m\tau+\kappa)^3(2m\bar{\tau}+\kappa)} + \frac{1}{(2m\tau+\kappa)^2(2m\bar{\tau}+\kappa)^2}\right)$$

$$- 18\left(2\Re\frac{1}{(3m\tau+\kappa)^3(3m\bar{\tau}+\kappa)} + \frac{1}{(3m\tau+\kappa)^2(3m\bar{\tau}+\kappa)^2}\right)$$

$$+ 288\left(2\Re\frac{1}{(6m\tau+\kappa)^3(6m\bar{\tau}+\kappa)} + \frac{1}{(6m\tau+\kappa)^2(6m\bar{\tau}+\kappa)^2}\right).$$

4.1. Key points in the proof of the previous result.

- More geometry is necessary since the polynomials Q_k define $K3$-surfaces X_k.
- Since X_k is a $K3$-surface, there is on X_k a unique (up to scalars) holomorphic 2-form.
- Since X_k is a $K3$-surface, one can define periods.
- The family of periods satisfies a Picard-Fuchs differential equation of order 3 [32].
- The derivative of the Mahler measure of the polynomial Q_k with respect to the parameter k, $\frac{dm(P_k)}{dk}$, is a period of X_k, hence satisfies the Picard-Fuchs equation.

- The family X_k is modular, so the solutions of the Picard-Fuchs equation can be expressed in terms of modular forms.
- By integrating and taking the development in Fourier series, we find the previous formulae.

For X_k singular, $m(P_k)$ is related to the L-series of the variety X_k [4]. In fact, one of the most important result on $K3$-surfaces is a theorem of Morrison: a $K3$-surface X, $\mathcal{M}$-polarized, with Picard number 19, has a Shioda-Inose structure, that is to say

$$
\begin{array}{ccc}
X & & A = E \times E/C_N \\
& \searrow{\scriptstyle \iota} \quad \swarrow & \\
& Y = \mathrm{Kum}(A/\pm) &
\end{array}
$$

where E is an elliptic curve, C_N is a cyclic group of isogeny, ι an involution and $X/\langle \iota \rangle$ is birationally isomorphic to Y. Moreover, if X is singular (Picard number 20), then E has complex multiplication.

So we get the following theorem.

Theorem 3. (2005)[3] *Let* $\mathbb{Q}(\sqrt{-3})$ *and* $R = (1, 2\sqrt{-3}) \subset R' = (1, \sqrt{-3})$ *two orders of discriminants* -48 *(resp.* -12*), with class numbers* 2 *(resp.* 1*). Let* Φ_R *(resp.* $\Phi_{R'}$*), the Hecke Grössencharacter of weight 3, be defined by*

$$
\Phi_R(\alpha R) = \alpha^2, \quad \Phi_R(P) = -3, \quad P = (3, 2\sqrt{-3}),
$$

$$
\Phi_{R'}(\beta R') = \beta^2.
$$

Then, the relation

$$
2m(Q_{-36}) = 4m(Q_{-6}) + m(Q_0)
$$

is equivalent to the relation

$$
\frac{9}{8} \sum_{(m,\kappa) \neq (0,0)} \frac{m^2 - 3\kappa^2}{(m^2 + 3\kappa^2)^3} = \sum_{(m,\kappa) \neq (0,0)} \left(\frac{4m^2 - 3\kappa^2}{(4m^2 + 3\kappa^2)^3} - \frac{12m^2 - \kappa^2}{(12m^2 + \kappa^2)^3} \right),
$$

which in turn can be expressed as a relation between Hecke L-series

$$
L_{R'}(\phi_{R'}, 3) = L_R(\phi_R, 3).
$$

Zagier [33] has proved that this is in fact a relation between the L-series of weight 3 modular forms for $\Gamma_0(4)$:

$$
(1 + 2 \cdot 4^{1-s})L(f, s) = L(f_1, s) + L(f_2, s),
$$

where $f = [\theta_1, \theta_3]$, $f_1 = [\theta_1, \theta_{12}]$ and $f_2 = [\theta_4, \theta_3]$ are Rankin-Cohen brackets. In our situation, if

$$\theta_a = \sum_{n \in \mathbb{Z}} q^{an^2},$$

the Rankin-Cohen bracket is

$$RC(g, h) = [g, h] = kgh' - lg'h$$

and is a modular form of weight $k + l + 2$ provided that g is of weight k and h of weight l.

5. FINAL REMARKS

Using the Zagier-Goncharov trilogarithm, Lalin [17] has generalized the wedge product to 3 variables, explaining for instance Smyth's second relation and many other relations such as

$$m\big((1 + x + y^{-1}) - (1 + x + y)z\big) = \tfrac{14}{3\pi^2}\zeta(3) \qquad \text{(Smyth) [28], [29],}$$

$$m\big((1 + x)(1 + y) - (1 - x)(1 - y)z)\big) = \tfrac{7}{3\pi^2}\zeta(3) \qquad \text{(Lalin) [18].}$$

For all these examples, the surfaces are rational of a certain type. So my last question is: what are the explicit formulae for rational elliptic surfaces, for instance the rational elliptic modular surface associated to $\Gamma_0(6)$ defined by

$$x(x - 1)(y - 1) = zy(x - y)?$$

Acknowledgement. I thank the referee for his pertinent suggestions concerning the presentation of this article.

REFERENCES

[1] A. Beilinson, Higher regulators of modular curves, Application of algebraic K-theory to algebraic geometry and number theory, Part I, II (Boulder, Colo., 1983), *Contemp. Math.*,**55**, Amer. Math. Soc., Providence, R. I., (1986), 1–34.

[2] M.J. Bertin, Mesure de Mahler d'une famille de polynômes, *J. reine angew. Math.* **569** (2004), 175–188.

[3] ———, *Mesure de Mahler d'hypersurfaces K3*, http://arxiv.org/pdf/math.NT/0501153 (preprint).

[4] ———, Mahler's measure and L-series of $K3$ hypersurfaces, *Mirror Symmetry V*, AMS/IP Studies in Advanced Mathematics **38** (2006), 3–18.

[5] M.J. Bertin, A. Decomps, M. Grandet-Hugot, M. Pathiaux-Delefosse, J.P. Schreiber, *Pisot and Salem numbers*, Birkhaüser (1992).

[6] S. Bloch & D. Grayson, K_2 and L-functions of elliptic curves computer calculations, *Contemp. Math.* **55**, Part I, (1986), 79–88.

[7] D.W. Boyd, Kronecker's theorem and Lehmer's problem for polynomials in several variables, *J. Number Theory* **13** (1981), 116–121.

[8] ———, Speculations concerning the Range of Mahler's Measure, *Canad. Math. Bull.* **24** (1981), 453–469.

[9] _____, Mahler's measure and special values of L-functions, *Experiment. Math.* **7** (1998), 37–82.

[10] _____, Mahler's measure and invariants of hyperbolic manifolds, *Number Theory for the millennium* **I** (Urbana, IL, 2000), 127-143, A K Peters, Natick, MA,2002.

[11] _____, Personal communication.

[12] D.W. Boyd, F. Rodriguez Villegas, Mahler's measure and the dilogarithm I *Canad. J. Math.* **54** (2002), no. 3, 468–492.

[13] D.W. Boyd, F. Rodriguez Villegas, N.M. Dunfield, *Mahler's measure and the dilogarithm II* http://arxiv.org/list/math.NT/0308041, 37 p.

[14] Breusch, Robert. On the distribution of the roots of a polynomial with integral coefficients. *Proc. Amer. Math. Soc.* **2** (1951), 939–941.

[15] C. Deninger, Deligne periods of mixed motives, K-theory and the entropy of certain $\mathbb{Z}^n$-actions, *J. Amer. Math. Soc.***10:2** (1997), 259–281.

[16] M. Lalin, Some examples of Mahler measures as multiple polylogarithms, *J. Number Theory* **103** (2003), no. 1, 85–108.

[17] _____, Mahler measure of some n-variable polynomial families, *J. Number Theory* **116** (2006), no. 1, 102–139.

[18] _____, *Some relations of Mahler measure with hyperbolic volumes and special values of L-functions*, Doctoral Dissertation, University of Texas at Austin (2005),

[19] _____, Mahler measure and volumes in hyperbolic space, *Geom. Dedicata* **107** (2004), 211–234.

[20] D. H. Lehmer, Factorization of certain cyclotomic functions, *Ann. of Math. (2)* **34** (1933), 461–479.

[21] V. Maillot, *Géométrie d'Arakelov des grassmanniennes, des variétés toriques et de certaines hypersurfaces*, Thèse Université Paris 7 (1997).

[22] _____, *Mesure de Mahler et cohomologie motivique*, preprint (2006).

[23] A. Néron, Modèles minimaux des variétés abéliennes sur les corps locaux et globaux, *Inst. Hautes Etudes Sci. Publ. Math.*,**21**, (1964), 88–128.

[24] F. Rodriguez Villegas, *Modular Mahler measures*, preprint (1996).

[25] _____, Modular Mahler measures I, *Topics in Number Theory* (S.D. Ahlgren, G.E. Andrews & K. Ono, ed.), Kluwer, Dordrecht (1999), 17–48.

[26] _____, Identities between Mahler measures, *Number Theory for the millennium*, **III** (Urbana, IL, 2000), 223–229, A K Peters, Natick, MA, 2002.

[27] C. J. Smyth, On the product of conjugates outside the unit circle of an algebraic integer, *Bull. London Math. Soc.* **3** (1971), 169–175.

[28] _____, Mahler measure of a family of 3-variable polynomials, *J. Théor. Nombres Bordeaux,* **14** (2002), 683–700.

[29] _____, *Explicit formulas for the Mahler measure of families of multivariate polynomials* (preprint).

[30] S. Vandervelde, A formula for the Mahler measure of $axy + bx + cy + d$, *J. Number Theory,* **100** (2003), no. 1, 184–202.

[31] _____, *The Mahler Measure of parametrizable polynomials*, http://www.mandelbrot.org/parampolys.pdf (preprint).

[32] H. Verrill, The *L*-series of certain rigid Calabi-Yau threefolds, *J. Number Theory,* **81** (2000), no. 2, 310–334.

[33] D. Zagier, Personal communication.

EXPLICIT CALCULATION OF ELLIPTIC FIBRATIONS OF $K3$-SURFACES AND THEIR BELYI-MAPS

FRITS BEUKERS AND HANS MONTANUS

ABSTRACT. In this paper we give an introduction to Belyi-maps and Grothendieck's dessins d'enfant. In addition we provide an explicit method to compute the Belyi-maps corresponding to all semi-stable families of elliptic curves with six singular fibers.

1. INTRODUCTION

In a paper by Miranda and Persson [5], the authors study semi-stable elliptic fibrations over $\mathbb{P}^1$ of K3-surfaces with 6 singular fibres. In their paper the authors give a list of possible fiber types for such fibrations. It turns out that there are 112 cases. The corresponding J-invariant is a so-called Belyi-function. More particularly, J is a rational function of degree 24, it ramifies of order 3 in every point above 0, it ramifies of order 2 in every point above 1, and the only other ramification occurs above infinity. To every such map we can associate a so-called 'dessin d'enfant' (a name coined by Grothendieck) which in its turn uniquely determines the Belyi map. If $f : C \to \mathbb{P}^1$ is a Belyi map, the dessin is the inverse image under f of the real segment [0,1].

Several papers, e.g. [3],[7],[13], have been devoted to the calculation of some of the rational J-invariants for the Miranda-Persson list. It turns out that explicit calculations quickly become too cumbersome (even for a computer) if one is not careful enough. The goal of this paper is to compute all J-invariants corresponding to the Miranda-Persson list. We use a trick which enables us to reduce the calculation to the solution of a set of three polynomial equations in three unknowns (see Section 7 for details). The results can be found on the website

http://www.math.uu.nl/people/beukers/mirandapersson/Dessins.html

On that website, an entry like 14-3-2-2-2-1 means that one finds there all dessins of J-functions with ramification orders 14,3,2,2,2,1 above infinity. Alternatively one can say that the special elliptic fibers are of type I_n with $n = 14, 3, 2, 2, 2, 1$. If to a partition there corresponds only one picture, this means that J is a rational function with coefficients in $\mathbb{Q}$. If there are several

---

2000 *Mathematics Subject Classification.* 14D05, 14H52.
Key words and phrases. Belyi-map, elliptic fibration.
33

pictures, the corresponding fields of definition are indicated. On this website one also finds the explicit J-invariants.

Besides giving the computation of explicit formulas for the Miranda-Persson list this paper also contains a brief introduction to Belyi maps and dessins d'enfant.

Acknowledgement. I would like to thank one of the referees in particular for valuable advice concerning the presentation of the tables and the computational results. Thanks are also due to Jeroen Sijsling who compiled Table 1 in the Appendix.

2. Dessins d'enfant

Let X be a smooth algebraic curve and $\phi : X \to \mathbb{P}^1$ a non-constant rational function, which can be considered as a morphism of curves. A point $P \in X$ is called a *point of ramification* if $d\phi(P) = 0$. The image under ϕ of the ramification points is called the *branched set*. Let S be a finite subset of $\mathbb{P}^1$. We say that ϕ is *unramified outside S* if the branched set is contained in S. We have the following theorem.

Theorem 2.1. *Let X be a smooth algebraic curve defined over $\mathbb{C}$ and $\phi : X \to \mathbb{P}^1$ a non-constant rational map. Suppose the branched set of ϕ is contained in $\mathbb{P}^1(\overline{\mathbb{Q}})$. Then both X and ϕ can be defined over $\overline{\mathbb{Q}}$.*

Although the theorem is quite well known, its proof is not easy to recover from the literature. In Serre's book [10, p. 71] on the Mordell-Weil theorem we are referred to SGA1 (Séminaire de Géometrie Algébrique, LNM 224).

The following remarkable theorem is crucial to the story of dessins d'enfant.

Theorem 2.2 (Belyi). *Let X be an algebraic curve defined over $\overline{\mathbb{Q}}$. Then there exists a non-constant rational map $\phi : X \to \mathbb{P}^1$ which is unramified outside $\{0, 1, \infty\}$.*

See [1] for a proof. Together with the previous theorem, this theorem characterises algebraic curves defined over $\overline{\mathbb{Q}}$ as algebraic curves that allow a rational map to $\mathbb{P}^1$ unramified outside $\{0, 1, \infty\}$. In other words, we have a geometric characterisation for curves defined over $\overline{\mathbb{Q}}$.

A pair (X, ϕ), where X is a smooth algebraic curve and $\phi : X \to \mathbb{P}^1$ a non-constant morphsim unramified outside $\{0, 1, \infty\}$, is called a *Belyi pair*. Two Belyi pairs (X, ϕ) and (X', ϕ') are considered equivalent if there is an isomorphism $\sigma : X \to X'$ such that $\phi = \phi' \circ \sigma$. From now on, when we speak of Belyi pairs, we mean their equivalence class.

The other surprise, due to an observation of Grothendieck, is that the geometrical criterion can be turned into a purely combinatorial description using certain graphs. To this end we shall use connected, bi-coloured, graphs with rotation.

 (1) A graph is called *connected* if every vertex is connected to any other vertex via a sequence of edges.

(2) A graph is called *bi-coloured* if every vertex is given one of two colours (say black and white) such that vertices of the same colour are not connected by an edge.

(3) A graph is called a *graph with rotation* if at every vertex there is a given cyclic ordering of the edges ending in the vertex (in mathematical physics literature they are referred to as *ribbon graphs*).

A graph satisfying all three of properties above is called a *dessin d'enfant*, or simply a *dessin*. These names were first coined by Grothendieck, see [6]. Two dessins are considered equivalent if there exists a graph isomorphism between them preserving the bi-colouring and the cyclic ordering at the vertices. From now on, when we speak of a dessin, we mean its equivalence class.

Before explaining the connection between Belyi pairs and dessins, we introduce two other categories of interest. An ordered triple of permutations $\sigma_0, \sigma_1, \sigma_\infty \in S_n$, the group of permutations on $\{1, 2, \ldots, n\}$, is called a *permutation triple* if $\sigma_0\sigma_1\sigma_\infty = \mathrm{Id}$ and the group generated by σ_0, σ_1 acts transitively on $\{1, 2, \ldots, n\}$. Two permutation triples σ_i $(i = 0, 1, \infty)$ and σ_i' $(i = 0, 1, \infty)$ are considered equivalent if there exists $\tau \in S_n$ such that $\sigma_i' = \tau\sigma_i\tau^{-1}$ for $i = 0, 1, \infty$. From now on, when we speak of a permutation triple, we mean its equivalence class.

Finally we consider finite extensions of $\overline{\mathbb{Q}}(z)$ unramified outside $0, 1, \infty$. Two such extensions K, K' are considered equivalent if there exists a field isomorphism $\psi : K \to K'$ fixing the subfield $\overline{\mathbb{Q}}(z)$. Again, we consider equivalence classes of such extensions.

Above we have defined four categories,

I Belyi pairs (X, ϕ).
II Dessins (d'enfant).
III Permutation triples.
IV Finite extensions of $\overline{\mathbb{Q}}(z)$, unramified outside of $0, 1, \infty$.

We would like to make these classes more refined by introducing the order n.

I(n) Belyi pairs (X, ϕ) where $\deg(\phi) = n$.
II(n) Dessins (d'enfant) with n edges.
III(n) Permutation triples in S_n.
IV(n) Finite extensions of $\overline{\mathbb{Q}}(z)$ of degree n, unramified outside $0, 1, \infty$.

We now give an explicit set of natural bijections

$$\mathrm{I}(n) \to \mathrm{IV}(n) \to \mathrm{III}(n) \to \mathrm{II}(n) \to \mathrm{I}(n)$$

whose composition is the identity map $\mathrm{I}(n) \to \mathrm{I}(n)$. The reader should verify that in each case the mapping is actually well defined on equivalence classes.

$\mathrm{I}(n) \to \mathrm{IV}(n)$. To a Belyi pair (X, ϕ) we associate the function field $\overline{\mathbb{Q}}(X)$ as an extension of $\overline{\mathbb{Q}}(\phi)$.

IV$(n) \to$ III(n). Let K be the degree n extension of $\overline{\mathbb{Q}}(z)$. Choose $y \in K$ such that $K = \overline{\mathbb{Q}}(z, y)$. Let $P(z, y)$ be the minimal polynomial of y over $\overline{\mathbb{Q}}(z)$. For the following consideration we need to consider an embedding $\overline{\mathbb{Q}} \hookrightarrow \mathbb{C}$ and then work over $\mathbb{C}$ with a monodromy argument.

Choose a point $z_0 \neq 0, 1, \infty$ and such that $P(z_0, y)$ is well defined and has n distinct zeros. Consider the n Taylor series solutions $y_1, y_2, \ldots, y_n$ of $P(z, y) = 0$ around z_0. Choose a closed path $\gamma_0 \in \pi_1(\mathbb{C} \setminus \{0, 1\}, z_0)$, which loops around 0 exactly once in the positive direction, and zero times around 1. Analytic continuation of the functions $y_1, \ldots, y_n$ permutes these functions. Denote this permutation by σ_0. Similarly we choose a loop γ_1 going around 1 exactly once in the positive direction. Analytic continuation along γ_1 generates a permutation σ_1. Together, the permutations $\sigma_0, \sigma_1, \sigma_\infty = (\sigma_0\sigma_1)^{-1}$ form a permutation triple. Transitivity of the group generated by σ_0, σ_1 follows from the irreducibility of $P(y)$.

III$(n) \to$ II(n). Let $\sigma_i \in S_n$ $(i = 0, 1, \infty)$ be a permutation triple. Take n line segments each with a black and a white endpoint. We now identify the black points and the white points in the following manner to obtain a dessin. We number the segments $1, 2, \ldots, n$. For each cycle in the cycle decomposition of σ_0 we identify the black points corresponding to the numbers in that cycle and choose the ordering of the cycle as ordering on the edges ending in the newly formed black vertex. We proceed in the same way with the white vertices using the cycle decomposition of σ_1. Because the group generated by σ_0, σ_1 acts transitively on $1, 2, \ldots, n$, the resulting graph is connected.

II$(n) \to$ I(n). Let D be a dessin. The argument used here is a topological one. Choose a closed compact oriented surface which allows an embedding $D \to S$ such that:

(1) the vertex orientation of D coincides with the positive orientation on S;
(2) $S \setminus D$ is a finite union of simply connected open sets $U_1, U_2, \ldots, U_r$ with piecewise smooth boundaries.

These properties determine the genus g of S uniquely. We now complete the embedded dessin into a triangulation of S. Choose in every open set U_i a grey point and connect it by edges to all vertices on D which are on the boundary of U_i. This gives a triangulation of S. There are two kinds of triangles: positive ones, in which the ordering of the vertices is black-white-grey, and negative ones, in which the ordering of the vertices is black-grey-white. Both types occur equally often. Now construct a continous map $\phi : S \to \mathbb{P}^1$ such that:

(1) the positive triangles are mapped homeomorphically onto the northern hemisphere (Im$(z) \geq 0$) with the black, white and grey vertices mapping to $0, 1, \infty$;

(2) the negative triangles are mapped homeomorphically onto the south-
   ern hemisphere (Im(z) ≤ 0) with the black, white and grey vertices
   mapping to $0, 1, \infty$.

Now pull back the complex structure of $\mathbb{P}^1$ to a complex structure on S via
ϕ. In this way S becomes a compact Riemann surface, hence an algebraic
curve.

Of course, any other mapping between these sets is a composition of one
or more of the above ones. There are a few useful shortcuts though.

I(n) → II(n). Given a Belyi pair (X, ϕ) the corresponding dessin is simply $\phi^{-1}([0, 1])$ where the two sets of vertices are $\phi^{-1}(0)$ and $\phi^{-1}(1)$. The
orientation around the vertices is induced by the positive orientation on X.

II(n) → III(n). Let n be the number of edges of the dessin. We number
these edges $1, 2, \ldots, n$. For σ_0 we take the permutation induced by the cyclic
ordering around the black vertices; for σ_1 we take the permutation induced
by the ordering around the white vertices. Because the dessin is connected,
the subgroup of S_n generated by σ_0, σ_1 acts transitively on $1, 2, \ldots, n$.

In each of the sets I, II, III, IV there are natural subsets which are also in
1-1 correspondence. They are

 I′ Connected graphs with n edges and a cyclic order at each vertex.
    These graphs arise if we take a dessin in which every white vertex has
    multiplicity 2 (i.e., there are two edges emanating from it) and where
    the white vertices are subsequently erased.
 II′ Belyi pairs (X, ϕ) with $\deg(\phi) = 2n$, such that every point in $\phi^{-1}(1)$
    is ramified of order two.
 III′ Permutation triples in S_{2n} such that σ_1 is a product of n disjoint cycle
    pairs.
 IV′ Finite extensions of $\overline{\mathbb{Q}}(z)$ of degree $2n$, unramified outside $0, 1, \infty$ and
    ramification order 2 in every place above 1.

Following Schnepps [6, p. 50] we call I′ the set of *clean dessins*. We can
always recover the original dessin from the clean one by putting a white vertex
in the middle of each edge. Although it seems like a restriction, any dessin
can be mapped to a clean dessin as follows. Let $\phi : X \to \mathbb{P}^1$ be the Belyi
map corresponding to the general dessin. Then the map $X \to \mathbb{P}^1$ given by
$P \mapsto 4\phi(P)(1 - \phi(P))$ is again a Belyi map which now corresponds to a clean
dessin. This is based on the idea that $x \mapsto 4x(1 - x)$ maps ∞ to ∞, the
points $0, 1$ to 0 and it ramifies of order 2 above 1. The clean dessin is simply
obtained from the original dessin by changing the colour of the white points
to black.

Theorem 2.3. *Let $\phi : X \to \mathbb{P}^1$ be a Belyi map of degree N. Let n_0, n_1, n_∞
be the number of distinct points in $\phi^{-1}(0), \phi^{-1}(1), \phi^{-1}(\infty)$ respectively. Then*

$$2g(X) - 2 = N - n_0 - n_1 - n_\infty,$$

where $g(X)$ denotes the genus of X.

Proof. We prove this theorem using the description of the Belyi map via dessins. The pre-image $\phi^{-1}([0,1]) \subset X$ defines a cell decomposition of X. The number of 0-cells is $n_0 + n_1$, the number of 2-cells is n_∞ and the number of 1-cells N. The theorem follows from the computation of the Euler characteristic of X,

$$2 - 2g(X) = n_0 + n_1 - N + n_\infty.$$

$\square$

3. COUNTING DESSINS

We address the following question. Given the ramification indices above $0, 1, \infty$, how many corresponding dessins are there?

Before we answer the question we need a slight extension of the concept of permutation triples and dessins. By a *generalised permutation triple* we mean any three $g_0, g_1, g_\infty \in S_n$ such that $g_0 g_1 g_\infty = \mathrm{Id}$. So we have dropped the transitivity condition. In the same way as before we shall consider equivalence classes of permutation triples. A *generalised dessin* is simply a bicoloured oriented graph, without the condition of connectivity. Again we consider only equivalence classes of generalised dessins.

Furthermore, we call $g \in S_n$ an automorphism of the triple g_i ($i = 0, 1, \infty$) if $g g_i g^{-1} = g_i$ for $i = 0, 1, \infty$. We denote the automorphism group by $\mathrm{Aut}(g_0, g_1, g_\infty)$. An automorphism of a generalised dessin is of course an automorphism as of a bicoloured oriented graph. We observe the following.

Remark 3.1. *The natural bijection between* II*(n) and* III*(n) extends to a natural bijection between generalised dessins and generalised permutation triples. Moreover, the automorphism group of a generalised dessin and the automorphism group of the corresponding generalised permutation triple are isomorphic.*

One can also extend the classes I(n) and IV(n) and their correspondence with generalised dessins. The class I(n) must be extended to sets of Belyi-pairs where the sum of the degrees of the maps is n. The class IV(n) would have to be extended to commutative $\overline{\mathbb{Q}}(z)$-algebras of dimension n. But we shall not pursue this here.

The answer to our question is based on the following theorem (see also [11]).

Theorem 3.2. *Let $n \in \mathbb{N}$. Let C_0, C_1, C_∞ be three conjugacy classes in S_n. The number of triples $g_0 \in C_0, g_1 \in C_1, g_\infty \in C_\infty$ such that $g_0 g_1 g_\infty = \mathrm{Id}$ is given by*

$$N(C_0, C_1, C_\infty) = \frac{|C_0||C_1||C_\infty|}{n!} \sum_\chi \frac{\chi(C_0)\chi(C_1)\chi(C_\infty)}{\dim(\chi)},$$

where the sum is over all irreducible characters of S_n and $\dim(\chi)$ *denotes the dimension of the representation corresponding to* χ.

Notice that for any solution g_0, g_1, g_∞ of the problem, the conjugates gg_ig^{-1} ($i = 0, 1, \infty$) are also solutions. So we arrive at the observation that the quantity $N(C_0, C_1, C_\infty)/n!$ counts the number of equivalence classes of $g_i \in C_i$ ($i = 0, 1, \infty$) with $g_0g_1g_\infty = \mathrm{Id}$ with a weight equal to $1/|\mathrm{Aut}(g_0, g_1, g_\infty)|$.

Notice that the conjugacy classes of S_n are in 1-1 correspondence with partitions of n, hence with sets of ramification indices that add up to n. So suppose we are given n, three partitions p_0, p_1, p_∞ of n corresponding to ramification indices above $0, 1, \infty$, and C_0, C_1, C_∞ their corresponding conjugacy classes. Then $N(C_0, C_1, C_\infty)/n!$ counts the number of generalised dessins with the given ramification data, where each dessin D is counted with weight $1/|\mathrm{Aut}(D)|$.

4. AN EXAMPLE: PLANAR DESSINS

As first examples we consider Belyi pairs $(\mathbb{P}^1, \phi)$, so we take $X = \mathbb{P}^1$. A Belyi map $\mathbb{P}^1 \to \mathbb{P}^1$ is called a *rational Belyi map*. The automorphism group of $\mathbb{P}^1$ is given by the fractional linear transforms $z \to \frac{az+b}{cz+d}$ where $ad - bc \neq 0$. So it is clear that if $\phi(z)$ is a rational Belyi map, then any equivalent one is given by $\phi\left(\frac{az+b}{cz+d}\right)$ with $ad - bc \neq 0$.

Consider a dessin in $\mathbb{P}^1$ corresponding to a rational Belyi map and suppose it does not contain ∞. After stereographic projection this dessin becomes a dessin in the complex plane, i.e., a two-coloured planar graph. The cyclic order of the edges around every vertex is induced by the positive orientation in the plane.

Now let $\phi : \mathbb{P}^1 \to \mathbb{P}^1$ be a Belyi map, which we write as $\phi(x) = \frac{A(x)}{B(x)}$ where $A(x), B(x)$ are polynomials with gcd equal to 1. Define $C(x) = A(x) - B(x)$. Then $\phi^{-1}(0), \phi^{-1}(\infty), \phi^{-1}(1)$ are the zeros of $A(x), B(x), C(x)$ respectively. Here ∞ is counted as a zero of $A(x)$ if $\deg(A) < \deg(B), \deg(C)$, and similarly for B, C. Let S be the set of distinct zeros of ABC (possibly including ∞) and let N be the degree of ϕ. Then our genus formula implies that $-2 = N - |S|$, and hence that

$$|S| = N + 2.$$

In fact, for any triple of polynomials $A(x), B(x), C(x)$ with $A(x) + B(x) + C(x) = 0$ and $\gcd(A, B, C) = 1$ we know that $|S| \geq N + 2$, where $N = \max\big(\deg(A), \deg(B), \deg(C)\big)$. This inequality is known as Mason-Stothers inequality or the ABC-theorem. So we see that Belyi maps provide us with optimal cases of Mason's inequality.

5. J-MAPS

In this paper we shall be interested in special Belyi-maps which we call J-maps: these have ramification order 3 in every point above 0 and unique ramification order 2 above 1. Because of this the degree of a J-map is always a multiple of 6, say $6m$. The corresponding dessin d'enfant now has vertices of multiplicity two in the points $\phi^{-1}(1)$. We might as well suppress these points and we are left with a graph with $3m$ edges and precisely $2m$ vertices, each with multiplicity 3. In the group theoretic description the element σ_0 now has cycle type $33\cdots 3$ ($2m$ threes) and σ_1 has cycle type $22\cdots 2$ ($3m$ twos).

Proposition 5.1. *Equivalence classes of J-maps are in 1-1 correspondence with torsion-free finite index subgroups Γ of $PSL(2,\mathbb{Z})$. The degree of the J-map equals the index $[PSL(2,\mathbb{Z}) : \Gamma]$.*

The proof of this Proposition is fairly straightforward and can be found in [12].

We remark that the subgroups in the above Proposition are in general not congruence subgroups. In fact the majority is not. However the congruence subgroups are of course of special interest. In recent work by Sebbar and McKay [9], [4] a complete classification is given of all torsion-free subgroups Γ that are congruence subgroups and such that $\mathcal{H}/\Gamma$ is a rational curve. For even more extensive computations see [2] and the references therein.

Proposition 5.2. *Let $J : X \to \mathbb{P}^1$ be a J-map. Let $n_J = \#J^{-1}(\infty)$, g the genus of X and $6m = \text{degree}(J)$. Then $2g - 2 = m - n_J$.*

Proof. This follows from Theorem 2.3 with $n_0 = 3m, n_1 = 2m, N = 6m$ and $n_\infty = n_J$. □

6. COUNTING RATIONAL J-MAPS

A J-map is called *rational* if $X = \mathbb{P}^1$. In this section we count the number of equivalence classes of rational J-maps of degree 24. One way to do this would be to draw all possible dessins d'enfant. However, this is a bit risky since it is easy to overlook possible graphs. That is why we shall do the drawing in conjunction with the group theoretic count using Theorem 3.2.

For a J-map of degree $6m$ the cycle types are $33\cdots 3$ above 0 ($2m$ threes, this is C_0), $22\cdots 2$ above 1 ($3m$ twos, this is C_1) and a partition of $6m$ above ∞ into $m + 2$ parts (this is C_∞). In the Appendix we have given the table of values for $N_{C_0 C_1 C_\infty}/(6m)!$ for these cycle types in the cases $m = 4$. We do not include the details of this computation in this paper, since the calculation of the character table of S_{24} has been rather cumbersome. Suffice it to say that we used the software package **Lie** with some special tweaks to find the table.

To enumerate all rational J-maps with $m = 4$ we consider each case in Table 1 from the Appendix in the following way. Suppose for example that C_∞ is given by the partition 14-6-1-1-1-1 of 24. In Table 1 we find the counting number $25/12$. Let us draw all dessins corresponding to these data, where we do not draw the vertices above 1 and the vertices above 0 are recognizable by the three-fold branchings in the picture.

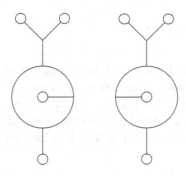

These are two mirror images (the orientations are induced by the orientation in the plane) and each has trivial automorphism group. So they contribute 2 to our character formula. The remaining 1/12 are accounted for by the following generalised dessin

The smaller dessin has opposite orientations at the vertices, so it is not a planar dessin. It has an automorphism group of order 6. The larger dessin is planar and has an automorphism group of order 2. So the total automorphism group has order 12, which explains the missing 1/12 in our character formula. Moreover, we are now certain that we have listed all possible dessins corresponding to the partition 14-6-1-1-1-1. On the website we have only pictured the connected dessins.

7. COMPUTATION OF RATIONAL J-MAPS

In this section we describe a method to compute rational J-maps efficiently, in the sense that it gives us answers in a reasonable time when $m \leq 4$. We have not explored the cases when $m \geq 5$. Moreover, due to condition (2) later on, we must require that the partition of $6m$ into $m + 2$ parts must contain at least one part $\geq m$. This is always true when $m \leq 5$.

Proposition 7.1. *Let ϕ be a rational J-map of degree $6m$ such that $\phi(\infty) = \infty$. Then there exist polynomials $c_4, c_6 \in \mathbb{C}[t]$ of degrees $2m, 3m$ respectively and a polynomial Δ of degree $< 6m$ such that:*

(1) $\gcd(c_4, c_6) = 1$,

(2) $c_4^3 - c_6^2 = \Delta$,
(3) Δ has $m + 1$ distinct zeros,
(4) $\phi = c_4^3/\Delta$.

Proof. Since ϕ is a rational function which ramifies of order 2 in every point above 1 and order 3 in every point above 0, there exist polynomials c_4, c_6, Δ such that $\phi(t) = c_4^3/\Delta = 1 + c_6^2/\Delta$. Hence $c_4^3 - c_6^2 = \Delta$. Since ϕ has degree $6m$, the degrees of c_4, c_6, Δ follow. From Proposition 5.2 it follows that Δ has precisely $n_J - 1$ distinct zeros (∞ is left out) which equals $(m + 2 - 2g) - 1 = m + 1$.

$\square$

Proposition 7.2. *Let the notation and assumptions be as in the previous Proposition, and Δ' be the derivative of Δ. Define $\delta = \gcd(\Delta, \Delta')$ and $\Delta = p\delta$, $\Delta' = q\delta$, where the leading coefficient of δ is chosen the same as that of Δ, i.e., p is monic. Then c_4, c_6, Δ can be normalised in such a way that*

$$
\begin{aligned}
c_6 &= c_4 q - 3c_4' p, \\
c_4^2 &= c_6 q - 2c_6' p, \\
\delta &= 3c_4' c_6 - 2c_6' c_4.
\end{aligned}
$$

Moreover, there exists a polynomial l of degree $\leq m-1$ such that $c_4 = q^2 + lp$.

Proof. The identity $c_4^3 - c_6^2 = p\delta$ and its derivative $3c_4^2 c_4' - 2c_6 c_6' = q\delta$ can be rewritten in vector notation as

$$
c_4^2 \begin{pmatrix} c_4 \\ 3c_4' \end{pmatrix} + c_6 \begin{pmatrix} -c_6 \\ -2c_6' \end{pmatrix} + \delta \begin{pmatrix} -p \\ -q \end{pmatrix} = 0.
$$

We consider this as a system of linear equations in the unknowns c_4^2, c_6. The coefficient determinant is $2c_4 c_6' - 3c_6 c_4'$. Suppose it vanishes identically. This implies $(c_6^2/c_4^3)' = 0$, hence c_6^2/c_4^3 is in $\mathbb{C}$, which cannot happen since c_4, c_6 are relatively prime, non-constant polynomials. We conclude that $2c_4 c_6' - 3c_6 c_4' \neq 0$ and solving the equation gives us the existence of $\lambda \in \mathbb{C}(z)^*$ such that

$$
\lambda \begin{pmatrix} c_6 \\ c_4^2 \\ \delta \end{pmatrix} = \begin{pmatrix} c_4 q - 3c_4' p \\ c_6 q - 2c_6' p \\ 3c_4' c_6 - 2c_4 c_6' \end{pmatrix}.
$$

Since $\gcd(c_4, c_6) = 1$ we can infer that $\lambda \in \mathbb{C}[z]$. We note that the degrees of p, q are $m + 1$ and m respectively. So from the first equation it follows that

$$
\deg(\lambda) + 3m \leq \deg(c_4) + \deg(q) = 2m + m = 3m.
$$

Hence $\deg(\lambda) \leq 0$ and so $\lambda \in \mathbb{C}^*$. By making the substitution $c_4 \mapsto \lambda^{-2}c_4$, $c_6 \mapsto \lambda^{-3}c_6$, $\delta \mapsto \lambda^{-6}\delta$ we can see to it that the new λ equals 1.

Finally notice that p and c_4 are relatively prime. Consider the first two equations modulo p. We obtain $c_6 \equiv c_4 q \pmod{p}$ and $c_4^2 \equiv c_6 q \pmod{p}$.

Elimination of c_6 yields $c_4^2 \equiv c_4 q^2$ (mod p). Since c_4 and p are relatively prime this implies $c_4 \equiv q^2$ (mod p) from which our last assertion follows.

$\square$

Suppose we wish to compute the rational J-maps with cycle type $n_1, n_2,$ $\ldots, n_{m+1}, n_{m+2}$ above ∞. In particular $n_i > 0$ for all i and $\sum_i n_i = 6m$. We assume that $J(\infty) = \infty$ and that ∞ has ramification order n_{m+2}. The polynomial Δ in Proposition 7.2 has the form

$$\Delta = \Delta_0 \prod_{i=1}^{m+1} (z - a_i)^{n_i}, \Delta_0 \in \overline{\mathbb{Q}}^*.$$

The polynomials p, q in Proposition then have the form

$$p = \prod_{i=1}^{m+1} (z - a_i), \qquad q = p \sum_{i=1}^{m+1} \frac{n_i}{z - a_i}. \tag{1}$$

Proposition 7.3. *Suppose we have polynomials p, q given as in (1). Suppose there exist polynomials c_4, c_6 of degrees $2m, 3m$ in z such that*

$$c_6 = c_4 q - 3 c_4' p \qquad c_4^2 = c_6 q - 2 c_6' p.$$

Then $c_4^3 - c_6^2$ is proportional to $\prod_{i=1}^{m+1} (z - a_i)^{n_i}$. Moreover, if there exists a polynomial l of degree $\leq m-1$ such that $c_4 = q^2 + lp$, then c_4, c_6 are relatively prime.

Proof. From the two equations for c_4, c_6 it easily follows that

$$c_4^2 (c_4 q - 3 c_4' p) = c_6 (c_6 q - 2 c_6' p).$$

Hence

$$(c_4^3 - c_6^2) q = (c_4^3 - c_6^2)' p$$

and thus

$$\frac{(c_4^3 - c_6^2)'}{(c_4^3 - c_6^2)} = \frac{q}{p} = \sum_{i=1}^{m+1} \frac{n_i}{z - a_i}.$$

In other words, $c_4^3 - c_6^2$ is a constant times $\prod_i (z - a_i)^{n_i}$.

Suppose $c_4 = q^2 + lp$. Since p, q have no common zeros, the same holds for c_4 and p. Suppose c_4, c_6 have a common zero ρ. Then, from the first equation it follows that $\mathrm{ord}_\rho(c_6) = \mathrm{ord}_\rho(c_4) - 1$. From the second it follows that $2\mathrm{ord}_\rho(c_4) = \mathrm{ord}_\rho(c_6) - 1$. This gives a contradiction. Hence c_4 and c_6 are relatively prime.

$\square$

Given $p(z), q(z)$ as above, we solve the system of equations

$$\begin{aligned} c_6 &= c_4 q - 3 c_4' p, \\ c_4^2 &= c_6 q - 2 c_6' p, \\ c_4 &= q^2 + lp, \end{aligned}$$

in polynomials c_4, c_6, l of degrees $2m, 3m, m-1$. We can eliminate c_4, c_6 from these equations to obtain the following single equation for $l(z)$:

$$
\begin{aligned}
0 = \ & p\big(pl^2 + q^2 l + 5p'ql - 6(p')^2 l + 2pq'l - 6pp''l \\
& + 5pql' - 18pp'l' + 12q^2 q' - 6p^2 l'' - 12p'qq' - 12p(q')^2 - 12pqq''\big)
\end{aligned}
$$

Note that we can divide by p on both sides. Denote by Q the polynomial on the right between parentheses. In principle Q has degree $3m - 1$. We now write $l = \sum_{i=0}^{m-1} l_i z^i$ and determine the coefficients l_i recursively for $i = m - 1, m - 2, \ldots$ by setting the coefficient of $z^{3m-1}, z^{3m-2}, \ldots$ in Q equal to zero.

Let us first compute the leading coefficient of Q. The leading coefficient of p equals 1, the leading coefficient of q equals $N := \sum_{i=1}^{m+1} n_i$. Then it is straightforward to compute the leading coefficient of Q. It reads

$$
(l_{m-1} + N^2)(l_{m-1} + 12mN - 36m^2).
$$

Since $l_{m-1} + N^2$ is also the leading coefficient of c_4, which we assumed to be non-zero, we conclude that

$$
l_{m-1} = 36m^2 - 12mN.
$$

Fix $k < m - 1$. The coefficient l_k occurs in the expansion of Q as a polynomial in z. However, closer examination reveals that l_k does not occur in the coefficients of z^r in Q when $r > 2m + k$. When $r = 2m + k$ the coefficient l_k occurs linearly with coefficient

$$
(N - 5m - k - 1)(N - 12m + 6k + 6),
$$

where we have used our evaluation of l_{m-1}. Setting the coefficient of z^{2m+k} in Q equal to zero allows us to compute l_k in terms of l_i with $i > k$ and the a_i, n_i. In particular we see by induction that l_k is polynomial in the a_i. Notice that the calculation of l_k is only possible if $(N - 5m - k - 1)(N - 12m + 6k + 6)$ is non-zero. The procedure will work whenever $N \leq 5m$. Add n_{m+2} on both sides and use $N + n_{m+2} = 6m$ to find that the latter condition is equivalent to

$$
n_{m+2} \geq m. \tag{2}
$$

When we restrict to $m \leq 4$, as we will do here, this does not give us problems.

Once we have have expressed the coefficients l_k in terms of the zeros a_i, we can substitute this in the equation $Q = 0$. By construction, the coefficients of z^j will be zero for $j = 2m, 2m + 1, \ldots, 3m - 1$. The requirement that the coefficients of z^j for $j = 0, 1, \ldots, 2m - 1$ should be zero provides us with a set of polynomial equations for the unknown points $a_1, a_2, \ldots, a_{m+1}$.

8. A SAMPLE COMPUTATION

As an illustration of the algorithm from the previous section we compute the J-map corresponding to the case $m = 4$ and the partition 19-1-1-1-1-1 above ∞.

We assume $n_6 = 19$ and take Δ proportional to $z^5 + az^4 + bz^3 + cz^2 + dz + e$. We do this rather than taking Δ proportional to $\prod_{i=1}^{5}(z - a_i)$. In the latter case the numbers a_i may turn out to be algebraic, which may complicate our calculations. Furthermore, by shifting z over a constant we can see to it that $a = 0$. So Δ is taken proportional to $z^5 + bz^3 + cz^2 + dz + e$.

We find easily that $p = z^5 + bz^3 + cz^2 + dz + e$ and $q = 5z^4 + 3bz^2 + 2cz + d$. We now use Proposition 7.2 to get $c_4 = q^2 + (l_3 z^3 + l_2 z^2 + l_1 z + l_0)p$ where $l_3 = -576$. Substitute this in $c_6 = c_4 q - 3c'_4 p$ and substitute that in $(c_4^2 - c_6 q)/p + 2c'_6 = 0$. We get the equation

$$450 l_2 z^{10} + (l_2^2 + 527 l_1 - 64872b)z^9 + \cdots = 0.$$

Setting the coefficient of z^{10} equal to zero gives us $l_2 = 0$. We are left with the equation

$$(527 l_1 - 64872b)z^9 + (592 l_0 - 43392c)z^8 + \cdots = 0.$$

Setting the coefficient of z^9 equal to zero gives us $l_1 = 64872b/527$. Notice also that setting the coefficient of z^8 equal to zero gives us $l_0 = 43392c/592$. We are left with an equation of the form

$$Q_7 z^7 + Q_6 z^6 + \cdots + Q_0 = 0,$$

where

$$
\begin{aligned}
Q_7 &= -25920d + 6486480b^2/961 \\
Q_6 &= 12096e - 13384224bc/1147 \\
Q_5 &= 855360bd/31 + 6486480c^2/1369 \\
&\quad + 6486480b^3/961
\end{aligned}
$$

$$\vdots \qquad \vdots$$

Setting Q_5, Q_6, Q_7 equal to zero gives us a set of polynomal equations in b, c, d, e whose solution reads

$$b = 62t^2, \quad c = 148t^3, \quad d = 1001t^4, \quad e = 8852t^5,$$

where t can be chosen arbitrarily. Surprisingly enough, all coefficients Q_i ($i = 0, \ldots, 7$) vanish for this choice of b, c, d, e. Hence we have found a solution which, choosing $t = 1$, reads

$$
\begin{aligned}
c_4 &= 19^2(z^8 + 84z^6 + 176z^5 + 2366z^4 + 13536z^3 \\
&\quad + 26884z^2 + 218864z + 268777), \\
c_6 &= 19^3(z^{12} + 126z^{10} + 264z^9 + 6195z^8 + 31392z^7 \\
&\quad + 163956z^6 + 1260528z^5 + 3531639z^4 \\
&\quad + 19770400z^3 + 62912622z^2 + 94024776z + 291742453), \\
\Delta &= c_4^3 - c_6^2 \\
&= -2^{38}3^3 19^6(z^5 + 62z^3 + 148z^2 + 1001z + 8852).
\end{aligned}
$$

Always, when the computation is done, one can reduce the complexity of the coefficients to a large extent by performing a variable substitution. In our example we substitute z by $4z - 1$ and obtain a new relation

$$
c_4^3 - c_6^2 = -16(4z^5 - 5z^4 + 18z^3 - 3z^2 + 14z + 31),
$$

where

$$
\begin{aligned}
c_4 &= 4(z^8 - 2z^7 + 7z^6 - 6z^5 + 11z^4 + 4z^3 + 12z + 1), \\
c_6 &= 4(2z^{12} - 6z^{11} + 24z^{10} - 38z^9 + 78z^8 \\
&\quad - 48z^7 + 60z^6 + 72z^5 - 30z^4 + 120z^3 + 27z^2 + 9z + 29).
\end{aligned}
$$

9. APPENDIX

In this section we list the values of $N(C_\infty) := N_{C_0 C_1 C_\infty}/24!$ for S_{24}, where C_0 has cycle type consisting of 8 three's, C_1 has cycle type consisting of 12 two's and where C_∞ consists of exactly six cycles. We list only those cases for which $N(C_\infty) \neq 0$. The first column lists the partition, the second the corresponding value of $N(C_\infty)$. In the third column we find the number of irreducible dessins. They are stated in the form $m\alpha$ or $m\alpha + n\beta$, where m, n indicate the length(s) of the Galois orbit(s) and α, β indicate the weights with which the dessin contributes to $N(C_\infty)$. Sometimes we find the entry 0 in the third column, meaning that only reducible dessins belong to that case. Also note that due to the occurrence of reducible dessins the number in the third column need not equal $N(C_\infty)$. We have not made any effort to describe the reducible (generalised) dessins. The polynomials in the fourth column are the defining polynomials for the field of moduli (equalling the field of definition in these cases) of the dessins enumerated in the third column.

The abbreviation n.a. stands for 'not applicable'.

Table 1: Counting dessins in the case $m = 4$

Partition						$N(C_\infty)$	# dessins	fields
19	1	1	1	1	1	1	$1 \cdot 1$	X
18	2	1	1	1	1	$5/2$	$1 \cdot 1/2 + 2 \cdot 1$	$X, X^2 + 3$
17	3	1	1	1	1	1	$1 \cdot 1$	X

Table 1: Counting dessins in the case $m = 4$, (continued)

Partition						$N(C_\infty)$	# dessins	fields
17	2	2	1	1	1	2	$2 \cdot 1$	$X^2 - 17$
16	4	1	1	1	1	11/4	$1 \cdot 1/4$	X
16	3	2	1	1	1	3	$1 \cdot 1 + 2 \cdot 1$	$X, X^2 + 2$
16	2	2	2	1	1	4/3	$1 \cdot 1/2$	X
15	5	1	1	1	1	2	$2 \cdot 1$	$X^2 + 15$
15	4	2	1	1	1	9/2	$2 \cdot 1$	$X^2 + 15$
15	3	3	1	1	1	1	$1 \cdot 1$	X
15	3	2	2	1	1	3	$1 \cdot 1 + 2 \cdot 1$	$X, X^2 + 15$
15	2	2	2	2	1	5/6	0	n.a.
14	6	1	1	1	1	25/12	$2 \cdot 1$	$X^2 + 3$
14	5	2	1	1	1	4	$1 \cdot 1 + 3 \cdot 1$	$X, X^3 + 2X + 2$
14	4	3	1	1	1	5/2	$1 \cdot 1$	X
14	4	2	2	1	1	5/2	$2 \cdot 1$	$X^2 + 7$
14	3	3	2	1	1	2	$1 \cdot 1 + 1 \cdot 1$	X, X
14	3	2	2	2	1	3/2	$1 \cdot 1$	X
14	2	2	2	2	2	1/6	0	n.a.
13	7	1	1	1	1	1	$1 \cdot 1$	X
13	6	2	1	1	1	13/6	$2 \cdot 1$	$X^2 + 3$
13	5	3	1	1	1	1	$1 \cdot 1$	X
13	5	2	2	1	1	2	$2 \cdot 1$	$X^2 - 65$
13	4	4	1	1	1	1	0	n.a.
13	4	3	2	1	1	2	$1 \cdot 1$	X
13	4	2	2	2	1	1/3	0	n.a.
13	3	3	2	2	1	1	$1 \cdot 1$	X
13	3	2	2	2	2	1/3	0	n.a.
12	7	2	1	1	1	2	$2 \cdot 1$	$X^2 + 3$
12	6	3	1	1	1	19/18	$1 \cdot 1$	X
12	6	2	2	1	1	19/12	$1 \cdot 1 + 1 \cdot 1/2$	X, X
12	5	4	1	1	1	1	0	n.a.
12	5	3	2	1	1	4	$4 \cdot 1$	$X^4 - 3X^2 + 6$
12	5	2	2	2	1	4/3	$1 \cdot 1$	X
12	4	4	2	1	1	7/4	$1 \cdot 1$	X
12	4	3	3	1	1	11/6	$1 \cdot 1 + 1 \cdot 1/2$	X, X
12	4	3	2	2	1	3	$1 \cdot 1 + 2 \cdot 1$	$X, X^2 + 3$
12	4	2	2	2	2	1/4	0	n.a.
12	3	3	3	2	1	1	$1 \cdot 1$	X
12	3	3	2	2	2	11/18	$1 \cdot 1/2$	X
11	9	1	1	1	1	4/3	$1 \cdot 1$	X
11	8	2	1	1	1	7/2	$3 \cdot 1$	$X^3 - X^2 + X + 1$
11	7	3	1	1	1	3	$3 \cdot 1$	$X^3 + X^2 - 3$

Table 1: Counting dessins in the case $m = 4$, (continued)

Partition						$N(C_\infty)$	# dessins	fields
11	7	2	2	1	1	2	$2 \cdot 1$	$X^2 + 7$
11	6	4	1	1	1	37/12	$2 \cdot 1$	$X^2 - 33$
11	6	3	2	1	1	25/6	$3 \cdot 1$	$X^3 + X^2 - X + 1$
11	6	2	2	2	1	13/36	0	n.a.
11	5	5	1	1	1	3/2	$1 \cdot 1$	X
11	5	4	2	1	1	7/2	$3 \cdot 1$	$X^3 + X^2 - 10X + 10$
11	5	3	3	1	1	2	$2 \cdot 1$	$X^2 - 5$
11	5	3	2	2	1	1	$1 \cdot 1$	X
11	5	2	2	2	2	1/6	0	n.a.
11	4	4	3	1	1	3/2	$1 \cdot 1$	X
11	4	4	2	2	1	1/4	0	n.a.
11	4	3	3	2	1	1	$1 \cdot 1$	X
11	4	3	2	2	2	1/6	0	n.a.
11	3	3	3	3	1	1/12	0	n.a.
10	10	1	1	1	1	5/4	$1 \cdot 1/4 + 2 \cdot 1/2$	$X, X^2 - 5$
10	9	2	1	1	1	13/6	$2 \cdot 1$	$X^2 - 5$
10	8	3	1	1	1	1	$1 \cdot 1$	X
10	8	2	2	1	1	1/4	0	n.a.
10	7	4	1	1	1	5/2	$1 \cdot 1$	X
10	7	3	2	1	1	2	$2 \cdot 1$	$X^2 - 21$
10	7	2	2	2	1	3/2	$1 \cdot 1$	X
10	6	5	1	1	1	25/6	$1 \cdot 1 + 3 \cdot 1$	$X, X^3 - X^2 - 3X - 3$
10	6	4	2	1	1	71/24	$1 \cdot 1$	X
10	6	3	3	1	1	1/12	0	n.a.
10	6	3	2	2	1	5/3	$1 \cdot 1$	X
10	6	2	2	2	2	19/72	0	n.a.
10	5	5	2	1	1	5/4	$2 \cdot 1/2$	$X^2 - 5$
10	5	4	3	1	1	1/2	0	n.a.
10	5	4	2	2	1	3	$3 \cdot 1$	$X^3 - X^2 + 2X + 2$
10	5	3	2	2	2	7/6	$1 \cdot 1$	X
10	4	4	4	1	1	1/4	0	n.a.
10	4	4	3	2	1	1	$1 \cdot 1$	X
10	4	4	2	2	2	5/24	0	n.a.
10	4	3	3	2	2	1	$1 \cdot 1$	X
10	3	3	3	3	2	1/24	0	n.a.
9	9	3	1	1	1	1/9	0	n.a.
9	9	2	2	1	1	2	$1 \cdot 1/2 + 3 \cdot 1/2$	$X, X^3 - 3X - 4$
9	8	4	1	1	1	7/12	0	n.a.
9	8	3	2	1	1	19/6	$3 \cdot 1$	$X^3 + 2$
9	8	2	2	2	1	1/6	0	n.a.

Table 1: Counting dessins in the case $m = 4$, (continued)

Partition						$N(C_\infty)$	# dessins	fields
9	7	5	1	1	1	1	$1 \cdot 1$	X
9	7	4	2	1	1	2	$2 \cdot 1$	$X^2 + 7$
9	7	3	2	2	1	3	$3 \cdot 1$	$X^3 - 6X + 12$
9	6	6	1	1	1	55/216	0	n.a.
9	6	5	2	1	1	19/6	$3 \cdot 1$	$X^3 - 3X + 3$
9	6	4	3	1	1	37/36	$1 \cdot 1$	X
9	6	3	3	2	1	10/3	$3 \cdot 1$	$X^3 + 2$
9	6	3	2	2	2	1/108	0	n.a.
9	5	5	3	1	1	1/6	0	n.a.
9	5	5	2	2	1	1	$1 \cdot 1$	X
9	5	4	3	2	1	3	$3 \cdot 1$	$X^3 + 3X - 6$
9	5	3	3	3	1	1	$1 \cdot 1$	X
9	4	4	3	2	2	1/12	0	n.a.
9	4	3	3	3	2	1	$1 \cdot 1$	X
9	3	3	3	3	3	1/36	0	n.a.
8	8	4	2	1	1	9/8	$1 \cdot 1/2$	X
8	8	3	3	1	1	3/2	$1 \cdot 1/2 + 2 \cdot 1/2$	$X, X^2 + 2$
8	8	2	2	2	2	7/24	$1 \cdot 1/8$	X
8	7	6	1	1	1	2	$2 \cdot 1$	$X^2 + 3$
8	7	5	2	1	1	3	$3 \cdot 1$	$X^3 + 2X - 2$
8	7	4	3	1	1	5/2	$2 \cdot 1$	$X^2 + 6$
8	7	4	2	2	1	2	$2 \cdot 1$	$X^2 + 7$
8	7	3	3	2	1	1	$1 \cdot 1$	X
8	7	3	2	2	2	1/6	0	n.a.
8	6	6	2	1	1	307/144	$2 \cdot 1$	$X^2 + 3$
8	6	5	3	1	1	2	$2 \cdot 1$	$X^2 - 5$
8	6	5	2	2	1	7/6	$1 \cdot 1$	X
8	6	4	4	1	1	13/48	0	n.a.
8	6	4	3	2	1	29/12	$2 \cdot 1$	$X^2 - 2$
8	6	4	2	2	2	157/144	$1 \cdot 1$	X
8	6	3	3	2	2	1/12	0	n.a.
8	5	5	4	1	1	3/8	0	n.a.
8	5	5	2	2	2	1/12	0	n.a.
8	5	4	3	3	1	2	$2 \cdot 1$	$X^2 - 10$
8	5	4	3	2	2	1	$1 \cdot 1$	X
8	4	4	4	3	1	1	$1 \cdot 1$	X
8	4	4	4	2	2	9/16	$1 \cdot 1/2$	X
8	4	3	3	3	3	1/48	0	n.a.
7	7	7	1	1	1	1/3	$1 \cdot 1/3$	X
7	7	6	2	1	1	1/6	0	n.a.

Table 1: Counting dessins in the case $m = 4$, (continued)

Partition						$N(C_\infty)$	# dessins	fields
7	7	5	3	1	1	2	$2 \cdot 1$	$X^2 - 21$
7	7	4	4	1	1	5/4	$2 \cdot 1/2$	$X^2 + 7$
7	7	4	2	2	2	1/12	0	n.a.
7	7	3	3	2	2	1	$2 \cdot 1/2$	$X^2 - 7$
7	6	6	3	1	1	1/6	0	n.a.
7	6	6	2	2	1	1	$1 \cdot 1$	X
7	6	5	4	1	1	1	$1 \cdot 1$	X
7	6	5	3	2	1	19/6	$3 \cdot 1$	$X^3 + 2X - 2$
7	6	4	4	2	1	1	$1 \cdot 1$	X
7	6	4	3	3	1	1/6	0	n.a.
7	5	5	4	2	1	2	$2 \cdot 1$	$X^2 - 2$
7	5	5	3	2	2	1	$1 \cdot 1$	X
7	5	4	4	3	1	1	$1 \cdot 1$	X
7	5	4	3	3	2	1	$1 \cdot 1$	X
6	6	6	4	1	1	1855/2592	$1 \cdot 1/2$	X
6	6	6	3	2	1	19/72	0	n.a.
6	6	6	2	2	2	1855/7776	$1 \cdot 1/6$	X
6	6	5	5	1	1	175/144	$2 \cdot 1/2$	$X^2 - 3$
6	6	5	4	2	1	7/6	$1 \cdot 1$	X
6	6	5	3	3	1	1	$1 \cdot 1$	X
6	6	4	4	2	2	187/288	$1 \cdot 1/2$	X
6	6	4	3	3	2	1/2	$1 \cdot 1/2$	X
6	6	3	3	3	3	235/864	$1 \cdot 1/4$	X
6	5	5	3	3	2	1/12	0	n.a.
6	5	4	4	3	2	1	$1 \cdot 1$	X
6	4	4	4	3	3	1/36	0	n.a.
5	5	5	5	2	2	1/4	$1 \cdot 1/4$	X
5	5	4	4	3	3	1/2	$1 \cdot 1/2$	X
4	4	4	4	4	4	1/24	$1 \cdot 1/24$	X

REFERENCES

[1] G. Belyi, On Galois extensions of a maximal cyclotomic field, *Izv. Akad. Nauk USSR, Ser. Mat.* **43:2** (1979), 269–276 (in Russian). English translation: *Math USSR Izv.* **14** (1979), 247–256.

[2] C.J. Cummins, Congruence subgroups of groups commensurable with PSL(2,Z) of genus 0 and 1, *Experimental Math.* **13** (2004), 361–382.

[3] Y. Iron, *An explicit presentation of a K3 surface that realizes [1,1,1,1,1,19]*, MSc. Thesis, Hebrew University of Jerusalem, 2003.

[4] J. McKay, A. Sebbar, Arithmetic semistable elliptic surfaces, *Proceedings on Moonshine and related topics (Montréal, QC, 1999)*, *CRM Proc. Lecture Notes* **30**, 119–130, Providence, RI, 2001.

[5] R. Miranda, U. Persson, Configurations of I_n fibers on elliptic $K3$ surfaces, *Math. Z.* **201** (1989), 339–361.

[6] L. Schneps (editor), *The Grothendieck theory of dessins d'enfants*, LMS Lecture Notes **200**, CUP 1994.

[7] M. Schütt, Elliptic fibrations of some extremal $K3$ surfaces, *Rocky Mountain J. of Math.* **37** (2007), 609–652.

[8] M. Schütt, J. Top, Arithmetic of the [19,1,1,1,1,1] fibration, *Comm. Math. Univ. St.Pauli* **55** (2006), 9–16.

[9] A. Sebbar, Classification of torsion-free genus zero congruence groups, *Proc. Amer. Math. Soc.* **129** (2001), 2517–2527 (electronic).

[10] J.-P. Serre, *Lectures on the Mordell-Weil Theorem*, Vieweg Verlag 1989.

[11] ———, *Topics in Galois theory*, A.K.Peters 1992.

[12] W.W. Stothers, Polynomial identities and hauptmoduln, *Quart.J. Math. Oxford (2)* **32** (1981), 349–370.

[13] J. Top and N. Yui, Explicit equations of some elliptic modular surfaces, *Rocky Mountain J. of Math.* **37** (2007), 663–688.

THE MERIT FACTOR PROBLEM

PETER BORWEIN, RON FERGUSON, AND JOSHUA KNAUER

ABSTRACT. The merit factor problem is of considerable practical interest
to communications engineers and theoretical interest to number theorists.
For binary sequences, although it is generally believed that the merit factor
is bounded, it still has not been completely established that the number
of even length Barker sequences, each with merit factor N, is bounded.
In this paper, we present an overview of the problem and results of quite
extensive searches we have conducted in lengths up to slightly beyond 200.

1. INTRODUCTION

For the sequence $A = [a_1, a_2, \ldots, a_N]$ the *kth acyclic autocorrelation coefficient*, or *kth shift sidelobe* ($0 \leq k \leq N - 1$), is given by

$$c_k = \sum_{j=1}^{N-k} a_j \overline{a_{j+k}},$$

where the superimposed bar indicates the complex conjugate, in the case where the sequence takes complex values. Of particular interest are the polyphase sequences, where the modulus of each coefficient is 1. In these cases the 0th coefficient, or main lobe to engineers, is simply the length of the sequence. The other coefficients, or positive shift sidelobes, measure self-interference of a signal based on this sequence. This points, for example, to the use of signals based on sequences with low autocorrelation in radar detection. The energy in the kth shift sidelobe is defined as $|c_k|^2$ and higher sidelobe values correspond to energy inefficiencies in the signal. The base energy of the sequence is the total of the energies in these sidelobes, i.e.,

$$E = \sum_{i=1}^{N-1} |c_k|^2.$$

The merit factor, F, of the sequence relates energy in the sidelobes to energy in the main lobe,

$$F = \frac{N^2}{2E}.$$

2000 *Mathematics Subject Classification.* 11B83, 11Y55.
Key words and phrases. Integer sequences, autocorrelation, polynomials, merit factor, flatness.

The *merit factor* is a measure of the quality of the sequence in terms of engineering applications.

For number theorists, these coefficients arise in the expression for the modulus of a polynomial on the unit circle. For $p(z) = \sum_{j=1}^{N-k} a_j z^{j-1}$ the L_α norm of $p(z)$ on the unit circle C is given by

$$L_\alpha(p) = \left(\frac{1}{2\pi} \int_0^{2\pi} |p(e^{i\theta})|^\alpha d\theta \right)^{1/\alpha}.$$

In particular, for polyphase sequences and $\alpha = 2$ and $z \in C$, we have

$$|p(z)|^2 = f(z)\overline{f(z)} = N + \sum_{k=1}^{N-1} \left(c_k z^{-k} + \overline{c_k} z^k \right),$$

so that

$$L_2(p) = \sqrt{N}. \tag{1}$$

For the L_4-norm we then obtain

$$L_4(p)^4 = N^2 + 2\sum_{k=1}^{N-1} c_k^2$$

$$= N^2 \left(1 + \frac{1}{F} \right),$$

and

$$L_2(|p|^2 - N)^2 = L_4(p)^4 - N^2 = \frac{N^2}{F}. \tag{2}$$

This equation relates a higher merit factor with less deviation of $|p|$ from its L_2 average value of $\sqrt{N}$.

1.1. Barker sequences.

Barker sequences are sequences for which $|c_k| \leq 1$ for $1 \leq k < N$, i.e., each autocorrelation coefficient has absolute value less than or equal one. For binary sequences, this implies that $|c_k| = 1$ for autocorrelation sums of odd length while $|c_k| = 0$ for sums of even length. Barker [1] was interested in their use for pulse compression of radar signals. They exist for lengths 2,3,4,5,7,11,13 and conjecturally for no longer length. Storer and Turin [45] proved that there are none for odd lengths greater than 13. For even lengths the conjecture has been proved for lengths up to 10^{22} by Leung and Schmidt [35].

For sequences consisting of 3rd, 4th, or 6th roots of unity, the condition $|c_k| = 0$ or 1 still applies. For sequences consisting of higher roots of unity or for more general polyphase sequences, we can have $0 < |c_k| < 1$. Borwein and Ferguson [8] have shown the existence of such sequences up to length 63.

1.2. Skew-symmetric sequences. A binary sequence $A = [a_1, a_2, \ldots, a_N]$ is *symmetric* (or *reciprocal*) if $a_j = a_{N+1-j}$ for each j and *antisymmetric* if $a_j = -a_{N+1-j}$ for each j. A skew-symmetric sequence is formed by interleaving an odd length symmetric sequence with an even antisymmetric sequence of length greater or less by 1. This means that for skew-symmetric $A = [a_1, a_2, \ldots, a_N]$, we have

$$a_j a_{j+k} + a_{N+1-j-k} a_{N+1-j} = 0$$

for odd k, implying that all even length autocorrelation sums are 0. Since half of the sidelobe energies are zero, it is natural to search for sequences of high merit factor among skew-symmetric sequences.

The equivalent condition with generalized or polyphase sequences would require a conjugate skew-symmetric sequence, i.e., a sequence which is conjugate reciprocal interleaved with a sequence which is the negative of its conjugate reciprocal. A difference here is that the odd length sequence may be either conjugate reciprocal or the negative of its conjugate reciprocal. In either case, we have

$$a_j \overline{a_{j+k}} + a_{N+1-j-k} \overline{a_{N+1-j}} = 0$$

for odd k so that again the even length autocorrelation sums are all 0. Thus we may expect to find high merit factor sequences among this class. In practice, however, we find that more optimal examples are obtained among reciprocal sequences of odd length. Then for any k,

$$a_j \overline{a_{j+k}} + a_{N+1-j-k} \overline{a_{N+1-j}} = a_j \overline{a_{j+k}} + a_{j+k} \overline{a_j}$$

is purely real. In these cases, the imaginary part of each autocorrelation disappears, so, in effect, half of the sidelobe energy expansion terms disappear for these sequences as well.

The square of the middle entry is either 1 or -1. For binary sequences, this implies that an odd length symmetric sequence is interleaved with an even length antisymmetric sequence, since $i = \sqrt{-1}$ cannot be an entry.

In all of these cases, specifying the entries up to and including the middle term is enough to determine the entire sequence, so searching unrestricted sequences at length N is comparable in complexity to searching symmetrics, antisymmetrics and skew-symmetrics at approximately double the length.

Thus, with binary sequences, we have evidence that, with high probability, we have found the optimal merit factor sequences for binary sequences up to length 85, and for skew-symmetrics up to length 165. For polyphase sequences, we believe the optimal results to be highly accurate to length 45. The best examples found at longer odd lengths are often from searches restricted to symmetric searches.

1.3. Sequence equivalence. There are a number of operations for which the sidelobe energies of a sequence remain unchanged and which generate a group under composition. These include:

1. Multiplication of all entries by a constant of modulus 1.
2. Taking the complex conjugate of all entries.
3. Sequence reversal.
4. Multiplication of successive entries by linearly increasing powers of a constant of modulus 1.

As operations on the space of binary sequences, the second is redundant while the remaining three generate a noncommutative group of order 8. Multiplying all entries by -1 or every second entry by -1 will give a new sequence, so these operations applied to a sequence of length at least 2 will produce at least 4 different sequences. In most cases, sequence reversal will add 4 more. However, for symmetric, antisymmetric or skew-symmetric sequences the number produced remains at 4.

Using the first and fourth operations, any sequence can be transformed to a sequence with 1's as the first two entries. Using sequence reversal, followed again by transformation to a sequence with 1's as the first two entries, we obtain a second sequence, not necessarily different. Applying complex conjugation to these gives us a total of 4 from which to choose a canonical representative of the orbit class of the original sequence. One method is to compare successive entries and prefer a sequence for which the entry has a smaller argument. Normalization to this canonical form allows us to identify equivalence between sequences with the same base energy.

1.4. **Flat polynomials.** 'Flatness' of a polynomial on the unit circle is a term used to describe closeness of the modulus of its values to the average value over the whole circle. Equation 2 shows how the merit factor may be used to provide a measure for flatness of ± 1 polynomials. This is further illustrated in Figures 1, 2, 3 below, showing the modulus of the polynomial $p(z) = \sum_{j=1}^{n} a_j z^{j-1}$ on the unit circle, where each $A = [a_1, a_2, \ldots, a_{63}]$ is a sequence of length 63.

Figure 1 arises from the randomly generated binary sequence

$$42F11C5DFFE24B8E$$

in hexadecimal notation with merit factor 1.4185. Here the leading 4 converts to 0100 in binary and $1, -1, -1$ in terms of ± 1 coefficients (dropping the leading 0), while the following 2 converts to 0010 in binary and $-1, -1, 1, -1$ in ± 1 coefficients.

Figure 2 represents the polynomial formed for the binary sequence

$$6C9B015052F14339$$

with merit factor $F = 9.5870$, which we believe is optimal for binary sequences of this length.

Figure 3 is a graph of the modulus of the polynomial formed with coefficients from the polyphase Barker sequence of length 63 with entries $a_j = e^{2\pi i \phi_j}$

The merit factor problem

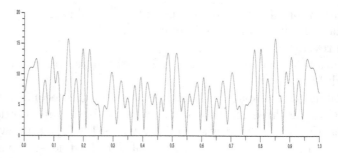

FIGURE 1. Modulus of $p(e^{2\pi it})$ for binary sequence with $F = 1.4185$ and $0 \le t \le 1$.

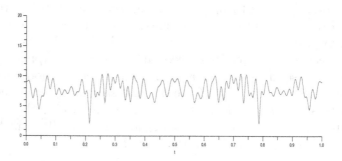

FIGURE 2. Modulus of $p(e^{2\pi it})$ for binary sequence with $F = 9.5870$.

with $\phi_1, \phi_2, \ldots, \phi_{63}$ having the values

0.000000, 0.000000, 0.044072, 0.100041, 0.124944, 0.044316, 0.915805,
0.834292, 0.896073, 0.072380, 0.153734, 0.145180, 0.264172, 0.409227,
0.678385, 0.779028, 0.703430, 0.582492, 0.464976, 0.434226, 0.137145,
0.048468, 0.004949, 0.928442, 0.365491, 0.394539, 0.867998, 0.074881,
0.666226, 0.614514, 0.194754, 0.471911, 0.761195, 0.956267, 0.323923,
0.119675, 0.556891, 0.854043, 0.099691, 0.332923, 0.935108, 0.561814,
0.731794, 0.132518, 0.422282, 0.875526, 0.519252, 0.026738, 0.368575,
0.879993, 0.399091, 0.939885, 0.425655, 0.919075, 0.551357, 0.209371,
0.855254, 0.577566, 0.272426, 0.992504, 0.662106, 0.376538, 0.022081,

which has $F = 37.5022$.

A high merit factor does not guarantee uniformity of closeness, since there may be narrow domains where spikes occur as illustrated in Figure 2. Still, the tendency toward a more uniform flatness with increasing merit factor is shown.

In his book [37], Littlewood introduced this notion of flatness and its expression in terms of the L_2 and L_4 norms. At that time, a plot of known optimal merit factor values would have had the appearance of Figure 4.

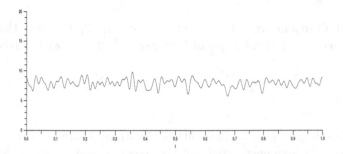

FIGURE 3. Modulus of $p(e^{2\pi it})$ for polyphase sequence with $F = 37.5022$.

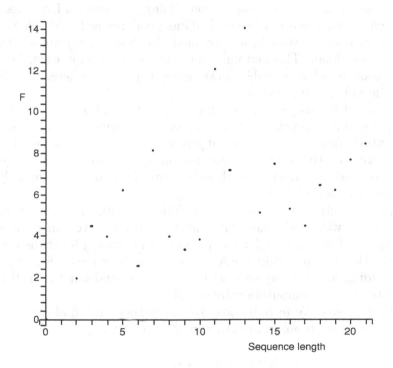

FIGURE 4. Optimal merit factor values for binary sequences to length 21.

The values for lengths 2,3,4,5,7,11, and 13 are in linear succession and correspond to Barker sequences. These lend considerable bias to the picture, the other points being more scattered. This may have been what led him to suggest the existence of an infinite sequence of polynomials p_{N_i} with merit factor of order $\sqrt{N_i}$ which would imply, in particular, that merit factors are unbounded. He further formulated the following conjecture:

Littlewood Conjecture: There exist constants C_1, C_2 such that we can find a sequence of polynomials p_N of increasing degree N with ± 1-coefficients such that

$$C_1 \sqrt{N} < |p_N(z)| < C_2 \sqrt{N}$$

for all z on C.

The Rudin-Shapiro polynomials satisfy the upper bound of this conjecture. The existence of a sequence satisfying the lower bound, however, has not yet been confirmed. In the extension to polyphase sequences, Kahane [29] has both confirmed the conjecture and the existence of sequences with unbounded merit factors.

In contrast to Littlewood's suggestion, Golay [21], using a hypothesis that sidelobe energies move toward statistical independence as the length of binary sequences increase, developed an argument to show an asymptotic limit of 12.32... as maximal. This certainly does not settle the question, but gives an expression to what probably most researchers now believe, i.e., that an asymptotic upper bound exists.

It was noted by Turyn and later proved [26] that Legendre sequences rotated by 1/4 of their lengths will provide sequences with merit factors having an asymptotic limit of 6. A few authors suggested that this might be optimal. More recently, a further construction applied to Legendre sequences appears to produce sequences with asymptotic limit of approximately 6.34 for the merit factor [6], [33].

Computationally, this is still a very difficult problem. From results in the ranges for which adequate data can be collected, we can project that an asymptotic limit of $F > 7$ for sequences of increasing length is certainly expected. There is good evidence for $F > 8$ as well and even $F > 9$ appears likely. Finding another sequence with $F > 10$ beyond length 13, if indeed such exists, may be computationally out of range.

More comprehensive introductions to the history and applications of the binary merit factor problem are given in [25] and [28].

2. SEARCH ALGORITHMS

Where a sequence has entries drawn from a finite alphabet, e.g., the Kth roots of unity, the number of sequences at a given length is finite. Thus, finding the optimal sequences of a given type up to a fixed length by checking the whole space is theoretically possible, but quickly becomes impractical as the length increases. More clever methods may be applied, which eliminate vast sections of the space from consideration as the search progresses in order to confirm optimal examples.

In contrast, at least beyond very short lengths, the search space for optimal polyphase sequences is not finite. Since the range of each coordinate is continuous, methods of calculus may be used to transform this into a finite problem. However, the number of local maxima and minima proliferates as the length increases, so this type of exhaustive search bogs down quite quickly.

Greater reach is achieved by using directed stochastic methods. Such algorithms include direct descent, simulated annealing, great deluge, genetic, and tabu search. Here optimality is not confirmed, but on the assumption that the number of locally optimal solutions within a given range is finite and the search method is not biased in locating these, statistical analysis using capture-recapture [40],[36] or the inverse collectors problem [34] can establish levels of confidence.

2.1. Exhaustive searches. For binary sequences, this is essentially a 2^{N-2} problem. Using a Gray code helps minimize recalculation through the iteration. To obtain precise information on base energy distributions we have conducted exhaustive searches up to length 44 for the general case and length 89 for skew-symmetrics.

2.2. Branch and bound. This is the method which has been used to confirm optimality of merit factor for general sequences up to length 60 [38], and for skew-symmetric sequences up to length 109 [9]. It uses the fact that the shorter and increasing length autocorrelation sums only involve terms progressing from the ends of a sequence to the middle. As a sequence is developed in this way, more sidelobe energy values become determined and better lower bounds for others are established. If this sum already exceeds some predetermined bound, then continued development can be aborted and the iteration passed to an earlier stage. The amount of truncation and thus the speed of the search space depends on this bound. A known sequence of low base energy supplies a good initial bound. This can be replaced when a better example is found.

2.3. Directed stochastic searches. A search starts either with a sequence in which all the coordinate values are randomly generated or a random selection of the coordinate values of an existing sequence are regenerated. A coordinate position is chosen either at random or by some directed method, a change in value is proposed and either accepted or rejected according to some selection criterion, which then returns either an altered or the same sequence accordingly. This second process repeats until either no further improvement is possible using this method or some other limit is reached. The process then either reverts to an earlier position or restarts. The decision to discontinue may be based on either time, computational resources, or an estimation that any improvement is unlikely.

For the descent method, changes are allowed only if there is an improvement in quality. Initially the coordinates for proposed changes are chosen at random, though a final stage may be added where coordinates are chosen iteratively until no further improvement is possible. Experience has shown, however, that better results are achievable if the demand for improvement at each stage is relaxed.

At the starting stage we have used the great deluge approach to locate a sequence of merit value above an initial bound. The method of descent was then used first by random and finally by consecutive (with wrap-around) choice of coordinates until either termination or a sequence with merit value exceeding a second bound was located.

Two levels of intensified search then followed for appropriate sequences in an approach derived from the genetic method.

1) For the same length sequence coordinates were chosen in order from the beginning of the sequence. Any sequence found exceeding another preset bound was chosen for further development. Otherwise the process backtracked to an earlier stage from which a further choice was possible.

2) An appropriate sequence from the first stage was then stripped back, dropping coordinate entries at both ends successively until either the merit factor dropped below an established bound or the last coordinate change was reached but not exceeded. This core sequence was then extended at either end iteratively while still retaining the merit factor bound.

These intensification methods ensure a comprehensive search of clusters of sequences of high merit factor at neighbouring lengths.

For binary sequences, there is a single option for changing a coordinate value or two choices for an extension of one unit. For polyphase sequences, after a coordinate position was chosen for investigation, the base energy was expressed as a function of this coordinate variable, keeping other coordinates at their established values. This function was found to have up to two minimum values. The coordinate value at the lower of these was chosen for the proposed change.

2.4. Comparison of methods. For determining optimal values for the merit factor of binary sequences, the resources required to conduct the branch and bound search at length 60 can be considered roughly the equivalent of those required to perform a fully exhaustive search at length 53, or a directed stochastic search at length 90 with a 99% confidence level of success. In terms of complexity, this translates to $O(2^N)$ for exhaustive methods, $O(1.84^N)$ for branch and bound and $O(1.5^N)$ for the directed stochastic algorithm used here.

3. RESULTS

3.1. Growth trends for merit factor values. Figures 5 and 6 show plots of best known merit factor values recorded for binary sequences and for the more general polyphase sequences. We believe Figure 6 shows probable optimal values for unrestricted binary sequences with lengths into the late 80's and for skew-symmetrics up to length 160. For polyphase sequences this confidence extends to length 45.

The horizontal line in Figure 5 is drawn for reference at $F = 9$. Up to length 90, there appears to be a trend for more of these values to approach or exceed 9. Beyond $N = 112$ all records are derived from skew-symmetrics, which appear to reflect trends for unrectricted sequences at half the lengths.

Figure 6 shows the much more rapid growth for polyphase sequences.

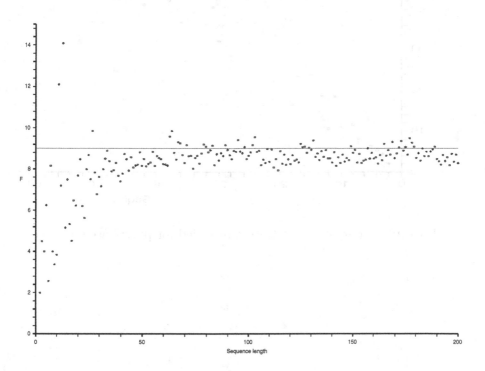

FIGURE 5. Largest merit factors recorded for binary sequences.

3.2. Distribution of base energies for binary sequences.

Figure 7 shows histograms of all 2^{39} values of the reciprocal of the merit factor, $1/F = 2E/N^2$, at length $N = 39$, normalized to have unit area. Each base energy value is congruent to 3 modulo 4. The graph on the left seems to be aligned with two smooth curves. A more careful analysis finds that it is better separated into four sections, each corresponding to a separate congruence class modulo sixteen for the base energy and each seeming to conform to a smooth curve. An alternative is to combine the congruences classes 3,7,11,15 modulo 16, giving the histogram on the right.

This shows a remarkable resemblance to the extreme value distribution. In fact, this was discovered through curve-fitting as illustrated in Figure 8.

The merit factor problem

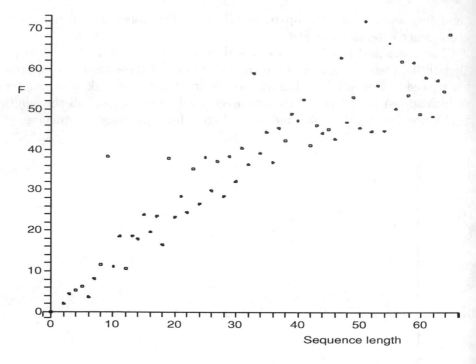

FIGURE 6. Largest merit factors recorded for polyphase sequences.

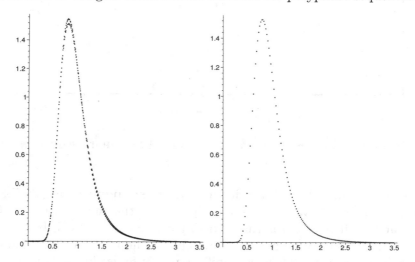

FIGURE 7. Full distribution of $1/F$ values at $N = 39$ and distribution combining counts from successive groups of 4.

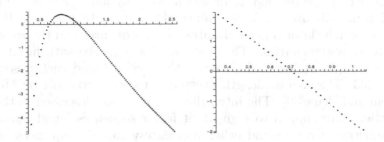

FIGURE 8. Plot of the logarithm of the values in the graph on the right in Figure 7 and and a plot of the logarithm of a linear fit to the right tail minus the values for this adjacent graph.

These graphs point to an exponential tail to the right but a doubly exponential growth on the left which are characteristic for the extreme value distribution. Figure 9 shows least squares fits of the extreme value distribution to histograms of $1/F$ values at lengths 39, 99, and 300. Since obtaining full distributions at lengths 39 and 300 is computationally out of range, these we obtained for 2^{37} and 2^{36} sequences generated by a random process.

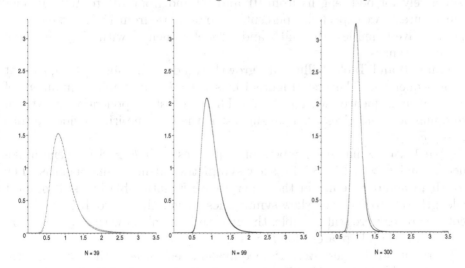

FIGURE 9. Least squares fit of extreme value distribution to $1/F$ distributions at lengths 39, 99 and 300.

Our primary interest is in the tail of the distributions on the left. The least squares fit is best at length 99. At 39, the curve overestimates the count of higher merit factor sequences while at 300 it gives an underestimate.

3.3. **High merit factor binary sequences.** Using our directed stochastic search method we have found approximately 2800 unequivalent sequences with merit factor greater than 8, of which probably more than 80% may be classified as new discoveries. This comprised two separate programs, the first designed to search through general unrestricted sequences and the second restricted to skew-symmetrics. The first found a sequence with merit factor greater than eight at length 115, while the second found such a sequence at length 233. The greatest length previously recorded was 161 by Militzer, Zamparelli and Beule[39]. The intensification methods described in the previous section were applied to high merit factor sequences found during the skew-symmetric search to find other non-skewsymmetric sequences of high merit in the surrounding cluster, finding sequences with $F \geq 8$ for all lengths between 100 and 200 except 112 and 114. A sequence with $F > 8$ at 112 was found using the first program.

We estimate the first program to have been close to exhaustive up to length 85, and the second to length 161 as outlined in the analysis below. Confidence that over 50% of the sequences with $F > 8$ have been found remains to lengths 90 and lengths 181 respectively. Comparing the percentage that the skew-symmetric sequences form of sequences found at odd lengths provides another rough yardstick. Between lengths 61 and 85 skew-symmetrics form 11% of sequences found at odd lengths with $F \geq 8$. These increase to 26% and 77% respectively for odd lengths from 91 and 99 and from 101 to 109. In fact from Figure 10 we expect this percentage to decrease from 11%, so we expect that we have found less than 40% and 12% of sequences with $F \geq 8$ in these respective ranges.

Figure 10 and Table 1 illustrate growth trends in numbers of high merit factor sequences. The first, Figure 10, is a log-linear graph of numbers of inequivalent sequences with $F \geq 7$. This suggests exponential growth in these numbers and gives what we suggest is the first empirical evidence that $\limsup F_{\max} > 7$.

Table 1 lists estimated numbers of sequences with $F > 8$ for general sequences and $F > 8$, $F > 8.5$ for skew-symmetrics at increasing lengths. The growth in numbers found in the general case is noticeable but still modest in lengths up to 85. For skew-symmetrics in lengths up to 181, where in depth searches were still possible, the growth in numbers with $F \geq 8$ is more apparent, and still apparent for $F > 8.5$.

Before our work, there were 15 inequivalent sequences with $F \geq 9$ known, one each at lengths 11, 13, 27, 64, 71, 83, 95, 105, 125, 127, 129, 131 and three at length 67. We have found 23 more which are listed in Table 2. There seems to be a trend for more of these to appear at longer lengths, but it is difficult to say more.

3.4. The inverse collector's problem.

In collecting data on sequences with $F \geq 7$, we continued to run our programs at lengths up to 85 to the point that we could have good expectation that the best examples were collected. In advance, we do not know the size of the sample space of sequences

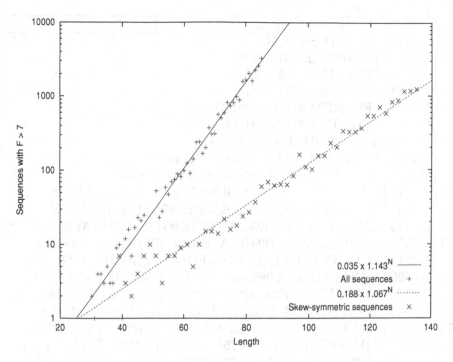

FIGURE 10. Growth in number of sequences with $F \geq 7$.

N	$F > 8$	$N^{\dagger}$	$F > 8$	$F > 8.5$
73	8	159	44	5
74	3	161	57	4
75	9	163	52	5
76	5	165	57	3
77	11	167	73	5
78	12	169	73	11
79	12	171	58	7
80	14	173	90	13
81	17	175	97	8
82	13	177	99	10
83	16	179	153	9
84	10	181	114	11
85	18	183	125	9

TABLE 1. Estimated numbers of inequivalent sequences with $F \geq 8$. Here † denotes skew-symmetric sequences.

N	F	Hexadecimal sequence
63	9.5870	64CBED0FAFEAC631
68	9.2480	FFD0B564E4D74798E
79	9.2050	7F36491D815A531AA871
80	9.0909	FFE81E89A8D1C665A9A5
89	9.1678	1FF924F246C19C2D4B8D454
95	9.3427	7FFC0FA333154B534DA71C69
98	9.0775	3FFAA55B45978719636C4F633
102	9.1746	383F0C38A4D5D6673480256A44
126	9.0720	3C7854315FE710B9990BB655FB2FE96D
149	9.0542	1C71C7AB46CDDABF9F82959501DCC6F016DB6D†
149	9.1137	1FE0003921C9CC3E4CBD0CE52CD8DA392AAA55†
157	9.0223	1F0600F83071FF993CC57ECD39955B6B294AA6B5†
165	9.2351	1D5B2B41689B1B24BAA6E846010E31B1887AF031FD†
169	9.3215	1C1C7C623B8EB1FD05DAFDD41DEBD5B0491226D6DAD†
172	9.0526	E03F9CF6030FF9EDBF293338351C5954B2A74D952A5
173	9.3179	18006FFE1FCF33F079C3D999D2D96B5334D5A5546AA9†
173	9.3645	1E03F9CF6030FF9EDBF293338351C5954B2A74D952A5†
175	9.0768	6AA32AF1A35998A5E530DAF8D30687D9983792FF37FE†
177	9.5052	1D3842C58FCB33401779175F7B977AAF330D49EC2E93D†
178	9.2915	3D3842C58FCB33401779175F7B977AAF330D49EC2E93D
179	9.0974	7A70858B1F9666802EF22EBEF72EF55E661A93D85D27A†
183	9.0073	6311C73B838E2A72BF958A85FD81ABF27F6DB5BB249136†
189	9.0847	1C39CE1FE1CBC67F3B7BF9002AB951713566D0DA55A4D92D

TABLE 2. New sequences found with merit factor > 9. Here †
indicates a skew-symmetric sequence.

with energies in this range. At some point, however, the continual repetition of previous examples suggests that we come close to exhausting this sample space. We use the statistical model described below to give substance to this observation. Part of this may be described as the inverse collector's problem as described in [11] and [34]. However, our problem is not specifically to establish the most probable size of the sample space, but to estimate the likelihood that we have found the example with the lowest base energy.

Let S be the size of the sample space, i.e., the total number of sequences, unique up to equivalence as described above, with $F \geq 7$. Let n be the number of trials in terms of sequences collected, and k the number of these which are different. Where M is the event that we have collected the optimal example, what we seek to evaluate is

$$P(M \mid n, k),$$

the probability that we have the best example, given the values n and k arising from our data. We make assumptions:
(1) the occurrence of the different examples are equally likely;
(2) there is no specific bias on the actual size of the sample space in the range where this is significant.

Then we use

$$P(M \mid n, k) = \sum_{i \geq 0} P(M \mid S = k + i, k) P(S = k + i \mid n, k).$$

From our assumptions, we have $P(M \mid S = k + i, k) = k/(k + i)$. Using Baysian probability theory, we then derive

$$P(S = k + i \mid n, k) = \frac{P(k \mid S = k + i, n)}{\sum_{j \geq 0} P(k \mid S = k + j, n)}.$$

Example: At length 80 we collected 9965 sequences of which 1636 were different. This calculates to a 0.998 probability that we have the optimal example [8], [9].

It remains, at present, computationally hard to verify optimality of merit factor values beyond length 60 for unrestricted binary sequences and about double this for skew-symmetrics. Other authors have published examples with 'high' merit factor without any further assessment of quality. This method of statistical analysis offers a way to estimate quality.

4. COMMENTS

Determining the maximal merit factor for binary sequences of length N is widely regarded as a difficult task in combinatorial optimization. Indeed it appears as the fifth problem on CSPLib [24], a library of test problems for use in benchmarking constraint solvers.

We have found good evidence that the upper limit for $\max_N F$ is greater than 8 and in fact greater than 8.5. These maximal values may routinely exceed 9 in lengths over 200, but it would be difficult to establish this computationally. With the information we have collected, it is difficult to project whether values will continue to grow slowly or level off.

The match of the distribution of $1/F$ to the extreme value distribution for lengths under 100 is intriguing. If this continued, it would suggest a sharp cutoff for merit value as lengths increased. However, this fit becomes less good as we go further, giving an underestimate for the tail containing high merit factor values. Perhaps a match to a gaussian distribution is a better choice for longer lengths, but this would require further investigation.

The upper limit for $\max_N F$ is known to be infinite for polyphase sequences. What growth rates can be established for $\max_N F$? Figure 6 suggests this might be close to linear for polyphase sequences. If it can be established that $\max_N F$ is bounded for binary sequences, is this true for other finite alphabet sequences as well?

REFERENCES

[1] R.H. Barker, Group synchronizing of binary digital sequences, *Communication Theory* (1953),273–287.

[2] H. Bauke, S. Mertens, Ground states of the Bernasconi model with open boundary conditions, http://itp.nat.uni-magdeburg.de/ mertens/bernasconi/open.dat, (2003).

[3] G.F.M. Beenker, T.A.C.M. Claasen, P.W.C. Hermens, Binary sequences with a maximally flat amplitude spectrum, *Philips Journal of Research* **40** (1985), 289–304.

[4] J. Bernasconi, Low autocorrelation binary sequences: statistical mechanics and configuration space analysis, *J. Phys.* **48** (1987), no. 4, 559–567.

[5] P. Borwein, *Computational excursions in analysis and number theory*, CMS Books in Mathematics **10**, Springer Verlag, New York, 2002.

[6] P. Borwein, K.-K.S. Choi, J. Jedwab, Binary sequences with merit factor greater than 6.34, *IEEE Trans. Inform. Theory* **50** (2004), no. 12, 3234–3249.

[7] P. Borwein and R. Ferguson, Polyphase sequences with low autocorrelation, 2003 (preprint).

[8] ——, ——, Polyphase sequences with low autocorrelation, *IEEE Trans. Inform. Theory* **51** (2005), no. 4, 1564–1567.

[9] P. Borwein, R. Ferguson, J. Knauer, The merit factor of binary sequences, (to appear).

[10] F. Brglez, M. Stallmann, X.Y. Li, B. Militzer, Reliable cost predictions for finding optimal solutions to LABS problem: evolutionary and alternative algorithms, *Proceedings of The Fifth International Workshop on Frontiers in Evolutionary Algorithms (FEA2003), Cary, NC, USA, September 26-30*, 2003.

[11] B. Dawkins, Siobhan's problem: the coupon collector revisited, *The American Statistician* **45** (1991), no. 1, 76–82.

[12] H. Deng, Synthesis of binary sequences with good autocorrelation and crosscorrelation properties by simulated annealing, *IEEE Transactions on Aerospace Electronics Systems* **32** (1996), no. 1, 98–107.

[13] X. Deng, P. Fan, New binary sequences with good aperiodic autocorrelations obtained by evolutionary algorithm, *IEEE Communications Letters* **3** (1999), no. 10, 288–290.

[14] G. Dueck, New optimization heuristics, The great deluge algorithm and the record-to-record travel, *Journal of Computation Physics* **104** (1993), 86–92.

[15] C. de Groot, D. Würtz, Statistical Mechanics of Low Autocorrelation Skew-symmetric Binary Sequences, *Helvetica Physica Acta* **64** (1991), 86–91.

[16] C. de Groot, D. Wurtz, K.H. Hoffman, Low autocorrelation binary sequences: exact enumeration and optimization by evolutionary algorithm, *Optimization* **23** (1992), 369–384.

[17] L. Eberhardt, A course in quantitative ecology, http://nmml.afsc.noaa.gov/Accessibility/AccLibQuantita.htm, 2003.

[18] F.F. Ferreira, J.F. Fontanari, P.F. Stadler, Landscape statistics of the low-autocorrelation binary string problem, *J. Phys. A* **33** (2000), no. 48, 8635–8647.

[19] M.J.E. Golay, A class of finite binary sequences with alternate autocorrelation values equal to zero, *IEEE Transactions on Information Theory* **IT-18** (1972), no. 3, 449–450.

[20] ——, Sieves for low autocorrelation binary sequences, *IEEE Transactions on Information Theory* **IT-23** (1977), no. 1, 43–51.

[21] ——, The merit factor of long low autocorrelation binary sequences, *IEEE Transactions on Information Theory* **IT-28** (1982), no. 3, 543–549.

[22] ——, The merit factor of Legendre sequences, *IEEE Transactions on Information Theory* **IT-29** (1983), no. 6, 934–936.

[23] M.J.E. Golay, D.B. Harris, A new search for skewsymmetric binary sequences with optimal merit factors, *IEEE Transactions on Information Theory* **36** (1990), no. 5, 1163–1166.

[24] I.P. Gent, T. Walsh, CSPLib: a benchmark library for constraints, Technical report APES-09-1999 (1999), available from http://csplib.cs.strath.ac.uk/. A shorter version appears in the Proceedings of the 5th International Conference on Principles and Practices of Constraint Programming (CP-99).

[25] T. Høholdt, The merit factor of binary sequences, *Difference sets,sequences and their correlation properties*, Series C: Mathematical and Physical Sciences **542**, 227–237, Kluwer, 1999.

[26] T. Høholdt, H.E. Jensen, Determination of the merit factor of Legendre sequences, *IEEE Transactions on Information Theory* **34** (1988), no. 1, 161–164.

[27] F. Hu, P.Z. Fan, M. Darnell, F. Jin, Binary sequences with good aperiodic autocorrelations obtained by evolutionary algorithm, *Electronics Letters* **33** (1997), no. 8, 688–690.

[28] J. Jedwab, A survey of the merit factor problem for binary sequences, *Sequences and Their Applications – Proceedings of SETA 2004*, Lecture Notes in Computer Science **3486**, 30–55, Springer-Verlag, Berlin, 2004.

[29] J.-P. Kahane, Sur les polynômes à coefficients unimodulaires, *Bull. London Math. Soc.* **12** (1980), no. 5, 321–342.

[30] A. Kirilusha, G. Narayanaswamy, Construction of new asymptotic classes of binary sequences based on existing asymptotic classes, Dept. Math. and Comput. Science, University of Richmond, Summer Science Program Technical Report, July 1999.

[31] Ş.E. Kocabaş, A. Atalar, Binary sequences with low aperiodic autocorrelation for synchronization purposes, *IEEE Communications Letters* **7** (2003), no. 1, 36–38.

[32] R. Kristiansen, M.G. Parker, Binary Ssequences with asymptotic aperiodic merit factor > 6.3, preprint (2003).

[33] ———, ———, Binary Sequences with merit factor > 6.3, *IEEE Trans. Inform. Theory* **50** (2004), no. 12, 3385–3389.

[34] E. Langford, R. Langford, Solution of the inverse coupon collector's problem, *Mathematical Scientist* **27** (2002), no. 1, 32–35.

[35] K.H. Leung, B. Schmidt, The field descent method, *Des. Codes Cryptogr.* **36** (2005), no. 2, 171–188.

[36] F.C. Lincoln, Calculating waterfowl abundance on the basis of banding returns, *U.S. Department of Agriculture Circulation* **18** (1930), 1–4.

[37] J.E. Littlewood, *Some problems in real and complex analysis*, D.C. Heath and Co. Raytheon Education Co., Lexington, Mass., 1968.

[38] S. Mertens, Exhaustive search for low-autocorrelation binary sequences, *Journal of Physics A: Mathematical and General* **29** (1996), no. 18, L473–L481. Updated results can be found at
http://itp.nat.uni-magdeburg.de/ mertens/bernasconi/open.dat.

[39] B. Militzer, M. Zamparelli, D. Beule, Evolutionary Search for Low Autocorrelated Binary Sequences, *IEEE Transactions on Evolutionary Computation* **2** (1998), no. 1, 34–39.

[40] C.G.J. Petersen, The yearly immigration of young plaice into the Limfjord from the German Sea, *Report of Danish Biological Station* **6** (1896), 1–48.

[41] S. Prestwich, A hybrid search architecture applied to hard random 3-SAT and low-autocorrelation binary sequences, *The Sixth international conference on principles and practice of constraint programming*, Lecture Notes in Computer Science **1894**, 337–352, Springer-Verlag, Berlin, 2000.

[42] A. Reinholz, *Ein paralleler genetischer Algorithmus zur Optimierung der binären Autokorrelations-Funktion*, Masters Thesis, Universität Bonn, Oct. 1993.

[43] B. Schmidt, Cyclotomic integers and finite geometry, *J. Amer. Math. Soc.* **12** (1999), 929–952.

[44] M.R. Schroeder, *Number theory in science and communication*, 3rd edn., Springer-Verlag, New York, 1997.

[45] J. Storer, R. Turyn, On binary sequences, *Proc. Amer. Math. Soc.* **12** (1961), 394–399.

BARKER SEQUENCES AND FLAT POLYNOMIALS

PETER BORWEIN AND MICHAEL J. MOSSINGHOFF

ABSTRACT. A Barker sequence is a finite sequence of integers, each ± 1, whose aperiodic autocorrelations are all as small as possible. It is widely conjectured that only finitely many Barker sequences exist. We describe connections between Barker sequences and several problems in analysis regarding the existence of polynomials with ± 1 coefficients that remain flat over the unit circle according to some criterion. First, we amend an argument of Saffari to show that a polynomial constructed from a Barker sequence remains within a constant factor of its L_2 norm over the unit circle, in connection with a problem of Littlewood. Second, we show that a Barker sequence produces a polynomial with very large Mahler's measure, in connection with a question of Mahler. Third, we optimize an argument of Newman to prove that any polynomial with ± 1 coefficients and positive degree $n-1$ has L_1 norm less than $\sqrt{n-.09}$, and note that a slightly stronger statement would imply that long Barker sequences do not exist. We also record polynomials with ± 1 coefficients having maximal L_1 norm or maximal Mahler's measure for each fixed degree up to 24. Finally, we show that if one could establish that the polynomials in a particular sequence are all irreducible over $\mathbb{Q}$, then an alternative proof that there are no long Barker sequences with odd length would follow.

1. INTRODUCTION

For a sequence of complex numbers a_0, a_1, ..., a_{n-1}, define its *aperiodic autocorrelation sequence* $\{c_k\}$ by

$$c_k := \sum_{j=0}^{n-1-k} a_j \overline{a}_{j+k}$$

for $0 \le k < n$ and

$$c_{-k} := \overline{c}_k .$$

We are interested here in the case when the a_j are all of unit modulus, in particular when each $a_j = \pm 1$. Thus the *peak autocorrelation* c_0 has the value $c_0 = n$, and in many applications it is of interest to minimize the *off-peak autocorrelations* $c_{\pm k}$ with $0 < k < n$. In the integer case, clearly the

2000 *Mathematics Subject Classification.* Primary: 11B83, 42A05; Secondary: 30C10, 94A55.

Key words and phrases. Barker sequences, flat polynomials, Littlewood polynomials, Mahler's measure, L_p norm.

Research of P. Borwein supported in part by NSERC of Canada and MITACS.

71

optimal situation occurs when $|c_k| \leq 1$ for each $k \neq 0$, so $c_k = 0$ if $2 \mid (n - k)$ and $c_k = \pm 1$ otherwise. A sequence achieving this for each k is called a *Barker sequence*. Barker first asked for ± 1 sequences with this property in 1953 [1]. (In fact, Barker asked for the stricter condition that $c_k \in \{0, -1\}$ for $k \neq 0$.) For the complex unimodular case, we say $\{a_k\}$ is a *generalized Barker sequence* if each off-peak autocorrelation satisfies $|c_k| \leq 1$.

Since negating every other term of a sequence $\{a_k\}$ does not disturb the magnitudes of its autocorrelations, we may assume that $a_0 = a_1 = 1$ in a Barker sequence. With this normalization, just eight Barker sequences are known, all with length at most 13. These are shown in Table 1. (Only three of these satisfy the more strict condition requested by Barker—the ones of length 3, 7, and 11.) It is widely conjectured that no additional Barker sequences exist, and in section 2 we survey some known restrictions on their existence. First however we describe a broader conjecture that arises in signal processing, and an equivalent problem in analysis regarding norms of polynomials.

Sequences with small off-peak autocorrelations are of interest in a number of applications in signal processing and communications (see [1, 18, 13]). In engineering applications, a common measure of the value of a sequence is the ratio of the square of the peak autocorrelation to the sum of the squares of the moduli of the off-peak values. This is called the *merit factor* of the sequence. For a sequence $A_n = \{a_j\}$ of length n, its merit factor is defined by

$$\mathrm{MF}(A_n) := \frac{n^2}{2 \sum_{k=1}^{n-1} |c_k|^2}.$$

Golay introduced this quantity in 1972 [16], and in [17] he conjectured that the merit factor of a binary sequence is bounded, presenting a heuristic argument that $\mathrm{MF}(A_n) < 12.32$ for large n. Several researchers in engineering, physics, and mathematics have made similar conjectures; see for instance [6] or [18]. It is clear, however, that a Barker sequence of length n has merit factor near n, so certainly Golay's merit factor conjecture contains the question of the existence of long Barker sequences as a special case.

TABLE 1. Barker sequences with $a_0 = a_1 = 1$.

n	Sequence	Merit factor
2	++	2.00
3	++-	4.50
4	+++-	4.00
4	++-+	4.00
5	+++-+	6.25
7	+++--+-	8.17
11	+++---+--+-	12.10
13	+++++--++-+-+	14.08

The merit factor problem may be restated as a question on polynomials. We first require some notation. Given a sequence $\{a_j\}_{j=0}^{n-1}$, define a polynomial $f(z)$ of degree $n-1$ by

$$f(z) = \sum_{j=0}^{n-1} a_j z^j.$$

For a positive real number p, let $\|f\|_p$ denote the value

$$\|f\|_p := \left(\int_0^1 |f(e(t))|^p \, dt \right)^{1/p},$$

where $e(t) := e^{2\pi i t}$. If $p \geq 1$, this is the usual L_p norm of f on the unit circle. We also let $\|f\|_\infty$ denote the supremum norm of f,

$$\|f\|_\infty := \lim_{p\to\infty} \|f\|_p = \sup_{|z|=1} |f(z)|,$$

and we let $\|f\|_0$ denote its geometric mean on the unit circle,

$$\|f\|_0 := \lim_{p\to 0+} \|f\|_p = \exp\left(\int_0^1 \log|f(e(t))| \, dt \right).$$

This is *Mahler's measure* of the polynomial. We recall that if $p < q$ are positive real numbers and f is not a monomial, then

$$\|f\|_0 < \|f\|_p < \|f\|_q < \|f\|_\infty.$$

Assuming that $|a_j| = 1$ for each j, we have $\|f\|_2^2 = n$ by Parseval's formula, and, since $\bar{z} = 1/z$ on the unit circle, it is easy to see that

$$\|f\|_4^4 = \|f(z)\overline{f(z)}\|_2^2 = \left\| \sum_{k=1-n}^{n-1} c_k z^k \right\|_2^2 = n^2 + 2\sum_{k=1}^{n-1} |c_k|^2. \qquad (1.1)$$

Thus, the merit factor of a sequence $\{a_j\}$ can be expressed in terms of certain L_p norms of its associated polynomial,

$$\mathrm{MF}(f) := \frac{\|f\|_2^4}{\|f\|_4^4 - \|f\|_2^4}.$$

Golay's problem on maximizing the merit factor of a family of sequences of fixed length is thus equivalent to minimizing the L_4 norm of a collection of polynomials of fixed degree. This latter problem is one instance of a family of questions regarding the existence of so-called *flat polynomials*.

For a positive integer n, let $\mathfrak{U}_n$ denote the set of polynomials in $\mathbb{C}[x]$ defined by

$$\mathfrak{U}_n := \left\{ f(z) = \sum_{j=0}^{n-1} a_j z^j : |a_j| = 1 \text{ for } 0 \leq j < n \right\},$$

and let $\mathfrak{L}_n$ denote the subset

$$\mathfrak{L}_n := \left\{ f(z) = \sum_{j=0}^{n-1} a_j z^j : a_j = \pm 1 \text{ for } 0 \leq j < n \right\}.$$

We call the first set the *unimodular polynomials* of degree $n - 1$, and the second set the *Littlewood polynomials* of fixed degree. In 1966, Littlewood [23] asked about the existence of polynomials in these sets with particular flatness properties. More precisely, he asked if there exist absolute positive constants α_1 and α_2 and arbitrarily large integers n such that there exists a polynomial $f_n \in \mathfrak{U}_n$ (or, more strictly, $f_n \in \mathfrak{L}_n$), where

$$\alpha_1 \sqrt{n} \leq |f_n(z)| \leq \alpha_2 \sqrt{n}$$

for all z with $|z| = 1$. Since each polynomial in such a sequence never strays far from its L_2 norm, we say such a sequence is *flat*. In 1980, Körner [21] established that flat sequences of unimodular polynomials exist, and in the same year Kahane [20] proved moreover that for any $\epsilon > 0$ there exists a flat sequence of unimodular polynomials with $\alpha_1 = 1 - \epsilon$ and $\alpha_2 = 1 + \epsilon$. Such sequences are often called *ultraflat*.

Much less is known regarding flat sequences of Littlewood polynomials. The Rudin-Shapiro polynomials [28, 31] satisfy the upper bound in the flatness condition with $\alpha_2 = \sqrt{2}$, but no sequence is known that satisfies the lower bound. In fact, the best known result here is due to Carrol, Eustice, and Figiel [8], who used the Barker sequence of length 13 to show that for sufficiently large n there exist polynomials $f_n \in \mathfrak{L}_n$ with $|f_n(z)| > n^{.431}$ on $|z| = 1$. Also, in 1962 Erdős [12] conjectured that ultraflat Littlewood polynomials do not exist, opining that there exists an absolute positive constant ϵ such that

$$\frac{\|f\|_\infty}{\|f\|_2} > 1 + \epsilon$$

for every Littlewood polynomial of positive degree. (Littlewood however [23, sec. 6; 24, prob. 19] in effect conjectured that no such ϵ exists.) Since $\|f\|_4 \leq \|f\|_\infty$, we see then that Golay's merit factor problem is in fact a stronger version of Erdős' conjecture. Further, from (1.1) it follows that if the coefficients of f form a Barker sequence of length n, then

$$\frac{\|f\|_4}{\sqrt{n}} \leq \left(1 + \frac{1}{n}\right)^{1/4} < 1 + \frac{1}{4n}.$$

Therefore, to show that long Barker sequences do not exist, it would suffice to prove that $\|f\|_4 \geq \sqrt{n} + \frac{1}{4\sqrt{n}}$ for $f \in \mathfrak{L}_n$ and n large. Similar observations occur for example in [4, Chap. 14] and [5].

In this paper, we describe some further connections between Barker sequences and flatness problems for polynomials. Section 2 summarizes some known results on Barker sequences. Section 3 shows that long Barker sequences provide an answer to Littlewood's question on flat polynomials, amending an argument of Saffari that connects these two problems. Section 4 then ties the existence of long Barker sequences to a problem of Mahler's concerning Littlewood polynomials with large measure. Section 5 connects the Barker sequence question to problems on the L_1 norm of Littlewood polynomials, and optimizes an argument of Newman to provide an improved restriction on the flatness of Littlewood polynomials with respect to this norm. Finally, Section 6 outlines a possible alternative method for establishing that there are no Barker sequences of certain lengths.

2. BARKER SEQUENCES

We first record some facts about Barker sequences. The following results are due to Turyn and Storer [35, 32]; we include the proof here for the reader's convenience.

Theorem 2.1. *Suppose* $a_0, a_1, \ldots, a_{n-1}$ *is a sequence of integers with each* $a_i = \pm 1$, *and let* $\{c_k\}$ *denote its aperiodic autocorrelations. Then*

$$c_k + c_{n-k} \equiv n \pmod 4.$$

If in addition the sequence $\{a_k\}$ *is a Barker sequence, then*

$$a_k a_{n-1-k} = (-1)^{n-1-k}.$$

If furthermore n *is even and* $n > 2$, *then* $n = 4m^2$ *for some integer* m, *and* $c_{n-k} = -c_k$ *for* $0 < k < n$. *If* n *is odd, then* $c_k + c_{n-k} = (-1)^{(n-1)/2}$ *for each* k.

Proof. Since c_k records the difference between the number of positive and negative terms in $\sum_{i=0}^{n-1-k} a_i a_{i+k}$, it follows that

$$\prod_{i=0}^{n-k-1} a_i a_{i+k} = (-1)^{(n-k-c_k)/2} \tag{2.1}$$

for $0 \le k < n$. Multiplying this product by the same expression with k replaced by $n - k$, we obtain

$$(-1)^{(n-c_k-c_{n-k})/2} = \prod_{i=0}^{k-1} a_i a_{i+n-k} \prod_{i=0}^{n-k-1} a_i a_{i+k} = 1,$$

so $c_k + c_{n-k} \equiv n \pmod 4$. Assume now that $\{a_k\}$ forms a Barker sequence of length n. Multiplying (2.1) by the same equation with k replaced by $k + 1$, we compute that

$$a_k a_{n-1-k} = (-1)^{n-1-k}.$$

Also, certainly $c_k = 0$ if $0 < k < n$ and $n \equiv k \pmod 2$, and $c_k = \pm 1$ for the other k in this range. In particular, if n is even and $n > 2$, then $c_2 + c_{n-2} = 0$, so $n \equiv 0 \pmod 4$. It follows then that $c_k + c_{n-k} = 0$ for $0 < k < n$ in this case. Last, since

$$\left(\sum_{i=0}^{n-1} a_i\right)^2 = c_0 + \sum_{k=1}^{n-1}(c_k + c_{n-k}) = n\,,$$

we see that n is a perfect square if $n \geq 4$ is even.   □

Recall that a polynomial $f(z)$ with integer coefficients is *skew-symmetric* if $f(z) = \pm z^{\deg f} f(-1/z)$. We remark that Theorem 2.1 then shows that every Barker sequence of odd length corresponds to a skew-symmetric Littlewood polynomial.

Much more is known about possible lengths of Barker sequences. Turyn and Storer [35] proved that if the length n of a Barker sequence is odd then $n \leq 13$, so the complete list for this case appears in Table 1. It also follows from this that no additional sequences satisfy Barker's original requirement for sequences whose off-peak autocorrelations are all 0 or -1, since Theorem 2.1 implies that any such sequence must have length $n \equiv 3 \pmod 4$. For the even case, we write $n = 4m^2$. In 1965 Turyn [33] showed in effect that m must be odd and cannot be a prime power (see also [2, sec. 2D and 4C], [9], [10]). In 1990, Eliahou, Kervaire, and Saffari [11] proved that if $p \mid m$ then $p \equiv 1 \pmod 4$; in 1992 Eliahou and Kervaire [9] and Jedwab and Lloyd [19] both used this constraint, together with some additional restrictions on m, to show that there are no Barker sequences with $1 < m < 689$. In 1999, Schmidt [30] obtained much stronger restrictions on m, determining that no Barker sequences exist with $m \leq 10^6$. This method was refined and extended by Leung and Schmidt in 2005 [22], who established that no Barker sequences exist with $1 < m \leq 5 \cdot 10^{10}$, that is, with even length n satisfying $4 < n \leq 10^{22}$. Another restriction was obtained in 1989 by Fredman, Saffari, and Smith [15], who proved that a Barker sequence may not be palindromic.

3. LITTLEWOOD'S PROBLEM

In 1990, Saffari [29] noted that if there are in fact infinitely many Barker sequences, then Littlewood's conjecture on the existence of flat polynomials with ± 1 coefficients follows. We present Saffari's proof here, in part because we require the result in Section 4, but also to correct an oversight in the original article. The correction here affects the values of the constants in the following theorem.

Theorem 3.1. *Suppose f is a Littlewood polynomial of degree $n-1$ whose sequence of coefficients $\{a_k\}$ forms a Barker sequence of length n. Then*

$$\alpha_1 + O\left(\frac{1}{n}\right) \leq \frac{|f_n(z)|}{\sqrt{n}} \leq \alpha_2 + O\left(\frac{1}{n}\right)$$

for each z of modulus 1, where $\alpha_1 = \sqrt{1-\theta} = 0.52477485\ldots$, $\alpha_2 = \sqrt{1+\theta} = 1.31324459\ldots$, and

$$\theta = \sup_{t>0} \frac{\sin^2 t}{t} = 0.7246113537\ldots .$$

Proof. Suppose $f \in \mathcal{L}_n$ with $n > 13$, and write $n = 4m$. Using the fact that the off-peak autocorrelations satisfy $c_{n-k} = -c_k$ from Theorem 2.1, and that $c_{2j} = 0$ for $j \geq 1$, we compute

$$\left|f\left(e^{it}\right)\right|^2 - n = 2\sum_{k=1}^{n-1} c_k \cos kt$$

$$= 2\sum_{k=1}^{2m-1} c_k\left(\cos kt - \cos((n-k)t)\right)$$

$$= 4\sin(2mt)\sum_{k=1}^{2m-1} c_k \sin\left((2m-k)t\right)$$

$$= 4\sin(2mt)\sum_{k=1}^{m} c_{2m-2k+1}\sin\left((2k-1)t\right).$$

Thus

$$\left|\frac{\left|f\left(e^{it}\right)\right|^2}{n} - 1\right| \leq \theta_m, \tag{3.1}$$

where θ_m is defined by

$$\theta_m := \max_{0 \leq t < 2\pi} \frac{|\sin(2mt)|}{m} \sum_{k=1}^{m} \left|\sin\left((2k-1)t\right)\right|.$$

Define ϕ_m and ψ_m by

$$\phi_m := \max_{0 \leq t \leq \pi/4} \frac{|\sin(2mt)|}{m} \sum_{k=1}^{m} \left|\sin\left((2k-1)t\right)\right|$$

and

$$\psi_m := \max_{0 \leq t \leq \pi/4} \frac{|\sin(2mt)|}{m} \sum_{k=1}^{m} \left|\cos\left((2k-1)t\right)\right|,$$

so that $\theta_m = \max\{\phi_m, \psi_m\}$. For ϕ_m, note first that the quantity

$$\frac{1}{m}\sum_{k=1}^{m} \left|\sin\left((2k-1)t\right)\right|$$

is the midpoint approximation over m subintervals of equal size for the integral

$$\int_0^1 |\sin(2mtx)|\, dx.$$

We consider the error incurred when approximating this integral with the sum over each interval $[(k-1)/m, k/m]$. If no cusp occurs in the interval, certainly the error is at most $1/24m^3$, so the total error incurred from these intervals is $O(1/m^2)$. If a cusp occurs in an interval, in the worst case it lies at the midpoint, and the ratio of the error incurred in this case to the error when the cusp occurs at an endpoint is $\tan(t/2)/(t - \sin t)$. If $\pi/\sqrt{m} \le t \le \pi/4$, this ratio is $3m/\pi^2 + O(1)$, so the total error incurred on the intervals with cusps is

$$O\left(\frac{3m}{\pi^2} \cdot \frac{1}{24m^3} \cdot \frac{m}{2}\right) = O\left(\frac{1}{m}\right).$$

If $0 \le t < \pi/\sqrt{m}$, then there are at most $2\sqrt{m}$ cusps, and the error in the worst case at each cusp is $(2 - \cos(\pi/\sqrt{m}))/2\pi\sqrt{m}$, so the total error in this case is also $O(1/m)$. Therefore,

$$
\begin{aligned}
\phi_m &= \max_{0 \le t \le \pi/4} |\sin(2mt)| \int_0^1 |\sin(2mtx)|\, dx + O\left(\frac{1}{m}\right) \\
&\le \sup_{\alpha \ge 0} |\sin \alpha| \int_0^1 |\sin(\alpha t)|\, dt + O\left(\frac{1}{m}\right) \\
&= \sup_{n \ge 0} \max_{0 \le x \le \pi} \frac{(2n + 1 - \cos x)\sin x}{n\pi + x} + O\left(\frac{1}{m}\right) \\
&= \max_{0 \le x \le \pi} \frac{(1 - \cos x)\sin x}{x} + O\left(\frac{1}{m}\right) \\
&= 0.6639534894\ldots + O\left(\frac{1}{m}\right).
\end{aligned}
\tag{3.2}
$$

In the same way,

$$
\begin{aligned}
\psi_m &= \max_{0 \le t \le \pi/4} |\sin(2mt)| \int_0^1 |\cos(2mtx)|\, dx + O\left(\frac{1}{m}\right) \\
&\le \sup_{\alpha \ge 0} |\sin \alpha| \int_0^1 |\cos(\alpha t)|\, dt + O\left(\frac{1}{m}\right) \\
&= \sup_{n \ge 0} \max_{-\frac{\pi}{2} \le x \le \frac{\pi}{2}} \frac{(2n + \sin x)|\sin x|}{n\pi + x} + O\left(\frac{1}{m}\right) \\
&= \max_{0 \le x \le \frac{\pi}{2}} \frac{\sin^2 x}{x} + O\left(\frac{1}{m}\right) \\
&= 0.7246113537\ldots + O\left(\frac{1}{m}\right).
\end{aligned}
\tag{3.3}
$$

The statement then follows from (3.1), (3.2), and (3.3).   □

We remark that Saffari computed the limiting value of θ_m to be $0.66395\ldots$ by considering only the computation of ϕ_m above for $0 \le t \le 2\pi$. However, this argument breaks down when t is very close to $\pi/2$ or $3\pi/2$.

4. MAHLER'S PROBLEM

In 1963, Mahler [25] posed the question of maximizing the normalized measure $\|f\|_0 / \|f\|_2$ of polynomials with complex coefficients and fixed degree. He proved that for each degree the maximum is attained by a unimodular polynomial, and Fielding [14] proved that there exist unimodular polynomials with normalized measure arbitrarily close to 1. Beller and Newman [3] proved further that there exists a positive constant c such that for each $n > 0$ there exists a polynomial $f_n \in \mathfrak{U}_n$ such that $\|f_n\|_0 > \sqrt{n} - c \log n$. The problem remains open for Littlewood polynomials; the largest known normalized measure in this case is $0.98636598\ldots$, achieved by the polynomial whose coefficients form the Barker sequence of length 13. We prove here that long Barker sequences would also provide an answer to Mahler's problem for the case of Littlewood polynomials.

Theorem 4.1. *Let f_n be a Littlewood polynomial whose coefficients form a Barker sequence of length n. Then*

$$\frac{\|f_n\|_0}{\sqrt{n}} > 1 - \frac{1}{\sqrt{n}}$$

for sufficiently large n.

Proof. Let $f_n(z) = \sum_{j=0}^{n-1} a_j z^j$, with $\{a_j\}$ a Barker sequence. Since the off-peak autocorrelation c_k is 0 if $n \equiv k \pmod{2}$ and ± 1 otherwise, it follows from (1.1) that

$$\|f_n\|_4^4 = n^2 + n - \epsilon(n),$$

where $\epsilon(n) = 0$ if n is even and 1 if n is odd. Thus

$$\int_0^1 \left(\frac{|f_n(e(t))|^2}{n} - 1 \right)^2 dt = \frac{\|f_n\|_4^4}{n^2} - 1 = \frac{n - \epsilon(n)}{n^2}.$$

Next, if $a > b > 0$ it is straightforward to verify that

$$\frac{a - b}{b} \geq \log a - \log b,$$

so setting

$$a(t) = \max\left\{1, \frac{|f_n(e(t))|^2}{n}\right\} \quad \text{and} \quad b(t) = \min\left\{1, \frac{|f_n(e(t))|^2}{n}\right\}$$

for t in $[0, 1]$, we obtain

$$\int_0^1 \left(\frac{|f_n(e(t))|^2}{n} - 1 \right)^2 dt$$

$$\geq \int_0^1 \min\left\{ \frac{|f_n(e(t))|^2}{n}, 1 \right\}^2 \left(2\log|f_n(e(t))| - \log n \right)^2 dt,$$

so

$$\int_0^1 \left(2\log|f_n(e(t))| - \log n\right)^2 dt \le \frac{1}{\alpha_1^2 n} + O\left(\frac{1}{n^2}\right),$$

where $\alpha_1 = 0.52477\ldots$ is the constant appearing in statement of Theorem 3.1. By the Schwarz inequality,

$$\int_0^1 \left|2\log|f_n(e(t))| - \log n\right| dt \le \frac{1}{\alpha_1\sqrt{n}} + O\left(\frac{1}{n^{3/2}}\right),$$

and so

$$\int_0^1 \log|f_n(e(t))| dt \ge \log\sqrt{n} - \frac{1}{2\alpha_1\sqrt{n}} + O\left(\frac{1}{n^{3/2}}\right).$$

Since $1/2\alpha_1 = 0.9527\ldots$, it follows then that

$$\frac{\|f_n\|_0}{\sqrt{n}} \ge 1 - \frac{1}{2\alpha_1\sqrt{n}} + O\left(\frac{1}{n^{3/2}}\right) > 1 - \frac{1}{\sqrt{n}}$$

for sufficiently large n. $\square$

For each $n \le 25$, Table 2 lists a Littlewood polynomial with degree $n - 1$ having maximal Mahler's measure over $\mathfrak{L}_n$. We remark that the coefficient sequences for $n = 2$, 3, 4, 5, 7, 11, and 13 are precisely the Barker sequences. (The two Barker sequences of length 4 correspond to polynomials with identical Mahler's measure.)

5. NEWMAN'S PROBLEM

One may also study flatness properties of polynomials by using the L_1 norm. In this case, again the problem is largely resolved for unimodular polynomials, and largely open for Littlewood polynomials. For the unimodular case, in 1965 Newman [27] proved that there exists a positive constant c so that for each $n \ge 2$ there exists a polynomial $f_n \in \mathfrak{U}_n$ such that $\|f\|_1 > \sqrt{n} - c$. In his proof, Newman first constructed a polynomial f_n whose L_4 norm satisfies $\|f_n\|_4/\sqrt{n} = 1 + O(1/\sqrt{n})$, then used Hölder's inequality to obtain a lower bound on $\|f_n\|_1$ of the desired form.

Much less is known for the Littlewood case. In [26], Newman mentioned a conjecture (without attribution) for the L_1 norm for these polynomials, similar to Erdős' conjecture for the supremum norm: There exists a positive constant $c < 1$ such that $\|f\|_1 < c\sqrt{n}$ whenever $f \in \mathfrak{L}_n$ and $n \ge 2$. This problem remains open, as does the weaker question of whether there exists a positive constant c such that $\|f\|_1 < \sqrt{n} - c$ for $f \in \mathfrak{L}_n$ of positive degree. Resolving a still weaker problem however suffices for answering the question of the existence of Barker sequences of large degree.

Theorem 5.1. *If $f(z) = \sum_{k=0}^{n-1} a_j z_j$ is a Littlewood polynomial whose coefficients form a Barker sequence of length n, then $\|f\|_1 > \sqrt{n-1}$.*

TABLE 2. Maximal Mahler's measure of Littlewood polynomials by degree.

n	Coefficients of f	$\|f\|_0$	$\|f\|_0/\sqrt{n}$	$\sqrt{n}-\|f\|_0$
2	++	1.00000	0.70711	0.41421
3	++-	1.61803	0.93417	0.11402
4	+++-	1.83929	0.91964	0.16071
5	+++-+	2.15372	0.96317	0.08235
6	++++-+	2.22769	0.90945	0.22180
7	+++--+-	2.49670	0.94366	0.14905
8	++++-+-	2.64209	0.93412	0.18634
9	+++-+--++	2.72501	0.90834	0.27499
10	++++-+--++	2.92076	0.92363	0.24152
11	+++---+--+-	3.16625	0.95466	0.15038
12	+++++--++-+-	3.33463	0.96262	0.12948
13	+++++--++-+-+	3.55639	0.98637	0.04916
14	++++++--++-+-+	3.57536	0.95556	0.16630
15	+++++--++--+-+-	3.74089	0.96589	0.13209
16	+++-+++---+-++-+	3.77645	0.94411	0.22355
17	++-++--++++-+-+--	3.87848	0.94067	0.24463
18	+++-+++---+-++-+--	4.01406	0.94612	0.22858
19	++-+---+-++++-++++-	4.16269	0.95499	0.19621
20	+++++-+---+-++---++-	4.30167	0.96188	0.17047
21	++--------++--+-+-+--+	4.39853	0.95984	0.18405
22	+++++-++--+-+-++---+++	4.47518	0.95411	0.21523
23	+++++++---++--+---+-+-+-	4.57183	0.95329	0.22400
24	++--+++------+-+-+--+--+	4.71462	0.96237	0.18436
25	+++---++++++-+-+--+--++-	4.83413	0.96683	0.16587

Proof. Suppose $f \in \mathfrak{L}_n$ has coefficients forming a Barker sequence. From (1.1) we see that

$$\|f\|_4^4 = n^2 + n - \epsilon(n),$$

where $\epsilon(n) = 1$ if n is odd and 0 if n is even. Using Hölder's inequality, we have

$$\|f\|_2^2 < \|f\|_1^{2/3} \|f\|_4^{4/3},$$

and so

$$\|f\|_1^2 > \frac{n^3}{n^2 + n - \epsilon(n)} = n - 1 + \frac{1}{n+1}\left(1 + \frac{\epsilon(n)n^2}{n^2 + n - 1}\right). \qquad \square$$

This statement in fact appears in the 1968 paper of Turyn [34], who attributes the observation to Newman.

Newman in fact proved a statement similar to Theorem 5.1 in 1960 [26], showing that $\|f\|_1 < \sqrt{n} - .03$ for $f \in \mathfrak{L}_n$ of positive degree. We revisit Newman's argument here, choosing parameters in an optimal way and employing the results of some computations on Littlewood polynomials to obtain an improved lower bound. It is clear from the proof however that a new approach is needed to obtain the constant 1, as Newman observed.

Theorem 5.2. *If f is a Littlewood polynomial of positive degree $n-1$, then*

$$\|f\|_1 < \sqrt{n} - .09 \,.$$

Proof. Let $f(z) = \sum_{k=0}^{n-1} a_k z^k$ with $a_k = \pm 1$ for $0 \le k < n$, and let $\alpha > 1$ be a real number whose value will be selected later. The argument splits into two cases, depending on the size of $\|f\|_\infty$.

Case 1: $\|f\|_\infty \le \alpha\sqrt{n}$. Let c_k denote the kth aperiodic autocorrelation of the sequence of coefficients of f. Since $\sum_{k=1}^{n-1} c_k^2 \ge \lfloor n/2 \rfloor$, using (1.1) we have

$$\|f\|_4^4 \ge n^2 + n - \epsilon(n) \,,$$

where $\epsilon(n) = 1$ if n is odd and 0 otherwise. Next, since

$$\int_0^1 \left(|f(e(t))|^2 - n \right)^2 dt = \|f\|_4^4 - 2n\|f\|_2^2 + n^2 \ge n - \epsilon(n) \,,$$

we compute

$$\int_0^1 \left(|f(e(t))| - \sqrt{n} \right)^2 dt = \int_0^1 \left(\frac{|f(e(t))|^2 - n}{|f(e(t))| + \sqrt{n}} \right)^2 dt \ge \frac{n - \epsilon(n)}{(\alpha+1)^2 n} \,.$$

However,

$$\int_0^1 \left(|f(e(t))| - \sqrt{n} \right)^2 dt = 2n - 2\sqrt{n}\,\|f\|_1 \,,$$

so

$$\begin{aligned}
\|f\|_1^2 &\le n - \frac{n - \epsilon(n)}{(\alpha+1)^2 n} + \frac{(n - \epsilon(n))^2}{4(\alpha+1)^4 n^3} \\
&= n - \frac{1}{(\alpha+1)^2} + O(1/n) \,.
\end{aligned} \tag{5.1}$$

Case 2: $\|f\|_\infty > \alpha\sqrt{n}$. Suppose $\max_{|z|=1} |f(z)| = A\sqrt{n}$, occurring at $z = e(t_0)$. By Bernstein's inequality, $|f'(z)| \le A(n-1)\sqrt{n}$, so for $0 \le t \le 1$, it follows that

$$|f(e(t))| \ge A\sqrt{n}\,(1 - 2\pi(n-1)\,|t - t_0|) \,.$$

Let β be a small positive number whose value will be selected later, let I denote the interval $[t_0 - \beta/n, t_0 + \beta/n]$, and let $B = \int_I |f(e(t))|^2 \, dt$. Then

$$B \geq 2\alpha^2 n \int_{t_0}^{t_0+\beta/n} (1 - 2\pi(n-1)(t-t_0))^2 \, dt$$

$$= 2\alpha^2 \beta \left(1 - 2\pi\beta \left(1 - \frac{1}{n} \right) + \frac{4\pi^2\beta^2}{3} \left(1 - \frac{1}{n} \right)^2 \right).$$

(5.2)

It follows that

$$B \geq 2\alpha^2 \beta \left(1 - 2\pi\beta + 4\pi^2\beta^2/3 \right)$$

(5.3)

if $\beta < 3/4\pi$.

Next, let J denote the complement of I (modulo 1) in $[0, 1]$ so that

$$\int_J |f(e(t))|^2 \, dt = n - B.$$

By the Schwarz inequality,

$$\left(\int_I |f(e(t))| \, dt \right)^2 \leq 2\beta B/n$$

and

$$\left(\int_J |f(e(t))| \, dt \right)^2 \leq (n - B)(1 - 2\beta/n),$$

so

$$\|f\|_1 \leq \sqrt{2\beta B/n} + \sqrt{(n - B)(1 - 2\beta/n)}.$$

(5.4)

The expression on the right is decreasing in B for $B \geq 2\beta$, so assuming that $\alpha^2(1 - 2\pi\beta + 4\pi^2\beta^2/3) \geq 1$, we may replace B in (5.4) with the expression in (5.3) to obtain

$$\|f\|_1^2 \leq \left(2\alpha\beta\sqrt{(1 - 2\pi\beta + 4\pi^2\beta^2/3)/n} \right.$$

$$\left. + \sqrt{(n - 2\alpha^2\beta(1 - 2\pi\beta + 4\pi^2\beta^2/3))(1 - 2\beta/n)} \right)^2$$

$$= n - 2\beta \left(1 + \alpha^2 - 2\alpha^2\beta\pi + 4\alpha^2\beta^2\pi^2/3 - 2\alpha\sqrt{1 - 2\beta\pi + 4\beta^2\pi^2/3} \right)$$

$$+ O(1/n).$$

(5.5)

Selecting parameters. Now we wish to choose α and β, subject to the identified constraints, so that the constant terms in the expressions (5.1) and (5.5) match and are as large as possible. (Newman uses $\alpha = 2\sqrt{\pi} \approx 3.54$, $\beta = 1/4\pi \approx .0796$, and $B \geq 1$, which yields .0484 in case 1 and .361 in case 2.) Selecting candidate values for β between 0 and $3/4\pi$ produces the values of $1/(\alpha + 1)^2$ shown in Figure 1. The optimal value is approximately .092347, occurring near $\alpha = 2.2907$ and $\beta = .064804$.

FIGURE 1. Optimal constant term.

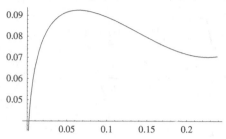

When n is even, we obtain from (5.1) that $\|f\|_1^2 \le n - .091281$, and it is straightforward to verify that the bound in (5.5) is slightly smaller for all n. However, for odd $n \ge 3$ we obtain from (5.1) that $\|f\|_1^2 \le n - .09$ only for $n \ge 41$. To obtain the inequality for all odd n, we first perform the analysis a bit more carefully to decrease this threshold, then we complete the proof by determining for small n the maximal value of the L_1 norm of a Littlewood polynomial of degree $n - 1$. To this end, we replace the parameter α with the expression $\alpha - \gamma/n$ and use the more precise lower bound from (5.2) for B in (5.4) in place of the bound (5.3). Choosing $\gamma = .899634$ to balance the $1/n$ terms in the respective asymptotic expansions, we verify that both (5.1) and (5.4) yield $\|f\|_1^2 < n - .09$ when n is odd for $n \ge 21$.

To complete the proof, we therefore need only check that every Littlewood polynomial with even degree $n - 1 \le 18$ satisfies $\|f\|_1 < \sqrt{n - .09}$. This is established in Table 3, which displays for each $n \le 25$ a Littlewood polynomial of degree $n - 1$ having maximal L_1 norm.   □

We remark that the last column of the table shows that the value of .09 in Theorem 5.2 cannot in general be replaced with any number larger than .1856 . . . . We also note that the extremal polynomials with respect to the L_1 norm in Table 3 are precisely the same as the extremal Littlewood polynomials with respect to Mahler's measure in Table 2. In particular, the coefficient sequences appearing in Table 3 for $n = 2, 3, 4, 5, 7, 11$, and 13 are Barker sequences. Again, the other Barker sequence of length 4 has the same L_1 norm as that of the $n = 4$ entry in the table.

6. AN IRREDUCIBILITY QUESTION

As we noted in Section 2, Turyn and Storer [35] proved that no Barker sequences of odd length n exist for $n > 13$. Their proof is elementary, though somewhat complicated, and relies on showing that long Barker sequences of odd length must exhibit certain patterns. We describe here a possible alternative route to proving this result, in the hope of spurring further research. The material in this section also appears in [7].

TABLE 3. Maximal L_1 norms of Littlewood polynomials by degree.

n	Coefficients of f	$\|f\|_1$	$\|f\|_1/\sqrt{n}$	$n - \|f\|_1^2$
2	++	1.27324	0.90032	0.37886
3	++-	1.67761	0.96857	0.18562
4	+++-	1.92555	0.96277	0.29227
5	+++-+	2.19412	0.98124	0.18583
6	++++-+	2.33899	0.95489	0.52912
7	+++--+-	2.58397	0.97665	0.32311
8	++++--+-	2.73681	0.96761	0.50989
9	+++-+--++	2.87385	0.95795	0.74097
10	++++-+--++	3.04989	0.96446	0.69817
11	+++---+--+-	3.25835	0.98243	0.38317
12	+++++--++-+-	3.40074	0.98171	0.43498
13	+++++--++-+-+	3.57946	0.99276	0.18749
14	++++++--++-+-+	3.65775	0.97757	0.62088
15	+++++--++--+-+-	3.80732	0.98305	0.50430
16	+++-+++---+-++-+	3.89389	0.97347	0.83764
17	++-++--++++-+-+--	4.00380	0.97106	0.96956
18	+++-+++---+-++-+--	4.13097	0.97368	0.93505
19	++-+---+-++++-++++-	4.26105	0.97755	0.84344
20	+++++-+----+-++---++-	4.39129	0.98192	0.71659
21	++--------++--+-+-+--+	4.50012	0.98201	0.74893
22	+++++--++--+-+-+-++---+++	4.58809	0.97818	0.94943
23	++++++++---++--+---+-+-+-	4.68409	0.97670	1.05934
24	++--+++------+-+-+---+--+	4.81295	0.98244	0.83550
25	+++---+++++++-+-+--+--++-	4.92189	0.98438	0.77497

For a polynomial $f(x)$, we define its *reciprocal polynomial* $f^*(x)$ by $f^*(x) := x^{\deg f} f(1/x)$. For $f(x) \in \mathbb{Z}[x]$, we say f is *reciprocal* if $f = \pm f^*$.

Theorem 6.1. *If the polynomial*

$$g_m(x) := \sum_{k=1}^{m} \left(x^{2m-2k} + x^{2m+2k} \right) + (-1)^m (2m+1)x^{2m}$$

is irreducible, then no Barker sequence of length $2m + 1$ exists.

Proof. Suppose $\{a_k\}$ is a Barker sequence of length $2m + 1$, and let $f_m(x) = \sum_{k=0}^{2m} a_k x^k$. By Theorem 2.1, the aperiodic autocorrelation c_k is 0 if k is odd and $(-1)^m$ if $k \neq 0$ and k is even. Thus

$$f_m(x)f_m^*(x) = \sum_{k=-m}^{m} c_{2k}x^{2k+2m}$$

$$= (2m+1)x^{2m} + \sum_{k=1}^{m}(-1)^m \left(x^{2m+2k} + x^{2m-2k}\right),$$

and so $g_m(x) = (-1)^m f_m(x)f_m^*(x)$.   □

The polynomials $g_m(x)$ are in fact irreducible for $6 < m \leq 900$, and it would be interesting if there is a short proof of this for large m. We note however that Erich Kaltofen has observed that the polynomials $g_m(x)$ are in fact always reducible mod p, for any prime p. With his permission, we include his proof of the following more general statement.

Theorem 6.2. *Suppose $f(x)$ is an even, reciprocal polynomial with integer coefficients and $\deg(f) \geq 4$. Then $f(x)$ is reducible mod p for every prime p.*

Proof. If $f = -f^*$ then $f(\pm 1) = 0$ so f is reducible over $\mathbb{Q}$. If $f = f^*$ and $\deg(f) = 4n+2$ then $f(\pm i) = 0$, so again f is reducible over $\mathbb{Q}$ for $n \geq 1$. Suppose then that $f = f^*$ and $\deg(f) = 4n$ with $n \geq 1$, and write $f(x) = g(x^2)$. Clearly $f(x) \equiv g(x)^2 \pmod 2$, so suppose p is an odd prime, and $g(x)$ is irreducible mod p. Let α be a root of g in its splitting field $\mathbb{F}_{p^{2n}}$ over $\mathbb{F}_p$, so that

$$g(x) = \prod_{k=0}^{2n-1} \left(x - \alpha^{p^k}\right).$$

Let γ be a primitive element of $\mathbb{F}_{p^{2n}}$, and let $\alpha = \gamma^t$ for some integer t. Since g is reciprocal, α^{-1} is also a root of g, so $\alpha^{-1} = \gamma^{-t} = \alpha^{p^j} = \gamma^{tp^j}$ for some positive integer $j < 2n$. Then $\gamma^{tp^{2j}} = \gamma^{-tp^j} = \gamma^t$, so $\alpha^{p^{2j}-1} = 1$, and consequently $j = n$. Therefore $\gamma^{t(p^n+1)} = 1$, so $(p^n - 1) \mid t$ and thus t is even. Let $\beta = \gamma^{t/2}$. Then

$$f(x) = \prod_{k=0}^{2n-1} \left(x + \beta^{p^k}\right) \cdot \prod_{k=0}^{2n-1} \left(x - \beta^{p^k}\right),$$

and each of these products lies in $\mathbb{F}_p[x]$.   □

ACKNOWLEDGMENTS

We thank the Heilbronn Institute for Mathematical Research for sponsoring this conference on number theory and polynomials. We also thank Erich Kaltofen for Theorem 6.2. In addition, we thank the centre for Interdisciplinary Research in the Mathematical and Computational Sciences (IRMACS) and the High Performance Computing Centre at Simon Fraser University for computational resources. Lastly, we thank the referee for careful reading of the manuscript, and helpful comments.

REFERENCES

[1] R.H. Barker, Group synchronizing of binary digital systems, *Communication theory*, 273–287, Butterworths Sci. Pub., London, 1953.

[2] L.D. Baumert, *Cyclic difference sets*, Lecture Notes in Mathematics **182**, Springer Verlag, Berlin, 1971.

[3] E. Beller, D.J. Newman, An extremal problem for the geometric mean of polynomials, *Proc. Amer. Math. Soc.* **39** (1973), 313–317.

[4] P. Borwein, *Computational excursions in analysis and number theory*, CMS Books Math./Ouvrages Math. SMC **10**, Springer Verlag, New York, 2002.

[5] _____, Paul Erdős and polynomials, *Paul Erdős and his mathematics, I (Budapest, 1999)*, Bolyai Soc. Math. Stud. **11**, 161–174, János Bolyai Math. Soc., Budapest, 2002.

[6] P. Borwein, E. Kaltofen, K.-K.S. Choi, J. Jedwab, Binary sequences with merit factor greater than 6.34, *IEEE Trans. Inform. Theory* **50** (2004), no. 12, 3234–3249.

[7] P. Borwein, M.J. Mossinghoff, A question of irreducibility, in preparation.

[8] F.W. Carroll, D. Eustice, T. Figiel, The minimum modulus of polynomials with coefficients of modulus one, *J. London Math. Soc. (2)* **16** (1977), no. 1, 76–82.

[9] S. Eliahou, M. Kervaire, Barker sequences and difference sets, *Enseign. Math. (2)* **38** (1992), no. 3–4, 345–382.

[10] _____, _____, Corrigendum to "Barker sequences and difference sets", *Enseign. Math. (2)* **40** (1994), no. 1–2, 109–111.

[11] S. Eliahou, M. Kervaire, B. Saffari, A new restriction on the lengths of Golay complementary sequences, *J. Combin. Theory Ser. A* **55** (1990), no. 1, 49–59.

[12] P. Erdős, An inequality for the maximum of trigonometric polynomials, *Ann. Polon. Math.* **12** (1962), 151–154.

[13] P. Fan, M. Darnell, *Sequence design for communications applications*, Research Studies Press, Somerset, England, 1996.

[14] G.T. Fielding, The expected value of the integral around the unit circle of a certain class of polynomials, *Bull. London Math. Soc.* **2** (1970), 301–306.

[15] M.L. Fredman, B. Saffari, B. Smith, Polynômes réciproques: conjecture d'Erdős en norme L^4, taille des autocorrélations et inexistence des codes de Barker, *C. R. Acad. Sci. Paris Sér. I Math.* **308** (1989), no. 15, 461–464.

[16] M.J.E. Golay, A class of finite binary sequences with alternate autocorrelation values equal to zero, *IEEE Trans. Inform. Theory* **18** (1972), 449–450.

[17] _____, The merit factor of long low autocorrelation binary sequences, *IEEE Trans. Inform. Theory* **28** (1982), no. 3, 543–549.

[18] J. Jedwab, A survey of the merit factor problem for binary sequences, *Sequences and Their Applications—Proceedings of SETA 2004*, Lecture Notes in Computer Science **3486**, 30–55, Springer Verlag, Berlin, 2005.

[19] J. Jedwab, S. Lloyd, A note on the nonexistence of Barker sequences, *Des. Codes Cryptogr.* **2** (1992), no. 1, 93–97.

[20] J.-P. Kahane, Sur les polynômes à coefficients unimodulaires, *Bull. London Math. Soc.* **12** (1980), 321–342.

[21] T.W. Körner, On a polynomial of Byrnes, *Bull. London Math. Soc.* **12** (1980), no. 3, 219–224.

[22] K.H. Leung, B. Schmidt, The field descent method, *Des. Codes Cryptogr.* **36** (2005), no. 2, 171–188.

[23] J.E. Littlewood, On polynomials $\sum^n \pm z^m$, $\sum^n e^{\alpha_m i} z^m$, $z = e^{\theta i}$, *J. London Math. Soc.* **41** (1966), 367–376.

[24] _____, *Some problems in real and complex analysis*, D.C. Heath and Co., Lexington, Mass., 1968.

[25] K. Mahler, On two extremum properties of polynomials, *Illinois J. Math.* **7** (1963), 681–701.

[26] D.J. Newman, Norms of polynomials, *Amer. Math. Monthly* **67** (1960), 778–779.

[27] _____, An L^1 extremal problem for polynomials, *Proc. Amer. Math. Soc.* **16** (1965), 1287–1290.

[28] W. Rudin, Some theorems on Fourier coefficients, *Proc. Amer. Math. Soc.* **10** (1959), 855–859.

[29] B. Saffari, Barker sequences and Littlewood's "two-sided conjectures" on polynomials with ±1 coefficients, *Séminaire d'Analyse Harmonique, Année 1989/90*, 139–151, Univ. Paris XI, Orsay, 1990.

[30] B. Schmidt, Cyclotomic integers and finite geometry, *J. Amer. Math. Soc.* **12** (1999), no. 4, 929–952.

[31] H.S. Shapiro, *Extremal problems for polynomials and power series*, Master's Thesis, Mass. Inst. of Technology, 1951.

[32] R. Turyn, On Barker codes of even length, *IEEE Trans. Inform. Theory* **51** (1963), no. 9, 1256.

[33] _____, Character sums and difference sets, *Pacific J. Math.* **15** (1965), 319–346.

[34] _____, Sequences with small correlation, *Error Correcting Codes (Proc. Sympos. Math. Res. Center, Madison, Wis.)*, 195–228, John Wiley, New York, 1968.

[35] R. Turyn, J. Storer, On binary sequences, *Proc. Amer. Math. Soc.* **12** (1961), 394–399.

THE HANSEN-MULLEN PRIMITIVITY CONJECTURE: COMPLETION OF PROOF

STEPHEN D. COHEN AND MATEJA PREŠERN

ABSTRACT. This paper completes an efficient proof of the Hansen-Mullen Primitivity Conjecture (HMPC) when $n = 5, 6, 7$ or 8. The HMPC (1992) asserts that, with some (mostly obvious) exceptions, there exists a primitive polynomial of degree n over any finite field with any coefficient arbitrarily prescribed. This has recently been proved whenever $n \geq 9$ or $n \leq 4$. We show that there exists a primitive polynomial of any degree $n \geq 5$ over any finite field with third coefficient, i.e., the coefficient of x^{n-3}, arbitrarily prescribed. This completes the HMPC when $n = 5$ or 6. For $n \geq 7$ we prove a stronger result, namely that the primitive polynomial may also have its constant term prescribed. This implies further cases of the HMPC and completes the HMPC when $n = 7$. We also show that there exists a primitive polynomial of degree $n \geq 8$ over any finite field with the coefficient of x^{n-4} arbitrarily prescribed, and this completes the HMPC when $n = 8$. A feature of the method, when the cardinality of the field is 2 or 3, is that 2-adic and 3-adic analysis is required for the proofs. The article is intended to provide the reader with an overview of the general approach to the solution of the HMPC without the weight of detail involved in unravelling the situation of arbitrary degree.

1. INTRODUCTION

For q a power of a prime p, let $\mathbb{F}_q$ be the finite field of order q. Its multiplicative group $\mathbb{F}_q^*$ is cyclic of order $q - 1$ and a generator of $\mathbb{F}_q^*$ is called a *primitive element* of $\mathbb{F}_q$. More generally, a primitive element γ of $\mathbb{F}_{q^n}$, the unique extension of degree n of $\mathbb{F}_q$, is the root of a (necessarily monic and automatically irreducible) *primitive polynomial* $f(x) \in \mathbb{F}_q[x]$ of degree n. Any root of f is a primitive element of $\mathbb{F}_{q^n}$ and so are its conjugates $\gamma^q, \ldots, \gamma^{q^{n-1}}$. In 1992, T. Hansen and G.L. Mullen [16] stated a conjecture on the existence of a primitive polynomial of degree n over $\mathbb{F}_q$ with an arbitrary coefficient prescribed. (See also [22] and [23].)

Conjecture 1.1 (Hansen and Mullen, 1992). *Let $a \in \mathbb{F}_q$ and let $n \geq 2$ be a positive integer. Fix an integer m with $0 < m < n$. Then there exists a primitive polynomial $f(x) = x^n + \sum_{j=1}^{n} a_j x^{n-j}$ of degree n over $\mathbb{F}_q$ with*

2000 *Mathematics Subject Classification.* 11T06, 11T30, 11T24, 11L40, 11S85.

Key words and phrases. primitive polynomial, finite field, prescribed coefficient, Hansen-Mullen conjecture.

$a_m = a$ *with (genuine) exceptions when*

$$(q, n, m, a) = (q, 2, 1, 0), \ (4, 3, 1, 0), \ (4, 3, 2, 0) \ or \ (2, 4, 2, 1).$$

We shall refer to Conjecture 1.1 as the Hansen-Mullen Primitivity Conjecture (HMPC). Substantial progress has already been made towards a complete proof of the HMPC. We outline some of these steps. (For a fuller bibliography consult Cohen's survey of the last decade's activity, [5].) When $m = 1$, it was demonstrated by Cohen, [1]. (See [9] for a self-contained exposition.) For $n = m - 1$, it follows from [2], [6], [17]. The papers of Han [15] and Cohen and Mills [8] cover most cases with $m = 2$ and $n \geq 5$ (although the situation when q is even and $n = 5$ or 6 is not altogether clear). For $m = 3$, the conjecture holds provided $n \geq 7$ by [13], [14], [21] and [7]. As remarked previously in [10], however, when $m = 2$ or 3, significant computer verification in a large number of cases was necessary to resolve these questions, particularly when $5 \leq n \leq 7$. Next, the HMPC follows from [3] whenever $m \leq \frac{n}{3}$ (except that for $q = 2$ the restriction is to $m \leq \frac{n}{4}$). For *even* prime powers q and *odd* degrees n it was shown by Fan and Han [12] provided $n \geq 7$. Recently, the HMPC has been established when $m = 2$ or, provided $n \geq 6$, when $m = n - 2$ [10]. Finally, the whole conjecture has been established by Cohen whenever $n \geq 9$, [4].

To resolve the HMPC for particular values of n and m, it is evidently more delicate when n is small and, less evidently perhaps, when m is around $\frac{n}{2}$ (see [4]). From the above summary, the outstanding cases all have $5 \leq n \leq 8$. In particular, the existence of a primitive quintic ($n = 5$), a primitive sextic ($n = 6$) and a primitive septic ($n = 7$) with the coefficient of x^3 prescribed ($m = 3$), as well as the existence of a primitive octic ($n = 8$) with the coefficient of x^4 prescribed ($m = 4$) has not been settled.

In this paper, we prove existence in all of the remaining cases of the HMPC listed above. In particular, with regard to $m = 3$ or 4, we give a self-contained proof of the following Theorems 1.2 and 1.4 with a minimal amount of computation.

Theorem 1.2. *Suppose $n \geq 5$. Let a be an arbitrary member of the finite field $\mathbb{F}_q$. Then there exists a primitive polynomial $f(x) \in \mathbb{F}_q[x]$ of degree n with third coefficient prescribed as a.*

Note that when $n = 5$ and $a = 0$ the conclusion of Theorem 1.2 is a consequence of Theorem 1.2 in [10]. A (difficult) case of the HMPC is an immediate consequence of Theorem 1.2.

Corollary 1.3. *Suppose $5 \leq n \leq 6$. Then the HMPC holds.*

Theorem 1.4. *Suppose $n \geq 8$. Let a be an arbitrary member of the finite field $\mathbb{F}_q$. Then there exists a primitive polynomial $f(x) \in \mathbb{F}_q[x]$ of degree n with fourth coefficient prescribed as a.*

Corollary 1.5. *Suppose $n = 8$ and $m = 4$. Then the HMPC holds.*

Moreover, when the degree $n \geq 7$, we prove a stronger version of Theorem 1.2 wherein additionally the constant term of the primitive polynomial is appropriately prescribed as $(-1)^n c \in \mathbb{F}_q$. Here, necessarily c must be a *primitive* element of $\mathbb{F}_q$, since this is the norm of a root of the polynomial.

Theorem 1.6. *Suppose $n \geq 7$. Let a be an arbitrary non-zero member of the finite field $\mathbb{F}_q$ and c be an arbitrary primitive element of $\mathbb{F}_q$. Then, there exists a primitive polynomial $f(x) \in \mathbb{F}_q[x]$ of degree n with third coefficient a and constant term $(-1)^n c$.*

In view of the fact that a monic polynomial $f(x) \in \mathbb{F}_q[x]$ of degree n with constant term $(-1)^n c$ is primitive if and only if the reciprocal polynomial $\frac{x^n}{(-1)^n c} \cdot f\left(\frac{1}{x}\right)$ is primitive, Theorem 1.2 (for $a = 0$) and Theorem 1.6 (for $a \neq 0$) imply further cases of the HMPC.

Corollary 1.7. *Suppose $n \geq 7$ and $a \in \mathbb{F}_q$. Then there exists a primitive polynomial of degree n over $\mathbb{F}_q$ with its coefficient of x^3 equal to a. In particular, the HMPC is established for $(n, m) = (7, 4)$ and $(8, 5)$.*

Corollary 1.8. *Suppose $7 \leq n \leq 8$. Then the HMPC holds.*

Generally, for the numerical aspects we can suppose $5 \leq n \leq 8$, though the calculations could easily be extended to larger values of the degree. (Of course, the working becomes easier as n increases).

In the proofs of Theorems 1.2, and 1.6 we will separately approach fields of orders $\equiv 0 \pmod{3}$ (the *ternary problem*) and orders $\not\equiv 0 \pmod{3}$. We shall refer to this as the *non-ternary problem*. This is because, when $q \equiv 0 \pmod{3}$, the criterion for prescribing the third coefficient has a different shape and 3-adic analysis is employed. (Recall that in [10], for prescribed second coefficient, 2-adic analysis was involved.)

Granted Theorem 1.6, for $a \neq 0$ we need only consider $n = 5$ or 6 in Theorem 1.2. When $n = 5$ in Theorem 1.2 and $n = 7, 8$ in Theorem 1.6, we only need give the proof for when $a \neq 0$ (the *non-zero problem*), but when $n = 6$, we shall distinguish two cases according to whether $a \neq 0$ or $a = 0$ (the *zero problem*). In particular, in the non-zero problem, when $n = 7, 8$ we also treat the case when the constant term is prescribed.

In the proof of Theorem 1.4, we have to treat fields of even and odd orders separately. Here in numerical work it suffices to take $n = 8$. When $q \equiv 0 \pmod{2}$ and 2-adic analysis is brought in, we will speak of the *even problem* and refer to everything else as the *odd problem*. In both problems we distinguish between the *zero problem*, when the prescribed coefficient is zero, and the *non-zero problem* (otherwise).

In every case careful work on expressing the number of desired primitive polynomials in terms of character sum expressions is required, as well as a sieving technique. However, in most cases for a number of values of q examples had to be found directly using the computer algebra package Maple. Full details will be given in [24].

2. BASIC NOTATION WITH APPLICATIONS

In order to render this account as self-contained as possible, we reproduce some basic material from [10].

Throughout take $Q_n = \frac{q^n-1}{q-1}$ and, for any integer r, denote by $\theta(r)$ the ratio $\frac{\phi(r)}{r}$, ϕ being Euler's function.

Observe that a primitive element of $\mathbb{F}_{q^n}$ is not a d-th power in $\mathbb{F}_{q^n}$ for any divisor d of $q^n - 1$ exceeding 1. More generally, for any divisor k of $q^n - 1$, call a (non-zero) element of $\mathbb{F}_{q^n}$ k-free if it is not a d-th power in $\mathbb{F}_{q^n}$ for any divisor d of k exceeding 1.

Given $a \in \mathbb{F}_q$, for a divisor k of $q^n - 1$ denote (temporarily) by $\pi_a(k)$ the number of k-free elements of $\mathbb{F}_q^n$ whose characteistic polynomial over $\mathbb{F}_q$ has specified coefficient a (here third or fourth). It is required to show that $\pi_a(q^n - 1)$ is positive. In particular, in the zero problem ($a = 0$), the number is $\pi_0(q^n - 1)$. Evidently, from the definition of k-free, the value of $\pi_a(k)$ depends only on the square-free part of k, that is, the product of all distinct primes dividing k. Accordingly, we replace k by its square-free part, whenever appropriate.

Lemma 2.1. *Suppose that an (irreducible) polynomial $f(x) \in \mathbb{F}_q[x]$ of degree n has specified coefficient 0 and a root $\gamma \in \mathbb{F}_{q^n}$ that is Q_n-free. Then there exists $b \in \mathbb{F}_q^*$, such that the minimal polynomial of $\gamma^* := b\gamma$ is primitive of degree n and also has specified coefficient 0.*

Proof. Since γ is Q_n-free, for a fixed primitive element $\xi \in \mathbb{F}_{q^n}$, $\gamma = \xi^e$, where $\gcd(e, Q_n) = 1$. Set $b = \xi^{jQ_n}$ (automatically in $\mathbb{F}_q$) for some j to be chosen. Then, for any choice of j, $\gamma* := b\gamma$ remains Q_n-free. Write $q - 1 = q_1 q_2$, where q_1 and q_2 are co-prime with q_1 the largest factor of $q - 1$ co-prime to Q_n. Thus, for any b, $b\gamma = \gamma^*$ is already q_2-free. It is additionally q_1-free (and so primitive) if j is chosen so that $e + jQ_n \equiv 1 \pmod{q_1}$. This is always possible. The result follows. $\square$

Consequently, from Lemma 2.1, in the zero problem in order to establish that $\pi_0(q^n - 1)$ is positive, it suffices to show that $\pi_0(Q_n)$ is positive.

For Theorem 1.6 (wherein the constant term is also prescribed), introduce E_n, defined as the product of distinct primes in $q^n - 1$ that are *not* factors of $q - 1$. In particular, E_n is an *odd* divisor of Q_n. Further, for a ($\neq 0$) $\in \mathbb{F}_q$, c primitive in $\mathbb{F}_q$ and $k|(q^n - 1)$, define $\pi_{a,c}(k)$ to be the number of k-free $\gamma \in \mathbb{F}_{q^n}$ whose characteristic polynomial has third coefficient a and constant term $(-1)^n c$. We want to show that when $n \geq 6$, then $\pi_{a,c}(q^n - 1)$ is positive.

Lemma 2.2. *Suppose that $f(x) \in \mathbb{F}_q[x]$ is an irreducible polynomial of degree n with constant term $(-1)^n c$, where c is a primitive element of $\mathbb{F}_q$. Then f is primitive if and only if any root $\gamma \in \mathbb{F}_{q^n}$ is E_n-free.*

Proof. Since $\gamma^{\frac{q^n-1}{q-1}} = c$ is a primitive element of $\mathbb{F}_q$, γ is guaranteed to be $(q-1)$-free. To be primitive (in $\mathbb{F}_{q^n}$) it therefore suffices for γ to be E_n-free. $\square$

By Lemma 2.2, it suffices to show that $\pi_{a,c}(E_n)$ is positive.

The next items of notation relate to the characteristic function of the set of (non-zero) k-free elements of $\mathbb{F}_{q^n}$. For any $d|(q^n-1)$, write η_d for a typical (multiplicative) character in $\widehat{\mathbb{F}_{q^n}}^*$ of order d. Then η_d is extended to a function on $\mathbb{F}_{q^n}$ by setting $\eta_d(0) = 0$ (even when $d=1$). Thus η_1 is the trivial character. We shall however write $\eta = \mathbf{1}$ for the version of the trivial character for which $\eta(0) = 1$. As in other papers, adopt an 'integral' notation for weighted sums; namely, for $k|(q^n-1)$, set

$$\int_{d|k} \eta_d := \sum_{d|k} \frac{\mu(d)}{\phi(d)} \sum_{(d)} \eta_d,$$

where the inner sum runs over all $\phi(d)$ characters of order d. (Once again, only square-free divisors d have any influence.) Then the characteristic function for the subset of k-free elements of $\mathbb{F}_{q^n}$ is

$$\theta(k) \int_{d|k} \eta_d(\gamma), \quad \gamma \in \mathbb{F}_{q^n}, \tag{2.1}$$

with $\theta(k)$ as above.

The next batch of notation relates to the sieving technique. Given k (taken to be square-free), write $k = k_0 p_1 \cdots p_s$, $s \geq 1$, for some divisor k_0 and distinct primes $p_1, \ldots, p_s$. Then (k_0, s) is called a *decomposition* of k. To such a decomposition we associate a number

$$\delta := 1 - \sum_{i=1}^{s} \frac{1}{p_i}, \tag{2.2}$$

which is of special significance. To be useful it is essential that k_0 is selected so that δ is positive: it will always be assumed that this is so.

From here on, we write $\pi(k)$ for $\pi_a(k)$ whenever $a \neq 0$.

Lemma 2.3. *For any divisor d of $q^n - 1$, let $\pi(k)$ denote the number of k-free elements of $\mathbb{F}_{q^n}$ satisfying prescribed conditions. Suppose that (k_0, s) is a decomposition of k. Then*

$$\pi(k) \geq \left(\sum_{i=1}^{s} \pi(k_0 p_i) \right) - (s-1)\pi(k_0) \tag{2.3}$$

$$= \delta\pi(k_0) + \sum_{i=1}^{s} \left(\pi(k_0 p_i) - \left(1 - \frac{1}{p_i}\right)\pi(k_0) \right). \tag{2.4}$$

Proof. The results are trivial for $s = 1$. The basic sieving inequality (2.3) holds by induction on $s \geq 2$. When $s = 2$, $\mathcal{S}(k_0) \subseteq \mathcal{S}(k_0 p_1) \cup \mathcal{S}(k_0 p_2)$, where $\mathcal{S}(k)$ denotes the set of elements counted by $\pi(k)$.

The expression (2.4) is a useful rearrangement of the right side of (2.3). $\square$

In brief, for a given k (such as $q^n - 1$), one starts out by estimating $\pi(k)$ directly (i.e., take $s = 1$ in the above) for sufficiently large q. For smaller values of q, genuine applications of the sieve $(s > 1)$ become crucial.

For any positive integer r, denote by $W(r) = 2^{\omega(r)}$ the number of square-free divisors of r, where $\omega(r)$ is the number of distinct prime divisors of r. For a given decomposition (k_0, s) define

$$\Delta_{s,\delta} := \frac{s-1}{\delta} + 2.$$

When $s = 1$, then $\Delta_{s,\delta} = 2$ and $W(k) = 2W(k_0)$.

3. Prescribing the third coefficient

3.1. The non-ternary problem. Throughout this section we will only consider fields with characteristic *not* 3. First we recall a standard general fact.

Lemma 3.1 (Newton's formula). *For a field F, let $f(x) \in F[x]$ be a separable monic irreducible polynomial in $F[x]$ with a root $\gamma \in E$, say. For $1 \leq t \leq m$, denote by s_t the E/F-trace of γ^t. Then the m-th symmetric function σ_m of the roots of f satisfies*

$$(-1)^{m-1} m \sigma_m = s_m - s_{m-1}\sigma_1 + s_{m-2}\sigma_2 + \cdots + (-1)^{m-1} s_1 \sigma_{m-1}. \quad (3.1)$$

As it stands when $m = 3$, Lemma 3.1 is useful only when the characteristic of F is not 3. Suppose now that $q \not\equiv 0 \pmod 3$ and that $a \in \mathbb{F}_q$ is given. From (3.1), considering that $2\sigma_2 = s_1^2 - s_2$ and $\sigma_1 = s_1$, we have

$$6\sigma_3 = s_1^3 - 3s_1 s_2 + 2s_3.$$

As we want to assign the value a to the third coefficient, we put $\sigma_3 = -a$. Set $s_1 = 0$. Then we have to put $s_3 = -3a$. The characteristic function for the set of elements $\gamma \in \mathbb{F}_{q^n}$ for which $s_1 = 0$ and $s_3 = -3a$ is

$$\frac{1}{q^2} \sum_{\alpha,\beta \in \mathbb{F}_q} \chi_n(\alpha\gamma^3 + \beta\gamma)\chi(-3\alpha a).$$

Here χ is the canonical additive character on $\mathbb{F}_q$ (so that

$$\chi(b) = \exp\left(\frac{2\pi i T_u(b)}{p}\right),$$

where $q = p^u$) and χ_n is the canonical character on $\mathbb{F}_{q^n}$. Also $\bar{\chi}$ is the complex conjugate character to χ.

Therefore, for $k|q^n - 1$, redefining $\pi_a(k)$ to refer specifically to the number of primitive polynomials with $s_1 = 0$, $s_3 = -a$, we obtain

$$\frac{q^2 \pi_a(k)}{\theta(k)} = \int_{d|k} \sum_{\alpha,\beta \in \mathbb{F}_q} \bar{\chi}(\alpha a) S_n(\alpha, \beta; \eta_d), \qquad (3.2)$$

where $S_n(\alpha, \beta; \eta) = \sum_{\gamma \in \mathbb{F}_{q^n}} \chi_n(\alpha \gamma^3 + \beta \gamma) \eta(\gamma)$ and, for simplicity, $3a$ has been replaced by a .

More generally, suppose that (k_0, s) is a decomposition of k. Then, by the equivalence of (2.3) and (2.4),

$$\frac{q^2 \pi_a(k)}{\theta(k)} = \delta \int_{d|k_0} \sum_{\alpha,\beta \in \mathbb{F}_q} \bar{\chi}(\alpha a) S_n(\alpha, \beta; \eta_d)$$

$$+ \sum_{i=1}^{s} \left(1 - \frac{1}{p_i}\right) \int_{d|k_0} \sum_{\alpha,\beta \in \mathbb{F}_q} \bar{\chi}(\alpha a) S_n(\alpha, \beta; \eta_{dp_i}). \qquad (3.3)$$

For (3.2) use the equivalence of the right sides of (2.3) and (2.4). Of course, (3.2) is recovered from (3.3) by setting $s = 1$. Estimates for $S_n(\alpha, \beta; \eta_d)$ are standard as now described.

Lemma 3.2. *Suppose $\alpha, \beta \in \mathbb{F}_q$, not both 0.*
If $\alpha = 0$, then $S_n(0, \beta; \mathbf{1}) = 0$; otherwise

$$\left| S_n(\alpha, \beta; \mathbf{1}) \right| \leq q^{\frac{n}{2}}.$$

Suppose $d|q^n - 1$ with $d > 1$. Then

$$\left| S_n(\alpha, \beta; \eta_d) \right| \leq \begin{cases} 3q^{\frac{n}{2}}, & \text{if } \alpha \neq 0, \\ q^{\frac{n}{2}}, & \text{if } \alpha = 0. \end{cases}$$

Define $S_1(\kappa \beta^T, \eta) := \sum_{\beta \in \mathbb{F}_q} \chi(\kappa \beta^T) \eta(\beta)$. The following lemma gives the bounds for $S_1(\kappa \beta^T, \eta)$. It is a version of Lemma 9.5 in [4].

Lemma 3.3. *Suppose $p \nmid T$ and $T' := \gcd(m, q - 1)$. Assume that $\kappa \in \mathbb{F}_q^*$ and $\eta \in \widehat{\mathbb{F}}_q^*$. Then*

$$\left| S_1(\kappa \beta^T, \eta) + 1 \right| \leq (T' - 1) q^{\frac{1}{2}} \quad \text{if } \eta \text{ is trivial},$$
$$\left| S_1(\kappa \beta^T, \eta) \right| \leq T' q^{\frac{1}{2}} \qquad \text{otherwise}.$$

Proof. For any $\nu \in \widehat{\mathbb{F}}_q^*$, define the Gaussian sum $G(\nu) := \sum_{z \in \mathbb{F}_q} \psi(z) \nu(z)$. As usual, $G(\nu) = -1$ if ν is trivial and, otherwise, $|G(\nu)| = \sqrt{q}$. Now let ν be a *generator* of $\widehat{\mathbb{F}}_q^*$, so that ν has order $q - 1$. Then $\hat{\chi} = \nu^i$ for some $i \leq q - 2$: $i = 0$ if $\hat{\chi}$ is trivial.

Moreover, for $y \in \mathbb{F}_q$,

$$\psi(\kappa y) = \frac{1}{q-1} \sum_{j=0}^{q-2} G(\bar{\nu}^j) \, \nu^j(\kappa y).$$

Hence,

$$S_1(\kappa x^T, \hat{\chi}) = \frac{1}{q-1} \sum_{j=0}^{q-2} G(\bar{\nu}^j) \sum_{c \in \mathbb{F}_q} \nu^j(\kappa c^T) \nu^i(c)$$

$$= \frac{1}{q-1} \left(\sum_{j=1}^{q-2} G(\bar{\nu}^j) \nu^j(\kappa) \sum_{c \in \mathbb{F}_q} \nu^{jT+i}(c) - \sum_{c \in \mathbb{F}_q} \nu^i(c) \right).$$

Now, $\sum_{c \in \mathbb{F}_q} \nu^{jT+i}(c) = 0$ unless ν^{jT+i} is trivial (in which case the sum is $q-1$). The latter occurs precisely when $jT + i \equiv 0 \pmod{q-1}$. For this, necessarily $T' | i$, in which case there are T' solutions $j \pmod{q-1}$.

When $i = 0$, there are $T' - 1$ such solutions j with $1 \le j \le q-2$; otherwise there are T' solutions. $\square$

Lemma 3.4. *Suppose that $p \neq 3$ and that $a \in \mathbb{F}_q$ is non-zero and $k | P_{n,3} := \frac{q^n-1}{\gcd(3,q-1)}$. Suppose also that $k = k_0 p_1 \cdots p_s$, $s \ge 1$, $p_1, \ldots, p_s$ prime, with δ positive. Then $\pi_a(k)$ is positive whenever*

$$q^{\frac{n-3}{2}} > \begin{cases} 9W(k_0)\Delta_{s,\delta}, & \text{when } q \equiv 1 \pmod 3; \\[2mm] 3W(k_0)\Delta_{s,\delta}, & \text{when } q \equiv 2 \pmod 3. \end{cases} \tag{3.4}$$

Specifically, when $s = 1$ and $k = P_{n,3}$, the sufficient condition is

$$q^{\frac{n-3}{2}} > \begin{cases} 9W(q^n - 1), & \text{when } q \equiv 1 \pmod 3; \\[2mm] 3W(q^n - 1), & \text{when } q \equiv 2 \pmod 3. \end{cases} \tag{3.5}$$

Proof. In (3.3), aggregate the contributions to the right side relating to a specific multiplicative character η_d or η_{dp_i} (without the weighting factor implicit in the integral notation). Suppose $d | k_0$ and take η_d: similar reasoning applies to each η_{dp_i}. The contribution to the right-hand side of (3.3) attributable to values of $\alpha = \beta = 0$ is $\delta(q^n - 1)$, as $S_n(0, 0; \eta_d) = 0$ unless $d = 1$. This yields the main term. It remains to show that for each character η_d for any $d|k$ (with $d > 1$) the sum of terms with α, β not both zero is bounded absolutely by $9\delta q^{\frac{n+3}{2}}$ when $q \equiv 1 \pmod 3$ and by $3\delta q^{\frac{n+3}{2}}$ when $q \equiv 2 \pmod 3$.

Suppose $d \nmid Q_n$. Then the restriction $\tilde{\eta}_d$ (of $\eta_d \in \widehat{\mathbb{F}}_{q^n}$ to $\widehat{\mathbb{F}}_q$) is nontrivial. For the terms with $\beta \neq 0$, we can replace α by $\alpha\beta^3$ and γ by $\frac{\gamma}{\beta}$ to yield $\sum_{\alpha \in \mathbb{F}_q} \bar{S}_1(\alpha a; \tilde{\eta}_d) S_n(\alpha, 1, \eta_d)$, where $S_1(x, \tilde{\eta}) = \sum_{\beta \in \mathbb{F}_q} \chi(x\beta)\tilde{\eta}(\beta)$. Because $q \nmid Q_n$, when $\alpha = 0$, then $S_1(\alpha a; \tilde{\eta}_d) = 0$. On the other hand, when $\alpha \neq 0$, then, as usual, $|S_n(\alpha, 1, \eta_d)| \le 3q^{\frac{n}{2}}$, whereas $|S_1(\alpha a; \tilde{\eta}_d)| \le m' q^{\frac{1}{2}}$ by Lemma

3.3. So the total contribution from terms with $\beta \neq 0$ is bounded absolutely by $9\delta(q-1)q^{\frac{n+1}{2}}$ when $q \equiv 1 \pmod 3$ and by $3\delta(q-1)q^{\frac{n+1}{2}}$ when $q \equiv 2 \pmod 3$.

For the remaining contribution in this case, consider terms with $\beta = 0$ (and hence $\alpha \neq 0$): $\sum_{\alpha \in \mathbb{F}_q} \chi(-\alpha a) S_n(\alpha, 0; \eta_d)$. Here, if $3 \nmid (q-1)$, replace α by α^3 and γ by $\frac{\gamma}{\alpha}$ to obtain $\bar{S}_1(a, \tilde{\eta}_d) S_n(1, 0; \eta_d)$. Again $|S_n| \leq 3q^{\frac{n}{2}}$ and $|S_1| \leq q^{\frac{1}{2}}$ so that the total contribution is $3\delta q^{\frac{n+1}{2}}$. On the other hand, if $3|q-1$, one has to split S_n into three sums (each with weight $\frac{1}{3}$) by replacing α by $g^i\alpha^3$ ($i = 0, 1, 2$) and γ by $\frac{\gamma}{\alpha}$ for a fixed non-cube in $\mathbb{F}_q$. Each S_3 is bounded as before whereas, now $|S_1| \leq 3q^{\frac{1}{2}}$. Now the total contribution is bounded by $9\delta q^{\frac{n+1}{2}}$.

Thus, adding the contributions we obtain the required absolute bound $3\delta \gcd(3, q-1)q^{\frac{n+3}{2}}$.

<u>Suppose $d|Q_n$.</u> Now $\tilde{\eta}_d$ is trivial. For the terms with $\beta \neq 0$ proceed as before. This time, provided $\alpha \neq 0$, we can conclude that $\left|S_1(\alpha a; \tilde{\eta}_d)\right| \leq (\gcd(3, q-1) - 1)q^{\frac{1}{2}} + 1$. On the other hand, when $\alpha = 0$, we can only use the trivial bound $|S_1| \leq (q-1)$, though $\left|S_n(0, 1; \eta_d)\right| \leq q^{\frac{n}{2}}$ in this case. Hence for these terms (with $\beta \neq 0$) we have a total absolute bound of

$$3\delta(q-1)\left((\gcd(3, q-1) - 1)q^{\frac{1}{2}} + 1\right)q^{\frac{n}{2}} + \delta(q-1)q^{\frac{n}{2}}$$

$$= 3\delta(q-1)q^{\frac{n}{2}}\left((\gcd(3, q-1) - 1)q^{\frac{1}{2}} + \tfrac{4}{3}\right)$$

$$\leq 3\delta \gcd(3, q-1)(q-1)q^{\frac{n+1}{2}}.$$

The contribution from terms with $\beta = 0$ (and so $\alpha \neq 0$) is certainly bounded by $3\delta \gcd(3, q-1)q^{\frac{n+1}{2}}$ as before. Indeed, the factor $\gcd(3, q-1)q^{\frac{1}{2}}$ could be reduced to $(\gcd(3, q-1) - 1)q^{\frac{1}{2}} + 1$.

The remaining terms on the right side of (3.3) (involving characters like η_{dp_i}) are estimated in the same way: we have used no special properties for $d|k_0$. Taking into account that there are $\phi(d)$ characters of order d for each divisor d we deduce that numerically the right side of (3.3) exceeds

$$\delta\left(q^n - 3\gcd(3, q-1)q^{\frac{n+3}{2}}\Delta_{s,\delta}\right),$$

with $\Delta_{s,\delta}$ as in Section 2, since $\sum_{i=1}^{s}\left(1 - \frac{1}{p_i}\right) = s - 1 + \delta$.   □

To prescribe additionally the constant term of the polynomial, we use the condition of the following lemma.

Lemma 3.5. *Suppose that $a \in \mathbb{F}_q$ is non-zero, c is a primitive element of $\mathbb{F}_q$ and $k|E_n$. Suppose also $k = k_0 p_1 \cdots p_s$, $s \geq 1$, $p_1, \ldots, p_s$ prime, with δ*

positive. Then $\pi_{a,c}(k)$ *is positive whenever*

$$q^{\frac{n-5}{2}} > \begin{cases} 9W(k_0)\Delta_{s,\delta}, & \text{when } q \equiv 1 \ (\text{mod } 3)\,; \\[2mm] 3W(k_0)\Delta_{s,\delta}, & \text{when } q \equiv 2 \ (\text{mod } 3)\,. \end{cases} \tag{3.6}$$

Specifically, when $s = 1$, $W(k_0)\Delta_{s,\delta}$ *is replaced by* $W(E_n)$ *in* (3.6).

In particular, the norm case of the HM-problem is solved whenever (3.6) holds with $k = E_n$.

Proof. The characteristic function for the subset of $\mathbb{F}_{q^n}^*$ comprising elements with $\mathbb{F}_q$-norm c (i.e. $N_n(\gamma) = c$) is $\dfrac{1}{q-1} \displaystyle\sum_{\nu \in \widehat{\mathbb{F}_{q^n}^*}} \nu(N_n(\gamma)c^{-1})$ $\left(\widehat{\mathbb{F}_{q^n}^*}\right.$ being the group of multiplicative characters of $\mathbb{F}_{q^n}^*$). Redefining $\pi_{a,c}$ to refer to the number of primitive polynomials with $s_1 = 0$, $s_3 = -a$, we obtain the following modification of the condition (3.3), where $\hat{\nu}$ denotes the lift of ν to $\widehat{\mathbb{F}_{q^n}^*}$ (so that $\hat{\nu}(\gamma) = \nu(N_n(\gamma))$):

$$\frac{(q-1)q^2\pi_{a,c}(k)}{\theta(k)} = \delta \int_{d|k_0} \sum_{\alpha,\beta \in \mathbb{F}_q} \sum_{\nu \in \widehat{\mathbb{F}_{q^n}^*}} \bar{\nu}(c)\bar{\chi}(3\alpha a)S_n(\alpha,\beta;\eta_d\hat{\nu})$$

$$\tag{3.7}$$

$$+ \sum_{i=1}^{s}\left(1 - \frac{1}{p_i}\right)\int_{d|k_0} \sum_{\alpha,\beta \in \mathbb{F}_q} \sum_{\nu \in \widehat{\mathbb{F}_{q^n}^*}} \bar{\nu}(c)\bar{\chi}(3\alpha a)S_n(\alpha,\beta;\eta_{dp_i}\hat{\nu}).$$

The result is then obtained by the same methods as (3.4). $\qquad\square$

We will distinguish between the cases of $q \equiv 1$ or 2 (mod 3) for smaller values of q, when this gives us a useful saving. In general, however, we will use only the condition for $q \equiv 1$ (mod 3), which applies to all cases.

3.1.1. *Quintics.* To assist in the application of Lemma 3.4 we employ some auxiliary results. The first is an easy fact that was also quoted as Lemma 4.2 in [9].

Lemma 3.6. *Suppose n and l are odd primes such that* $l|Q_n = \frac{q^n-1}{q-1}$. *Then either* $l = n$ *or* $l \in L_{2n}$ *(with* $l \nmid (q-1)$*). Here* L_{2n} *denotes the set of primes congruent to* 1 (mod $2n$).

Remark. We will use l to denote a prime number throughout.

Throughout we use some explicit bounds for the number of square-free divisors of an integer h.

Here is an example to illustrate how to obtain such a bound. Let h be a positive integer, such that $\omega(h) \geq 13$. Then the number of square-free divisors of h is bounded by $W(h) < h^{\frac{3}{11}}$. This holds because $l^{\frac{3}{11}} > 2$ when $l \geq 41$ (the 13-th prime) and so, by calculation,

$$\frac{2^{\omega(h)}}{h^{\frac{3}{11}}} \leq \prod_{l \mid h} \frac{2}{l^{\frac{3}{11}}} \leq \prod_{l \leq 41} \frac{2}{l^{\frac{3}{11}}} < 1.$$

Other bounds we use are obtained analogously.

Express the product of distinct primes in $q^5 - 1$ as $K_1 \cdot K_2$, where K_1 (a factor of $q - 1$) is the product of all distinct prime divisors of $q - 1$ and K_2 (a factor of Q_5) is the product of distinct prime divisors of Q_5 that do not divide $q - 1$. Observe that $5 \mid (q - 1)$ if and only if $5 \mid Q_5$ and therefore all prime divisors of K_2 are $\equiv 1 \pmod{10}$. Denote $\omega(K_1)$ by ω_1 and $\omega(K_2)$ by ω_2.

Lemma 3.7. *Suppose that $n = 5$, $\omega_1 \geq 13$ or $\omega_2 \geq 26$. Let a $(\neq 0) \in \mathbb{F}_q$. Then there exists a primitive polynomial of degree 5 over $\mathbb{F}_q$ with the coefficient of x^2 prescribed as a.*

Proof. First suppose $\omega_1 \geq 13$ and $\omega_2 \geq 26$. The number of square-free divisors of h, an integer with $\omega(h) \geq 13$, is bounded by $W(h) < h^{\frac{3}{11}}$. Therefore $W(K_1) < (q - 1)^{\frac{3}{11}} < q^{\frac{3}{11}}$. Also, when integer h is a product of primes $l \equiv 1 \pmod{10}$ and $\omega(h) \geq 26$, then $W(h) < h^{\frac{13}{99}}$. That yields $W(K_2) < (Q_5)^{\frac{13}{99}} < (q^5)^{\frac{13}{99}} = q^{\frac{65}{99}}$. It follows that $W(q^5 - 1) < q^{\frac{92}{99}}$. Consequently, by (3.5), to show existence it suffices that $q > 9q^{\frac{92}{99}}$, i.e., $q \geq 9^{\frac{99}{7}} \approx 3.131 \cdot 10^{13}$, which holds as $\omega_1 \geq 13$ and $\omega_2 \geq 26$ both yield $q > 10^{14}$.

Next, suppose $\omega_1 \leq 12$ and $\omega_2 \geq 26$. Set k_0 to be the product of K_2 and the least three primes in K_1. Thus $s \leq 9$, $\delta \geq 1 - \frac{1}{7} - \ldots - \frac{1}{37} > 0.440$ and $\Delta_{s,\delta} < 20.19$. By the above, $W(k_0) < 8q^{\frac{65}{99}}$ and (3.4) is satisfied whenever $q \geq 1615716202$. This is the case since $\omega_2 \geq 26$, whence $q > 10^{14}$.

Finally, suppose $\omega_1 \geq 13$ and $\omega_2 \leq 25$. Put $k_0 = K_1$. Then $s \leq 25$, $\delta \geq 1 - \sum_{\substack{l \leq 571 \\ l \equiv 1 (\text{mod } 10)}} \frac{1}{l} > 0.743$ and $\Delta_{s,\delta} < 34.31$. Now (3.4) is satisfied whenever $q \geq 2651$. This completes the proof since $\omega_1 \geq 13$ implies $q > 10^{14}$. $\square$

Following Lemma 3.7, we assume $\omega_1 \leq 12$, $\omega_2 \leq 25$ and run the full sieving process. The sieving steps are shown in the table below, where $q_{\min}$ denotes the minimum integer q satisfying (3.4) numerically.

#	q	$\omega_1 \leq$	$\omega_2 \leq$	k_0	$\omega(k_0)$	$s \leq$	$\delta \geq$	$\Delta_{s,\delta} <$	$q_{\min}$
1		12	25	30	3	34	0.263	127.48	9179
2	≤ 9178	5	8	6	2	11	0.378	28.46	1025
3	≤ 1024	4	7	6	2	9	0.461	19.36	697
4	≤ 696	4	6	6	2	8	0.469	16.93	610
5	≤ 609	4	5	6	2	7	0.479	14.53	524

In line 5, $q \leq 609$ yields $\omega_2 \leq 6$, but as there are no values of q in this range with $\omega_2 = 6$, we can suppose $\omega_2 \leq 5$. (Similar reductions apply to subsequent tables.)

At this point, it is appropriate to separate the calculations regarding mod 3. Firstly, criterion (3.4) takes a milder form for values of $q \equiv 2 \pmod 3$ and line 5 of the table above gives $q_{\min} = 175$. We may then suppose $q \leq 174$, $\omega_1 \leq 3$ and $\omega_2 \leq 5$. Putting $k_0 = 6$ gives $s \leq 6$, $\delta \geq 0.621$, $\Delta_{s,\delta} \leq 10.06$ and $q_{\min} = 121$.

Secondly, when $q \equiv 1 \pmod 3$ but $q \not\equiv 1 \pmod 9$, then 3 is not a factor of $P_{n,m}$. In line 5 we obtain $\omega_1 \leq 3$ and $\omega_2 \leq 5$. Putting $k_0 = 2$ yields $s \leq 7$, $\delta \geq 0.479$, $\Delta_{s,\delta} \leq 14.53$ and $q_{\min} = 262$. We may then suppose $q \leq 261$, $\omega_1 \leq 3$ and, as there are no values of q in this range with $\omega_2 = 5$, we proceed assuming $\omega_2 \leq 4$. Again we set $k_0 = 2$. Now $\delta \geq 0.493$, $\Delta_{s,\delta} \leq 12.15$ and $q_{\min} = 219$.

The remaining prime powers are checked separately. In fact, by individual consideration, all remaining prime powers ≥ 121, other than $q = 163$, satisfy (3.4) with k_0 the least prime factor of $P_{n,m}$. These comprise 361, 343 and 289 $(\equiv 1 \pmod 9)$ together with 211, 199, 196, 193, 181, 169, 157, 151, 139 and 127 $(\equiv 1 \pmod 3)$. Indeed, lower values of q similarly satisfy (3.4) except for $q = 109, 67, 64, 61, 49, 43, 37, 31, 25, 19, 16, 13, 11, 8, 7, 5, 4$ and 2. For these and for $q = 163$ a primitive polynomial had to be found explicitly in each case, using Maple.

3.1.2. Sextics: the non-zero problem. Write the product of distinct primes in $q^6 - 1$ as $K_1 \cdot K_2$, where K_1 is the product of all distinct prime divisors of $q^2 - 1$ and K_2 is the product of distinct prime divisors of $\frac{q^6-1}{q^2-1}$ that do not divide $q^2 - 1$. Notice that 3 cannot be a factor of K_2 and so (by an analogue of Lemma 3.6) any prime divisor l of K_2 is $\equiv 1 \pmod 6$, i.e., $l \in L_6$. Denote $\omega(K_1)$ by ω_1 and $\omega(K_2)$ by ω_2.

Lemma 3.8. *Suppose that $n = 6$, $\omega_1 \geq 13$ or $\omega_2 \geq 15$. Let $a \, (\neq 0) \in \mathbb{F}_q$. Then there exists a primitive polynomial of degree 6 over $\mathbb{F}_q$ with the coefficient of x^3 prescribed as a.*

The proof of Lemma 3.8 is similar to that of Lemma 3.7.

Consequently to Lemma 3.8, we now assume $\omega_1 \leq 12$, $\omega_2 \leq 14$ and run the full sieving process: steps are shown in the following table. Here $q_{\min}$ denotes the minimal integral value of q for which (3.4) holds with the displayed value of δ.

#	$q \leq$	$\omega_1 \leq$	$\omega_2 \leq$	k_0	$\omega(k_0)$	$s \leq$	$\delta \geq$	$\Delta_{s,\delta} <$	$q_{\min}$
1		12	14	30	3	24	0.241	93.29	356
2	355	5	7	6	2	11	0.235	44.56	138
3	137	5	6	6	2	9	0.295	29.12	104
4	103	5	5	6	2	8	0.318	24.02	91

Here, as in Lemma 3.7, in the last line, $\omega_2 \leq 6$, has been reduced to $\omega_2 \leq 5$.

We now assume $q \leq 89$ separately treat $q \equiv 1$ and $q \equiv 2 \pmod 3$. When $q \equiv 2 \pmod 3$, $\omega_1, \omega_2 \leq 5$ and setting $k_0 = 6$ results in s, δ and $\Delta_{s,\delta}$ as in row 4 of the table above, and $q_{\min} = 44$. Assuming $q \leq 41$ and putting $k_0 = 6$, gives $s \leq 7$, $\delta \geq 0.377$, $\Delta_{s,\delta} \leq 17.92$ and $q_{\min} = 36$.

When $q \equiv 1 \pmod 3$ but $q \not\equiv 1 \pmod 9$, then $\omega_1 \leq 4$ and $\omega_2 \leq 5$. We put $k_0 = 2$ and obtain $s \leq 8$, $\delta \geq 0.318$, $\Delta_{s,\delta} \leq 24.04$ and $q_{\min} = 58$.

There are two values of $q \equiv 1 \pmod 9$ in between 89 and 36, namely $q = 73$ and $q = 64$, and both satisfy condition (3.4) when we set k_0 to be the smallest divisor of $P_{n,m}$.

Of all the other values of q left out by the sieve procedure, $q = 49$, 43, 32, 31, 29, 25, 23 and 17, satisfy condition (3.4) with $\omega(k_0) = 1$. Primitive polynomials for $q = 37$, 19, 16, 13, 11, 8, 7, 5, 4 and 2 had to be found explicitly, using Maple.

3.1.3. Sextics: the zero problem.

Suppose thet the prescribed coefficient a is zero. By Lemma 2.1, it suffices to prove that $\pi_0(Q_n)$ is positive. Now, putting $a = 0$ in (3.3) yields the following proposition. Note that, by comparison with the proof of Lemma 3.4 we always have to use the trivial bound $|S_1(0, \bar{\bar{\eta}}_d)| \leq q - 1$ since $d | Q_n$.

Proposition 3.9. *Let $k | Q_n$ and let (k_0, s) be a decomposition of k. Suppose $q \not\equiv 0 \pmod 3$ and*

$$q^{\frac{n-4}{2}} > 9W(k_0)\Delta_{s,\delta}. \tag{3.8}$$

Then $\pi_0(k)$ is positive.
Specifically, when $s = 1$ and $k_0 = Q_n$, the sufficient condition is

$$q^{\frac{n-4}{2}} > 9W(Q_n). \tag{3.9}$$

Now take $n = 6$. Write the product of distinct primes in Q_6 as $K_1 \cdot K_2$, where K_1 is the product of all distinct prime divisors of $q+1$ and K_2 is the product of distinct prime divisors of $\frac{q^6-1}{q^2-1}$ that do not divide $q^2 - 1$. As in the previous section, any prime divisor l of K_2 is $\equiv 1 \pmod 6$. Denote $\omega(K_1)$ by ω_1 and $\omega(K_2)$ by ω_2.

Lemma 3.10. *Suppose that $n = 6$, $\omega_1 \geq 12$ or $\omega_2 \geq 25$. Then there exists a primitive polynomial of degree 6 over $\mathbb{F}_q$ with the coefficient of x^3 equal to zero.*

The proof parallels that of Lemma 3.7, giving lower bounds for $W(K_1)$, $W(K_2)$: though here $K_1 \cdot K_2$ is taken to be a divisor of $\frac{q^n-1}{q-1}$ and not $q^n - 1$.

Following Lemma 3.10, we assume $\omega_1 \leq 11$, $\omega_2 \leq 24$ and run the full sieving process. The sieving steps are shown in the table below.

#	$q \le$	$\omega_1 \le$	$\omega_2 \le$	k_0	$\omega(k_0)$	$s \le$	$\delta \ge$	$\Delta_{s,\delta} <$	q_{min}
1		11	24	30	3	32	0.174	180.17	12973
2	12972	5	10	30	3	12	0.437	27.18	1957
3	1956	4	8	6	2	10	0.322	29.96	1079
4	1078	4	7	6	2	9	0.337	25.74	927
5	926	4	6	6	2	8	0.354	21.78	785
6	784	3	6	6	2	7	0.445	15.49	558
7	557	3	5	6	2	6	0.468	12.69	468

Here there have been reductions to ω_1 and ω_2 in lines 5–7. When $q \le 167$, polynomials were found explicitly, using Maple.

Remark. Proposition 3.9 can also be used easily to treat the zero problem for $m = 3$ when $n = 7, 8$. This would avoid relying on the calculations used in [13], [14], [21] and [7].

3.1.4. *Septics.* For degrees $n = 7$ and $n = 8$ we prove a stronger result and prove the existence of primitive polynomials over $\mathbb{F}_q$ with prescribed third coefficient *and* constant term (necessarily a primitive element of $\mathbb{F}_q$). The main tool here is Lemma 3.5.

Lemma 3.11. *Suppose that $n = 7$ and $\omega(E_7) \ge 20$. Let $a \ (\ne 0) \in \mathbb{F}_q$ and c be a primitive element of $\mathbb{F}_q$. Then there exist a primitive polynomial of degree 7 over $\mathbb{F}_q$ with the coefficient of x^4 and the constant term specified as a and $-c$, respectively.*

Proof. Suppose $\omega(E_7) \ge 20$. Then $W(E_7) < (E_7)^{\frac{1}{8}} < (q^7)^{\frac{1}{8}} < q^{\frac{7}{8}}$. Then, by (3.6), to show existence it suffices that $q > 9 \cdot q^{\frac{7}{8}}$, or $q > 9^8$, which obviously holds as $\omega(E_7) \ge 20$ yields $q > 10^8$. $\qquad\qquad\square$

We can now assume $\omega(E_7) \le 19$ and sieve: the outcome is displayed in the table below.

#	$q \le$	$\omega(E_7) \le$	k_0	$\omega(k_0) \le$	s	$\delta \ge$	$\Delta_{s,\delta} <$	q_{min}
1		19	1	0	19	0.874	22.60	204
2	203	7	1	0	7	0.901	8.66	78
3	77	5	1	0	5	0.911	6.40	58

Next, $q = 53, 49, 47, 43, 41, 37, 32, 31, 29, 23, 17$ and 11 all satisfy criterion (3.6) when k_0 is set to be 1. But for $q = 25, 19, 16, 13, 8, 7, 5, 4$ and 2, suitable primitive polynomials had to be searched for with Maple.

3.1.5. *Octics.* Express the product of distinct primes in E_8 as $K_1 \cdot K_2$, where K_1 is the product of all distinct prime divisors of $(q+1)(q^2+1)$ and K_2 is the product of distinct prime divisors of $q^4 + 1$ that do not divide $q^4 - 1$. By an analogue of Lemma 3.6, any prime divisor l of K_2 is $\equiv 1 \pmod 8$, i.e., $l \in L_8$. Denote $\omega(K_1)$ by ω_1 and $\omega(K_2)$ by ω_2. Note that 2 is never a factor of E_8.

Lemma 3.12. *Suppose that $n = 8$, $\omega_1 \geq 16$ or $\omega_2 \geq 13$. Let a $(\neq 0) \in \mathbb{F}_q$ and c be a primitive element of $\mathbb{F}_q$. Then there exists a primitive polynomial of degree 8 over $\mathbb{F}_q$ with the coefficient of x^5 prescribed as a and constant term c.*

The proof is analogous to that of Lemma 3.7. Consequently, we now assume $\omega_1 \leq 15$, $\omega_2 \leq 12$ and sieve. The sieving steps are shown in the table below.

#	$q \leq$	$\omega_1 \leq$	$\omega_2 \leq$	k_0	$w(k_0)$	$s \leq$	$\delta \geq$	$\Delta_{s,\delta} <$	q_{min}
1		15	12	15	2	25	0.247	99.17	234
2	233	7	5	3	1	11	0.274	38.50	79
3	78	6	4	3	1	9	0.328	26.40	61
4	60	5	4	3	1	8	0.381	20.38	52
5	51	5	3	3	1	7	0.392	17.31	46
6	45	3	3	3	1	5	0.560	9.15	31

Here there are reductions in ω_1, ω_2 in lines 5–6.

Each prime power q, $29 \geq q \geq 16$, satisfies (3.6) with k_0 the least prime in E_8. When $q = 13, 11, 8, 7, 5, 4$ or 2, however, the primitive polynomials had to be found explicitly by Maple.

3.2. The ternary problem.

In this section, we give a summary of appropriate p-adic analysis. It will later be applied first to the ternary problem (when $p = 3$) and then to the even problem (when $p = 2$).

The fields $\mathbb{F}_q$ and $\mathbb{F}_{q^n}$ will be identified with subsets (or finite quotient rings) of an extension of the field $\mathbb{Q}_p$ (the completion of the rational field with respect to the p-adic metric).

Introduce definitions and notation as follows.

- K_n is the splitting field of the polynomial $x^{q^n} - x$ over $\mathbb{Q}_p$.
- Γ_n ($\subseteq K_n$) is the set of roots of the polynomial above (the Teichmüller points of K). The non-zero elements of Γ_n form a cyclic group of order $q^n - 1$.
- R_n denotes the ring of integers of K_n. Then

$$\Gamma_n \subseteq R_n = \left\{ \sum_{i=0}^{\infty} p^i \gamma_i, \ \gamma_i \in \Gamma_n \right\}.$$

  Moreover, R_n is a local ring with unique maximal ideal pR_n and $R_n/pR_n \cong \mathbb{F}_{q^n}$.
- Distinct elements of Γ_n are already distinct modulo p. For a set isomorphic to $\mathbb{F}_{q^n}$, temporarily denoted by $\mathcal{G}_n$, all q^n members of Γ_n can be expressed uniquely in the form $\sum_{i=0}^{\infty} p^i \gamma_i$, $\gamma_i \in \mathcal{G}_n$, where $\gamma \in \Gamma_n$ is already fixed by specifying γ_0. For any integer $e \geq 1$, $\Gamma_{n,e}$ is the set (of cardinality q^n) of elements of Γ_n mod p^e, i.e., $\Gamma_{n,e} =$

$\{\sum_{i=0}^{e-1} p^i \gamma_i, \, \gamma_i \in \mathcal{G}_n\}$, where we retain the notation γ for the member associated with $\gamma \in \Gamma_{n,e}$. In particular, $\gamma^{q^n} = \gamma$ for $\gamma \in \Gamma_{n,e}$. Moreover, $\mathcal{G}_n = \Gamma_{n,1} \cong \mathbb{F}_{q^n}$.

- $R_{n,e} = \{\sum_{i=0}^{e-1} p^i \gamma_i, \gamma_i \in \Gamma_{n,e}\} \cong R_n/p^e R_n$, so that $R_{n,e}$ has cardinality q^{ne}. (Thus $R_{n,e}$ is a *Galois ring*.) Observe that here $R_{n,e}/pR_{n,e} \cong \mathbb{F}_{q^n}$ also. Moreover, $R_{n,1} = \Gamma_{n,1}$, which can be identified with $\mathbb{F}_{q^n}$. Conversely, each $\gamma \in \Gamma_{n,1}$ yields a unique lift, also denoted by γ, to every $\Gamma_{n,e}$ and to Γ_n itself. An element of (multiplicative) order r in $\Gamma_{n,1}$ lifts to an element of the same order in each $\Gamma_{n,e}$ and in Γ_n; in particular, a primitive element lifts to a primitive element.

Next, consider objects relating to the extension $\mathbb{F}_{q^n}/\mathbb{F}_q$. Note that K_1 is a subfield of K_n, with $\Gamma_1 \subseteq \Gamma_n$, and R_1 is a subring of R_n. Similar relationships apply to the Galois rings. Further, note that the Galois group of K_n/K_1 is isomorphic to that of $\mathbb{F}_{q^n}/\mathbb{F}_q$, being cyclic of order n and generated by the Frobenius automorphism τ_n, where $\tau_n(\gamma) = \gamma^q, \gamma \in \Gamma_n$. More generally, on R_n, $\tau_n\left(\sum_{i=0}^{\infty} p^i \gamma_i\right) = \sum_{i=0}^{\infty} p^i \gamma_i^q$ (where each $\gamma_i \in \Gamma_n$). This induces a ring homomorphism τ_n on $R_{n,e}$ such that $\tau_n\left(\sum_{i=0}^{e-1} p^i \gamma_i\right) = \sum_{i=0}^{e-1} p^i \gamma_i^q$ (where now each $\gamma_i \in \Gamma_{n,e}$).

Now we discuss polynomials. The polynomial $x^{q^n} - x$ over $\mathbb{F}_q$ (and so over R_1) is the product of all monic irreducible polynomials of degree a divisor of n. A typical monic irreducible polynomial $f(x)$ of degree d (a divisor of n) in $R_{1,1}[x]$ has the form

$$f(x) = (x - \gamma)(x - \gamma^q) \cdots (x - \gamma^{q^{d-1}}) = x^d - \sigma_1 x^{d-1} + \cdots + (-1)^d \sigma_d, \quad (3.10)$$

where $\gamma \in \Gamma_{n,1}$ and each $\sigma_j \in \Gamma_{1,1}$. The polynomial f *lifts* to a (unique) irreducible polynomial of degree d over each $R_{1,e}$, and over R_1 having the same form, except that γ is the corresponding lifted element of $\Gamma_{1,e}$ or Γ_1. But note that, in general, the coefficients σ_j in (3.10) lie in $R_{1,e}$ (or R_1), but may not be in $\Gamma_{1,e}$ (or Γ_1). From the above, the *order* of the polynomial f (which equals the order of any of its roots) or any of its lifts has the same value (a divisor of $q^n - 1$). In particular, f is *primitive* if it is irreducible of degree n and has order $q^n - 1$: this holds if and only if any of its lifts is primitive.

For any $\gamma \in \Gamma_n$, define its trace (over R_1) as $T_n(\gamma) := \gamma + \tau_n(\gamma) + \cdots + \tau_n^{n-1}(\gamma) = \gamma + \gamma^q + \cdots + \gamma^{q^{n-1}} \in R_1$. Observe that $T_n(c\gamma) = cT_n(\gamma)$, $c \in \Gamma_1$. A trace function T_n with similar properties is induced on $\Gamma_{n,e}$.

Next, let $\gamma \in \Gamma_n$ be a root of a lifted irreducible polynomial $f(x) \in R_1[x]$. Eventually, we can suppose γ is *primitive*: for the moment it suffices that f has degree n. Thus, (3.10) holds with $d = n$. Here σ_i denotes the i-th symmetric function of the roots $\gamma, \gamma^q, \ldots, \gamma^{q^{n-1}}$. Employing the trace, we have that s_i, the sum of the i-th powers of the roots of f, is given by $s_i = T_n(\gamma^i) \in R$. Of course, each s_i depends only on f and not on the specific root γ: moreover, all this translates to the expansion of f as a polynomial in

$R_e[x]$. For our purposes, we require an expression for the p-adic expansion of s_i.

We proceed to work with a lifted irreducible polynomial f of degree n in $R_1[x]$ and eventually its reduction to $R_{1,2}$. Henceforth, the letter t is reserved for an positive integer $\not\equiv 0 \pmod p$. Note from above that, for any such t, the value of s_{tp^i} for any $i \geq 0$ is already determined by s_t, and is given by $s_t^{(i)} := \tau^i(s_t)$. For any t, write $s_t = \sum_{j=0}^{\infty} s_{t,j} p^j$, $s_{t,j} \in \Gamma_1$, whence $s_t^{(i)} = \sum_{j=0}^{\infty} s_{t,j}^{2^i} p^j$. Since each positive integer L can be uniquely expressed as $L = tp^j$, then any *component* $s_{t,j}$ is uniquely associated with the integer tp^j.

Now assume $p = 3$. In the context of 3-adic analysis Lemma 3.1 assumes the following shape.

Lemma 3.13. *Let* $f(x) = x^n - \sigma_1 x^{n-1} + \cdots + (-1)^n \sigma_n \in R_1[x]$ *be a (lifted) irreducible polynomial with* σ_i *being a symmetric function of the roots of* f, $\sigma_1, \ldots, \sigma_n \in \Gamma_1$. *Let* s_i *be the sum of the i-th powers of the roots of* f. *Then*

$$3\sigma_3 = \sigma_2 s_1 - \sigma_1 s_2 + s_3. \tag{3.11}$$

Lemma 3.14. *Let* f, σ_i *and* s_i *be as in Lemma* 3.13. *Then*

$$2\sigma_3 = s_{1,0}^3 - s_{1,0} s_{2,0} + 2s_{1,1}^3 \pmod 3.$$

Proof. Over $R_{1,2}$, equality (3.11) translates to

$$
\begin{aligned}
6\sigma_3 &= s_1^3 - 3s_1 s_2 + 2s_3 \\
&= (s_{1,0} + 3s_{1,1})^3 - 3(s_{1,0} + 3s_{1,1})(s_{2,0} + 3s_{2,1}) + 2(s_{1,0}^3 + 3s_{1,1}^3)
\end{aligned}
$$

which, modulo 9, is congruent to $3s_{1,0}^3 - 3s_{1,0}s_{2,0} + 6s_{1,1}^3$. Hence $2\sigma_3 = s_{1,0}^3 - s_{1,0}s_{2,0} + 2s_{1,1}^3 \pmod 3$. $\square$

We wish to assign the value a to the third coefficient, i.e., set $\sigma_3 = -a$. In characteristic 3, we can write $-a = A^3$, $A \in \mathbb{F}_q$. To achieve this, set $s_{1,0} = 0$ and $s_{1,1} = A$.

The characteristic function for the set of elements $\gamma \in \Gamma_{n,2}$ for which $s_{1,0} = 0$ and $s_{1,1} = A$ is, with $\xi = \alpha_0 + 3\alpha_1$,

$$\frac{1}{q^2} \sum_{\xi \in R_{1,2}} \chi\big(\xi(T_n(\gamma) - 3A)\big) = \frac{1}{q^2} \sum_{\alpha_0, \alpha_1 \in \Gamma_{1,1}} \chi_{(n)}\big((\alpha_0 + 3\alpha_1)(\gamma)\big)\chi(-3\alpha_0 A),$$

It follows that, for k a divisor of $q^n - 1$,

$$\frac{q^2 \pi_a(k)}{\theta(k)} = \int_{d|k} \sum_{\alpha_0, \alpha_1 \in \Gamma_{1,1}} \bar\chi(3\alpha_0 A)\, S_n(\xi; \eta_d), \tag{3.12}$$

where $S_n(\xi; \eta) = \sum_{\gamma \in \Gamma_{n,2}} \chi(\xi\gamma)\eta(\gamma)$.

Of course, $S_n(\xi; \eta_d) = 0$ unless $d = 1$. Hence the 'main term' (corresponding to $\xi = 0$) is $q^n - 1$.

The next lemma summarises bounds for $S_n(\xi; \eta_d) = 0$ implied by [17] when $\xi \neq 0$.

Lemma 3.15. *Suppose* $\xi = \alpha_0 + 3\alpha_1 \in R_{1,2}^*$. *Then the following hold:*

- $S_n(\xi; 1) = 0$;
- $|S_n(3\alpha_1; \eta_d)| \le q^{\frac{n}{2}}$, $\alpha_1 \ne 0$;
- $|S_n(\alpha_0; \eta_d)| \le q^{\frac{n}{2}}$, $\alpha_0 \ne 0$;
- $|S_n(\alpha_0 + 3\alpha_1; \eta_d)| \le 3q^{\frac{n}{2}}$, $\alpha_0\alpha_1 \ne 0$.

We consider the contribution to the right side of (3.12) from terms with $\alpha_0 \ne 0$. Replace α_1 by $\alpha_0\alpha_1$ and γ by $\frac{\gamma}{\alpha_0}$. The contribution is

$$\int_{1 < d|k} \sum_{\alpha_1 \in \Gamma_{1,1}} \sum_{\alpha_0 \in \Gamma_{1,1}} \bar{\chi}(3\alpha_0 A)\bar{\bar{\eta}}_d(\alpha_0) S_n(1 + 3\alpha_1; \eta_d)$$

$$= \int_{1 < d|k} \sum_{\alpha_1 \in \Gamma_{1,1}} S_1(3A; \bar{\bar{\eta}}_d) S_n(1 + 3\alpha_1; \eta_d). \qquad (3.13)$$

For the terms with $\alpha_0 = 0$, $\alpha_1 \ne 0$, replace γ by $\frac{\gamma}{\alpha_1}$ to yield the contribution

$$\int_{1 < d|k} \sum_{\alpha_1 \in \Gamma_{1,1}^*} \bar{\bar{\eta}}_d(\alpha_1) S_n(3; \eta_d). \qquad (3.14)$$

Note that in (3.14), the sum over α_1 is zero unless $d|Q_n$ (as $\bar{\bar{\eta}}_d$ is trivial).

It is time to split the discussion into zero and non-zero cases according to the value of A. The zero case will only be needed when $n = 6$, and is therefore treated in Section 3.2.3. Here we proceed supposing $A \ne 0$. Then by Lemma 3.15, the sum $S_1(3A; \bar{\bar{\eta}}_d)$ in absolute value does not exceed 1 if $d|Q_n$ and does not exceed $\sqrt{q}$ otherwise. Hence, for each character η_d $(1 < d|k)$ with $d \nmid Q_n$, by Lemma 3.15, we obtain a total contribution of $3(q-1)q^{\frac{n+1}{2}} = 3(1 - \frac{1}{q})q^{\frac{n+3}{2}}$.

On the other hand, for each character η_d $(1 < d|k)$ with $d|Q_n$, we obtain a bound $3(q-1)q^{\frac{n}{2}}$ from contributions with $\alpha_0 \ne 0$ (governed by (3.13)) and $(q-1)q^{\frac{n}{2}}$ from contributions with $\alpha_0 = 0$, $\alpha_1 \ne 0$ (governed by (3.14)), a total of $4(1 - \frac{1}{q})q^{\frac{n}{2}+1}$. Since this is less than $3(1 - \frac{1}{q})q^{\frac{n+3}{2}}$, for simplicity we use the latter as a bound for the contribution for every η_d, $d > 1$.

Thus the right hand side of (3.12) is bounded by $3W(k)\left(1 - \frac{1}{q}\right)q^{\frac{n+3}{2}}$. This yields a sufficient condition (in the non-sieve case) of

$$q^{\frac{n-3}{2}} > 3\left(1 - \frac{1}{q}\right)W(k).$$

More generally:

Proposition 3.16. *Assume that* $q \equiv 0 \pmod{3}$ *and* $a \in \mathbb{F}$ *is non-zero. Assume also that* $k|q^n - 1$ *and that* (k_0, s) *is a decomposition of* k. *Suppose also that*

$$q^{\frac{n-3}{2}} > 3\left(1 - \frac{1}{q}\right)W(k_0)\Delta_{s,\delta}. \qquad (3.15)$$

Then $\pi_a(k)$ *is positive.*

Multiplicatively, $\mathbb{F}_{q^n}^* \cong \Gamma_{n,3}^*$. Take c to be a primitive element of $\mathbb{F}^* \cong \Gamma_{1,1}^*$ as well as $a \neq 0$ ($\in \mathbb{F} \cong \Gamma_{1,1}$). Then, with $k | E_n$ (by Lemma 2.2), there is an analogous expression for $\dfrac{(q-1)q\pi_{a,c}}{\theta(k_0)}$ to (3.12) comparable to the relationship Lemma 3.5 bears to Lemma 3.4. In particular, each 'integral' on the right side is also over a sum over characters $\nu \in \widehat{\Gamma_{1,2}^*}$ and each character such as η_d or η_{dp_i} replaced by a product $\eta_d \hat{\nu} \, \eta_{dp_i} \hat{\nu}$, where $\hat{\nu}$ is the lift of ν to $\Gamma_{n,2}^*$.

Proposition 3.17. *Assume that $q \equiv 0 \pmod 3$, $a \in \mathbb{F}$ is non-zero and c is a primitive element of $\mathbb{F}$. Assume also that $k | E_n$ and that (k_0, s) is a decomposition of k. Suppose also that*

$$q^{\frac{n-5}{2}} > 3 \left(1 - \frac{1}{q} \right)^2 W(k_0) \Delta_{s,\delta} . \qquad (3.16)$$

Then $\pi_{a,c}(k)$ is positive.

3.2.1. *Quintics.* Express the product of distinct primes in $q^5 - 1$ as $K_1 \cdot K_2$ and define ω_1, ω_2 as in Section 3.1.1 and note again that all prime divisors of K_2 are $\equiv 1 \pmod{10}$. Observe that, throughout Section 3.2, 3 is not a factor of $q^n - 1$.

Lemma 3.18. *Suppose that $n = 5$, $q \equiv 0 \pmod 3$ and $\omega_1 \geq 11$ or $\omega_2 \geq 22$. Let a ($\neq 0$) $\in \mathbb{F}_q$. Then there exists a primitive polynomial of degree 5 over $\mathbb{F}_q$ with the coefficient of x^2 prescribed as a.*

Even though the characteristic of the field is different, the proof of Lemma 3.18 is analogous to that of Lemma 3.7 and is not outlined here. Similarly, we will omit proofs to Lemmas 3.19, 3.20, 3.21 and 3.22.

Following Lemma 3.18, we assume $\omega_1 \leq 10$, $\omega_2 \leq 21$ and obtain the following table.

#	q	ω_1	ω_2	k_0	$\omega(k_0)$	s	$\delta \geq$	$\Delta_{s,\delta} <$	$q_{\min}$
1		≤ 10	≤ 21	10	2	≤ 29	0.291	98.22	1179
2	≤ 729	≤ 3	≤ 6	2	1	≤ 8	0.469	16.93	102
3	81	2	3	2	1	4	0.691	6.35	37.62...
4	27	2	2	2	1	3	0.831	4.41	25.48

When $q = 9$ or 3, primitive quintics with third coefficient prescribed have to be found explicitly, using Maple. They are stated in Section 5.

3.2.2. *Sextics: the non-zero problem.* Express the product of distinct primes in $q^6 - 1$ as $K_1 \cdot K_2$ and define ω_1, ω_2 as in Section 3.1.2.

Lemma 3.19. *Suppose that $n = 6$, $q \equiv 0 \pmod 3$ and $\omega_1 \geq 9$ or $\omega_2 \geq 12$. Let a ($\neq 0$) $\in \mathbb{F}_q$. Then there exists a primitive polynomial of degree 6 over $\mathbb{F}_q$ with the coefficient of x^3 prescribed as a.*

Consequently to the above Lemma, we now assume $\omega_1 \le 8$, $\omega_2 \le 11$ and run the full sieving process: steps are shown in the following table.

#	q	ω_1	ω_2	k_0	$\omega(k_0)$	s	$\delta \ge$	$\Delta_{s,\delta} <$	q_{min}
1		≤ 8	≤ 11	10	2	≤ 17	0.298	55.70	77
2	≤ 27	≤ 3	≤ 4	2	1	≤ 6	0.404	14.38	20

For primitive sextics when $q = 9$ or 3, see Section 5.

3.2.3. *Sextics: the zero problem.*

An isolated case of the ternary problem where considering the possibility of the third coefficient being prescribed zero is needed, is when $n = 6$. Therefore we treat it in this separate subsection, complete with the estimates necessary to establish the criterion for the existence of desired primitive polynomials.

Now, wishing to fix the third coefficient as zero, following Lemma 3.14 we suppose $A = 0$. We may therefore assume also that $k|Q_n$. Suppose $1 < d|Q_n$. In this case, the sum $S_1(3A; \bar{\eta}_d)$ is trivially $q - 1$ and we obtain a contribution for each η_d bounded by the sum $3(q - 1)q^{\frac{n}{2}+1}$ (from (3.13)) and $(q - 1)q^{\frac{n}{2}}$ (from (3.14)). In total this is less than $3\left(1 - \frac{2}{3q}\right)q^{\frac{n+4}{2}}$. Thus the right hand side of (3.12) is bounded by $3W(k)\left(1 - \frac{2}{3q}\right)q^{\frac{n+4}{2}}$. This yields a sufficient condition of

$$q^{\frac{n-4}{2}} > 3\left(1 - \frac{2}{3q}\right)W(k_0)\Delta_{s,\delta}. \tag{3.17}$$

As always, when the coefficient is prescribed to be zero, we only need to consider $W(Q_n)$. Therefore we express the product of distinct primes in Q_6 as $K_1 \cdot K_2$ and define ω_1, ω_2 as in Section 3.1.3.

Lemma 3.20. *Suppose that $n = 6$, $q \equiv 0 \pmod 3$ and $\omega_1 \ge 10$ or $\omega_2 \ge 23$. Then there exists a primitive polynomial of degree 6 over $\mathbb{F}_q$ with the coefficient of x^3 prescribed as zero.*

Here is the sieving table.

#	q	ω_1	ω_2	k_0	$\omega(k_0)$	s	$\delta \ge$	$\Delta_{s,\delta} <$	q_{min}
1		≤ 9	≤ 22	70	3	≤ 28	0.368	75.37	1809
2	≤ 729	≤ 3	≤ 7	2	1	≤ 9	0.337	25.74	155
3	81	2	4	2	1	5	0.741	7.40	44.03...

When $q = 27$, 9 or 3, primitive sextics with zero third coefficient are stated in Section 5.

Remark. Similarly, (3.17) could be used to treat the zero problem when $n = 7, 8$.

3.2.4. Septics.

Lemma 3.21. *Suppose that $n = 7$, $q \equiv 0 \pmod 3$ and $\omega(E_7) \geq 16$. Let $a \, (\neq 0) \in \mathbb{F}_q$ and c be a primitive element of $\mathbb{F}_q$. Then there exists a primitive polynomial of degree 7 over $\mathbb{F}_q$ with the coefficient of x^4 and the constant term specified as a and $-c$, respectively.*

We can now assume $\omega(E_7) \leq 15$. The sieving steps are shown in the table below.

#	q	$\omega(E_7)$	k_0	$\omega(k_0)$	s	$\delta \geq$	$\Delta_{s,\delta} <$	$q_{\min}$
1		≤ 15	1	0	≤ 15	0.879	17.93	54
2	≤ 27	≤ 4	1	0	≤ 4	0.919	5.27	16
3	9	2	1	0	3	0.997	3.01	7.13...

The primitive septics with prescribed third coefficient are in Section 5.

3.2.5. Octics.

Express the product of distinct primes in E_8 as $K_1 \cdot K_2$ and define ω_1, ω_2 as in Section 3.1.5.

Lemma 3.22. *Suppose that $n = 8$, $q \equiv 0 \pmod 3$ and $\omega_1 \geq 11$ or $\omega_2 \geq 10$. Let $a \, (\neq 0) \in \mathbb{F}_q$ and c be a primitive element of $\mathbb{F}_q$. Then there exists a primitive polynomial of degree 8 over $\mathbb{F}_q$ with the coefficient of x^5 and the constant term specified as a and c, respectively.*

Consequently to Lemma 3.22, we now assume $\omega_1 \leq 10$, $\omega_2 \leq 9$ and sieve:

#	q	ω_1	ω_2	k_0	$\omega(k_0)$	s	$\delta \geq$	$\Delta_{s,\delta} <$	$q_{\min}$
1		≤ 10	≤ 9	5	1	≤ 18	0.331	53.36	47
2	≤ 27	≤ 4	≤ 3	1	0	≤ 7	0.392	17.31	14
3	≤ 9	≤ 3	≤ 2	1	0	≤ 5	0.483	10.29	8.41...
4	3	1	1	1	0	2	0.775	3.30	2.68...

4. PRESCRIBING THE FOURTH COEFFICIENT

This section deals with $m = 4$. It achieves the final goal of settling the existence of primitive polynomials with $(n, m) = (8, 4)$, thereby completing the proof of the Hansen-Mullen conjecture. Consider separately the fields of characteristic 2 (the *even* problem) and those with odd characteristic (the *odd* problem).

4.1. The odd problem.

Throughout this section $\mathbb{F}_q$ is a field of characteristic *not* 2. Putting $m = 4$ in Lemma 3.1 and setting $s_1 = 0$ (already) yields

$$4\sigma_4 = -\sigma_2 s_2 - s_4 . \tag{4.1}$$

(Working in this order avoids any difficulties in characteristic 3.)

Condition (4.1) can be further expressed as

$$8\sigma_4 = s_2^2 - 2s_4.$$

For $a \in \mathbb{F}_q$ and for any $z \in \mathbb{F}_q$, setting $s_2 = z$ and $2s_4 = z^2 - 8a$ fixes the fourth coefficient of the polynomial as a. Now, using the characteristic functions defined in Section 2, we can deduce a basic formula for $\pi_a(k)$.

Lemma 4.1. *Suppose q is odd, $a \in \mathbb{F}_q$ is given and $k|q^n - 1$. Then*

$$q^3 \pi_a(k) = \theta(k) \int_{d|k} \sum_{\alpha,\beta,z\in\mathbb{F}_q} \bar{\chi}(\alpha(z^2 - 8a) + \beta z) \, S_n(2\alpha, \beta; \eta_d), \qquad (4.2)$$

where $S_n(\alpha, \beta; \eta)$ denotes the character sum $\sum_{\gamma\in\mathbb{F}_{q^n}} \chi_n(\alpha\gamma^2 + \beta\gamma)\eta(\gamma)$.
More generally, suppose that (k_0, s) is a decomposition of k. Then

$$\frac{q^3 \pi_a(k)}{\theta(k_0)} = \delta \int_{d|k_0} \sum_{\alpha,\beta,z\in\mathbb{F}_q} \bar{\chi}\big(\alpha(z^2 - 8a) + \beta z\big) \, S_n(2\alpha, \beta; \eta_d)$$

$$\qquad (4.3)$$

$$+ \sum_{i=1}^{s} \left(1 - \frac{1}{p_i}\right) \int_{d|k_0} \sum_{\alpha,\beta,z\in\mathbb{F}_q} \bar{\chi}\big(\alpha(z^2 - 8a) + \beta z\big) S_n(2\alpha, \beta; \eta_{dp_i}).$$

In particular, the contribution to the right side of (4.3) attributable to values of $\alpha = \beta = 0$ (the 'main term') is $\delta q(q^n - 1)$.

Proof. For (4.3) use the equivalence of the right sides of (2.3) and (2.4).
For the main term, observe that $S_n(0, 0; \eta_d)$ is zero unless $d = 1$ when the value is $q^n - 1$. Then summing over $z \in \mathbb{F}_q$ we obtain the 'main term' in (4.3). Of course, (4.2) is recovered from (4.3) by setting $s = 1$. □

Estimates for $S_n(\alpha, \beta; \eta_d)$ are standard:

Lemma 4.2. *Suppose $\alpha, \beta \in \mathbb{F}_q$, not both 0.*
If $\alpha = 0$, then $S_n(0, \beta; \mathbf{1}) = 0$; otherwise

$$\big|S_n(\alpha, \beta; \mathbf{1})\big| \leq q^{\frac{n}{2}}.$$

Suppose $d|q^n - 1$ with $d > 1$. Then

$$\big|S_n(\alpha, \beta; \eta_d)\big| \leq \begin{cases} 2q^{\frac{n}{2}}, & \text{if } \alpha \neq 0, \\ q^{\frac{n}{2}}, & \text{if } \alpha = 0. \end{cases}$$

Of course, now $S_1(\alpha, \beta; \eta_d)$ is not the same function as $S_1(\kappa\beta^T, \eta)$ in Section 3.

At this point it is convenient to split the discussion in two parts: the zero and the non-zero problem. First consider the case when the prescribed coefficient a is *non-zero*.

4.1.1. The odd non-zero problem.

Proposition 4.3. *Suppose q is odd and $a \neq 0$. Let $k|q^n - 1$ and (k_0, s) be a decomposition of k. Suppose*

$$q^{\frac{n-4}{2}} > 4W(k_0)\Delta_{s,\delta}. \tag{4.4}$$

Then $\pi_a(k)$ is positive.

Proof. Consider the expression (4.3). We aggregate the contributions to the right side relating to a specific multiplicative character η_d or η_{dp_i} (without the weighting factor implicit in the integral notation). Denote by $\tilde{\eta}_d$ the restriction of η_d to $\mathbb{F}_q$, the significance being that $\tilde{\eta}_d$ has order $\frac{d}{\gcd(d,Q_n)}$.

Suppose $d|k_0$ and take η_d (similar reasoning applies to each η_{dp_i}). Consider the contribution of terms with $\beta \neq 0$. Replace $\gamma \in \mathbb{F}_{q^n}$ by $\frac{\gamma}{\beta} \in \mathbb{F}_{q^n}$, $\alpha \in \mathbb{F}_q$ by $\alpha\beta^2 \in \mathbb{F}_q$, and $z \in \mathbb{F}_q$ by $\frac{z}{\beta} \in \mathbb{F}_q$. We obtain

$$\delta \sum_{\alpha \in \mathbb{F}_q} \sum_{\beta \in \mathbb{F}_q^*} \chi(8a\alpha\beta^2)\bar{\tilde{\eta}}_d(\beta) \sum_{z \in \mathbb{F}_q} \bar{\chi}(\alpha z^2 + z) \, S_n(2\alpha, 1; \eta_d),$$

which is the same as

$$\delta \sum_{\alpha \in \mathbb{F}_q} S_1(8a\alpha, 0; \bar{\tilde{\eta}}_d) \, \overline{S_1(\alpha, 1; 1)} \, S_n(2\alpha, 1; \eta_d). \tag{4.5}$$

The expression (4.5) is essentially the same as in the proof of Proposition 4.1 in [10]. Similarly, we obtain an analogous expression when considering the contribution from terms with $\beta = 0$. We therefore do not give a detailed discussion here; summarising, we obtain an absolute bound of $4\delta q^{\frac{n}{2}+3}$ for the (non-weighted) contribution of all terms corresponding to a character η_d.

The remaining terms on the right side of (4.3) (involving characters like η_{dp_i}) are estimated in the same way: we have used no special properties for $d|k_0$. Taking into account that there are $\phi(d)$ characters of order d for each divisor d we deduce that numerically the right side of (4.3) exceeds

$$\delta\left(q^{n+1} - 4q^{\frac{n}{2}+3}\Delta_{s,\delta}\right),$$

with $\Delta_{s,\delta}$ as in Section 2, since $\sum_{i=1}^{s}\left(1 - \frac{1}{p_i}\right) = s - 1 + \delta$.

$\square$

Now consider $n = 8$. Criterion (4.4) then takes form

$$q^2 > 4W(k_0)\Delta_{s,\delta}. \tag{4.6}$$

Express the product of distinct primes in $q^8 - 1$ as $K_1 \cdot K_2$ where K_1 is the product of all distinct prime divisors of $(q^2 + 1)(q^2 + 1)$ and K_2 is the product of distinct prime divisors of $q^4 + 1$ that do not divide $q^4 - 1$. Remember that any prime divisor l of K_2 is an element of L_8. Denote $\omega(K_1)$ by ω_1 and $\omega(K_2)$ by ω_2. Note that $16|q^4 - 1$ and therefore $\omega_1 = \omega(\frac{q^4-1}{8})$.

Lemma 4.4. *Suppose that* $n = 8$, *a* $(\neq 0) \in \mathbb{F}_q$, *q odd and* $\omega_1 \geq 13$ *or* $\omega_2 \geq 7$. *Then there exists a primitive polynomial of degree 8 over* $\mathbb{F}_q$ *with the coefficient of* x^5 *specified as a.*

Though the characteristic of the field is now different, the proof of this lemma parallels those of analogous lemmas in Section 3 and is omitted. Also the proofs of Lemmas 4.6, 4.13 and 4.14 will be omitted.

Now suppose $\omega_1 \leq 12$ or $\omega_2 \leq 6$. Employing the sieve, the existence of primitive octics with forth coefficient prescribed as $a \neq 0$ is proved in just two steps for $q \geq 17$. First, taking k_0 to be the product of three least primes in K_1 yields $s \leq 14$, $\delta > 0.371$, $\Delta_{s,\delta} < 37.05$ and condition (4.6) holds for $q \geq 35$. Now take k_0 to be the product of two least primes in K_1. Hence $s \leq 7$, $\delta > 0.393$, $\Delta_{s,\delta} < 17.31$ and (4.6) is satisfied when $q \geq 17$. Applying (4.6) to $q = 13$ and 11 with $k_0 = 6$ also proves existence for these two values. Only $q = 7$, 5 and 3 need direct verification with Maple.

4.1.2. *The odd zero problem.* When the prescribed coefficient is zero, it suffices to prove that $\pi_0(Q_n)$ is positive. Considering expression (4.5) with $a = 0$ and proceeding as before, the following proposition is derived.

Proposition 4.5. *Suppose q is odd,* $a = 0$ *and* $n = 8$. *Let* $k | Q_8$ *and* (k_0, s) *be a decomposition of k. Suppose*

$$q^{\frac{3}{2}} > 4W(k_0)\Delta_{s,\delta} .\qquad\qquad (4.7)$$

Then $\pi_0(k)$ *is positive.*

Express the product of distinct primes in Q_8 as $K_1 \cdot K_2$, where K_1 is the product of all distinct prime divisors of $(q+1)(q^2+1)$ (and so even) and K_2 is the product of distinct odd prime divisors of $q^4 + 1$. By an analogue of Lemma 3.6 any prime divisor l of K_2 is $\equiv 1 \pmod 8$, i.e., $l \in L_8$. Denote $\omega(K_1)$ by ω_1 and $\omega(K_2)$ by ω_2.

Lemma 4.6. *Suppose that* $n = 8$, $a = 0$, *q odd and* $\omega_1 \geq 16$ *or* $\omega_2 \geq 12$. *Then there exists a primitive polynomial of degree 8 over* $\mathbb{F}_q$ *with the coefficient of* x^5 *specified as 0.*

Now assume $\omega_1 \leq 15$ and $\omega_2 \leq 11$ and begin the sieve. In the first step, take k_0 to be the product of three least primes in K_1. Then $s \leq 23$, $\delta > 0.279$, $\Delta_{s,\delta} < 80.86$ and criterion (4.7) holds for $q \geq 99$. Now, taking k_0 to be the product of two least primes in K_1 yields $s \leq 8$, $\delta > 0.381$, $\Delta_{s,\delta} < 20.38$ and (4.7) is satisfied for $q \geq 48$. It also holds for $q = 47, 43, 41, 37, 31, 29, 27, 25, 23$ and 19, but smaller values (17, 13, 11, 9, 7, 5, 3) need direct verification with Maple.

4.2. The even problem.

Lemma 4.7. *Let* $f(x) = x^n - \sigma_1 x^{n-1} + \cdots + (-1)^n \sigma_n \in R_1[x]$ *be a (lifted) irreducible polynomial with* σ_i *being a symmetric function of the roots of* f, $\sigma_1, \ldots, \sigma_n \in \Gamma_1$. *Let* s_i *be the sum of the* i-*th powers of the roots of* f. *Then*

$$4\sigma_4 = \sigma_3 s_1 - \sigma_2 s_2 + \sigma_1 s_3 - s_4. \tag{4.8}$$

Lemma 4.8. *Let* f, σ_i *and* s_i *be as in Lemma 4.7. Suppose* $s_{1,0} = 0$. *Then* $\sigma_4 \equiv s_{1,2}^4 \pmod 2$.

Proof. Over $R_{1,3}$, s_i $i = 1, \ldots, 4$ expand to $s_1 = s_{1,0} + 2s_{1,1} + 4s_{1,2}$, $s_2 = s_{1,0}^2 + 2s_{1,1}^2 + 4s_{1,2}^2$, $s_3 = s_{3,0} + 2s_{3,1} + 4s_{3,2}$ and $s_4 = s_{1,0}^4 + 2s_{1,1}^4 + 4s_{1,2}^4$. Accordingly, (4.8) translates to

$$4\sigma_4 \equiv 4s_{1,0}^4 + 4s_{1,0}(s_{3,0} + 2s_{3,1} + 4s_{3,2}) + 4s_{1,2}^4 \pmod 8,$$

$$\sigma_4 \equiv s_{1,0}^4 + s_{1,0}(s_{3,0} + 2s_{3,1} + 4s_{3,2}) + s_{1,2}^4 \pmod 2. \tag{4.9}$$

Setting $s_{1,0} = 0$ in (4.9) yields $\sigma_4 \equiv s_{1,2}^4 \pmod 2$. □

As a consequence of Lemma 4.8, for σ_4 to be prescribed modulo 2, it suffices to prescribe $s_{1,0} = 0$ and $s_{1,2} \in \Gamma_1$ appropriately. The value of $s_{1,1}$ appears to be irrelevant. Nevertheless, in practice we cannot prescribe $s_{1,2}$ without assigning a value (say $z \in \Gamma_{1,1}$) to $s_{1,1}$. In view of Lemma 4.8, given $a \in \mathbb{F}_q \cong \Gamma_{1,1}$, write $a = A^4$, $A \in \mathbb{F}_q$. We wish to prescribe $s_1 = s_{1,0} + 2s_{1,1} + 4s_{1,2} \in R_{1,2}$ as $2z + 4A$.

In order to apply Lemma 4.8, we require to work with the multiplicative characters of $\Gamma_{n,3}^*$, a cyclic group of order $q^n - 1$, and the additive characters of $R_{n,3}$. So now, for any divisor d of $q^n - 1$, η_d is a character of order d. It is extended to $\Gamma_{n,3}$ by setting $\eta_d(0) = 0$. In particular, η_1 is the trivial character: for an alternative version with $\eta(0) = 1$ we write $\eta = \mathbf{1}$. For additive characters, write $\chi_{(n)}$ for the canonical additive character of $R_{n,3}$: thus

$$\chi_{(n)}(\gamma) = \exp\left(\frac{2\pi T_{nu}(\gamma)}{8}\right), \quad q = 2^u, \quad \gamma \in R_{n,3}.$$

Here $T_{nu}(\gamma)$ yields the *absolute trace* of γ. In particular, set $\chi_{(1)} = \chi$. The characteristic function for the set of elements $\gamma \in \Gamma_{n,3}$ for which $s_1 (= s_{1,0} + 2s_{1,1} + 4s_{1,2}) = 2z + 4A$ is

$$\frac{1}{q^3} \sum_{\xi \in R_{1,3}} \chi\Big(\xi\big(T_n(\gamma) - (2z + 4A)\big)\Big), \tag{4.10}$$

which equals

$$\frac{1}{q^3} \sum_{\alpha_0, \alpha_1, \alpha_2 \in \Gamma_{1,1}} \chi_{(n)}\big((\alpha_0 + 2\alpha_1 + 4\alpha_2)(\gamma)\big)\, \chi\big(-2(\alpha_0 + 2\alpha_1)z - 4\alpha_0 A\big).$$

For the next lemma, note that, if $\xi = \alpha_0 + 2\alpha_1 + 4\alpha_2 \in R_{1,3}$ with $\alpha_0, \alpha_1, \alpha_2 \in \Gamma_{1,1}$, then $2\xi = 2\hat{\xi}$, where $\hat{\xi} = \alpha_0 + 2\alpha_1 \in R_{1,2}$.

Lemma 4.9. *Write $\xi \in R_{1,3}^*$ as $\hat{\xi} + 4\alpha_2$, where $\hat{\xi} \in R_{1,2}$ and $\alpha_2 \in \Gamma_{1,1}$. Set $U_\xi = \sum_{z \in \Gamma_{1,1}} \chi(2\xi z)$. Then $U_\xi = 0$ unless $\hat{\xi} = 0$ (and $\alpha_2 \neq 0$) in which case $U_\xi = q$, or $\hat{\xi} = \pm\alpha_0 \; (\neq 0)$, in which case $U_\xi = \frac{1 \pm i}{2} \cdot q$, respectively.*

Proof. If $\hat{\xi} = 0$ (and $\alpha_2 \neq 0$) the result is obvious. Otherwise $U_\xi = \sum_{z \in \Gamma_{1,1}} \chi(\hat{\xi} z)$ and the conclusion follows from Lemma 6.3 of [10]. $\qquad\square$

For $k | q^n - 1$ and $a \neq 0 \in \mathbb{F}_q$, write $\pi_a(k)$ for the number of k-free elements of $\mathbb{F}_{q^n}$ whose characteristic polynomial has first coefficient zero and fourth coefficient $a = A^4 \in \mathbb{F}_q$.

Lemma 4.10. *Assume q is even and $a = A^4 \in \mathbb{F}_q \cong \Gamma_{1,1}^*$. Let $k | q^n - 1$ and (k_0, s) be a decomposition of k. Then*

$$
\frac{q^2 \pi_a(k)}{\theta(k_0)} = \delta \Bigg(q^n - 1 + \int_{d|k_0} \sum_{\alpha \in \Gamma_{1,1}^*} \bar{\tilde{\eta}}_d(\alpha) S_n(4, \eta_d)
$$

$$
+ \frac{1}{2} \int_{d|k_0} \sum_{\beta \in \Gamma_{1,1}} \sum_{\alpha \in \Gamma_{1,1}^*} \bar{\chi}(4\alpha A) \bar{\tilde{\eta}}_d(\alpha) \Big\{ (1-i) S_n(1 + 4\beta; \eta_d) +
$$

$$
(1+i) S_n\big(-(1 + 4\beta); \eta_d\big) \Big\} \Bigg)
$$

$$
+ \sum_{i=1}^{s} \left(1 - \frac{1}{p_i} \right) \left(\sum_{\alpha \in \Gamma_{1,1}^*} \bar{\tilde{\eta}}_d(\alpha) S_n(4, \eta_{dp_i}) \right)
$$

$$
+ \frac{1}{2} \int_{d|k_0} \sum_{\beta \in \Gamma_{1,1}} \sum_{\alpha \in \Gamma_{1,1}^*} \bar{\chi}(4\alpha A) \bar{\tilde{\eta}}_{dp_i}(\alpha) \Big\{ (1-i) S_n(1 + 4\beta; \eta_{dp_i})
$$

$$
+ (1+i) S_n\big(-(1 + 4\beta); \eta_{dp_i}\big) \Big\} \Bigg),
$$

where, for $\xi \in R_{1,3}$, $S_n(\xi; \eta_d) := \sum_{\gamma \in \Gamma_{n,3}} \chi_{(n)}(\xi\gamma) \eta_d(\gamma)$ and $\tilde{\eta}_d$ is the restriction of η_d to $\Gamma_{1,1}$.

Proof. For notational simplicity consider only the trivial decomposition of k with $s = 1$. Write $\xi = \alpha_0 + 2\alpha_1 + 4\alpha_2 = \hat{\xi} + 4\alpha_2$ for a typical element of $R_{1,3}$.

From the characteristic functions (in particular (4.10)) one obtains

$$
\frac{q^3 \pi_a(k)}{\theta(k)} = \int_{d|k} \sum_{\xi \in R_{1,3}} \chi(4\alpha_0 A) \, U_\xi \, S_n(\xi; \eta_d), \tag{4.11}
$$

with U_ξ as in Lemma 4.9. Since $S_n(0; \eta_d) = 0$ unless $d = 1$, the contribution to (4.11) from $\xi = 0$ (the 'main term') is $q(q^n - 1)$.

From Lemma 4.9, $U_\xi = 0$ unless $\hat{\xi} = 0$ (i.e., $\xi = 4\alpha_2 \neq 0$), or $\xi = \pm\alpha_0$. Note that, when U_ξ is non-zero, its value can be expressed as cq and, ultimately, a factor of q is cancelled from (4.11). We consider the contributions from these excepted ξ.

First suppose $\hat{\xi} = 0$, i.e., $\alpha_0 = \alpha_1 = 0$ but $\alpha_d \neq 0$. Then $U_\xi = q$. Replacing γ by $\frac{\gamma}{\alpha_2}$ in the expression for $S_n(4\alpha_2; \eta_d)$, we obtain the sum $q \sum_{\alpha_2 \in \Gamma_{1,1}^*} \bar{\bar{\eta}}_d(\alpha_2) S_n(4; \eta_d)$, which is equivalent to the expression shown.

Now consider the contribution from $\hat{\xi} = \pm \alpha_0 \neq 0$, i.e., $\alpha_0 \neq 0$, $\alpha_1 = 0$, α_2 arbitrary. Replace $\gamma \in \Gamma_{n,3}$ by $\frac{\gamma}{\alpha_0} \in \Gamma_{n,3}$ and $\alpha_2 \in \Gamma_{1,1}$ by $\alpha_0 \beta \in \Gamma_{1,1}$ to obtain the remainder of the displayed identity. □

The relevant bounds for $|S_n(\xi; \eta_d)|$ are as follows.

Lemma 4.11. *Suppose $\xi \in R_{1,3}^*$. Then $S_n(\xi; 1) = 0$. Further, if $d \ (> 1)$ divides $q^n - 1$, then one has $|S_n(\xi; \eta_d)| \leq 4q^{\frac{n}{2}}$. Indeed, if $\beta \in \Gamma_{1,1}$ then $|S_n(4; \eta_d)| \leq q^{\frac{n}{2}}$.*

Proof. This follows from Corollary 6.1 of [19]. The significant point is that the polynomial $(\alpha_0 + 2\alpha_1 + 4\alpha_2)x \in R_{1,3}^*[x]$ has *weighted degree* 4 (if $\alpha_0 \neq 0$) or 1 (if $\alpha_0 = \alpha_1 = 0$). □

Again it is now convenient to split the discussion into the non-zero or zero problems.

4.2.1. *The even non-zero problem.* Suppose that q is even and that the prescribed coefficient $a \in \mathbb{F} \cong \Gamma_{1,1}$ is non-zero.

Proposition 4.12. *Assume that q is even and $a \in \mathbb{F}$ is non-zero. Assume also that $k | q^n - 1$ and that (k_0, s) is a decomposition of k. Suppose also that*

$$q^{\frac{n-3}{2}} > 4\sqrt{2}\, W(k_0) \Delta_{s,\delta} . \tag{4.12}$$

Then $\pi_a(k)$ is positive.

Proof. Again for notational simplicity focus on the situation when the decomposition of k is trivial, i.e., $k_0 = k$ and only the first two lines of the identity of Lemma 4.10 are relevant. Indeed, apart from the main term $q^n - 1$, for any divisor d of k the balance of the contributions relating to multiplicative characters η_d arise from the first line and the second line is different, according to the value of d.

Specifically, suppose $d | Q_n$. Then $\bar{\bar{\eta}}_d$ is trivial and so $\sum_{\alpha \in \Gamma_{1,1}^*} \bar{\bar{\eta}}_d(\alpha) = q - 1$. Hence, by Lemma 4.11, the characters of order d contribute $(q-1)q^{\frac{n}{2}}$ from the first line. On the other hand, from the second line $\sum_{\alpha \in \Gamma_{1,1}^*} \bar{\chi}(4\alpha A)\bar{\bar{\eta}}_d(\alpha) = \sum_{\alpha \in \mathbb{F}_q^*} \bar{\chi}_0(\alpha A)$, where χ_0 is the canonical additive character on $\mathbb{F}_q$, and this is ≤ 1 in absolute value. Hence, by Lemma 4.11, the contributions from the second line of multiplicative characters of order d do not exceed $q^{\frac{n+1}{2}}$. Accordingly, the total contributions from characters or order d is certainly bounded by $2q^{\frac{n}{2}+1}$.

Finally, suppose, $d \nmid Q_0$. Then, in the first line, $\sum_{\alpha \in \Gamma_{1,1}^*} \bar{\bar{\eta}}_d(\alpha) = 0$. By the same lemma (with $n = 1$), in the second line $\sum_{\alpha \in \Gamma_{1,1}^*} \bar{\chi}(4\alpha A)\bar{\bar{\eta}}_d(\alpha) \leq q^{\frac{1}{2}}$ and, of course $\left| S_n(\pm(1 + 4\beta); \eta_d) \right| \leq 4q^{\frac{n}{2}}$. Accordingly, the contribution of characters of order d is bounded by $4\sqrt{2}q^{\frac{n+3}{2}}$. Since this easily exceeds the contribution for a divisor d of Q_n, we can use this as a global bound for any $d \mid k$ and the result follows.

$\square$

Express the product of distinct primes in $q^8 - 1$ as $K_1 \cdot K_2$, where K_1 is the product of all distinct odd prime divisors of $q^4 - 1$ and K_2 is the product of distinct odd prime divisors of $q^4 + 1$. By an analogue of Lemma 3.6 any prime divisor l of K_2 is $\equiv 1 \pmod 8$, i.e., $l \in L_8$. Denote $\omega(K_1)$ by ω_1 and $\omega(K_2)$ by ω_2.

Lemma 4.13. *Suppose that $n = 8$, q odd, $\omega_1 \geq 8$ or $\omega_2 \geq 5$. Let a ($\neq 0$) $\in \mathbb{F}_q$. Then there exists a primitive polynomial of degree 8 over $\mathbb{F}_q$ with the coefficient of x^4 prescribed as a.*

After Lemma 4.13 we can assume $\omega_1 \leq 7$ and $\omega_2 \leq 4$ and start the sieving process. It turns out, however, that one sieving step is enough. Taking $k_0 = 3 \cdot 5$ yields $s \leq 9$, $\delta \geq 0.285$ and $\Delta_{s,\delta} < 18.50$ whence (4.12) is satisfied for $q \geq 12$. Next, putting $k_0 = 3$, $q = 8$ satisfies (4.12), but the appropriate primitive polynomials over fields $\mathbb{F}_4$ and $\mathbb{F}_2$ have to be found explicitly.

4.2.2. The even zero problem. To fix the fourth coefficient as zero, following Lemma 4.8 we suppose $A = 0$. We may therefore assume also that $k \mid Q_n$. Suppose $1 < d \mid Q_n$ and take $n = 8$. A sufficient condition is then

$$q^2 > 4\sqrt{2}\, W(k_0)\Delta_{s,\delta}. \tag{4.13}$$

Define K_1, K_2 as in Section 4.1.2 and note that both K_1 and K_2 are odd. Denote $\omega(K_1)$ by ω_1 and $\omega(K_2)$ by ω_2.

Lemma 4.14. *Suppose that $n = 8$, q odd, $\omega_1 \geq 10$ or $\omega_2 \geq 8$. Then there exists a primitive polynomial of degree 8 over $\mathbb{F}_q$ with the coefficient of x^4 prescribed as 0.*

Consequently to Lemma 4.14, assume $\omega_1 \leq 9$ and $\omega_2 \leq 7$ and begin the sieve. Taking k_0 to be the product of least three primes in K_1 yields $s \leq 13$, $\delta > 0.450$, $\Delta_{s,\delta} < 28.67$ and condition (4.13) holds for $q \geq 37$. Hence assume $\omega_1 \leq 5$, $\omega_2 \leq 3$ and set $k_0 = 3$. Then $s \leq 7$, $\delta > 0.392$, $\Delta_{s,\delta} < 17.31$ and condition (4.13) holds for $q \geq 14$. For the three remaining values of q, primitive polynomials in $\mathbb{F}_8[x]$, $\mathbb{F}_4[x]$ and $\mathbb{F}_2[x]$ have to be found explicitly.

5. TABLES OF POLYNOMIALS

Here we give a primitive polynomial, found with Maple, for each case (n, q), where the existence of a primitive polynomial of degree n over $\mathbb{F}_q$ cannot be confirmed by criteria derived in Sections 3.2 and 4. In Section 3.1, however, there are too many polynomials to list and, as an illustration, we only give primitive polynomials of degree 5 over the largest field $\mathbb{F}_{163}$.

Note that the total amount of computation is small and was not needed for every degree n. As before, a is the prescribed third coefficient of a polynomial.

We begin with polynomials with prescribed third coefficient. First, consider the standard problem, that is, when $q \not\equiv 0 \pmod 3$. We list 3-tuples (a, b, c), $1 \le a \le 162$, representing primitive polynomials of degree 5 over $\mathbb{F}_{163}$ of the form $x^5 + ax^2 + bx + c$:

$(1, 0, 14)$, $(2, 0, 15)$, $(3, 0, 14)$, $(4, 0, 4)$, $(5, 0, 16)$, $(6, 0, 9)$, $(7, 0, 10)$, $(8, 0, 9)$, $(9, 0, 15)$,
$(10, 0, 15)$, $(11, 0, 15)$, $(12, 0, 10)$, $(13, 0, 43)$, $(14, 0, 15)$, $(15, 0, 35)$, $(16, 0, 4)$,
$(17, 0, 9)$, $(18, 0, 24)$, $(19, 0, 10)$, $(20, 0, 4)$, $(21, 0, 9)$, $(22, 0, 16)$, $(23, 0, 24)$, $(24, 0, 4)$,
$(25, 0, 10)$, $(26, 0, 47)$, $(27, 0, 26)$, $(28, 0, 10)$, $(29, 0, 4)$, $(30, 0, 9)$, $(31, 0, 9)$, $(32, 0, 39)$,
$(33, 0, 10)$, $(34, 0, 35)$, $(35, 0, 14)$, $(36, 1, 33)$, $(37, 0, 15)$, $(38, 0, 81)$, $(39, 0, 9)$,
$(40, 0, 15)$, $(41, 0, 83)$, $(42, 0, 14)$, $(43, 0, 4)$, $(44, 0, 24)$, $(45, 0, 9)$, $(46, 0, 4)$, $(47, 0, 43)$,
$(48, 0, 14)$, $(49, 0, 14)$, $(50, 0, 14)$, $(51, 0, 9)$, $(52, 0, 10)$, $(53, 0, 4)$, $(54, 0, 4)$, $(55, 0, 9)$,
$(56, 0, 16)$, $(57, 0, 35)$, $(58, 0, 24)$, $(59, 0, 10)$, $(60, 0, 10)$, $(61, 1, 4)$, $(62, 0, 16)$,
$(63, 0, 9)$, $(64, 1, 14)$, $(65, 0, 35)$, $(66, 0, 4)$, $(67, 0, 9)$, $(68, 0, 16)$, $(69, 0, 14)$, $(70, 0, 46)$,
$(71, 0, 39)$, $(72, 0, 9)$, $(73, 0, 9)$, $(74, 0, 26)$, $(75, 0, 9)$, $(76, 0, 4)$, $(77, 0, 10)$, $(78, 0, 14)$,
$(79, 0, 14)$, $(80, 0, 9)$, $(81, 0, 9)$, $(82, 0, 35)$, $(83, 0, 9)$, $(84, 0, 10)$, $(85, 0, 39)$, $(86, 0, 39)$,
$(87, 0, 4)$, $(88, 0, 14)$, $(89, 0, 15)$, $(90, 0, 43)$, $(91, 0, 9)$, $(92, 0, 4)$, $(93, 0, 10)$, $(94, 0, 15)$,
$(95, 0, 10)$, $(96, 0, 24)$, $(97, 0, 14)$, $(98, 0, 9)$, $(99, 0, 10)$, $(100, 0, 4)$, $(101, 0, 4)$,
$(102, 0, 4)$, $(103, 0, 9)$, $(104, 0, 4)$, $(105, 0, 4)$, $(106, 0, 16)$, $(107, 0, 26)$, $(108, 0, 15)$,
$(109, 0, 4)$, $(110, 0, 15)$, $(111, 0, 26)$, $(112, 0, 4)$, $(113, 0, 10)$, $(114, 0, 9)$, $(115, 1, 4)$,
$(116, 0, 9)$, $(117, 0, 24)$, $(118, 0, 4)$, $(119, 0, 9)$, $(120, 0, 14)$, $(121, 0, 9)$, $(122, 0, 4)$,
$(123, 0, 4)$, $(124, 0, 24)$, $(125, 0, 4)$, $(126, 1, 10)$, $(127, 0, 24)$, $(128, 0, 14)$, $(129, 0, 43)$,
$(130, 0, 4)$, $(131, 0, 24)$, $(132, 1, 10)$, $(133, 0, 43)$, $(134, 0, 16)$, $(135, 0, 46)$, $(136, 1, 26)$,
$(137, 0, 15)$, $(138, 0, 4)$, $(139, 0, 14)$, $(140, 0, 14)$, $(141, 0, 26)$, $(142, 0, 16)$, $(143, 0, 10)$,
$(144, 0, 4)$, $(145, 0, 4)$, $(146, 0, 35)$, $(147, 0, 4)$, $(148, 0, 9)$, $(149, 0, 4)$, $(150, 1, 56)$,
$(151, 0, 10)$, $(152, 0, 24)$, $(153, 0, 4)$, $(154, 0, 15)$, $(155, 0, 26)$, $(156, 0, 14)$, $(157, 0, 10)$,
$(158, 1, 33)$, $(159, 0, 10)$, $(160, 0, 10)$, $(161, 0, 14)$, $(162, 0, 15)$.

Next, we give a table of primitive polynomials over $\mathbb{F}_q$, with prescribed third coefficient, where $q \equiv 0 \pmod 3$. Here $\mathbb{F}_9$ is defined as $\mathbb{F}_3(\alpha)$ and $\mathbb{F}_{27}$ is defined as $\mathbb{F}_3(\beta)$, where α and β are roots of $x^2 + 2x + 2 \in \mathbb{F}_3$ and $x^3 + 2x + 1 \in \mathbb{F}_3$, respectively. Note that when $n = 7$, also the constant term of the polynomial is required to be prescribed, but it necessarily equals 1.

n	a	$q=3$	$q=9$	$q=27$
5	1	$x^5 + x^4 + x^2 + 1$	$x^5 + x^2 + \alpha$	—
	2	$x^5 + 2x^2 + x + 1$	$x^5 + 2x^2 + \alpha + 2$	—
	α	—	$x^5 + \alpha x^2 + \alpha$	—
	$\alpha + 1$	—	$x^5 + (\alpha + 1)x^2 + \alpha$	—
	$2\alpha + 1$	—	$x^5 + (2\alpha + 1)x^2 + 2\alpha + 1$	—
	$2\alpha + 2$	—	$x^5 + (2\alpha + 2)x^2 + 2\alpha + 1$	—
	$\alpha + 2$	—	$x^5 + (\alpha + 2)x^2 + 2\alpha + 1$	—
	2α	—	$x^5 + 2\alpha x^2 + \alpha$	—
6	1	$x^6 + x^3 + x + 2$	$x^6 + x^3 + x^2 + x + \alpha$	—
	2	$x^6 + 2x^3 + 2x + 2$	$x^6 + 2x^3 + x^2 + 2x + \alpha$	—
	α	—	$x^6 + \alpha x^3 + x^2 + 2\alpha + 2$	—
	$\alpha + 1$	—	$x^6 + (\alpha + 1)x^3 + x^2 + \alpha$	—
	$2\alpha + 1$	—	$x^6 + (2\alpha + 1)x^3 + x^2 + (\alpha + 2)x + \alpha + 1$	—
	$2\alpha + 2$	—	$x^6 + (2\alpha + 2)x^3 + x^2 + \alpha$	—
	$\alpha + 2$	—	$x^6 + (\alpha + 2)x^3 + x^2 + (2\alpha + 1)x + \alpha + 1$	—
	2α	—	$x^6 + 2\alpha x^3 + x^2 + 2\alpha + 2$	—
	0	$x^6 + x + 2$	$x^6 + x^2 + \alpha x + \alpha$	$x^6 + x + \beta$
7	1	$x^7 + x^6 + x^3 + 2x^2 + x + 1$	—	—
	2	$x^7 + x^6 + 2x^3 + x^2 + 2x + 1$	—	—

Now we give a complete list of polynomials with fourth coefficient prescribed, as required. The tables relating to fields of odd characteristic follow. Here are the tables:

$a \neq 0$

a	$q=3$	$q=5$	$q=7$
1	$x^8 + x^5 + x^4 + 2x^2 + 2$	$x^8 + x^5 + x^4 + 3x + 3$	$x^8 + x^5 + x^4 + x + 5$
2	$x^8 + 2x^5 + x^4 + 2x^2 + 2$	$x^8 + 2x^5 + x^4 + x + 3$	$x^8 + 2x^5 + x^4 + 2x + 5$
3	—	$x^8 + 3x^5 + x^4 + 4x + 3$	$x^8 + 3x^5 + x^4 + 5$
4	—	$x^8 + 4x^5 + x^4 + 2x + 3$	$x^8 + 4x^5 + x^4 + 5$
5	—	—	$x^8 + 5x^5 + x^4 + 5x + 5$
6	—	—	$x^8 + 6x^5 + x^4 + 6x + 5$

$a = 0$

$q = 3$	$x^8 + x^3 + 2$
$q = 5$	$x^8 + x^3 + 2x^2 + 2$
$q = 7$	$x^8 + x^3 + 3x^2 + 3$
$q = 9$	$x^8 + (2\alpha + 2)x^4 + (\alpha + 2)x + 2\alpha + 2$
$q = 11$	$x^8 + 2x^3 + x^2 + 2$
$q = 13$	$x^8 + x^3 + x^2 + 2$
$q = 17$	$x^8 + 2x^3 + x^2 + 3$

Lastly, here is the corresponding table for fields of even characteristic. The field $\mathbb{F}_4$ is defined as $\mathbb{F}_2(\alpha)$ and $\mathbb{F}_8$ is defined as $\mathbb{F}_2(\beta)$, where α and β are roots of $x^2 + x + 1 \in \mathbb{F}_2$ and $x^3 + x^2 + 1 \in \mathbb{F}_2$, respectively.

a	$q = 2$	$q = 4$	$q = 8$
1	$x^8 + x^5 + x^3 + x^2 + x + 1$	$x^8 + x^5 + x^2 + x + \alpha$	—
α	—	$x^8 + \alpha x^5 + (\alpha + 1)x^2 + \alpha$	—
$\alpha + 1$	—	$x^8 + (\alpha + 1)x^5 + \alpha x^2 + \alpha$	—
0	$x^8 + x^7 + x^2 + x + 1$	$x^8 + x^7 + x^2 + x + \alpha$	$x^8 + (\beta^2 + \beta)x^6$ $+(\beta^2 + \beta + 1)x^4$ $+(\beta^2 + \beta)x^3$ $+\beta^2 x^2 + \beta x + \beta$

REFERENCES

[1] S.D. Cohen, Primitive elements and polynomials with arbitrary trace, *Discr. Math.* **83** (1990), 1–7.

[2] _____, Gauss sums and a sieve for generators of Galois fields, *Publ. Math. Debrecen* **56** (2000), 293–312.

[3] _____, Primitive polynomials over small fields, *Finite Fields Appl. 7th International Conference, Toulouse* (2003), Lecture Notes in Computer Science **2948**, 197–214, Springer Verlag, Berlin, 2004.

[4] _____, Primitive polynomials with a prescribed coefficient, *Finite Fields Appl.* **12** (2006), 425–491.

[5] _____, Explicit results on generator polynomials, *Finite Fields Appl.* **11** (2005), 337–357.

[6] S.D. Cohen, S. Huczynska, Primitive free quartics with specified norm and trace, *Acta Arith.* **109** (2003), 359–385.

[7] S.D. Cohen, C. King, The three fixed coefficient primitive polynomial theorem, *JP J. Algebra Number Theory Appl.* **4** (2004), 79–87.

[8] S.D. Cohen, D. Mills, Primitive polynomials with first and second coefficients prescribed, *Finite Fields Appl.* **9** (2003), 334–350.

[9] S.D. Cohen, M. Prešern, Primitive finite field elements with prescribed trace, *Southeast Asian Bulletin of Mathematics* **29** (2005), 283–300.

[10] _____, _____, Primitive polynomials with prescribed second coefficient, *Glasgow Math. J.* **48** (2006), 281–307.

[11] S.-Q. Fan, W.-B. Han, p-adic formal series and primitive polynomials over finite fields, *Proc. Amer. Math. Soc.* **132** (2004), 15–31.

[12] _____, _____, Primitive polynomials over finite fields of characteristic two, *Appl. Algebra Engrg Comm. Comput.* **14** (2004), 381–395.

[13] _____, _____, Character sums over Galois rings and primitive polynomials over finite fields, *Finite Fields Appl.* **10** (2004), 36–52.

[14] _____, _____, Primitive polynomials with three coefficients prescribed, *Finite Fields Appl.* **10** (2004), 506–521.

[15] W.-B. Han, The coefficients of primitive polynomials over finite fields, *Math. Comp.* **65** (1996), 331–340.

[16] T. Hansen, G.L. Mullen, Primitive polynomials over finite fields, *Math. Comp.* **59** (1992), 639–643, S47–S50.

[17] S. Huczynska, S.D. Cohen, Primitive free cubics with specified norm and trace, *Trans. Amer. Math. Soc.* **355** (2003), 3099–3116.

[18] N. Koblitz, *p-adic numbers, p-adic analysis, and zeta-functions*, Springer, New York, 1984.

[19] W-C.W. Li, Character sums over p-adic fields, *J. Number Theory* **74** (1999), 181–229.

[20] R. Lidl, H. Niederreiter, *Finite Fields*, Addison-Wesley, Reading, Massachusetts (1983); second edition: Cambridge University Press, Cambridge (1997).

[21] D. Mills, Existence of primitive polynomials with three coefficients prescribed, *JP J. Algebra Number Theory Appl.* **4** (2004), 36–52.

[22] G.L. Mullen, Open problems in finite fields, *Congr. Numer.* **111** (1995), 97–103.

[23] G.L. Mullen, I. Shparlinski, Open problems and conjectures in finite fields, *Finite fields and applications (Glasgow, 1995)*, London Math. Soc. Lecture Note Ser. **233**, 243–268, Cambridge University Press, Cambridge, 1996.

[24] M. Prešern, *Existence problems of primitive polynomials over finite fields*, PhD thesis (University of Glasgow), in preparation.

AN INEQUALITY FOR THE MULTIPLICITY OF THE ROOTS OF A POLYNOMIAL

ARTŪRAS DUBICKAS

ABSTRACT. We give two inequalities in terms of the coefficients of a complex polynomial which imply an upper bound on the maximal multiplicity of its roots. In particular, for $f(x) = a_0 + a_1 x + \cdots + a_n x^n \in \mathbb{C}[x]$, where $a_0 a_n \neq 0$, our result implies that if $n|a_0| > \sum_{i=1}^{n-1} i|a_{n-i}|$ and $n|a_n| \geqslant \sum_{i=1}^{n-1} i|a_i|$ then f is a separable polynomial. This condition is best possible in the sense that the conclusion on f is no longer true if the first inequality is not strict. More generally, if there is a positive integer $k < n$ such that $\binom{n}{k}|a_0| > \sum_{i=k}^{n-1} \binom{i}{k}|a_{n-i}|$ and $\binom{n}{k}|a_n| \geqslant \sum_{i=k}^{n-1} \binom{i}{k}|a_i|$ then the multiplicity of each root of f is $\leqslant k$. This result sharpens a corresponding result of A.I. Bonciocat, N.C. Bonciocat and A. Zaharescu.

1. INTRODUCTION

Let $f(x) = a_0 + a_1 x + \cdots + a_{n-1} x^{n-1} + a_n x^n$, where $a_0 a_n \neq 0$ be a polynomial with complex coefficients. Suppose that

$$f(x) = a_n (x - \alpha_1)^{n_1} (x - \alpha_2)^{n_2} \ldots (x - \alpha_m)^{n_m},$$

where $\alpha_1, \ldots, \alpha_m$ are distinct roots of f and $n_1, \ldots, n_m$ are positive integers satisfying $n_1 + \cdots + n_m = n$. Set $e(f) = \max\{n_1, \ldots, n_m\}$. Recently, A.I. Bonciocat, N.C. Bonciocat and A. Zaharescu [1] proved that if j and k are two nonnegative integers satisfying $0 \leqslant j < k < n$,

$$|a_j| > \sum_{i=j+1}^{n} \binom{i}{j}|a_i| \quad \text{and} \quad \binom{n}{k}|a_n| > \sum_{i=k}^{n-1} \binom{i}{k}|a_i|, \tag{1}$$

then $e(f) \leqslant k$. The strongest version of their theorem is obtained when $j = 0$: if

$$|a_0| > \sum_{i=1}^{n} |a_i| \quad \text{and} \quad \binom{n}{k}|a_n| > \sum_{i=k}^{n-1} \binom{i}{k}|a_i|, \tag{2}$$

then $e(f) \leqslant k$. Finally, putting $k = 1$ into (2), they observed (see Corollary 1 in [1]) that the conditions

$$|a_0| > \sum_{i=1}^{n} |a_i| \quad \text{and} \quad n|a_n| > \sum_{i=1}^{n-1} i|a_i| \tag{3}$$

2000 *Mathematics Subject Classification.* 11C08, 12D10.
Key words and phrases. Complex polynomial, multiple roots.

imply that the polynomial f is separable.

In the proof of these and other, more technical, results, the authors of [1] use certain Hadamard's and Ostrowski's conditions for nonvanishing of determinants which represent certain resultants of the derivatives of f. We remark that one can give a very simple proof of the results quoted above as follows.

Suppose that $e(f) > k$. Then there is a complex number $\alpha \neq 0$ such that $f(\alpha) = f'(\alpha) = \cdots = f^{(k)}(\alpha) = 0$. For $j \leqslant k$, we have

$$\frac{f^{(j)}(\alpha)}{j!} = a_j + \binom{j+1}{j} a_{j+1}\alpha + \binom{j+2}{j} a_{j+2}\alpha^2 + \cdots + \binom{n}{j} a_n \alpha^{n-j} = 0.$$

So, by the first inequality of (1), we deduce that

$$\sum_{i=j+1}^{n} \binom{i}{j}|a_i| < |a_j| = \left| -\sum_{i=j+1}^{n} \binom{i}{j} a_i \alpha^{i-j} \right| \leqslant \max\{1,|\alpha|\}^{n-j} \sum_{i=j+1}^{n} \binom{i}{j}|a_i|.$$

Since $\sum_{i=j+1}^{n} \binom{i}{j}|a_i| \geqslant |a_n| > 0$, we obtain that $\max\{1,|\alpha|\} > 1$, so $|\alpha| > 1$. Similarly, $f^{(k)}(\alpha)/k! = \sum_{i=k}^{n} \binom{i}{k} a_i \alpha^{i-k} = 0$ implies that

$$\binom{n}{k} a_n = -\sum_{i=k}^{n-1} \binom{i}{k} a_i \alpha^{i-n},$$

so at least one of the numbers $a_k, \ldots, a_{n-1}$ is nonzero. Using the second inequality of (1), we derive that

$$\sum_{i=k}^{n-1} \binom{i}{k}|a_i| \; < \; \binom{n}{k}|a_n| = \left| -\textstyle\sum_{i=k}^{n-1} \binom{i}{k} a_i \alpha^{i-n} \right|$$

$$\leqslant \; \max\{|\alpha|^{-1}, |\alpha|^{k-n}\} \sum_{i=k}^{n-1} \binom{i}{k}|a_i|.$$

As $\sum_{i=k}^{n-1} \binom{i}{k}|a_i| > 0$, this yields $\max\{|\alpha|^{-1}, |\alpha|^{k-n}\} > 1$, i.e., $|\alpha| < 1$, contrary to $|\alpha| > 1$. Since the assumption $e(f) > k$ leads to a contradiction, we conclude that $e(f) \leqslant k$.

Of course, $e(f) = e(f^*)$, where $f^*(x) = a_n + a_{n-1}x + \cdots + a_0 x^n$ is the polynomial *reciprocal* to f (whose coefficients are written in reverse order). However, none of the conditions (1)–(3) is 'symmetric' with respect to the coefficients $a_0, \ldots, a_n$ of f and the coefficients $a_n, \ldots, a_0$ of f^*. Moreover, for instance, the second inequality of (3), i.e., $n|a_n| > \sum_{i=1}^{n-1} i|a_i|$ implies that $|a_n|$ is 'large', so, by the first inequality of (3), $|a_0| > |a_1| + \cdots + |a_n|$, the first coefficient $|a_0|$ is even greater than $|a_n|$. It seems very likely that this signals the fact that the first inequality of (3) is not sharp, so that it can be replaced by a weaker condition. This turns out to be the case.

The aim of this note is to settle both problems ('symmetry' and 'sharpness') described above. The following theorem gives the same conclusion $e(f) \leqslant k$ under a weaker assumption. This assumption is given by two symmetric inequalities. It is not only less restrictive than the corresponding condition of [1], but is also best possible, at least when $k = 1$ and $k = n - 1$.

Theorem 1. *Let* $f(x) = a_0 + a_1 x + \cdots + a_n x^n \in \mathbb{C}[x]$, *where* $a_0 a_n \neq 0$, *and let* $k < n$ *be a positive integer. If*

$$\binom{n}{k}|a_0| > \sum_{i=k}^{n-1} \binom{i}{k}|a_{n-i}| \quad and \quad \binom{n}{k}|a_n| > \sum_{i=k}^{n-1} \binom{i}{k}|a_i| \qquad (4)$$

then $e(f) \leqslant k$. *Moreover, one of the inequalities* $>$ *in* (4) *(but, for* $k \in \{1, n-1\}$, *not both!) can be replaced by* $\geqslant$.

For $k = 1$ the condition (4) becomes

$$\begin{aligned} n|a_0| &> (n-1)|a_1| + \cdots + 2|a_{n-2}| + |a_{n-1}| \\ \text{and} \quad n|a_n| &> (n-1)|a_{n-1}| + \cdots + 2|a_2| + |a_1| \,. \end{aligned} \qquad (5)$$

Obviously, the first inequality of (5) is a weaker restriction on $|a_0|$ then the one given by the first inequality of (3). So Theorem 1 with $k = 1$ strengthens Corollaries 1 and 2 of [1].

Theorem 1 asserts that the inequalities in (5) cannot be replaced by equalities. Indeed, put

$$f(x) = 1 - x - x^{n-1} + x^n = (1 - x)(1 - x^{n-1})$$

for any integer $n \geqslant 2$. It is easy to see that then

$$n|a_0| = n|a_n| = \sum_{i=1}^{n-1} i|a_i| = \sum_{i=1}^{n-1} i|a_{n-i}| = n \,,$$

but $e(f) = 2$, because f has a double root at $x = 1$ and $n - 2$ other distinct roots on the unit circle. Thus the inequality $e(f) \leqslant 1$, i.e., the conclusion that f is separable, does not follow from the inequalities

$$n|a_0| \geqslant \sum_{i=1}^{n-1} i|a_{n-i}| \quad and \quad n|a_n| \geqslant \sum_{i=1}^{n-1} i|a_i|.$$

Similarly, if $k = n - 1$, put $f(x) = (1 - x)^n$. Then (4) becomes $n|a_0| > |a_1|$ and $n|a_n| > |a_{n-1}|$. For $f(x) = (1 - x)^n$, we have

$$n|a_0| = n|a_n| = |a_1| = |a_{n-1}| = n$$

and $e(f) = n > k$. This proves that the inequalities $n|a_0| \geqslant |a_1|$ and $n|a_n| \geqslant |a_{n-1}|$ do not imply that $e(f) \leqslant n - 1$. The remaining part of the proof of Theorem 1 will be given in the next section.

Note that

$$c_{i,k,n} = \frac{\binom{i}{k}}{\binom{n}{k}} = \frac{i(i-1)\ldots(i-k+1)}{n(n-1)\ldots(n-k+1)} < 1,$$

so the first inequality of (4),

$$|a_0| > \sum_{i=1}^{n-1} c_{i,k,n} |a_{n-i}|,$$

is a weaker restriction on $|a_0|$ than the corresponding inequality of (2), namely, $|a_0| > \sum_{i=1}^{n} |a_i|$. In fact, the weights $c_{i,k,n}$ are 'close' to zero if i is 'small'.

2. PROOF OF THEOREM 1

We shall summarize our argument given in the introduction as follows:

Lemma 2. *Let* $f(x) = a_0 + a_1 x + \cdots + a_n x^n \in \mathbb{C}[x]$, *where* $a_0 a_n \neq 0$. *Suppose that* $k < n$ *is a positive integer and* α *is a complex number such that* $(x - \alpha)^{k+1} \mid f(x)$. *If* $\binom{n}{k} |a_n| \geq \sum_{i=k}^{n-1} \binom{i}{k} |a_i|$ *then* $|\alpha| \leq 1$. *Similarly, if* $\binom{n}{k} |a_n| > \sum_{i=k}^{n-1} \binom{i}{k} |a_i|$ *then* $|\alpha| < 1$.

Proof of the lemma: It is clear first of all that $\alpha \neq 0$. Thus the equation $f^{(k)}(\alpha)/k! = \sum_{i=k}^{n} \binom{i}{k} a_i \alpha^{i-k} = 0$ yields

$$\binom{n}{k} |a_n| = \left| -\sum_{i=k}^{n-1} \binom{i}{k} a_i \alpha^{i-n} \right| \leq \max\{|\alpha|^{-1}, |\alpha|^{k-n}\} \sum_{i=k}^{n-1} \binom{i}{k} |a_i|.$$

Moreover, at least one of the numbers $a_k, \ldots, a_{n-1}$ is nonzero, as $a_n \neq 0$. Now, $\binom{n}{k} |a_n| \geq \sum_{i=k}^{n-1} \binom{i}{k} |a_i| > 0$ implies that $\max\{|\alpha|^{-1}, |\alpha|^{k-n}\} \geq 1$, giving $|\alpha| \leq 1$. Similarly, strict inequality $\binom{n}{k} |a_n| > \sum_{i=k}^{n-1} \binom{i}{k} |a_i|$ implies that $\max\{|\alpha|^{-1}, |\alpha|^{k-n}\} > 1$, i.e., $|\alpha| < 1$. $\square$

Proof of Theorem 1: Suppose that $e(f) \geq k + 1$. Then there is a nonzero complex number α such that $(x - \alpha)^{k+1} | f(x)$. Suppose that both inequalities $\binom{n}{k} |a_n| \geq \sum_{i=k}^{n-1} \binom{i}{k} |a_i|$ and $\binom{n}{k} |a_0| > \sum_{i=k}^{n-1} \binom{i}{k} |a_{n-i}|$ hold. Then Lemma 2 implies that $|\alpha| \leq 1$.

On the other hand $(x - \beta)^{k+1} \mid f^*(x)$, where $\beta = \alpha^{-1} \neq 0$ and

$$f^*(x) = b_0 + b_1 x + \cdots + b_n x^n = a_n + a_{n-1} x + \cdots + a_0 x^n.$$

Here, $b_i = a_{n-i}$ for $i = 0, \ldots, n$. Now, by Lemma 2 and the inequality

$$\binom{n}{k} |b_n| = \binom{n}{k} |a_0| > \sum_{i=k}^{n-1} \binom{i}{k} |a_{n-i}| = \sum_{i=k}^{n-1} \binom{i}{k} |b_i|,$$

we obtain that $|\beta| < 1$. Hence $|\alpha| = |\beta^{-1}| > 1$, a contradiction with $|\alpha| \leq 1$.

In the case when $\binom{n}{k} |a_n| > \sum_{i=k}^{n-1} \binom{i}{k} |a_i|$ and $\binom{n}{k} |a_0| \geq \sum_{i=k}^{n-1} \binom{i}{k} |a_{n-i}|$ one obtains that $|\alpha| < 1$ and $|\alpha| \geq 1$, respectively, a contradiction again. This completes the proof of the theorem. $\square$

3. POLYNOMIALS WITH INTEGER COEFFICIENTS

In conclusion, by the same, but p-adic argument, we shall give a p-adic version of Theorem 1 for polynomials with integer coefficients.

Let p be a rational prime. For each nonzero $r \in \mathbb{Q}$, let $\nu_p(r)$ be the exponent of p in the prime decomposition of r. Put $\nu_p(0) = \infty$. Suppose that $f(x) = a_0 + a_1 x + \cdots + a_n x^n \in \mathbb{Z}[x]$, where $a_0 a_n \neq 0$. Proposition 5 of [1] implies that if there is a prime number p such that

$$\nu_p(a_0) \leqslant \min_{1 \leqslant i \leqslant n} \nu_p(a_i) \quad \text{and} \quad \nu_p\left(\binom{n}{k}a_n\right) < \min_{k \leqslant i \leqslant n-1} \nu_p\left(\binom{i}{k}a_i\right),$$

then $e(f) \leqslant k$. This condition is a p-adic analogue of (2). As above, it is not symmetric. We give a corresponding symmetric condition which implies the same conclusion $e(f) \leqslant k$.

Theorem 3. *Let $f(x) = a_0 + a_1 x + \cdots + a_n x^n \in \mathbb{Z}[x]$, where $a_0 a_n \neq 0$, and let $k < n$ be a positive integer. If*

$$\nu_p\left(\binom{n}{k}a_0\right) \leqslant \min_{k \leqslant i \leqslant n-1} \nu_p\left(\binom{i}{k}a_{n-i}\right)$$

$$(6)$$

$$\text{and} \quad \nu_p\left(\binom{n}{k}a_n\right) < \min_{k \leqslant i \leqslant n-1} \nu_p\left(\binom{i}{k}a_i\right),$$

then $e(f) \leqslant k$.

Proof: Let us define the absolute value $|\cdot|_p$ of a rational number r as $|r|_p = p^{-\nu_p(r)}$. In contrast to the usual absolute value, $|\cdot|_p$ is called an *ultrametric absolute value*. It can be extended to a number field K. (See, for instance, Ch. 12 in [2] or Ch. 3 in [3].) It satisfies $|\beta\beta'|_p = |\beta|_p|\beta'|_p$ and

$$|\beta + \beta'|_p \leqslant \max\{|\beta|_p, |\beta'|_p\},$$

where $\beta, \beta' \in K$. (Compare with the inequality $|\beta + \beta'| \leqslant |\beta| + |\beta'|$ for the usual absolute value $|\cdot|$.)

Let K be a number field containing all the roots of $f(x) \in \mathbb{Z}[x]$. Suppose that $e(f) > k$. Then, as above, there is an $\alpha \in K$ such that

$$\binom{n}{k}a_n = -\sum_{i=k}^{n-1} \binom{i}{k}a_i \alpha^{i-n}.$$

Hence

$$\left|\binom{n}{k}a_n\right|_p \leqslant \max_{k \leqslant i \leqslant n-1}\left|\binom{i}{k}a_i \alpha^{i-n}\right|_p \leqslant \max_{k \leqslant i \leqslant n-1}\left|\binom{i}{k}a_i\right|_p \max\{|\alpha^{k-n}|_p, |\alpha^{-1}|_p\}.$$

Using $|r|_p = p^{-\nu_p(r)}$, by the second inequality of (6), we obtain that

$$1 < \max\{|\alpha^{k-n}|_p, |\alpha^{-1}|_p\}.$$

Hence $|\alpha|_p < 1$. Similarly, by considering the reciprocal polynomial $f^*(x) = a_n + a_{n-1}x + \cdots + a_0x^d$ with the root $\alpha^{-1} \in K$ of multiplicity $\geqslant k+1$, from the first inequality in (6) we obtain that $1 \leqslant |\alpha|_p$, a contradiction. $\square$

We remark that the inequalities $\geqslant$ and $>$ in (6) can be replaced by $>$ and $\geqslant$, respectively, with the same conclusion $e(f) \leqslant k$.

Finally, suppose that a_0, a_n and n are all odd. Then, taking $p = 2$, we obtain from Theorem 3 that if the numbers $a_1, a_3, \ldots, a_{n-2}$ are all even then the polynomial $f(x) = a_0 + a_1x + \cdots + a_nx^n \in \mathbb{Z}[x]$ is separable.

This research was supported in part by the Lithuanian State Studies and Science Foundation.

References

[1] A.I. Bonciocat, N.C. Bonciocat, A. Zaharescu, Bounds for the multiplicities of the roots for some classes of complex polynomials, *Math. Inequal. Appl.* **9** (2006), 11–22.

[2] S. Lang, *Algebra*, Graduate texts in mathematics **211**, Springer, Berlin, New York, 2002.

[3] M. Waldschmidt, *Diophantine approximation on linear algebraic groups. Transcendence properties of the exponential function in several variables*, Springer, Berlin, New York, 2000.

NEWMAN'S INEQUALITY FOR INCREASING EXPONENTIAL SUMS

TAMÁS ERDÉLYI
DEDICATED TO THE MEMORY OF GEORGE G. LORENTZ

ABSTRACT. Let $\Lambda_n := \{\lambda_0 < \lambda_1 < \cdots < \lambda_n\}$ be a set of real numbers. The collection of all linear combinations of $e^{\lambda_0 t}, e^{\lambda_1 t}, \ldots, e^{\lambda_n t}$ over $\mathbb{R}$ will be denoted by

$$E(\Lambda_n) := \operatorname{span}\{e^{\lambda_0 t}, e^{\lambda_1 t}, \ldots, e^{\lambda_n t}\}.$$

Elements of $E(\Lambda_n)$ are called exponential sums of $n+1$ terms. Let $\|f\|_{[a,b]}$ denote the uniform norm of a real valued function f defined on $[a,b]$. We prove the following results.

Theorem 1. *Let $n \geq 2$ be an integer. Let $\Lambda_n := \{\lambda_0 < \lambda_1 < \cdots < \lambda_n\}$ be a set of nonnegative real numbers, $b \in \mathbb{R}$. There is an absolute constant $c_1 > 0$ such that*

$$\frac{c_1}{\log n} \sum_{j=0}^{n} \lambda_j \leq \sup_{P} \frac{\|P'\|_{(-\infty,b]}}{\|P\|_{(-\infty,b]}} \leq 9 \sum_{j=0}^{n} \lambda_j\,,$$

where the supremum is taken for all $0 \neq P \in E(\Lambda_n)$ increasing on $(-\infty, \infty)$.

Theorem 2. *Let $n \geq 2$ be an integer. Let $\Lambda_n := \{\lambda_0 < \lambda_1 < \cdots < \lambda_n\}$ be a set of real numbers. Let $[a,b]$ be a finite interval with length $b - a > 0$. There are positive constants $c_2 = c_2(a,b)$ and $c_3 = c_3(a,b)$ depending only on a and b such that*

$$c_2 \left(n^2 + \frac{1}{\log n} \sum_{j=0}^{n} |\lambda_j| \right) \leq \sup_{P} \frac{\|P'\|_{[a,b]}}{\|P\|_{[a,b]}} \leq c_3 \left(n^2 + \sum_{j=0}^{n} |\lambda_j| \right),$$

where the supremum is taken for all $0 \neq P \in E(\Lambda_n)$ increasing on $(-\infty, \infty)$.

It is expected that the factor $1/\log n$ in the above theorems can be dropped.

1. INTRODUCTION AND NOTATION

Throughout the paper $[a,b]$ denotes a finite interval of length $b - a > 0$. The Markov inequality asserts that

$$\|p'\|_{[-1,1]} \leq n^2 \|p\|_{[-1,1]}$$

2000 *Mathematics Subject Classification.* Primary 41A17.

Key words and phrases. Markov inequality, Newman inequality, increasing exponential sums .

for all polynomials of degree at most n (with real coefficients). Here, and in what follows $\|f\|_{[a,b]}$ denotes the uniform norm of a real valued function f defined on $[a,b]$.

It has been observed by Bernstein that Markov's inequality for monotone polynomials is not essentially better than that for all polynomials. He proved that

$$\sup_p \frac{\|p'\|_{[-1,1]}}{\|p\|_{[-1,1]}} = \begin{cases} \frac{1}{4}(n+1)^2, & \text{if } n \text{ is odd,} \\ \frac{1}{4}n(n+2), & \text{if } n \text{ is even,} \end{cases}$$

where the supremum is taken for all polynomials $0 \neq p$ of degree at most n that are monotone on $[-1,1]$. See [16, p. 607], for instance.

In his book [2] Braess writes "The rational functions and exponential sums belong to those concrete families of functions which are the most frequently used in nonlinear approximation theory. The starting point of consideration of exponential sums is an approximation problem often encountered for the analysis of decay processes in natural sciences. A given empirical function on a real interval is to be approximated by sums of the form

$$\sum_{j=1}^n a_j e^{\lambda_j t},$$

where the parameters a_j and λ_j are to be determined, while n is fixed."

Let

$$E_n := \left\{ f : f(t) = a_0 + \sum_{j=1}^n a_j e^{\lambda_j t}, \quad a_j, \lambda_j \in \mathbb{R} \right\}.$$

So E_n is the collection of all $n+1$ term exponential sums with constant first term. Schmidt [17] proved that there is a constant $c(n)$ depending only on n so that

$$\|f'\|_{[a+\delta,b-\delta]} \leq c(n)\delta^{-1}\|f\|_{[a,b]}$$

for every $f \in E_n$ and $\delta \in \left(0, \frac{1}{2}(b-a)\right)$. The main result, Theorem 3.2, of [5] shows that Schmidt's inequality holds with $c(n) = 2n-1$. That is,

$$\sup_{0 \neq f \in E_n} \frac{|f'(y)|}{\|f\|_{[a,b]}} \leq \frac{2n-1}{\min\{y-a, b-y\}}, \qquad y \in (a,b). \tag{1}$$

In this Bernstein-type inequality even the point-wise factor is sharp up to a multiplicative absolute constant; the inequality

$$\frac{1}{e-1}\frac{n-1}{\min\{y-a, b-y\}} \leq \sup_{0 \neq f \in E_n} \frac{|f'(y)|}{\|f\|_{[a,b]}}, \qquad y \in (a,b),$$

is established by Theorem 3.3 in [5].

Bernstein-type inequalities play a central role in approximation theory via a machinery developed by Bernstein, which turns Bernstein-type inequalities into inverse theorems of approximation. See, for example, the books by Lorentz [14] and by DeVore and Lorentz [9]. From (1) one can deduce in a standard fashion that if there is a sequence $(f_n)_{n=1}^{\infty}$ of exponential sums with $f_n \in E_n$ that approximates g on an interval $[a, b]$ uniformly with errors

$$\|g - f_n\|_{[a,b]} = O\left(n^{-m}(\log n)^{-2}\right), \qquad n = 2, 3, \ldots,$$

where $m \in \mathbb{N}$ is a fixed integer, then g is m times continuously differentiable on (a, b). Let $\mathcal{P}_n$ be the collection of all polynomials of degree at most n with real coefficients. Inequality (1) can be extended to E_n replaced by

$$\widetilde{E}_n := \left\{ f : f(t) = a_0 + \sum_{j=1}^{N} P_{m_j}(t) e^{\lambda_j t}, \ P_{m_j} \in \mathcal{P}_{m_j}, \ \sum_{j=1}^{N} (m_j + 1) \leq n \right\},$$

where $a_0, \lambda_j \in \mathbb{R}$. In fact, it is well-known that $\widetilde{E}_n$ is the uniform closure of E_n on any finite subinterval of the real number line. For a real valued function f defined on a set A let

$$\|f\|_A := \|f\|_{L_\infty A} := \|f\|_{L_\infty(A)} := \sup_{x \in A} |f(x)|,$$

and let

$$\|f\|_{L_p A} := \|f\|_{L_p(A)} := \left(\int_A |f(x)|^p \, dx \right)^{1/p}, \qquad p > 0,$$

whenever the Lebesgue integral exists.

In this paper we make an effort to show that Newman's inequality (Theorem 2.1) for exponential sums on $(-\infty, b]$ and its extension to finite intervals $[a, b]$ (the case $p = \infty$ in Theorem 2.3) remain essentially sharp even if we consider only increasing exponential sums on the real number line.

2. SOME RECENT RESULTS

Let $\Lambda_n := \{\lambda_0 < \lambda_1 < \cdots < \lambda_n\}$ be a set of real numbers. The collection of all linear combinations of $e^{\lambda_0 t}, e^{\lambda_1 t}, \ldots, e^{\lambda_n t}$ over $\mathbb{R}$ will be denoted by

$$E(\Lambda_n) := \mathrm{span}\{e^{\lambda_0 t}, e^{\lambda_1 t}, \ldots, e^{\lambda_n t}\}.$$

Elements of $E(\Lambda_n)$ are called exponential sums of $n + 1$ terms. Newman's inequality (see [3] and [15]) is an essentially sharp Markov-type inequality for $E(\Lambda_n)$ on $(-\infty, 0]$ in the case when each λ_j is nonnegative.

Theorem 2.1. *(Newman's Inequality) Let* $\Lambda_n := \{\lambda_0 < \lambda_1 < \cdots < \lambda_n\}$ *be a set of nonnegative real numbers. Let* $b \in \mathbb{R}$. *We have*

$$\frac{2}{3} \sum_{j=0}^{n} \lambda_j \leq \sup_{0 \neq P \in E(\Lambda_n)} \frac{\|P'\|_{(-\infty, b]}}{\|P\|_{(-\infty, b]}} \leq 9 \sum_{j=0}^{n} \lambda_j.$$

An L_p version of this is established in [3], [6], [8], and [10].

Theorem 2.2. *Let $\Lambda_n := \{\lambda_0 < \lambda_1 < \cdots < \lambda_n\}$ be a set of nonnegative real numbers. Let $1 \leq p \leq \infty$. Let $b \in \mathbb{R}$. Then*

$$\|P'\|_{L_p(-\infty,b]} \leq 9 \left(\sum_{j=0}^{n} \lambda_j \right) \|P\|_{L_p(-\infty,b]}$$

for every $P \in E(\Lambda_n)$.

Note that in the above theorems the case $b = 0$ represents the general case. This can be seen by the substitution $u = t - b$.

The following $L_p[a,b]$ $(1 \leq p \leq \infty)$ analogue of Theorem 2.2 has been established in [1].

Theorem 2.3. *Let $n \geq 1$ be an integer. Let $\Lambda_n := \{\lambda_0 < \lambda_1 < \cdots < \lambda_n\}$ be a set of real numbers. Let $1 \leq p \leq \infty$. There is a positive constant $c_4 = c_4(a,b)$ depending only on a and b such that*

$$\sup_{0 \neq P \in E(\Lambda_n)} \frac{\|P'\|_{L_p[a,b]}}{\|P\|_{L_p[a,b]}} \leq c_4 \left(n^2 + \sum_{j=0}^{n} |\lambda_j| \right).$$

Theorem 2.3 was proved earlier in [4] and [10] under the additional assumptions that $\lambda_j \geq \delta j$ for each j with a constant $\delta > 0$ and with $c_4 = c_4(a,b)$ replaced by $c_4 = c_4(a,b,\delta)$ depending only on a, b, and δ. The novelty of Theorem 2.3 was the fact that $\Lambda_n := \{\lambda_0 < \lambda_1 < \cdots < \lambda_n\}$ is an arbitrary set of real numbers, not even the nonnegativity of the exponents λ_j is needed.

In [11] the following Nikolskii-Markov type inequality has been proved for $E(\Lambda_n)$ on $(-\infty, 0]$.

Theorem 2.4. *Suppose $0 < q \leq p \leq \infty$. Let $\Lambda_n := \{\lambda_0 < \lambda_1 < \cdots < \lambda_n\}$ be a set of nonnegative real numbers. Let μ be a nonnegative integer. Let $b \in \mathbb{R}$. There are constants $c_5 = c_5(p,q,\mu) > 0$ and $c_6 = c_6(p,q,\mu) > 0$ depending only on p, q, and μ such that*

$$c_5 \left(\sum_{j=1}^{n} \lambda_j \right)^{\mu+\frac{1}{q}-\frac{1}{p}} \leq \sup_{0 \neq P \in E(\Lambda_n)} \frac{\|P^{(\mu)}\|_{L_p(-\infty,b]}}{\|P\|_{L_q(-\infty,b]}} \leq c_6 \left(\sum_{j=1}^{n} \lambda_j \right)^{\mu+\frac{1}{q}-\frac{1}{p}},$$

where the lower bound holds for all $0 < q \leq p \leq \infty$ and for all $\mu \geq 0$, while the upper bound holds when $\mu = 0$ and $0 < q \leq p \leq \infty$, and when $\mu \geq 1$, $p \geq 1$, and $0 < q \leq p \leq \infty$. Also, there are constants $c_5 = c_5(q,\mu) > 0$ and $c_6 = c_6(q,\mu) > 0$ depending only on q and μ such that

$$c_5 \left(\sum_{j=1}^{n} \lambda_j \right)^{\mu+\frac{1}{q}} \leq \sup_{0 \neq P \in E(\Lambda_n)} \frac{|P^{(\mu)}(y)|}{\|P\|_{L_q(-\infty,y]}} \leq c_6 \left(\sum_{j=1}^{n} \lambda_j \right)^{\mu+\frac{1}{q}}$$

holds for every $y \in \mathbb{R}$.

Motivated by a question of Michel Weber (Strasbourg) in [13] we proved the following couple of theorems.

Theorem 2.5. *Let* $\Lambda_n := \{\lambda_0 < \lambda_1 < \cdots < \lambda_n\}$ *be a set of real numbers. Suppose* $0 < q \leq p \leq \infty$. *There are constants* $c_7 = c_7(p, q, a, b) > 0$ *and* $c_8 = c_8(p, q, a, b) > 0$ *depending only on* p, q, a, *and* b *such that*

$$c_7 \left(n^2 + \sum_{j=1}^{n} |\lambda_j| \right)^{\frac{1}{q}-\frac{1}{p}} \leq \sup_{0 \neq P \in E(\Lambda_n)} \frac{\|P\|_{L_p[a,b]}}{\|P\|_{L_q[a,b]}} \leq c_8 \left(n^2 + \sum_{j=1}^{n} |\lambda_j| \right)^{\frac{1}{q}-\frac{1}{p}}.$$

Theorem 2.6. *Let* $\Lambda_n := \{\lambda_0 < \lambda_1 < \cdots < \lambda_n\}$ *be a set of real numbers. Suppose* $0 < q \leq p \leq \infty$. *There are constants* $c_9 = c_9(p, q, a, b) > 0$ *and* $c_{10} = c_{10}(p, q, a, b) > 0$ *depending only on* p, q, a, *and* b *such that*

$$c_9 \left(n^2 + \sum_{j=1}^{n} |\lambda_j| \right)^{1+\frac{1}{q}-\frac{1}{p}} \leq \sup_{0 \neq P \in E(\Lambda_n)} \frac{\|P'\|_{L_p[a,b]}}{\|P\|_{L_q[a,b]}} \leq c_{10} \left(n^2 + \sum_{j=1}^{n} |\lambda_j| \right)^{1+\frac{1}{q}-\frac{1}{p}},$$

where the lower bound holds for all $0 < q \leq p \leq \infty$, *while the upper bound holds when* $p \geq 1$ *and* $0 < q \leq p \leq \infty$.

The lower bounds in these inequalities were shown by a method with the Pinkus-Smith Improvement Theorem in the centre. We formulate the useful lemmas applied in the proofs of these lower bounds. To emphasize the power of the technique of interpolation, we present the short proofs of these lemmas, versions of which will be used in the proofs of our new results. We also note that essentially sharp Bernstein-type inequalities for linear combinations of shifted Gaussians are proved in [12].

In fact, a closer look at the proof of Theorems 2.5 and 2.6 presented in [13] gives the following results.

Theorem 2.5*. *Suppose* $0 \leq \lambda_0 < \lambda_1 < \cdots < \lambda_n$, $0 < q \leq \infty$. *There are constants* $c_7 = c_7(q, a, b) > 0$ *and* $c_8 = c_8(q, a, b) > 0$ *such that*

$$c_7 \left(n^2 + \sum_{j=1}^{n} \lambda_j \right)^{\frac{1}{q}} \leq \sup_{0 \neq P \in E(\Lambda_n)} \frac{|P(b)|}{\|P\|_{L_q[a,b]}} \leq c_8 \left(n^2 + \sum_{j=1}^{n} \lambda_j \right)^{\frac{1}{q}}.$$

Theorem 2.5**. *Suppose* $\lambda_0 < \lambda_1 < \cdots < \lambda_n \leq 0$, $0 < q \leq \infty$. *There are constants* $c_7 = c_7(q, a, b) > 0$ *and* $c_8 = c_8(q, a, b) > 0$ *such that*

$$c_7 \left(n^2 + \sum_{j=1}^{n} |\lambda_j| \right)^{\frac{1}{q}} \leq \sup_{0 \neq P \in E(\Lambda_n)} \frac{|P(a)|}{\|P\|_{L_q[a,b]}} \leq c_8 \left(n^2 + \sum_{j=1}^{n} |\lambda_j| \right)^{\frac{1}{q}}.$$

Theorem 2.6*. *Suppose* $0 \leq \lambda_0 < \lambda_1 < \cdots < \lambda_n$, $0 < q \leq \infty$. *There are constants* $c_9 = c_9(q, a, b) > 0$ *and* $c_{10} = c_{10}(q, a, b) > 0$ *such that*

$$c_9 \left(n^2 + \sum_{j=1}^{n} \lambda_j \right)^{1+\frac{1}{q}} \leq \sup_{0 \neq P \in E(\Lambda_n)} \frac{|P'(b)|}{\|P\|_{L_q[a,b]}} \leq c_{10} \left(n^2 + \sum_{j=1}^{n} |\lambda_j| \right)^{1+\frac{1}{q}}.$$

Theorem 2.6**. *Suppose* $\lambda_0 < \lambda_1 < \cdots < \lambda_n \leq 0$, $0 < q \leq \infty$. *There are constants* $c_9 = c_9(q, a, b) > 0$ *and* $c_{10} = c_{10}(q, a, b) > 0$ *such that*

$$c_9 \left(n^2 + \sum_{j=1}^{n} |\lambda_j| \right)^{1+\frac{1}{q}} \leq \sup_{0 \neq P \in E(\Lambda_n)} \frac{|P'(a)|}{\|P\|_{L_q[a,b]}} \leq c_{10} \left(n^2 + \sum_{j=1}^{n} |\lambda_j| \right)^{1+\frac{1}{q}}.$$

3. New Results

We make an effort to show that Newman's inequality (Theorem 2.1) on $(-\infty, b]$ and its extension to finite intervals $[a, b]$ with length $b - a > 0$ (the case $p = \infty$ in Theorem 2.3) remain essentially sharp even if we consider only increasing exponential sums on the real number line.

Theorem 3.1. *Let* $n \geq 2$ *be an integer. Let* $\Lambda_n := \{\lambda_0 < \lambda_1 < \cdots < \lambda_n\}$ *be a set of positive real numbers,* $b \in \mathbb{R}$. *There is an absolute constant* $c_1 > 0$ *such that*

$$\frac{c_1}{\log n} \sum_{j=0}^{n} \lambda_j \leq \sup_{P} \frac{\|P'\|_{(-\infty,b]}}{\|P\|_{(-\infty,b]}} \leq 9 \sum_{j=0}^{n} \lambda_j,$$

where the supremum is taken for all $0 \neq P \in E(\Lambda_n)$ *increasing on* $(-\infty, \infty)$.

Theorem 3.2. *Let* $n \geq 2$ *be an integer. Let* $\Lambda_n := \{\lambda_0 < \lambda_1 < \cdots < \lambda_n\}$ *be a set of real numbers. There are positive constants* $c_2 = c_2(a, b)$ *and* $c_3 = c_3(a, b)$ *depending only on* a *and* b *such that*

$$c_2 \left(n^2 + \frac{1}{\log n} \sum_{j=0}^{n} |\lambda_j| \right) \leq \sup_{P} \frac{\|P'\|_{[a,b]}}{\|P\|_{[a,b]}} \leq c_3 \left(n^2 + \sum_{j=0}^{n} |\lambda_j| \right),$$

where the supremum is taken for all $0 \neq P \in E(\Lambda_n)$ *increasing on* $(-\infty, \infty)$.

It is expected that the factor $1/\log n$ in the above theorems can be dropped.

4. Lemmas

Let $q \in (0, \infty]$ and let w be a not identically zero continuous function defined on $[a, b]$. Our first lemma can be proved by a simple compactness argument and may be viewed as a simple exercise.

Lemma 4.1. *Let* $\Delta_n := \{\delta_0 < \delta_1 < \cdots < \delta_n\}$ *be a set of real numbers. Let* $c \in [b, \infty)$. *There exists a* $0 \neq T \in E(\Delta_n)$ *such that*

$$\frac{|T(c)|}{\|Tw\|_{L_q[a,b]}} = \sup_{0 \neq P \in E(\Delta_n)} \frac{|P(c)|}{\|Pw\|_{L_q[a,b]}},$$

and there exists a $0 \neq S \in E(\Delta_n)$ *such that*

$$\frac{|S'(c)|}{\|Sw\|_{L_q[a,b]}} = \sup_{0 \neq P \in E(\Delta_n)} \frac{|P'(c)|}{\|Pw\|_{L_q[a,b]}}.$$

Our next lemma is an essential tool in proving our key lemmas, Lemmas 4.3 and 4.4.

Lemma 4.2. *Let* $\Delta_n := \{\delta_0 < \delta_1 < \cdots < \delta_n\}$ *be a set of real numbers. Let* $c \in (b, \infty)$. *Let* T *and* S *be the same as in Lemma 4.1. Then* T *has exactly* n *zeros in* $[a, b]$, *counted with multiplicity. Under the additional assumption* $\delta_n \geq 0$, S *also has exactly* n *zeros in* $[a, b]$, *counted with multiplicity.*

The heart of the proof of our theorems is the following pair of comparison lemmas. The proof of the next couple of lemmas is based on basic properties of Descartes systems, in particular on Descartes' Rule of Signs, and on a technique used earlier by P.W. Smith and Pinkus. Lorentz ascribes this result to Pinkus, although it was P.W. Smith [18] who published it. I have learned about the method of proof of these lemmas from Peter Borwein, who also ascribes it to Pinkus. This is the proof we present here. Section 3.2 of [3], for instance, gives an introduction to Descartes systems. Descartes' Rule of Signs is stated and proved on page 102 of [3].

Lemma 4.3. *Let* $\Delta_n := \{\delta_0 < \delta_1 < \cdots < \delta_n\}$ *and* $\Gamma_n := \{\gamma_0 < \gamma_1 < \cdots < \gamma_n\}$ *be sets of real numbers satisfying* $\delta_j \leq \gamma_j$ *for each* $j = 0, 1, \ldots, n$. *Let* $c \in [b, \infty)$. *We have*

$$\sup_{0 \neq P \in E(\Delta_n)} \frac{|P(c)|}{\|Pw\|_{L_q[a,b]}} \leq \sup_{0 \neq P \in E(\Gamma_n)} \frac{|P(c)|}{\|Pw\|_{L_q[a,b]}}.$$

Under the additional assumption $\delta_n \geq 0$ *we also have*

$$\sup_{0 \neq P \in E(\Delta_n)} \frac{|P'(c)|}{\|Pw\|_{L_q[a,b]}} \leq \sup_{0 \neq P \in E(\Gamma_n)} \frac{|P'(c)|}{\|Pw\|_{L_q[a,b]}}.$$

In addition, the above inequalities hold if the supremums are taken over all nonnegative not identically zero $P \in E(\Delta_n)$ *and* $P \in E(\Gamma_n)$, *respectively.*

The result below follows from Lemma 4.3 by a standard compactness argument.

Lemma 4.3*. *The statements of Lemma 4.3 remain valid if* $\delta_0 > 0$, *the interval* $[a, b]$ *is replaced by* $(-\infty, b]$, *and* w *is a not identically zero, continuous, and bounded function on* $(-\infty, b]$.

Lemma 4.4. *Let* $\Delta_n := \{\delta_0 < \delta_1 < \cdots < \delta_n\}$ *and* $\Gamma_n := \{\gamma_0 < \gamma_1 < \cdots < \gamma_n\}$ *be sets of real numbers satisfying* $\delta_j \leq \gamma_j$ *for each* $j = 0, 1, \ldots, n$. *Let* $c \in (-\infty, a]$. *We have*

$$\sup_{0 \neq P \in E(\Delta_n)} \frac{|P(c)|}{\|Pw\|_{L_q[a,b]}} \geq \sup_{0 \neq P \in E(\Gamma_n)} \frac{|P(c)|}{\|Pw\|_{L_q[a,b]}}.$$

Under the additional assumption $\gamma_0 \leq 0$ *we also have*

$$\sup_{0 \neq P \in E(\Delta_n)} \frac{|Q'(c)|}{\|Qw\|_{L_q[a,b]}} \geq \sup_{0 \neq P \in E(\Gamma_n)} \frac{|Q'(c)|}{\|Qw\|_{L_q[a,b]}}.$$

In addition, the above inequalities hold if the supremums are taken over all nonnegative not identically zero $P \in E(\Delta_n)$ and $P \in E(\Gamma_n)$, respectively.

The result below follows from Lemma 4.4 by a standard compactness argument.

Lemma 4.4*. *The statements of Lemma 4.4 remain valid if $\gamma_n < 0$, the interval $[a, b]$ is replaced by $[a, \infty)$, and w is a not identically zero, continuous, and bounded function on $[a, \infty)$.*

5. Proofs of the Lemmas

Proof of Lemma 4.1. Since Δ_n is fixed, the proof is a standard compactness argument. We omit the details. $\qquad\square$

To prove Lemma 4.2 we need the following two facts:

(a) every $0 \neq f \in E(\Delta_n)$ has at most n real zeros, counted with multiplicity;

(b) if $t_1 < t_2 < \cdots < t_m$ are real numbers and $k_1, k_2, \ldots, k_m$ are positive integers such that $\sum_{j=1}^{m} k_j = n$, then there is a $0 \neq f \in E(\Delta_n)$ having a zero at t_j with multiplicity k_j for each $j = 1, 2, \ldots, m$.

Proof of Lemma 4.2. We prove the statement for T first. Suppose to the contrary that $t_1 < t_2 < \cdots < t_m$ are real numbers in $[a, b]$ such that t_j is a zero of T with multiplicity k_j for each $j = 1, 2, \ldots, m$, $k := \sum_{j=1}^{m} k_j < n$, and T has no other zeros in $[a, b]$ different from $t_1, t_2, \ldots, t_m$. Let $t_{m+1} := c$ and $k_{m+1} := n - k \geq 1$. Choose $0 \neq R \in E(\Delta_n)$ such that R has a zero at t_j with multiplicity k_j for each $j = 1, 2, \ldots, m + 1$, and normalize so that $T(t)$ and $R(t)$ have the same sign at every $t \in [a, b]$. Let $T_\varepsilon := T - \varepsilon R$. Note that T and R are of the form

$$T(t) = \widetilde{T}(t) \prod_{j=1}^{m} (t - t_j)^{k_j} \qquad \text{and} \qquad R(t) = \widetilde{R}(t) \prod_{j=1}^{m} (t - t_j)^{k_j},$$

where both $\widetilde{T}$ and $\widetilde{R}$ are continuous functions on $[a, b]$ having no zeros on $[a, b]$. Hence, if $\varepsilon > 0$ is sufficiently small, then $|T_\varepsilon(t)| < |T(t)|$ at every $t \in [a, b] \setminus \{t_1, t_2, \ldots, t_m\}$, so

$$\|T_\varepsilon w\|_{L_q[a,b]} < \|Tw\|_{L_q[a,b]}.$$

This, together with $T_\varepsilon(c) = T(c)$, contradicts the maximality of T.

Now we prove the statement for S. Without loss of generality we may assume that $S'(c) > 0$. Suppose to the contrary that $t_1 < t_2 < \cdots < t_m$ are real numbers in $[a, b]$ such that t_j is a zero of S with multiplicity k_j for each $j = 1, 2, \ldots, m$, $k := \sum_{j=1}^{m} k_j < n$, and S has no other zeros in $[a, b]$ different from $t_1, t_2, \ldots, t_m$. Choose

$$0 \neq Q \in \text{span}\{e^{\delta_{n-k}t}, e^{\delta_{n-k+1}t}, \ldots, e^{\delta_n t}\} \subset E(\Delta_n)$$

such that Q has a zero at t_j with multiplicity k_j for each $j = 1, 2, \ldots, m$, and normalize so that $S(t)$ and $Q(t)$ have the same sign at every $t \in [a, b]$. Note that S and Q are of the form

$$S(t) = \widetilde{S}(t) \prod_{j=1}^{m} (t - t_j)^{k_j} \quad \text{and} \quad Q(t) = \widetilde{Q}(t) \prod_{j=1}^{m} (t - t_j)^{k_j} ,$$

where both $\widetilde{S}$ and $\widetilde{Q}$ are continuous functions on $[a, b]$ having no zeros on $[a, b]$. Let $t_{m+1} := c$ and $k_{m+1} := 1$. Choose

$$0 \neq R \in \text{span}\{e^{\delta_{n-k-1}t}, e^{\delta_{n-k}t}, \ldots, e^{\delta_n t}\} \subset E(\Delta_n)$$

such that R has a zero at t_j with multiplicity k_j for each $j = 1, 2, \ldots, m + 1$, and normalize so that $S(t)$ and $R(t)$ have the same sign at every $t \in [a, b]$. Note that S and R are of the form

$$S(t) = \widetilde{S}(t) \prod_{j=1}^{m} (t - t_j)^{k_j} \quad \text{and} \quad R(t) = \widetilde{R}(t) \prod_{j=1}^{m} (t - t_j)^{k_j} ,$$

where both $\widetilde{S}$ and $\widetilde{R}$ are continuous functions on $[a, b]$ having no zeros on $[a, b]$. It can be easily seen that $\delta_n \geq 0$ implies that $Q'(t)$ does not vanish on (t_m, ∞) (divide by $e^{\delta_n t}$ and then use Rolle's Theorem). Similarly, since $\delta_n \geq 0$, it is easy to see that if Q' is positive on (t_m, ∞), then R' is negative on (c, ∞). Hence $Q'(c)R'(c) < 0$, so the sign of $Q'(c)$ is different from the sign of $R'(c)$. Let $U := Q$ if $Q'(c) < 0$ and let $U := R$ if $R'(c) < 0$. Let $S_\varepsilon := S - \varepsilon U$. Hence, if $\varepsilon > 0$ is sufficiently small, we have $|S_\varepsilon(t)| < |T(t)|$ at every $t \in [a, b] \setminus \{t_1, t_2, \ldots, t_m\}$, so

$$\|S_\varepsilon w\|_{L_q[a,b]} < \|S w\|_{L_q[a,b]} .$$

This, together with $S'_\varepsilon(c) > S'(c) > 0$, contradicts the maximality of S. $\square$

Proof of Lemma 4.3. We begin with the first inequality. We may assume that $a < b < c$. The general case when $a < b \leq c$ follows by a standard continuity argument. Let $k \in \{0, 1, \ldots, n\}$ be fixed and let

$$\gamma_0 < \gamma_1 < \cdots < \gamma_n , \qquad \gamma_j = \delta_j , \quad j \neq k , \qquad \text{and} \qquad \delta_k < \gamma_k < \delta_{k+1}$$

(let $\delta_{n+1} := \infty$). To prove the lemma it is sufficient to study the above cases since the general case follows from this by a finite number of pairwise comparisons. By Lemmas 4.1 and 4.2, there exists $0 \neq T \in E(\Delta_n)$ such that

$$\frac{|T(c)|}{\|Tw\|_{L_q[a,b]}} = \sup_{0 \neq P \in E(\Delta_n)} \frac{|P(c)|}{\|Pw\|_{L_q[a,b]}} ,$$

where T has exactly n zeros in $[a, b]$, counted with multiplicity. Denote the distinct zeros of T in $[a, b]$ by $t_1 < t_2 < \cdots < t_m$, where t_j is a zero of T with multiplicity k_j for each $j = 1, 2, \ldots, m$, and $\sum_{j=1}^{m} k_j = n$. Then T has no other zeros in $\mathbb{R}$ different from $t_1, t_2, \ldots, t_m$. Let

$$T(t) =: \sum_{j=0}^{n} a_j e^{\delta_j t}, \qquad a_j \in \mathbb{R}.$$

Without loss of generality we may assume that $T(c) > 0$. Now $T(t) > 0$ for every $t > c$, otherwise, in addition to its n zeros in $[a, b]$ (counted with multiplicity), T would have at least one more zero in (c, ∞), which is impossible. Hence

$$a_n := \lim_{t \to \infty} T(t) e^{-\delta_n t} \geq 0.$$

Since $E(\Delta_n)$ is the span of a Descartes system on $(-\infty, \infty)$, it follows from Descartes' Rule of Signs that

$$(-1)^{n-j} a_j > 0, \qquad j = 0, 1, \ldots, n.$$

So, in particular, $a_n > 0$. Choose $R \in E(\Gamma_n)$ of the form

$$R(t) = \sum_{j=0}^{n} b_j e^{\gamma_j t}, \qquad b_j \in \mathbb{R},$$

so that R has a zero at each t_j with multiplicity k_j for each $j = 1, 2, \ldots, m$, and normalize so that $R(c) = T(c)(> 0)$ (this $R \in E(\Gamma_n)$ is uniquely determined). Similarly to $a_n \geq 0$ we have $b_n \geq 0$. Since $E(\Gamma_n)$ is the span of a Descartes system on $(-\infty, \infty)$, Descartes' Rule of Signs yields,

$$(-1)^{n-j} b_j > 0, \qquad j = 0, 1, \ldots, n.$$

So, in particular, $b_n > 0$. We have

$$(T - R)(t) = a_k e^{\delta_k t} - b_k e^{\gamma_k t} + \sum_{\substack{j=0 \\ j \neq k}}^{n} (a_j - b_j) e^{\delta_j t}.$$

Since $T - R$ has altogether at least $n+1$ zeros at $t_1, t_2, \ldots, t_m$, and c (counted with multiplicity), it does not have any zero on $\mathbb{R}$ different from $t_1, t_2, \ldots, t_m$, and c. Since

$$(e^{\delta_0 t}, e^{\delta_1 t}, \ldots, e^{\delta_k t}, e^{\gamma_k t}, e^{\delta_{k+1} t}, \ldots, e^{\delta_n t})$$

is a Descartes system on $(-\infty, \infty)$, Descartes' Rule of Signs implies that the sequence

$$(a_0 - b_0, \ a_1 - b_1, \ \ldots, \ a_{k-1} - b_{k-1}, \ a_k, \ -b_k, \ a_{k+1} - b_{k+1}, \ \ldots, \ a_n - b_n)$$

strictly alternates in sign. Since $(-1)^{n-k} a_k > 0$, this implies that $a_n - b_n < 0$ if $k < n$, and $-b_n < 0$ if $k = n$, so

$$(T - R)(t) < 0, \qquad t > c.$$

This can be seen by dividing the left hand side by $e^{\gamma_n t}$ and taking the limit as $t \to \infty$. Since each of T, R, and $T - R$ has a zero at t_j with multiplicity k_j for each $j = 1, 2, \ldots, m$; $\sum_{j=1}^{m} k_j = n$, and $T - R$ has a sign change (a zero with multiplicity 1) at c, we can deduce that each of T, R, and $T - R$ has the same sign on each of the intervals (t_j, t_{j+1}) for every $j = 0, 1, \ldots, m$ with $t_0 := -\infty$ and $t_{m+1} := c$. Hence $|R(t)| \le |T(t)|$ holds for all $t \in [a, b] \subset [a, c]$ with strict inequality at every t different from $t_1, t_2, \ldots, t_m$. Combining this with $R(c) = T(c)$, we obtain

$$\frac{|R(c)|}{\|Rw\|_{L_q[a,b]}} \ge \frac{|T(c)|}{\|Tw\|_{L_q[a,b]}} = \sup_{0 \neq P \in E(\Delta_n)} \frac{|P(c)|}{\|Pw\|_{L_q[a,b]}} .$$

Since $R \in E(\Gamma_n)$, the first conclusion of the lemma follows from this.

Now we start the proof of the second inequality of the lemma. Although it is quite similar to that of the first inequality, we present the details. We may assume that $a < b < c$ and $\delta_n > 0$. The general case when $a < b \le c$ and $\delta_n \ge 0$ follows by a standard continuity argument. Let $k \in \{0, 1, \ldots, n\}$ be fixed and let

$$\gamma_0 < \gamma_1 < \cdots < \gamma_n , \qquad \gamma_j = \delta_j , \quad j \neq k , \qquad \text{and} \qquad \delta_k < \gamma_k < \delta_{k+1}$$

(let $\delta_{n+1} := \infty$). To prove the lemma it is sufficient to study the above cases since the general case follows from this by a finite number of pairwise comparisons. By Lemmas 4.1 and 4.2, there exists $0 \neq S \in E(\Delta_n)$ such that

$$\frac{|S'(c)|}{\|Sw\|_{L_q[a,b]}} = \sup_{0 \neq P \in E(\Delta_n)} \frac{|P'(c)|}{\|Pw\|_{L_q[a,b]}} ,$$

where S has exactly n zeros in $[a, b]$, counted with multiplicity. Denote the distinct zeros of S in $[a, b]$ by $t_1 < t_2 < \cdots < t_m$, where t_j is a zero of S with multiplicity k_j for each $j = 1, 2, \ldots, m$, and $\sum_{j=1}^m k_j = n$. Then S has no other zeros in $\mathbb{R}$ different from $t_1, t_2, \ldots, t_m$. Let

$$S(t) =: \sum_{j=0}^{n} a_j e^{\delta_j t} , \qquad a_j \in \mathbb{R} .$$

Without loss of generality we may assume that $S(c) > 0$. Since $\delta_n > 0$, we have $\lim_{t \to \infty} S(t) = \infty$, otherwise, in addition to its n zeros in (a, b), S would have at least one more zero in (c, ∞), which is impossible.

Because of the extremal property of S, $S'(c) \neq 0$. We show that $S'(c) > 0$. To see this observe that Rolle's Theorem implies that $S' \in E(\Delta_n)$ has at least $n - 1$ zeros in $[t_1, t_m]$ (counted with multiplicity). If $S'(c) < 0$, then $S(t_m) = 0$ and $\lim_{t \to \infty} S(t) = \infty$ imply that S' has at least 2 more zeros in (t_m, ∞). Thus $S'(c) < 0$ would imply that S' has at least $n + 1$ zeros in $[a, \infty)$, which is impossible. Hence $S'(c) > 0$, indeed. Also $a_n := \lim_{t \to \infty} S(t) e^{-\delta_n t} \ge 0$. Since $E(\Delta_n)$ is the span of a Descartes system on $(-\infty.\infty)$, it follows from Descartes' Rule of Signs that

$$(-1)^{n-j} a_j > 0 , \qquad j = 0, 1, \ldots, n .$$

So, in particular, $a_n > 0$. Choose $R \in E(\Gamma_n)$ of the form

$$R(t) = \sum_{j=0}^{n} b_j e^{\gamma_j t}, \qquad b_j \in \mathbb{R},$$

so that R has a zero at each t_j with multiplicity k_j for each $j = 1, 2, \ldots, m$, and normalize so that $R(c) = S(c) (> 0)$ (this $R \in E(\Gamma_n)$ is uniquely determined). Similarly to $a_n \geq 0$ we have $b_n \geq 0$. Since $E(\Gamma_n)$ is the span of a Descartes system on $[a, b]$, Descartes' Rule of Signs implies that

$$(-1)^{n-j} b_j > 0, \qquad j = 0, 1, \ldots, n.$$

So, in particular, $b_n > 0$. We have

$$(S - R)(t) = a_k e^{\delta_k t} - b_k e^{\gamma_k t} + \sum_{\substack{j=0 \\ j \neq k}}^{n} (a_j - b_j) e^{\delta_j t}.$$

Since $S - R$ has altogether at least $n+1$ zeros at $t_1, t_2, \ldots, t_m$, and c (counted with multiplicity), it does not have any zero on $\mathbb{R}$ different from $t_1, t_2, \ldots, t_m$, and c. Since

$$(e^{\delta_0 t}, e^{\delta_1 t}, \ldots, e^{\delta_k t}, e^{\gamma_k t}, e^{\delta_{k+1} t}, \ldots, e^{\delta_n t})$$

is a Descartes system on $(-\infty, \infty)$, Descartes' Rule of Signs implies that the sequence

$$(a_0 - b_0, \ a_1 - b_1, \ \ldots, \ a_{k-1} - b_{k-1}, \ a_k, \ -b_k, \ a_{k+1} - b_{k+1}, \ \ldots, \ a_n - b_n)$$

strictly alternates in sign. Since $(-1)^{n-k} a_k > 0$, this implies that $a_n - b_n < 0$ if $k < n$ and $-b_n < 0$ if $k = n$, so

$$(S - R)(t) < 0, \qquad t > c.$$

Since each of S, R, and $S - R$ has a zero at t_j with multiplicity k_j for each $j = 1, 2, \ldots, m$; $\sum_{j=1}^{m} k_j = n$, and $S - R$ has a sign change (a zero with multiplicity 1) at c, we can deduce that each of S, R, and $S - R$ has the same sign on each of the intervals (t_j, t_{j+1}) for every $j = 0, 1, \ldots, m$ with $t_0 := -\infty$ and $t_{m+1} := c$. Hence $|R(t)| \leq |S(t)|$ holds for all $t \in [a, b] \subset [a, c]$ with strict inequality at every t different from $t_1, t_2, \ldots, t_m$. Combining this with $0 < S'(c) \leq R'(c)$ (recall that $R(c) = S(c) > 0$), we obtain

$$\frac{|R'(c)|}{\|Rw\|_{L_q[a,b]}} \geq \frac{|S'(c)|}{\|Sw\|_{L_q[a,b]}} = \sup_{0 \neq P \in E(\Delta_n)} \frac{|P'(c)|}{\|Pw\|_{L_q[a,b]}}.$$

Since $R \in E(\Gamma_n)$, the second conclusion of the lemma follows from this. The proof of the last statement of the lemma is very similar. We omit the details. □

Proof of Lemma 4.4. The lemma follows from Lemma 4.3 by the substitution $u = -t$. □

6. PROOFS OF THE THEOREMS

Proof of Theorem 3.1. In the light of Theorem 2.1 we need to prove only the lower bound. Moreover, it is sufficient to prove only that for every $a < b$ there exists $0 \neq Q \in E(\Lambda_n)$ increasing on $(-\infty, \infty)$ such that

$$\frac{|Q'(b)|}{\|Q\|_{[a,b]}} \geq \frac{c_1}{\log n} \sum_{j=0}^{n} \lambda_j$$

with an absolute constant $c_1 > 0$, and the lower bound of the theorem follows by a standard compactness argument. Let

$$\lambda := \frac{1}{3 \log n} \sum_{j=0}^{n} \lambda_j.$$

Then there exists $k \in \{0, 1, \dots, n\}$ such that

$$\lambda_k \geq \frac{\lambda}{n - k + 1}.$$

Let $\varepsilon > 0$, $m := \lfloor \frac{n-k}{2} \rfloor$,

$$\widetilde{\delta}_j := \frac{\lambda}{2(n - k + 1)} + j\varepsilon, \qquad j = 0, 1, \dots, m.$$

Let $\widetilde{\Delta}_m := \{\widetilde{\delta}_0 < \widetilde{\delta}_1 < \dots < \widetilde{\delta}_m\}$. By Theorem 2.4 there exists $0 \neq R_m \in E(\widetilde{\Delta}_m)$ such that

$$\frac{|R_m(b)|}{\|R_m\|_{L_2[a,b]}} \geq \frac{|R_m(b)|}{\|R_m\|_{L_2(-\infty,b]}} \geq c_5 \left(\sum_{j=0}^{m} \widetilde{\delta}_j \right)^{1/2}$$

$$\geq c_5 \left(\frac{(m+1)\lambda}{2(n - k + 1)} \right)^{1/2} \geq \frac{c_5}{2} \lambda^{1/2}.$$

Moreover, by Lemma 4.2 we may assume that R_m has m zeros in $[a, b]$. Now let

$$\gamma_j := \lambda_{j+k} \quad \text{and} \quad \delta_j := \frac{\lambda}{n - k + 1} + j\varepsilon, \qquad j = 0, 1, \dots, 2m,$$

$$\Delta_{2m} := \{\delta_0 < \delta_1 < \dots < \delta_{2m}\} \quad \text{and} \quad \Gamma_{2m} := \{\gamma_0 < \gamma_1 < \dots < \gamma_{2m}\}.$$

Then

$$P_{2m} = R_m^2 \in E(\Delta_{2m})$$

is nonnegative on $(-\infty, \infty)$ having $2m$ zeros in $(-\infty, b]$, counted with multiplicity. Now, by Lemma 4.3* (if $\varepsilon > 0$ is sufficiently small, then the assumptions are satisfied), there exists $0 \neq Q_{2m} \in E(\Gamma_{2m}) \subset E(\Lambda_n)$ such that

$$\frac{|Q_{2m}(b)|}{\|Q_{2m}\|_{L_1(-\infty,b]}} \geq \frac{|P_{2m}(b)|}{\|P_{2m}\|_{L_1(-\infty,b]}} = \frac{|R_m(b)|^2}{\|R_m\|_{L_2(-\infty,b]}^2} \geq \frac{c_5^2}{4} \lambda.$$

Now let $S_{2m} \in E(\Gamma_{2m}) \subset E(\Lambda_n)$ be defined by

$$S_{2m}(x) = \int_{-\infty}^{x} Q_{2m}(t)\, dt\,.$$

Then S_{2m} is increasing on $(-\infty, \infty)$ and

$$\frac{|S'_{2m}(b)|}{\|S_{2m}\|_{[a,b]}} \geq \frac{|S'_{2m}(b)|}{\|S_{2m}\|_{(-\infty,b]}} \geq \frac{|Q_{2m}(b)|}{\|Q_{2m}\|_{L_1(-\infty,b]}} \geq \frac{c_5^2}{4}\lambda\,.$$

Note that the constant $c_5^2/4$ above is absolute and as such it is independent of a as well. Hence the standard compactness argument in the beginning of the proof can be implemented. $\qquad\square$

Proof of Theorem 3.2. The upper bound of the theorem follows from Theorem 2.5. Now we turn to the proof of the lower bound. Assume that

$$\lambda_0 < \lambda_1 < \cdots < \lambda_m < 0 \leq \lambda_{m+1} < \lambda_{m+2} < \cdots < \lambda_n\,.$$

We distinguish four cases.

Case 1: $\sum_{j=m+1}^{n} |\lambda_j| \geq \frac{1}{2}\sum_{j=0}^{n} |\lambda_j| \geq n^2 \log n$. In this case the lower bound of Theorem 3.1 gives the lower bound of the theorem.

Case 2: $\sum_{j=0}^{m} |\lambda_j| \geq \frac{1}{2}\sum_{j=0}^{n} |\lambda_j| \geq n^2 \log n$. In this case the lower bound of Theorem 3.1 gives the lower bound of the theorem after the substitution $u = -t$.

Case 3: $\frac{1}{2}\sum_{j=0}^{n} |\lambda_j| \leq n^2 \log n$ and $m < n/2$. Let $k = \lfloor n/4 \rfloor - 1$. Without loss of generality we may assume that $n \geq 8$, hence $k \geq 1$. Let

$$\Delta_k := \{\delta_0 < \delta_1 < \cdots < \delta_k\}\,, \qquad \delta_j := j\varepsilon\,, \quad j = 0, 1, \ldots, k\,.$$

By Theorem 2.5* there exists $0 \neq R_k \in E(\Delta_k)$ such that

$$\frac{|R_k(b)|}{\|R_k\|_{L_2[a,b]}} \geq c_7 n\,.$$

Moreover, by Lemma 4.2 we may assume that R_k has k zeros in $[a, b]$. Now let

$$\Delta_{2k} := \{\delta_0 < \delta_1 < \cdots < \delta_{2k}\}\,, \qquad \delta_j := j\varepsilon\,, \quad j = 0, 1, \ldots, 2k\,,$$

and

$$\Gamma_{2k} := \{\gamma_0 < \gamma_1 < \cdots < \gamma_{2k}\} := \{\lambda_{n-2k} < \lambda_{n-2k+1} < \cdots < \lambda_n\}\,.$$

Then

$$P_{2k} = R_k^2 \in E(\Delta_{2k})$$

is nonnegative on $(-\infty, \infty)$ and has $2k$ zeros in $[a, b]$, counted with multiplicity. Now, by Lemma 4.3* (if $\varepsilon > 0$ is sufficiently small, then the assumptions are satisfied), there exists $0 \neq Q_{2k} \in E(\Gamma_{2k}) \subset E(\Lambda_n)$ such that

$$\frac{|Q_{2k}(b)|}{\|Q_{2k}\|_{L_1[a,b]}} \geq \frac{|P_{2k}(b)|}{\|P_{2k}\|_{L_1[a,b]}} = \frac{|R_k(b)|^2}{\|R_k\|_{L_2[a,b]}^2} \geq c_7^2 n^2\,.$$

Now let $S_{2k} \in E(\Gamma_{2k}) \subset E(\Lambda_n)$ be defined by

$$S_{2k}(x) = \int_{-\infty}^{x} Q_{2k}(t)\, dt .$$

Then S_{2k} is increasing on $(-\infty, \infty)$ and

$$\frac{|S'_{2k}(b)|}{\|S_{2k}\|_{[a,b]}} \geq \frac{|Q_{2k}(b)|}{\|Q_{2k}\|_{L_1[a,b]}} \geq c_7^2 n^2 .$$

Case 4: $\frac{1}{2}\sum_{j=0}^{n} |\lambda_j| \leq n^2 \log n$ and $m \geq n/2$. The proof follows from that in Case 3 by the substitution $u = -t$. $\qquad\square$

REFERENCES

[1] D. Benko, T. Erdélyi, J. Szabados, The full Markov-Newman inequality for Müntz polynomials on positive intervals, *Proc. Amer. Math. Soc.* **131** (2003), 2385-2391.

[2] D. Braess, *Nonlinear approximation theory*, Springer-Verlag, Berlin, 1986.

[3] P. Borwein, T. Erdélyi, *Polynomials and polynomial inequalities*, Graduate Texts in Mathematics **161**, Springer Verlag, New York, NY, 1995.

[4] _____, Newman's inequality for Müntz polynomials on positive intervals, *J. Approx. Theory* **85** (1996), 132–139.

[5] _____, A sharp Bernstein-type inequality for exponential sums, *J. Reine Angew. Math.* **476** (1996), 127–141.

[6] _____, The L_p version of Newman's inequality for lacunary polynomials, *Proc. Amer. Math. Soc.* **124** (1996), 101–109.

[7] _____, Pointwise Remez- and Nikolskii-type inequalities for exponential sums, *Math. Ann.* **316** (2000), 39–60.

[8] P. Borwein, T. Erdélyi, J. Zhang, Müntz systems and orthogonal Müntz-Legendre polynomials, *Trans. Amer. Math. Soc.* **342** (1994), 523–542.

[9] R.A. DeVore, G.G. Lorentz, *Constructive approximation*, Springer Verlag, Berlin, 1993.

[10] T. Erdélyi, Markov- and Bernstein-type inequalities for Müntz polynomials and exponential sums in L_p, *J. Approx. Theory* **104** (2000), 142–152.

[11] _____, Extremal properties of the derivatives of the Newman polynomials, *Proc. Amer. Math. Soc.* **131** (2003), 3129-3134.

[12] _____, Bernstein-type inequalities for linear combinations of shifted Gaussians, *Bull. London Math. Soc.* **38** (2006), 124–138.

[13] _____, Markov-Nikolskii type inequalities for exponential sums on finite intervals, *Adv. Math.* **208** (2007), 135–146.

[14] G.G. Lorentz, *Approximation of functions*, 2nd edition., Chelsea, New York, 1986.

[15] D.J. Newman, Derivative bounds for Müntz polynomials, *J. Approx. Theory* **18** (1976), 360–362.

[16] Q.I. Rahman, G. Schmeisser, *Analytic theory of polynomials*, Clarendon Press, Oxford, 2002.

[17] E. Schmidt, Zur Kompaktheit der Exponentialsummen, *J. Approx. Theory* **3** (1970), 445–459.

[18] P.W. Smith, An improvement theorem for Descartes systems, *Proc. Amer. Math. Soc.* **70** (1978), 26–30.

ON PRIMITIVE DIVISORS OF $n^2 + b$

GRAHAM EVEREST AND GLYN HARMAN

ABSTRACT. We study primitive divisors of terms of the sequence $n^2 + b$, for a fixed integer b that is not a negative square. It seems likely that the number of terms with a primitive divisor has a natural density. This seems to be a difficult problem. We survey some results about divisors of this sequence as well as provide upper and lower growth estimates for the number of terms which have a primitive divisor.

1. PRIMITIVE PRIME DIVISORS

Given b, an integer that is not a negative square, consider the integer sequence with nth term $P_n = n^2 + b$. It seems likely that infinitely many of the terms are prime but a proof seems elusive. Perhaps this mirrors the status of the Mersenne Prime Conjecture, which predicts that the sequence with nth term $M_n = 2^n - 1$ contains infinitely many prime terms. At least with the Mersenne sequence, an old result shows that new primes are produced in a less restrictive sense.

Definition 1.1. Let (A_n) denote a sequence with integer terms. We say an integer $d > 1$ is a *primitive divisor* of A_n if

(1) $d \mid A_n$ and
(2) $\gcd(d, A_m) = 1$ for all non-zero terms A_m with $m < n$.

In 1886 Bang [2] showed that if a is any fixed integer with $a > 1$ then the sequence with nth term $a^n - 1$ has a primitive divisor for any index $n > 6$. This is remarkable because the number 6 is uniform across all a and it is small. Before we say any more about polynomials, a short survey follows indicating the incredible influence of Bang's Theorem.

1.1. Primitive divisor theorems.
In 1892 Zsigmondy obtained the generalization that for any choice of coprime a and b with $a > b > 0$, the term $a^n - b^n$ has a primitive divisor for any index $n > 6$. This lovely result was rediscovered several times in the early 20th century and it has turned out to be widely applicable. For example, the order of the group $GL_n(\mathbb{F}_q)$ has a primitive divisor for all large n, so Sylow's Theorem can be invoked

2000 *Mathematics Subject Classification.* 11A41, 11B32, 11N36.
Key words and phrases. Prime, primitive divisor, quadratic polynomial.
The first author wishes to thank the organizers and the sponsors for a great conference.

to deduce information about the structure of the group. See [22] and the references therein where applications to Group Theory are discussed.

The next major theoretical advance was made by Carmichael. Let u and v denote conjugate quadratic integers; in other words, zeros of a monic irreducible polynomial with integer coefficients. Consider the Lucas sequence defined by

$$U_n = (u^n - v^n)/(u - v).$$

The Fibonacci sequence (F_n) arises by taking the roots of the polynomial $x^2 - x - 1$. Carmichael [5] showed that if u and v are real then U_n has a primitive divisor for $n > 12$. This is a sharp result because F_{12} does not have a primitive divisor. Less is currently known about the corresponding Lehmer-Pierce sequence

$$V_n = (u^n - 1)(v^n - 1).$$

Kálmán Győry pointed out to the first author that if $uv = 1$ then V_n has a primitive divisor for all n beyond some bound; if $uv = -1$ on the other hand, then V_{2k} does not have a primitive divisor if k is odd, because of the identity $V_{2k} = -V_k^2$. In fact, when $uv = 1$, the bound mentioned is uniform. Győry's second observation is germane to this paper: when $uv = -1$, the set of terms with a primitive divisor has natural density equal to $\frac{3}{4}$ (see Theorem 1.2 and Conjecture 1.5). At the conference, Richard Pinch remarked that certain Lehmer-Pierce sequences count orders of groups: this time the groups are $E(\mathbb{F}_{p^n})$, where E denotes an elliptic curve.

Bilu, Hanrot and Voutier [4] used powerful methods from Diophantine analysis to prove, in the general case, that U_n has a primitive divisor for any $n > 30$. Again this is a sharp result as the sequence generated by the polynomial $x^2 - x + 2$ illustrates. Finally, Silverman [25] obtained a primitive divisor theorem for Elliptic Divisibility Sequences and a uniform version appears in [11] for a certain class of sequences.

1.2. Primitive divisors of $n^2 + b$.

Theorem 1.2. *Infinitely many terms of the sequence $n^2 + b$ do not have a primitive divisor.*

The proof of Theorem 1.2 follows very easily from [23] and will be discussed shortly. Schinzel's proof can be used to construct only a very thin set of terms with no primitive divisor. Dartyge [7] has improved Schinzel's result for $n^2 + 1$ (and in principle the method works for $n^2 + b$ also). The aim of this paper is to obtain a better grasp on the set of terms with no primitive divisor. We will also consider whether the set of indices n for which P_n has a primitive divisor has a natural density. Apparently this lies quite deep. For other interesting approaches to the study of divisors of quadratic integral polynomials; consult [7], [9], [10], [16], [20], [21] and [27]. For higher order polynomials see [8].

1.3. The greatest prime factor.

Let $P^+(m)$ denote the greatest prime factor of the integer $m > 1$. There is a wealth of literature about $P^+(n^2 + b)$ concerned with the fact that $P^+(n^2 + b) \to \infty$ as $n \to \infty$, see [24, Chapter 7]. In a slightly different direction, Luca [19] has recently revived an old method of Lehmer's [18] to show that, given B, the set of indices for which $P^+(n^2 + 1) < B$ is efficiently computable. Carmichael's result mentioned earlier for Lucas sequences plays a key role. Luca illustrates the method by showing that when $B = 101$, $n \leq 24208144$.

The following is an easy proposition, see [6] or [12], which relates $P^+(n^2 + b)$ to the existence of a primitive divisor.

Proposition 1.3. *For all $n > |b|$, the term $P_n = n^2 + b$ has a primitive divisor if and only if $P^+(n^2 + b) > 2n$. For all $n > |b|$, if P_n has a primitive divisor then that primitive divisor is a prime and it is unique.*

Proof of Theorem 1.2. Results of Schinzel [23, Theorem 13] show that for any $\alpha > 0$, $P^+(n^2 + b)$ is bounded above by n^α for infinitely many n. Taking $\alpha = \frac{1}{2}$, Proposition 1.3 shows that $P_n = n^2 + b$ fails to have a primitive divisor infinitely often. $\square$

Given $x > 1$, Schinzel's method constructs fewer than $\log x$ terms P_n with $n < x$ having no primitive divisor. For $\alpha > \frac{149}{179}$, Dartyge [7] showed that

$$\left|\{n \leq x : P^+(n^2 + 1) < x^\alpha\}\right| \gg x.$$

It should be noted that the implied constant is very small, involving a term $2^{-\delta^{-2}}$ where δ "est extrêment petit" [7, p.3 line 10]. In this paper we prove the following, which provides good upper and lower estimates for the number of terms with a primitive divisor.

Theorem 1.4. *Supposing $-b$ is not an integer square, define*

$$\rho_b(x) = \left|\{n \leq x : n^2 + b \text{ has a primitive divisor }\}\right|.$$

For all sufficiently large x we have

$$0.5324 < \frac{\rho_b(x)}{x} < 0.905.$$

1.4. Natural density.

Integers m with the property $P^+(m) > 2\sqrt{m}$ were studied by Chowla and Todd [6]. They proved that the set of these numbers has natural density $\log 2$. Perhaps this suggests the following:

Conjecture 1.5. *If $-b$ is not an integer square then $\rho_b(x) \sim x \log 2$.*

With the availability and power of modern computers, one would usually resort to some computational evidence in support of such a conjecture. The authors of [12] looked for such evidence. Whilst they found nothing to clearly contradict the conjecture, neither did they find overwhelming evidence to support it. The problem is that the convergence to the natural density is very slow.

The reason for this might best be explained as follows. Chowla and Todd's proof uses Mertens' Theorem about the asymptotic formula for the sum of inverse primes:

$$\sum_{p<x} \frac{1}{p} = \log\log x + C + O\left(\frac{1}{\log x}\right).$$

The main term of this formula grows very slowly and the error term shrinks very slowly as well. Perhaps, somehow, this lies behind the extremely slow convergence to the natural density of terms with primitive divisor, as in Conjecture 1.5. In addition, the arithmetical nature of the sequence $n^2 + b$ plays a significant rôle when discussing its very large prime divisors (see (11) below) and this will affect what happens for 'small' x. Our paper concludes with an explanation as to why we are not holding our breath about a proof of Conjecture 1.5.

2. Simple bounds

The article [12] gives some simple estimates for $\rho_b(x)$ which are sketched below. These are recalled here as a way in to the harder methods. The first bound in (1) counts indices which produce no primitive divisor. It is much better than the bound obtained from [23] but the set of indices still has density zero and perhaps indicates the limit of elementary methods. The second bound in (1) is very easy but already gives a good estimate for the density of terms with a primitive divisor if it exists.

Theorem 2.1. *For all sufficiently large* x,

$$\frac{x}{\log x} \ll x - \rho_b(x) \quad and \quad \tfrac{1}{2}x - \rho_b(x) \ll \frac{x}{\log x}. \tag{1}$$

The proofs use little apart from well-known estimates for sums over primes, which can be found in the book of Apostol [1]. Both begin with an old idea of Chebyshev which is used frequently as the starting point of investigating the greatest prime factor of certain sequences (see [17, Chapter 2] for example).

Apart from a finite number of primes, any prime p that divides $n^2 + b$ has the property that $-b$ is a quadratic residue modulo p. Let $\mathcal{R}$ denote the set of odd primes for which $-b$ is a quadratic residue; notice that $\mathcal{R}$ comprises the intersection of a finite union of arithmetic progressions with the set of primes. Write

$$Q_x = \prod_{n=1}^{x} |P_n|,$$

and denote by $\omega(Q_x)$ the number of prime divisors of Q_x. By Proposition 1.3 it is sufficient to bound $\omega(Q_x)$ because, with finitely many exceptions, a primitive divisor is unique.

2.1. Sketch proof of Theorem 2.1. Define

$$\mathcal{S} = \{p \in \mathcal{R} : p | Q_x, p < 2x\} \text{ and } \mathcal{S}' = \{p \in \mathcal{R} : p | Q_x, p \geq 2x\}.$$

Let $s = |\mathcal{S}|$ and $s' = |\mathcal{S}'|$. We seek bounds for $s + s'$. By Dirichlet's Theorem on primes in arithmetic progression it is sufficient to estimate s'. Following Chebyshev's method, use Stirling's Formula to obtain

$$\sum_{p | Q_x} e_p \log p = \log Q_x = 2x \log x + O(x), \qquad (2)$$

where the left-hand side corresponds to the prime decomposition of Q_x, for positive integers e_p. The sum on the left-hand side of (2) decomposes according to the definitions of $\mathcal{S}$ and $\mathcal{S}'$ to give

$$\sum_{p \in \mathcal{S}} e_p \log p + \sum_{p \in \mathcal{S}'} \log p = \log Q_x, \qquad (3)$$

noting that $e_p = 1$ whenever $p \geq 2x$. It is easy to show that

$$\sum_{p \in \mathcal{S}} e_p \log p = x \log x + O(x). \qquad (4)$$

Combining (2), (3) and (4) gives

$$x \log x + O(x) = \sum_{p \in \mathcal{S}'} \log p. \qquad (5)$$

The right hand side is bounded above by $s' \log(x^2 + 1)$ yielding a lower bound for s'.

The second bound in (1) arises similarly using a finer partition of the set $\mathcal{S}'$. For $K > 2$, define:

$$\begin{aligned} \mathcal{T} &= \{p \in \mathcal{R} : p | Q_x, 2x < p < Kx\}; \\ \mathcal{U} &= \{p \in \mathcal{R} \mid p | Q_x, Kx < p\}. \end{aligned}$$

Write $t = |\mathcal{T}|$ and $u = |\mathcal{U}|$; we seek an upper bound for the expression $t + u$. Using the definitions of $\mathcal{T}$ and $\mathcal{U}$ as well as equation (4) shows that

$$\sum_{p \in \mathcal{T}} \log p + \sum_{p \in \mathcal{U}} \log p = x \log x + O(x).$$

The extra leverage comes because the left-hand side is greater than

$$t \log x + u \log(Kx).$$

Now K can be chosen judiciously to beat the other O-constants. An upper bound for $t + u$ follows easily and hence the second bound in (1).

Note Actually K can be taken as large as $\log x$ which yields

$$\frac{x \log \log x}{\log x} < x - \rho_b(x)$$

for all large x. But this still fails to produce a positive density set.

3. BETTER BOUNDING

It is the aim of this section to prove Theorem 1.4. Take $b = 1$ for simplicity, so we can drop the subscript b on ρ; as with [9] the arguments in [17] can be used to generalise to $b \neq 1$. We then prove the following.

Theorem 3.1. *For all sufficiently large x we have*

$$0.5324 < \frac{\rho(x)}{x} < 0.905 .$$

Let

$$N_x(p) = \sum_{\substack{x \leq n < 2x \\ p | n^2 + 1}} 1 .$$

The previous section shows it is sufficient to estimate

$$\sum_{p \geq 2x} N_x(p) .$$

Recasting (5) using this definition:

$$\sum_{p \geq 2x} N_x(p) \log p = x \log x + O(x) . \tag{6}$$

The extreme cases arise if most of the contribution to this sum comes from p around $2x$ in size, or around $4x^2$ in size. In the former case the bound $\log p \geq \log x$ gives the trivial bound

$$\sum_{p \geq 2x} N_x(p) < x, \tag{7}$$

which is weaker than the first bound of the last section. On the other hand, $\log p \leq 2 \log x + O(1)$ gives

$$\sum_{p \geq 2x} N_x(p) > \tfrac{1}{2}x + o(x) , \tag{8}$$

which is essentially the second bound of the last section.

We could obtain improved results if we had better information about the following expression:

$$V_x(v) = \sum_{v < p \leq ev} N_x(p) .$$

It is a good exercise to show that

$$V_x(v) \sim \frac{x}{\log v} \tag{9}$$

implies the conjecture. Unfortunately, the asymptotic formula (9) is not expected to be true for very large v, in view of the arithmetic nature of $n^2 + 1$ (see below). However, it is expected that (9) will be true for $v < x^{2-\epsilon}$ for any $\epsilon > 0$ and this suffices to prove the conjecture.

3.1. A better upper bound for $\rho(x)$. We begin by modifying the definitions to allow us to use the Deshouillers-Iwaniec method in [9]. To be precise
we must use smooth functions in order to apply the mean-value estimates in
[9] for Kloostermann sums. Let ϵ, η be two small positive quantities. Let $b(u)$
be a function satisfying $b(u) \in [0,1]$ for all $u \in \mathbb{R}$, with

$$b(u) = \begin{cases} 1 & \text{if } (1+\epsilon)x \le u \le (2-\epsilon)x, \\ 0 & \text{if } u \le x \text{ or } u \ge 2x, \end{cases}$$

$$\frac{d^r b(u)}{du^r} \ll_{r,\epsilon} u^{-r} \text{ for all } r \in \mathbb{N}.$$

We redefine $N_x(p)$ to be

$$N_x(p) = \sum_{\substack{x \le n < 2x \\ p | n^2 + 1}} b(n).$$

An upper bound for this summed over p will give us an upper bound for the
original problem, since the two quantities will differ by at most

$$\tfrac{3}{2}\epsilon x.$$

Now write

$$X = \int_x^{2x} b(u)\,du, \quad |\mathcal{A}_d| = \sum_{n^2+1 \equiv 0 \,(\text{mod } d)} b(n).$$

By the working on [9, p.2] we can modify the Chebyshev argument to give

$$\sum_p |\mathcal{A}_p| \log p = 2X \log x + O(x).$$

Also, as shown in [9], we have

$$\sum_{p \le x} |\mathcal{A}_p| = X \log x + O(x).$$

Let

$$P_x = \max_{|\mathcal{A}_p| \ne 0} p = x^\sigma, \quad \text{say.}$$

We therefore have

$$\sum_{x \le p \le P_x} |\mathcal{A}_p| \log p = X \log x + O(x).$$

Deshouillers and Iwaniec then estimate this sum as

$$\sum_{1 \le j \le J} S(X, V_j) + O(x),$$

where $V_j = 2^j x$ and

$$S(x, V_j) = \sum_{V_j < p \le 4V_j} C_j(p) \log p.$$

Here the infinitely differentiable functions $C_j(u) \in [0,1]$ are supported in $[V_j, 4V_j)$, with

$$\sum_{1 \le j \le J} C_j(u) = \begin{cases} 1 & \text{if } 2x < u \le P_x, \\ 0 & \text{if } u < x \text{ or } u > P_x. \end{cases}$$

After several transformations and an application of the Rosser-Iwaniec sieve in tandem with their own sophisticated mean-value estimate for averages of Kloostermann sums, they prove that

$$S(x, V_j) \le \frac{2}{\log D_j} \int C_j(u) \frac{\log u}{u} \, du \left(1 + O\left(\frac{1}{\log D} \right) \right). \tag{10}$$

Here $D_j = x^{1-\eta} V_j^{-\frac{1}{2}}$. From this they deduce that σ is not less than the solution to

$$2 - \sigma - 2\log(2 - \sigma) = \tfrac{5}{4}.$$

That is, $\sigma = 1.202468\ldots$.

Now, the worst case scenario for the upper bound (7) is if (10) holds with equality for each V_j. This gives

$$\sum_{p \ge 2x} N_x(p) \le \sum_{1 \le j \le J} S(X, V_j)(\log V_j)^{-1} + O(x(\log x)^{-1})$$

$$= x \left(1 + O(\log x)^{-1} \right) \int_1^{\sigma} \frac{2}{1 - t/2} \, dt$$

$$= (2\sigma - \tfrac{3}{2})x \left(1 + O((\log x)^{-1}) \right) < 0.905x.$$

3.2. A better lower bound for $\rho(x)$.

Now we need to show that not all the contribution comes from primes near x^2. This is a relatively simple application of an upper bound sieve to the set

$$\{m : m\ell = n^2 + b, x < n \le 2x\} \quad \text{for } \ell \text{ in some range.}$$

In this case we can apply the sieve with *distribution level*

$$D_\ell = \frac{x}{\ell(\log x)^A},$$

for some A, by an elementary argument: this corresponds to V_j/x in the last section. Of course, this is why the elementary argument is no good for V_j near x in size. The crossover point between the two methods is at $V_j = x^{\frac{4}{3}}$, but we can get nowhere near this value for the problem discussed in [9]. For $\ell = 1$ the problem is the well-known one of representing almost-primes by values of $n^2 + 1$ and giving an upper bound for the number of prime values of this polynomial. By [14, Theorem 5.3] (or see [13, p.66]) we have

$$\sum_{\substack{x \le n \le 2x \\ n^2 + 1 = p}} 1 \le \frac{2x}{\log x} \prod_p \left(1 - \frac{\chi(p)}{p - 1} \right) \left(1 + O\left(\frac{\log \log 3x}{\log x} \right) \right). \tag{11}$$

Here $\chi(n)$ is the non-trivial character $(\mathrm{mod}\,4)$. Note the important product over primes above which encodes arithmetical information relevant to the polynomial $n^2 + 1$. This did not arise in the previous section since summing over a sufficiently long range for ℓ smooths out this factor (compare [9, §8]). It is expected that (11) holds with equality if the factor 2 is replaced by $\frac{1}{2}$ on the right-hand side, see [15, 3].

We obtain our desired bound by first considering

$$W(L, x) = \sum_{L \leq \ell \leq eL} \sum_{\substack{x \leq n \leq 2x \\ n^2 + 1 = \ell p}} 1 \, .$$

Write $\omega(d)$ for the number of solutions to $n^2 + 1 \equiv 0 \,(\mathrm{mod}\,d)$ and let $\{\lambda_d\}_{d \leq D}$ be the Rosser upper bound sieve of level $D_L = x(L(\log x)^A)^{-1}$ as described in [9, §4] and explicitly constructed in [13, Chapter 4]. We then have

$$W(L, x) \leq \sum_{L \leq \ell \leq eL} \sum_{d \leq D_L} \lambda_d \sum_{\substack{x \leq n \leq 2x \\ n^2 + 1 \equiv 0 \,(\mathrm{mod}\,d\ell)}} 1$$

$$= \sum_{L \leq \ell \leq eL} \sum_{d \leq D_L} \lambda_d \omega(d\ell) \left(\frac{x}{d\ell} + O(1) \right)$$

$$= x \sum_{L \leq \ell \leq eL} \sum_{d \leq D_L} \lambda_d \frac{\omega(d\ell)}{d\ell} + O\left(LD_L (\log x)^2 \right) .$$

In the above we have noted that $\omega(d\ell) \leq \tau(d)\tau(\ell)$ and used the well-known average value of the divisor function $\tau(n)$ to give a bound for the error term. We then use a similar analysis to that in [9, §8] to produce the 'main term'. We give all the details that differ from [9] here for completeness.

Firstly write

$$\sum_{L \leq \ell \leq eL} \sum_{d \leq D_L} \lambda_d \frac{\omega(d\ell)}{d\ell} = \sum_{d \leq D_L} \lambda_d \frac{\omega(d)}{d} J(d, L) \, .$$

Now put

$$L(s, d) = \sum_{m=1}^{\infty} \frac{\omega(dm)}{\omega(d)m^s} \, .$$

Note by [9, Lemma 4] that

$$L(s, d) = \frac{\zeta(s)L(s, \chi)}{\zeta(2s)} \prod_{p \mid d} \left(1 + \frac{1}{p^s} \right)^{-1} .$$

Using Perron's formula ([26, Theorem 3.12]) with $T = x, c = (\log x)^{-1}$ we have

$$J(d, L) = \frac{1}{2\pi i} \int_{c - ix}^{c + ix} L(s + 1, d) \frac{(eL)^s - L^s}{s} \, ds + O\left(x^{-\frac{1}{2}} \right) .$$

The final term is negligible. (Actually, this has been estimated very crudely, in reality it is $O(x^{\epsilon-1})$). Now take the contour of integration back to $\operatorname{Re} s = -\frac{1}{2}$. The pole at $s = 0$ gives a term

$$\frac{L(1,\chi)}{\zeta(2)} \prod_{p|d} \left(1 + \frac{1}{p}\right)^{-1}.$$

The pair of integrals on $\operatorname{Im} s = \pm x$ give a negligible contribution $(O(x^{\epsilon-\frac{2}{3}}))$, using

$$\max\left(|\zeta(s)|, |L(s,\chi)|\right) \ll T^{\frac{1}{6}} \text{ for } 1 \le |\operatorname{Im} s| \le T, \ \operatorname{Re} s \ge \frac{1}{2}.$$

The integral on the new contour can be estimated using:

$$\int_{-T}^{T} \frac{|\zeta(\frac{1}{2}+it)|^2}{1+|t|} dt \ll (\log T)^2,$$

with the same bound applying when $\zeta(s)$ is replaced by $L(s,\chi)$, together with ([26, p.135])

$$\frac{1}{\zeta(1+it)} \ll \log T \text{ for } |t| \le T,$$

and

$$\left| \prod_{p|d} \left(1 + \frac{1}{p^{s+1}}\right)^{-1} \right| \le \prod_{p|d} \left(1 - \frac{1}{p^{\frac{1}{2}}}\right)^{-1} < \tau(d).$$

This gives a bound for the integral which is

$$\ll \frac{\tau(d)(\log x)^3}{L^{\frac{1}{2}}}.$$

Thus

$$\sum_{L \le \ell \le eL} \sum_{d \le D_L} \lambda_d \frac{\omega(d\ell)}{d\ell} = \sum_{d \le D_L} \lambda_d \frac{\omega'(d)}{d} + O\left(\sum_{d \le D_L} \frac{\omega(d)\tau(d)(\log x)^3}{L^{\frac{1}{2}}d}\right).$$

Here

$$\omega'(d) = \omega(d) \prod_{p|d} \left(1 + \frac{1}{p}\right)^{-1}.$$

The rest of the working follows *mutatis mutandis* from [9, p.10]. Hence

$$x \sum_{L \le \ell \le eL} \sum_{d \le D_L} \lambda_d \frac{\omega(d\ell)}{d\ell} = \frac{2x}{\log D_L}\left(1 + O\left(\frac{1}{\log D_L}\right)\right) + O\left(\frac{x(\log x)^7}{L^{\frac{1}{2}}}\right).$$

The reader can thus see that the extra error term $O(x(\log x)^7 L^{-\frac{1}{2}})$ (the log power could be reduced here by more careful working) corresponds to the averaging over ℓ smoothing out the influence of the product in (11), and this must dominate the main term for small L since the 'main term' will be incorrect in this case. Assuming that $D_L = xL^{-1}(\log x)^{-4}$ and also that $x^{\frac{3}{4}} > L > (\log x)^{18}$ we obtain

$$W(N, L) \leq \frac{2x}{\log D_L} + O\left(\frac{x}{(\log x)^2}\right). \tag{12}$$

For $L \leq (\log x)^{18}$ we can establish a slightly cruder upper bound as follows. For each value of ℓ we do not sieve by primes dividing ℓ. This makes the λ_d depend on ℓ, but we have $\lambda_d = 0$ if $(d, \ell) > 1$. Hence we can write $\omega(d\ell) = \omega(d)\omega(\ell)$. Following the analysis above, the remainder term remains $O(LD_L(\log x)^2)$. The 'main term' for the upper bound is now

$$\frac{2x}{\log D_L} \sum_{L \leq \ell \leq eL} \frac{\omega(\ell)}{\phi(\ell)} \prod_{p \nmid \ell} \left(1 - \frac{\chi(p)}{p - 1}\right) \leq \frac{Kx}{\log D_L},$$

for some absolute constant K. The contribution from the terms for which $L \leq (\log x)^{18}$ is thus $\ll (\log \log x)(\log x)^{-1}$ times the total contribution for larger L. These terms may therefore be neglected asymptotically.

Now the worst case scenario for (7) has equality in (12) for

$$x^{2-\theta} \geq L \geq (\log x)^{18}, \quad \text{where} \quad \int_{\theta}^{2} \frac{2t}{t - 1} \, dt = 1.$$

In other words, θ is the solution to

$$2(2 - \theta) - 2\log(\theta - 1) = 1.$$

This is the limit of the sequence

$$a_1 = 2, \quad a_{n+1} = \frac{1}{2}\left(\frac{3}{2} + a_n - \log(a_n - 1)\right) \ (n \geq 1),$$

quickly giving the value $1.766249\ldots$. We then calculate

$$\int_{\theta}^{2} \frac{2}{t - 1} \, dt = 2\theta - 3 > 0.5324\ldots.$$

4. Some implications of Conjecture 1.5

The following argument shows that we do not expect Conjecture 1.5 to be settled in the near future. In the previous section we have used the tools that have been developed for the investigation of the greatest prime factor of $n^2 + 1$ to obtain (rather weak) approximations to the conjecture. Now we assume the conjecture and demonstrate that it would lead to a phenomenal improvement for the greatest prime factor problem.

The conjecture leads to

$$\sum_{p \geq 2x} N_x(p) \sim x \log 2.$$

By the Chebyshev argument (6), on average in these sums,

$$\frac{\log p}{\log x} \sim \frac{1}{\log 2} = \sigma \text{ (say)} = 1.4416\ldots.$$

Hence the greatest prime factor of $n^2 + 1$ infinitely often exceeds n^σ. This more than doubles the improvement of Deshouillers-Iwaniec over the trivial estimate! However, we can do still better using the elementary bound from the last section. The worst case scenario now has all the contribution to the left hand side of (6) coming from p close to x^σ. Since the bounds of the last section must hold (and they are better than the Deshouillers-Iwaniec estimates in this region), this corresponds to finding $\alpha < \sigma < \beta$ with

$$\int_\alpha^\beta \frac{2}{t-1}\, dt = \log 2, \qquad \int_\alpha^\beta \frac{2t}{t-1}\, dt = 1.$$

A little bit of manipulation gives the solution to be

$$\beta = 1 + \frac{1 - \log 2}{2 - \sqrt{2}} = 1.52383\ldots. \tag{13}$$

This gives the following result.

Theorem 4.1. *If Conjecture 1.5 is true, then infinitely often the greatest prime factor of $n^2 + 1$ exceeds n^β where β is given by (13).*

References

[1] T.M. Apostol, *Introduction to analytic number theory*, Undergraduate Texts in Mathematics, Springer Verlag, New York, 1976.

[2] A.S. Bang, Taltheoretiske Undersølgelser, *Tidskrift f. Math.* **5** (1886), 70–80 and 130–137.

[3] P.T. Bateman, R.A. Horn, A heuristic asymptotic formula concerning the distribution of prime numbers, *Math. Comp.* **16** (1962), 363–367.

[4] Y. Bilu, G. Hanrot, P.M. Voutier, Existence of primitive divisors of Lucas and Lehmer numbers, *J. Reine Angew. Math.* **539** (2001), 75–122, with an appendix by M. Mignotte.

[5] R.D. Carmichael, On the numerical factors of the arithmetic forms $a^n \pm b^n$, *Ann. of Math.* (2) **15** (1914), 49–70.

[6] S.D. Chowla, J. Todd, The density of reducible integers, *Canadian J. Math.* **1** (1949), 297–299.

[7] C. Dartyge, Entiers de la forme n^2+1 sans grand facteur premier, *Acta. Math. Hungar.* **72** (1996), 1–34.

[8] C. Dartyge, G. Martin, G. Tenenbaum, Polynomial values free of a large prime factors, *Periodica Math. Hung.* **43** (2001), 111–119.

[9] J.-M. Deshouillers, H. Iwaniec, On the greatest prime factor of $n^2 + 1$, *Annales de L'Institut Fourier* **32** (1982), 1–11.

[10] W. Duke, J.B. Friedlander, H. Iwaniec, Equidistribution of roots of a quadratic congruence to prime moduli, *Ann. of Math.* (2) **141** (1995), no. 2, 423–441.

[11] G. Everest, G. McLaren, T. Ward, Primitive divisors of elliptic divisibility sequences, *J. Number Theory* **118**, no. 1, (2006), 71–89.

[12] G.R. Everest, S. Stevens, D. Tamsett, T. Ward, Primes generated by recurrence sequences, *American Mathematical Monthly*, May 2007.

[13] G. Greaves, *Sieves in number theory*, Springer Verlag, Berlin, 2001.

[14] H. Halberstam, H.E. Richert, *Sieve methods*, London Mathematical Society Monographs 4, Academic Press, London, 1974.

[15] G.H. Hardy, J.E. Littlewood, Some problems of "partitio numerorum": III. On the expression of a number as a sum of primes, *Acta Math.* **44** (1922), 1–70.

[16] C. Hooley, On the greatest prime factor of a quadratic polynomial, *Acta Math.* **117** (1967), 281–299.

[17] _____, *Applications of sieve methods*, Cambridge Tracts in Mathematics **70**, Cambridge University Press, 1976.

[18] D.H. Lehmer, On a problem of Störmer, *Illinois J. Math.* **8** (1995), 57–59.

[19] F. Luca, Primitive divisors of Lucas sequences and prime factors of $x^2 + 1$ and $x^4 + 1$, *Acta Acad. Paedagog. Agriensis, Sect. Mat. (N.S.)*, **31** (2004), 19–24.

[20] J. McKee, On the average number of divisors of quadratic polynomials, *Math. Proc. Camb. Phil. Soc.* **117** (1995), 389–392.

[21] _____, The average number of divisors of an irreducible quadratic polynomial, *Math. Proc. Camb. Phil. Soc.* **126** (1999), 17–22.

[22] C. Praeger, Primitive prime divisor elements in finite classical groups, *Groups St. Andrews 1997 in Bath II*, 605–623, Cambridge University Press, 1999.

[23] A. Schinzel, On two theorems of Gelfond and some of their applications, *Acta Arith.* **13** (1978), 177–236.

[24] T. Shorey, R. Tijdeman, *Exponential diophantine equations*, Cambridge Tracts in Mathematics **87**, Cambridge University Press, 1986.

[25] J.H. Silverman, Wieferich's criterion and the *abc*-conjecture, *J. Number Theory* **30** (1988), no. 2, 226–237.

[26] E.C. Titchmarsh (revised by D.R. Heath-Brown), *The theory of the Riemann zeta-function*, Clarendon Press, Oxford, 1986.

[27] Á. Tóth, Roots of quadratic congruences, *Internat. Math. Res. Notices* (2000), no. 14, 719–739.

[28] K. Zsigmondy, Zur Theorie der Potenzreste, *Monatsh. Math.* **3** (1892), 265–284.

IRREDUCIBILITY AND GREATEST COMMON DIVISOR ALGORITHMS FOR SPARSE POLYNOMIALS

MICHAEL FILASETA, ANDREW GRANVILLE, AND ANDRZEJ SCHINZEL

ABSTRACT. We obtain algorithms for determining whether a nonreciprocal polynomial with integer coefficients is irreducible, for calculating the
greatest common divisor of two polynomials with integer coefficients provided one is free of cyclotomic factors, and for determining whether a given
polynomial with integer coefficients has a cyclotomic factor. Each algorithm has a running time that is linear or almost linear in the logarithm
of the maximal degree of the input polynomial(s). The dependence in the
running time on the number of terms and the height of the input varies
depending on the algorithm.

1. INTRODUCTION

Let $f(x) = \sum_{j=0}^{r} a_j x^{d_j} \in \mathbb{Z}[x]$ with each a_j nonzero and with $d_r > d_{r-1} > \cdots > d_1 > d_0 = 0$. For simplicity, we refer to the degree d_r of $f(x)$ as n. Observe that $r + 1$ is the number of terms of $f(x)$. For convenience, we suppose both $n > 1$ and $r > 0$. The height H, as usual, denotes the maximum of the absolute values of the a_j.

The lattice base reduction algorithm of A.K. Lenstra, H.W. Lenstra, Jr., and L. Lovasz [7] gives a factoring algorithm for $f(x)$ that runs in time that depends polynomially on $\log H$ and n. This clearly serves also as an irreducibility test for $f(x)$. One problem we address in this paper is the somewhat
different issue of describing an irreducibility algorithm for sparse polynomials,
that is where r is small compared to n. We view the input as being the list
of $r + 1$ coefficients a_j together with the list of $r + 1$ exponents d_j. With this
in mind, the input is of size $O(r(\log H + \log n))$. We give an algorithm for
this problem that runs in time that is polynomial in $\log n$ (but note that the
dependence on r and $\log H$ in our arguments is not polynomial).

For $f(x) \in \mathbb{Q}[x]$, we define $\tilde{f} = x^n f(1/x)$. We say that $f(x)$ is *reciprocal* if
$f(x) = \pm \tilde{f}(x)$. Otherwise, we say that $f(x)$ is *nonreciprocal*. We note that
$f(x)$ is reciprocal if and only if the condition $f(\alpha) = 0$ for $\alpha \in \mathbb{C}$ implies that

---

2000 *Mathematics Subject Classification.* 11Y16, 12Y05, 68W30, 11C08, 11R09.

Key words and phrases. Algorithm, cyclotomic, greatest common divisor, irreducibility,
sparse.

The first author was supported by the National Science Foundation and the National
Security Agency and the second author by the Natural Sciences and Engineering Research
Council of Canada.

155

$\alpha \neq 0$ and $f(1/\alpha) = 0$. Our methods require the additional assumption that $f(x)$ is nonreciprocal. We establish the following.

Theorem A. *There is a constant $c_1 = c_1(r, H)$ such that an algorithm exists for determining whether a given nonreciprocal polynomial $f(x) \in \mathbb{Z}[x]$ as above is irreducible and that runs in time $O\big(c_1 \log n \, (\log \log n)^2 \log \log \log n\big)$.*

The result relies heavily on some recent work by E. Bombieri and U. Zannier described by the latter in an appendix of [11]. Alternatively, we can make use of [1], work by these same authors and D. Masser, which describes a new simplified approach to the previous work. The other main ingredients are the third author's application of the work of Bombieri and Zannier, given originally in [10], and an improvement on the the first and third authors' joint work in [4].

The constant c_1 can be made explicit. We note though that c_1 depends on some effectively computable constants that are not explicitly given in the appendix of [11] or in [1]. We therefore do not address this issue further here.

The algorithm will give, with the same running time, some information on the factorization of $f(x)$ in the case that $f(x)$ is reducible. Specifically, we have the following:

(i) If $f(x)$ has a cyclotomic factor, then the algorithm will detect this and output an $m \in \mathbb{Z}^+$ such that the cyclotomic polynomial $\Phi_m(x)$ divides $f(x)$.

(ii) If $f(x)$ does not have a cyclotomic factor but has a non-constant reciprocal factor, then the algorithm will produce such a factor. In fact, the algorithm will produce a reciprocal factor of $f(x)$ of maximal degree.

(iii) Otherwise, if $f(x)$ is reducible, then the algorithm outputs a complete factorization of $f(x)$ as a product of irreducible polynomials over $\mathbb{Q}$.

The algorithm for Theorem A will follow along the lines given above. First, we will check if $f(x)$ has a cyclotomic factor. If it does, then the algorithm will produce m as in (i) and stop. If it does not, then the algorithm will check if $f(x)$ has a non-cyclotomic non-constant reciprocal factor. If it does, then the algorithm will produce such a factor as in (ii) and stop. If it does not, then the algorithm will output a complete factorization of $f(x)$ as indicated in (iii).

Our approach to (i) will allow us to obtain additional information about the complete set of cyclotomic factors of $f(x)$. In particular, we are able to describe, in the same running time given for the algorithm in Theorem A, the factor of $f(x)$ which has largest degree and only cyclotomic divisors. Details are given in the next section.

Our approach can be modified to show that if $f(x) \in \mathbb{Z}[x]$ is nonreciprocal and reducible, then $f(x)$ has a non-trivial factor in $\mathbb{Z}[x]$ containing $O(c_2)$ terms where $c_2 = c_2(r, H)$. We note that the results of [9] imply that if

$f(x)$ also does not have a reciprocal factor, then every factor of $f(x)$ in $\mathbb{Z}[x]$ contains $O(c_2)$ terms.

In the case that $f(x) \in \mathbb{Z}[x]$ is reciprocal, one can modify our approach to obtain some information on the factorization of $f(x)$. Define the nonreciprocal part of $f(x)$ to be the polynomial $f(x)$ removed of its irreducible reciprocal factors in $\mathbb{Z}[x]$ with positive leading coefficients. Then in the case that $f(x)$ is reciprocal, one can still determine whether the nonreciprocal part of $f(x)$ is irreducible in time $O(c_1 \log n (\log \log n)^2 \log \log \log n)$. Furthermore, in this same time, one can determine whether $f(x)$ has a cyclotomic factor and, if so, an integer m for which $\Phi_m(x)$ divides $f(x)$.

In addition, we address the problem of computing the greatest common divisor of two sparse polynomials. For nonzero $f(x)$ and $g(x)$ in $\mathbb{Z}[x]$, we use the notation $\gcd_{\mathbb{Z}}(f(x), g(x))$ to denote the polynomial in $\mathbb{Z}[x]$ of largest degree and largest positive leading coefficient that divides $f(x)$ and $g(x)$ in $\mathbb{Z}[x]$. Later in the paper, we will also make use of an analogous definition for $\gcd_{\mathbb{Z}}(f, g)$ where f and g are in $\mathbb{Z}[x_1, \ldots, x_r]$. In this case, we interpret the leading coefficient as the coefficient of the expression $x_1^{e_1} x_2^{e_2} \ldots x_r^{e_r}$ with e_1 maximal, then e_2 maximal given e_1, and so on. Our main result for the greatest common divisor of two sparse polynomials is the following.

Theorem B. *There is an algorithm which takes as input two polynomials $f(x)$ and $g(x)$ in $\mathbb{Z}[x]$, each of degree $\leq n$ and height $\leq H$ and having $\leq r+1$ nonzero terms, with at least one of $f(x)$ and $g(x)$ free of cyclotomic factors, and outputs the value of $\gcd_{\mathbb{Z}}(f(x), g(x))$ and runs in time $O(c_3 \log n)$ for some constant $c_3 = c_3(r, H)$.*

Our approach will imply that if $f(x), g(x) \in \mathbb{Z}[x]$ are as above with $f(x)$ or $g(x)$ not divisible by a cyclotomic polynomial, then $\gcd_{\mathbb{Z}}(f(x), g(x))$ has $O(c_4)$ terms where $c_4 = c_4(r, H)$. The same conclusion does not hold if one omits the assumption that either $f(x)$ or $g(x)$ is not divisible by a cyclotomic polynomial. The following example, demonstrating this, was originally noted in the related work of the third author [12]. Let a and b be relatively prime positive integers. Then

$$\gcd\left(x^{ab} - 1, (x^a - 1)(x^b - 1)\right) = \frac{(x^a - 1)(x^b - 1)}{x - 1}.$$

In connection with Theorem B, we note that D.A. Plaisted [8] has shown that computing $\gcd_{\mathbb{Z}}(f(x), g(x))$ for general sparse polynomials $f(x)$ and $g(x)$ in $\mathbb{Z}[x]$ is at least as hard as any problem in NP. On the other hand, his proof relies heavily on considering polynomials $f(x)$ and $g(x)$ that have cyclotomic factors. By contrast, our proof of Theorem B will rest heavily on the fact that one of $f(x)$ or $g(x)$ does not have any cyclotomic factors.

Our proof of Theorem A will rely on Theorem B. In fact, Theorem B is where we make use of the work of Bombieri and Zannier already cited. It is possible to prove Theorem A in a slightly more direct way, for exampl by making use of Theorem 80 in [11] instead of Theorem B and Theorem 1 below. This does not avoid the use of the work of Bombieri and Zannier since Theorem 80 of [11] is based on this work. We have chosen the presentation here, however, because it clarifies that parts of the algorithm in Theorem A can rest on ideas that have been around for over forty years. In addition, we want the added information given by (i), (ii) and (iii) above as well as Theorem B itself.

To aid in our discussions, we have used letters for labelling theorems that establish the existence of an algorithm and will refer to the algorithms using the corresponding format. As examples, Algorithm A will refer to the algorithm given by Theorem A, and Algorithm B will refer to the algorithm given by Theorem B. Also, we make use of the notation $O_{r,H}\big(w(n)\big)$ to denote a function with absolute value bounded by $w(n)$ times a function of r and H, for n sufficiently large. We note, however, that n being sufficiently large is for convenience to accommodate expressions appearing in the big-oh notation; the algorithms described are for all integers $n > 1$. Thus, the running time for Algorithm A and Algorithm B can be expressed as $O_{r,H}\big(\log n\,(\log\log n)^2\log\log\log n\big)$ and $O_{r,H}(\log n)$, respectively.

2. The Proof of Theorem A

We begin with the following result, which improves on the main result in [4].

Theorem C. *There is an algorithm that has the following property: given* $f(x) = \sum_{j=0}^{r} a_j x^{d_j} \in \mathbb{Z}[x]$ *of degree* $n > 1$ *and with* $r + 1 > 1$ *terms, the algorithm determines whether* $f(x)$ *has a cyclotomic factor in running time* $O_{r,H}\big(\log n\,(\log\log n)^2\log\log\log n\big)$, *where* H *denotes the height of* $f(x)$. *Furthermore, with the same running time, if* $f(x)$ *is divisible by a cyclotomic polynomial, then the algorithm outputs a positive integer* m *for which* $\Phi_m(x)$ *divides* $f(x)$.

Proof. We begin as in the proof of Theorem 2 of [4] and initially give an argument for the existence of an algorithm as in the theorem with running time $O_{r,H}\big((\log n)^2\big)$. We then explain how the algorithm can be sped up to produce the running time given in the statement of the theorem.

We describe and make use of Theorem 5 from [2]. For k a positive integer, define $\gamma(k) = 2 + \sum_{p|k}(p-2)$. Following [2], we call a vanishing sum S minimal if no proper subsum of S vanishes. We will be interested in sums $S = \sum_{j=1}^{t} a_j\omega_j$ where t is a positive integer, each a_j is a nonzero rational number and each ω_j is a root of unity. We refer to the reduced exponent of such an S as the least positive integer k for which $(\omega_i/\omega_1)^k = 1$ for all $i \in \{1, 2, \ldots, t\}$. Theorem 5 of [2] asserts then that if $S = \sum_{j=1}^{t} a_j\omega_j$ is a minimal vanishing sum, then $t \geq \gamma(k)$ where k is the reduced exponent of S.

Also, note that Theorem 5 of [2] implies that the reduced exponent k of a minimal vanishing sum is necessarily squarefree.

To explain our algorithm, suppose first that $f(x)$ has a cyclotomic factor $\Phi_m(x)$, and that we can write $f(x) = \sum_{i=1}^{s} f_i(x)$, where each $f_i(x)$ is a nonzero polynomial divisible by $\Phi_m(x)$, no two $f_i(x)$ have terms involving x to the same power, and s is maximal. Observe that each $f_i(x)$ necessarily has at least two terms. Setting $\zeta_m = e^{2\pi i/m}$, we see that each $f_i(\zeta_m)$ is a minimal vanishing sum. For each $i \in \{1, 2, \ldots, s\}$, we write $f_i(x) = x^{b_i} g_i(x^{e_i})$ where $g_i(x) \in \mathbb{Z}[x]$, b_i and e_i are nonnegative integers chosen so that $g_i(0) \neq 0$ and the greatest common divisor of the exponents appearing in $g_i(x)$ is 1. Then $g_i(\zeta_m^{e_i})$ is a minimal vanishing sum with reduced exponent $m_i = m/\gcd(m, e_i)$. Necessarily, we have $g_i(\zeta_{m_i}) = 0$ and m_i is squarefree. Also, if t_i denotes the number of nonzero terms of $g_i(x)$, we have

$$t_i \geq \gamma(m_i) = 2 + \sum_{p|m_i}(p - 2),$$

which implies each prime divisor of m_i is $\leq t_i$. Define

$$M_i = \{\ell \in \mathbb{Z}^+ : \Phi_\ell(x) \mid g_i(x), \ell \text{ is squarefree, and } \gamma(\ell) \leq t_i\}.$$

In particular, $m_i \in M_i$. In other words,

$$\frac{m}{\gcd(m, e_i)} \in M_i \quad \text{for all } i \in \{1, 2, \ldots, s\}. \tag{1}$$

We have not explained how we can write $f(x) = \sum_{i=1}^{s} f_i(x)$ as above. In particular, even if we know m exists with $\Phi_m(x)$ dividing $f(x)$, we do not know what m is. We circumvent this issue by considering every possible partition of the set $\{0, 1, \ldots, r\}$ as a disjoint union of sets $J_1, J_2, \ldots, J_s$ with each set J_i containing at least two elements. For each partition, we consider the polynomials

$$f_i(x) = \sum_{j \in J_i} a_j x^{d_j} = x^{b_i} g_i(x^{e_i}), \quad 1 \leq i \leq s,$$

where as before b_i and e_i are nonnegative integers chosen so that $g_i(0) \neq 0$ and the greatest common divisor of the exponents appearing in $g_i(x)$ is 1. Defining t_i and M_i as above, depending on the partition of $\{0, 1, \ldots, r\}$, we see then that if $f(x)$ is divisible by some $\Phi_m(x)$, then there is a partition for which (1) holds. On the other hand, if (1) holds for some positive integer m and some partition of $\{0, 1, \ldots, r\}$ as above, then we have $f_i(\zeta_m) = 0$ for each $i \in \{1, 2, \ldots, s\}$, which implies $f(\zeta_m) = 0$ and hence $\Phi_m(x) \mid f(x)$. Thus, (1) holding for some m and some partition of $\{0, 1, \ldots, r\}$ as above is a necessary and sufficient condition for $f(x)$ to be divisible by a cyclotomic polynomial.

With the above in mind, we describe the algorithm for determining whether $f(x)$ has a cyclotomic factor, give further justification that the algorithm works and give a proof that its running time is as claimed. The algorithm is as follows. We go through every partition of the set $\{0, 1, \ldots, r\}$ into disjoint non-empty sets $J_1, J_2, \ldots, J_s$ with each set J_i containing at least two elements. Observe that there are $O_r(1)$ such partitions. For each such partition and each $i \in \{1, 2, \ldots, s\}$, we set $u = u(i)$ to be the element of J_i for which d_u is minimal. In terms of our definition of $f_i(x)$ and $g_i(x)$, this means $b_i = d_u$ and e_i is the greatest common divisor of the degrees of the terms of the polynomial $f_i(x)/x^{d_u}$. We compute e_i by taking the greatest common divisor of the numbers $d_v - d_u$ where $v \in J_i$. In terms of the complexity of the algorithm, given J_i, determining d_u can be done in $O_r(\log n)$ bit operations and computing e_i takes at most $O_r\big((\log n)^2\big)$ bit operations (cf. the discussion of Euclid's algorithm in [3, p. 79]). We can in fact obtain a running time of $O_r\big(\log n \, (\log\log n)^2 \log\log\log n\big)$ using a recursive gcd computation for large integers [3, p. 428] leading to the running time stated in Theorem C, but for the moment we use the $O_r\big((\log n)^2\big)$ estimate. The number of these computations that are needed as we vary over the partitions of $\{0, 1, \ldots, r\}$ and vary over the sets J_i making up the partitions is $O_r(1)$. The computations have therefore thus far taken at most $O_r\big((\log n)^2\big)$ bit operations.

Next, for each partition $J_1, J_2, \ldots, J_s$ of $\{0, 1, \ldots, r\}$ as above, we compute the sets M_i as follows. Observe that t_i is the number of elements of J_i and is necessarily $\leq r + 1$. Thus, we can construct a list of the ℓ that are squarefree positive integers and such that $\gamma(\ell) \leq t_i$ in time $O_r(1)$. For each such ℓ, we want to check if $\Phi_\ell(x)$ divides $g_i(x)$. An algorithm that works well here and in more generality as well is given as Algorithm A in [4]. For our purposes, we can simply take each term $a_v x^{(d_v - d_u)/e_i}$ in $g_i(x)$, where $v \in J_i$, and replace it with $a_v x^{d'_v}$ where $d'_v \in \{0, 1, \ldots, \ell - 1\}$ and

$$d'_v \equiv \frac{d_v - d_u}{e_i} \pmod{\ell}.$$

If we call the resulting polynomial $h_i(x)$, then $g_i(x)$ is divisible by $\Phi_\ell(x)$ if and only if $h_i(x)$ is divisible by $\Phi_\ell(x)$. Observe that the degree of $h_i(x)$ is $\leq \ell \leq (r+1)^r$. Also, the height of $h_i(x)$ is $\leq (r+1)H$. Hence, one can check directly if $h_i(x)$ is divisible by $\Phi_\ell(x)$ in time $O_{r,H}(1)$. The construction of each $h_i(x)$ takes time no more than $O_{r,H}\big(\log n \, (\log\log n)^2\big)$, where the main contribution of the time required comes from the division of $d_v - d_u$ by e_i above. Hence, the total time spent on constructing the various M_i as we vary over the partitions $J_1, J_2, \ldots, J_s$ of $\{0, 1, \ldots, r\}$ and $i \in \{1, 2, \ldots, s\}$ is $O_{r,H}\big(\log n \, (\log\log n)^2\big)$.

For the algorithm, we consider each partition $J_1, J_2, \ldots, J_s$ of $\{0, 1, \ldots, r\}$ as above one at a time. We construct the numbers e_i and the sets M_i as indicated. Next, we want to determine for a fixed partition whether (1) holds for some positive integer m. In other words, we want to know whether there is an m and $m_i \in M_i$ for which

$$m = m_i \gcd(m, e_i) \quad \text{for } i \in \{1, 2, \ldots, s\}. \tag{2}$$

For a positive integer k, we use the notation $\nu_p(k)$ to denote the positive integer u such that $p^u \| k$. Then (2) holds if and only if each of the following is true:

- If $p \mid m_1 \ldots m_s$, then $\nu_p(m) \leq \nu_p(m_i e_i)$ for all i with equality whenever p divides m_i.
- If $p \nmid m_1 \ldots m_s$, then $\nu_p(m) \leq \nu_p(e_0)$, where $e_0 = \gcd(e_1, \ldots, e_s)$.

Defining

$$D = \prod_{\substack{p^t \| e_0 \\ p \nmid m_1 \cdots m_s}} p^t = e_0 \Big/ \prod_{\substack{p^t \| e_0 \\ p \mid m_1 \cdots m_s}} p^t \quad \text{and} \quad m_0 = \gcd(m_1 e_1, \ldots, m_s e_s)/D,$$

we see that a solution to (2) exists if and only if there exist m_i in M_i such that for every prime p dividing some m_i, the exact power of p dividing m_0 is the same as the exact power of p dividing $m_i e_i$. Furthermore, the set of m satisfying (2) in this case is precisely the set of $m = m_0 d$, where $d \mid D$. Observe that m_0 is the unique m satisfying (2) (if such m exist) with the property that every prime divisor of m is a divisor of $m_1 m_2 \cdots m_s$. Furthermore, every prime divisor of $m_1 m_2 \cdots m_s$ is a divisor of m_0. We are interested in knowing whether there exist m and m_i satisfying (2), so we simply restrict our attention to determining whether there exist m_i in M_i such that

$$m_0 = m_i \gcd(m_0, e_i) \quad \text{for } i \in \{1, 2, \ldots, s\}. \tag{3}$$

Recall that the numbers e_i and all elements of M_i have been computed (for each $i = 1, 2, \ldots, s$). Also, as the partitions vary, the number of different e_i and m_i in M_i that arise is $O_r(1)$. We go through all these possibilities and compute $\mathcal{P}$, the set of primes dividing $m_1 m_2 \cdots m_s$. There are $O_r(1)$ such primes and it takes $O_r(1)$ time to compute them. We compute e_0, D and m_0 as defined above and check whether (3) holds. Note that the second formula for D involves removing the prime divisors from e_0 that are in $\mathcal{P}$, which is a fixed set of primes of size $O_r(1)$. Thus, both e_0 and D can be computed in time $O_r\big((\log n)^2\big)$. We also compute m_0 and check (3) with the same bound on the running time. If an m_0 is obtained for which (3) holds, then we output that $f(x)$ has a cyclotomic factor, indicate that the choice of $m = m_0$ is such that $\Phi_m(x)$ divides $f(x)$ and end the algorithm. If no m_0 is obtained for which (3) holds, then we output that $f(x)$ does not have a cyclotomic factor. As there are $O_r(1)$ different m_0 each of size $O_r(n)$, the running time estimate is not affected by going through the various m_0 and outputting the result. Hence, the proof of the theorem, but with running time only $O_{r,H}\big((\log n)^2\big)$, has been explained.

We improve the running time as follows. For the algorithm above, we made use of a few different greatest common divisor computations. These were done to construct e_i for $i \in \{1, 2, \ldots, s\}$, to calculate $e_0 = \gcd(e_1, \ldots, e_s)$ and $m_0 = \gcd(m_1 e_1, \ldots, m_s e_s)/D$, and to determine the value of the right-hand side of (3). As noted earlier, we can apply known algorithms for gcd computations [3, p. 428] that would allow us to reduce the running time to that required by the theorem. However, it is also worth noting that these gcd computations can be circumvented and the required running time obtained in a different manner. We explain this approach now.

Let $J_1, J_2, \ldots, J_s$ be a partition of $\{0, 1, \ldots, r\}$ as in the argument above. Write $e_i = e_i' e_i''$ where every prime divisor of e_i' is $\leq r + 1$ and every prime divisor of e_i'' is $> r + 1$. Recall that $u = u(i) \in J_i$ is chosen so that d_u is minimal. One can compute e_i' without computing e_i from the formula

$$e_i' = \prod_{p \leq t_i} p^{\min_{v \in J_i} \{\nu_p(d_v - d_u)\}} .$$

In other words, for each $p \leq t_i$, we can calculate the minimum of $\nu_p(d_v - d_u)$ as v runs through the elements of J_i and then form the product above to get e_i'. As we shall see momentarily, the numbers e_i' can be calculated in time $O_r(\log n \, (\log \log n)^2 \log \log \log n)$.

We note now that

$$g_i(x^{e_i''}) = \sum_{v \in J_i} a_v x^{(d_v - d_u)/e_i'} ,$$

so we can compute $g_i(x^{e_i''})$ without computing $g_i(x)$, e_i or e_i''. Define

$$M_i' = \{\ell \in \mathbb{Z}^+ : \Phi_\ell(x) \mid g_i(x^{e_i''}), \ell \text{ is squarefree, and } \gamma(\ell) \leq t_i\} .$$

The set M_i' can be computed in the same manner that we computed M_i but with $g_i(x)$ replaced by $g_i(x^{e_i''})$. Thus, computing M_i', given the polynomials $g_i(x^{e_i''})$, takes time $O_{r,H}(\log n \, (\log \log n)^2)$. Recall that the prime divisors of e_i'' are all $> r + 1 \geq t_i$. We deduce that the numbers ℓ in the definition of M_i and M_i' are relatively prime to e_i''. It follows that $M_i = M_i'$. Thus, the above analysis allows us to compute M_i without explicitly computing the numbers e_i and with running time $O_{r,H}(\log n \, (\log \log n)^2 \log \log \log n)$.

Next, we address how to determine whether (3) holds. Recall that $\mathcal{P}$ is the set of prime divisors of $m_1 m_2 \cdots m_s$, and note that these primes are $\leq r + 1$. The prime divisors of m_0 are precisely the primes in $\mathcal{P}$. We deduce that (3) holds if and only if

$$\nu_p(m_0) = \nu_p(m_i) + \min\{\nu_p(m_0), \nu_p(e_i)\} \tag{4}$$

for each $i \in \{1, 2, \ldots, s\}$ and for each $p \in \mathcal{P}$. For each prime $p \in \mathcal{P}$, we compute the values of $\nu_p(e_i)$, for $i \in \{1, 2, \ldots, s\}$, by using that $\nu_p(e_i) = \nu_p(e_i')$. Next, we compute

$$\nu_p(m_0) = \min_{1 \leq i \leq s} \{\nu_p(m_i) + \nu_p(e_i)\} .$$

Then we check if (4) holds. Observe that each $\nu_p(m_i)$ is either 0 or 1, so $\nu_p(m_i)$ can be computed by a simple division. We want also a method to compute $\nu_p(e_i) = \nu_p(e_i')$, for $i \in \{1, 2, \ldots, s\}$. We further need to explain the computation of $\nu_p(d_v - d_u)$ to obtain e_i' above. For U a positive integer and p a prime $\leq r + 1$, the value of $\nu_p(U)$ can be computed as follows. We compute the values of p^{2^j} successively for $j \geq 0$ by squaring until we arrive at a positive integer t for which $p^{2^t} > U$. Observe that $t = O(\log \log U)$. We set $k_0 = 0$. For $j \in \{1, 2, \ldots, t\}$, we successively check if $p^{2^{t-j}} \mid U$ and, if so, set $k_j = k_{j-1} + 2^{t-j}$ and replace U with $U/p^{2^{t-j}}$. If $p^{2^{t-j}} \nmid U$, then we set $k_j = k_{j-1}$. Then $k_t = \nu_p(U)$. Using this procedure, we can compute $\nu_p(U)$ in time $O_r\big(\log U \,(\log \log U)^2 \log \log \log U\big)$. The theorem follows. $\quad\square$

Although it does not affect our main results, it is of some value to note that the running time of the algorithm can be shown to be

$$O_r\big(\log n \,(\log \log n)^2 \log \log \log n + \log H\big).$$

Indeed, the coefficients of $f(x)$ only take part in the algorithm when we form the polynomials $h_i(x)$ and when we check their divisibility by $\Phi_\ell(x)$. Forming the polynomials involves $O_r(1)$ additions of these coefficients and checking the divisibility of an $h_i(x)$ by $\Phi_\ell(x)$ takes time $O_r\big(\log(H+1)\big)$. Note that these divisions do not depend on n since the degrees and the coefficients of the polynomials are $O_r(1)$ and $O_r(H)$, respectively.

As it may be of interest in other contexts, we explain briefly how we can get a bit more out of the algorithm. More precisely, we explain how to obtain the largest monic factor $g(x)$ of $f(x)$ with each irreducible factor of $g(x)$ cyclotomic and in time $O_{r,H}\big(\log n \,(\log \log n)^2 \log \log \log n\big)$. We begin with determining the product of the distinct cyclotomic divisors of $f(x)$. We note, however, that the representations of $g(x)$ and the product of the distinct cyclotomic divisors of $f(x)$ as polynomials cannot be the obvious ones as it is not difficult to show that for $a \geq 2$, the cyclotomic factors of $x^{(a-1)^2} + x^a - x - 1$ are distinct and their product contains exactly $2a - 2$ terms. In other words, explicitly writing out $g(x)$, for example, can take time considerably more than any power of $\log n$.

For given positive integers u, v, define the set $C(u, v) = \{ud : d \mid v\}$. In the algorithm above, we determined values m_0 and D such that $\Phi_m(x)$ divides $f(x)$ whenever $m \in C(m_0, D)$. Let S be the set of all such pairs $\{m_0, D\}$ that can arise as a solution to (2) in Algorithm C. We proved that $\Phi_m(x)$ divides $f(x)$ if and only if m is in the set

$$C_S = \bigcup_{\{m_0, D\} \in S} C(m_0, D).$$

We want to determine

$$\Phi_S(x) = \prod_{m \in C_S} \Phi_m(x).$$

The obvious way to do this is by determining each $C(m_0, D)$ explicitly, but that would involve factoring D which, for complexity issues, should be avoided. However, we can get around determining $C(m_0, D)$ explicitly by taking advantage of the fact that

$$\prod_{m \in C(u,v)} \Phi_m(x) = \Phi_u(x^v)$$

as follows.

We make a few observations about the sets $C(u,v)$:

- One has $C(U,V) \subseteq C(u,v)$ if and only if UV divides uv, u divides U and, as a consequence, V divides v.
- Given positive integers u, v, u', v' with $\gcd(u,v) = \gcd(u',v') = 1$, define $U = \mathrm{lcm}(u,u')$, and let $V = \gcd(v,v')$. Note that $\gcd(U,V) = 1$. Then

$$C(u,v) \cap C(u',v') = \begin{cases} C(U,V) & \text{if } UV \text{ divides } \gcd(uv, u'v'), \\ \emptyset & \text{otherwise.} \end{cases}$$

- There is a natural ordering on the pairs $\{u,v\}$ where $u,v \in \mathbb{N}$, taking $\{U,V\} < \{u,v\}$ if $UV < uv$, or if $UV = uv$ and $V < v$. We see that if $C(U,V) \subset C(u,v)$ then $\{U,V\} < \{u,v\}$.

Now $|S| = O_r(1)$. Given S we create a new set T. We start with $T_0 = S$, and then recursively construct T_{k+1} as the union of T_k and

$$\left\{ \{U,V\} : \ C(U,V) = C(u,v) \cap C(u',v') \text{ for some } \{u,v\}, \{u',v'\} \in T_k \right\}.$$

One can show that $T_{k+1} = T_k$ for some $k = O_r(1)$. When $T_{k+1} = T_k$, we set $T = T_k$. Note that $|T| = O_r(1)$ and $\gcd(u,v) = 1$ for all $\{u,v\} \in T$. For each $\{u,v\} \in T$, beginning with uv minimal and v minimal given uv, we define the polynomials

$$\Phi_{\{u,v\}}(x) = \Phi_u(x^v) \Big/ \prod_{\substack{\{U,V\} \in T \\ \{U,V\} < \{u,v\} \\ C(U,V) \subset C(u,v)}} \Phi_{\{U,V\}}(x) \ \in \mathbb{Z}[x].$$

We do not compute these polynomials explicitly but can give their values as the quotient above where $\{u,v\}$ and each $\{U,V\}$ in the product are given explicitly. Then we have

$$\Phi_S(x) = \prod_{\{u,v\} \in T} \Phi_{\{u,v\}}(x).$$

Obtaining this description of $\Phi_S(x)$ takes $O_r\big(\log n \, (\log \log n)^2 \log \log \log n \big)$ bit operations.

The polynomial $\Phi_S(x)$ is the product of all the distinct cyclotomic factors of $f(x)$. To deal with cyclotomic factors to higher multiplicities, we make use of the following lemma due to G. Hajós [5] (also, see [11, p. 187]).

Lemma 1. *If $(x - \alpha)^k$ divides $f(x)$, then $k \leq r$.*

Recall that S was defined as the set of $\{m_0, D\}$ that gave rise to solutions of (2) corresponding to cyclotomic factors of $f(x)$. We construct similar sets S_j corresponding to cyclotomic factors of $f^{(j)}(x)$ for every $j \in \{0, 1, \ldots, r - 1\}$. Observe that the coefficients of $f^{(j)}(x)$ are bounded by $n^j H$, the degree of $f^{(j)}(x)$ is $n - j$ (assuming as we can that $n \geq r$) and the number of terms in $f^{(j)}(x)$ is $\leq r + 1$. Recalling that the running time of Algorithm C is $O_r\big(\log n\, (\log\log n)^2 \log\log\log n + \log H\big)$, it is not difficult to see that the running time for computing the various sets S_j is $O_{r,H}\big(\log n\, (\log\log n)^2 \log\log\log n\big)$. The exact multiplicity of a cyclotomic factor of $f(x)$ is k provided it divides $f^{(j)}(x)$ for $0 \leq j \leq k - 1$ and not $f^{(k)}(x)$. Lemma 1 further implies that if a cyclotomic polynomial divides $f^{(j)}(x)$ for every $j \in \{0, 1, \ldots, r - 1\}$, then the multiplicity of the factor is r (i.e., there is no need to check if the cyclotomic factor divides $f^{(r)}(x)$). However, we need to be able to determine the common cyclotomic factors determined by various sets S_j. To do this, we set $S_0^* = S_0$, and then construct recursively S_{k+1}^* as the set

$$\big\{\{U, V\} : C(U, V) = C(u, v) \cap C(u', v') \text{ for some } \{u, v\} \in S_k^*, \{u', v'\} \in S_{k+1}\big\}$$

for each $k \in \{1, 2, \ldots, r - 1\}$. One can then proceed by determining T_k^* from S_k^* as we constructed T from S above, and then compute $\Phi_{S_k^*}(x)$, the product of the distinct cyclotomic polynomials dividing $f(x)$ with multiplicity at least $k + 1$. The product of the polynomials $\Phi_{S_k^*}(x)$ for $k \in \{0, 1, \ldots, r - 1\}$ is therefore the largest degree factor of $f(x)$ that is a product of cyclotomic polynomials. The total running time is $O_{r,H}\big(\log n\, (\log\log n)^2 \log\log\log n\big)$ for describing this factor of $f(x)$.

We are now ready to return to our description of Algorithm A. Algorithm A begins by taking the input polynomial $f(x)$ and applying Algorithm C. If $f(x)$ has a cyclotomic factor, we obtain m as in (i). As $f(x)$ is not reciprocal, $f(x)$ cannot be a constant multiple of a cyclotomic polynomial. Hence, $f(x)$ is reducible and (i) holds.

This part of the algorithm does not actually depend on $f(x)$ being nonreciprocal. The proof of Algorithm C shows in fact that if $f(x)$ has a cyclotomic factor, then one can determine m as in (i) with every prime divisor of m being $\leq r + 1$. Thus, it would not be difficult to factor m and compute $\phi(m)$ in the running time required for Theorem A. Once $\phi(m)$ is computed, then one can determine if $f(x)$ is a constant multiple of the cyclotomic polynomial $\Phi_m(x)$ by comparing $\phi(m)$ with n.

We suppose now that $f(x)$ does not have a cyclotomic factor. The next step in Algorithm A is to determine whether $f(x)$ has a reciprocal factor. We shall do this by making use of Theorem B, which we establish in the next section.

We compute

$$\tilde{f}(x) = x^n f(1/x) = \sum_{j=0}^{r} a_j x^{n-d_j}.$$

Since $f(x)$ does not have a cyclotomic factor, we can apply Algorithm B to compute $h(x) = \gcd_{\mathbb{Z}}(f(x), \tilde{f}(x))$. Observe that $h(x)$ is reciprocal and each reciprocal factor of $f(x)$ divides $h(x)$. As $f(x)$ is not reciprocal, we must have $\deg h < \deg f$. If $h(x)$ is not constant, then $f(x)$ is reducible, $h(x)$ is a non-constant reciprocal factor of $f(x)$ and (ii) holds as $h(x)$ is a reciprocal polynomial of largest possible degree dividing $f(x)$. Otherwise, $f(x)$ does not have a non-constant reciprocal factor. Theorem B implies that this part of Algorithm A has running time $O_{r,H}(\log n)$.

We are now left with considering the case that $f(x)$ does not have any non-constant reciprocal factor. The basic idea here is to make use of the third author's work in [9] (see also Theorem 74 in [11]). For a polynomial $F(x_1, \ldots, x_r, x_1^{-1}, \ldots, x_r^{-1})$, in the variables $x_1, \ldots, x_r$ and their reciprocals $x_1^{-1}, \ldots, x_r^{-1}$, we define

$$J F = x_1^{u_1} \cdots x_r^{u_r} F(x_1, \ldots, x_r, x_1^{-1}, \ldots, x_r^{-1}),$$

where each u_j is an integer chosen as small as possible so that $J F$ is a polynomial in $x_1, \ldots, x_r$. In the way of examples, if

$$F = x^2 + 4x^{-1}y + y^3 \quad \text{and} \quad G = 2xyw - x^2 z^{-3} w - 12w,$$

then

$$J F = x^3 + 4y + xy^3 \quad \text{and} \quad J G = 2xyz^3 - x^2 - 12z^3.$$

In particular, note that although w is a variable in G, the polynomial $J G$ does not involve w. We call a multi-variable polynomial $F(x_1, \ldots, x_r) \in \mathbb{Q}[x_1, \ldots, x_r]$ *reciprocal* if

$$J F(x_1^{-1}, \ldots, x_r^{-1}) = \pm F(x_1, \ldots, x_r).$$

For example, $x_1 x_2 - x_1 - x_2 + 1$ and $x_1 x_2 - x_3 x_4$ are reciprocal. Note that this is consistent with our definition of a reciprocal polynomial $f(x) \in \mathbb{Z}[x]$.

To motivate the next result and begin our approach, we set

$$F(x_1, \ldots, x_r) = a_r x_r + \cdots + a_1 x_1 + a_0 \in \mathbb{Z}[x_1, \ldots, x_r].$$

The plan is to associate the factorization of $f(x) = F(x^{d_1}, x^{d_2}, \ldots, x^{d_r})$ with the factorization of a multi-variable polynomial of the form

$$J F(y_1^{m_{11}} \cdots y_t^{m_{1t}}, \ldots, y_1^{m_{r1}} \cdots y_t^{m_{rt}}),$$

where the number of variables t is $\leq r$ and $m_{ij} \in \mathbb{Z}$ for $1 \leq i \leq r$ and $1 \leq j \leq t$. The above multi-variable polynomial can be expressed as

$$y_1^{u_1} \cdots y_t^{u_t} F(y_1^{m_{11}} \cdots y_t^{m_{1t}}, \ldots, y_1^{m_{r1}} \cdots y_t^{m_{rt}}),$$

where

$$u_j = -\min\{m_{1j}, m_{2j}, \ldots, m_{rj}\} \quad \text{for } 1 \leq j \leq t. \tag{5}$$

To make the connection with the factorization of $f(x)$, we want the matrix $M = (m_{ij})$ to be such that

$$
\begin{pmatrix} d_1 \\ \vdots \\ d_r \end{pmatrix} = M \begin{pmatrix} v_1 \\ \vdots \\ v_t \end{pmatrix}
\tag{6}
$$

for some integers $v_1, v_2, \ldots, v_t$. In this way, the substitution $y_j = x^{v_j}$ for $1 \leq j \leq t$ takes any factorization

$$
y_1^{u_1} \cdots y_t^{u_t} F\left(y_1^{m_{11}} \cdots y_t^{m_{1t}}, \ldots, y_1^{m_{r1}} \cdots y_t^{m_{rt}}\right) = F_1(y_1, \ldots, y_t) \cdots F_s(y_1, \ldots, y_t)
\tag{7}
$$

in $\mathbb{Z}[y_1, \ldots, y_t]$ into the form

$$
x^{u_1 v_1 + \cdots + u_t v_t} F\left(x^{d_1}, x^{d_2}, \ldots, x^{d_r}\right) = F_1\left(x^{v_1}, \ldots, x^{v_t}\right) \cdots F_s\left(x^{v_1}, \ldots, x^{v_t}\right).
\tag{8}
$$

We restrict our attention to factorizations in (7) where the $F_i(y_1, \ldots, y_t)$ are non-constant. We will be interested in the case that s is maximal; in other words, we will want the right-hand side of (7) to be a complete factorization of the left-hand side of (7) into irreducibles over $\mathbb{Q}$. For achieving the results in this paper, we want some algorithm for obtaining such a complete factorization of multi-variable polynomials; among the various sources for this, we note that A.K. Lenstra's work in [6] provides such an algorithm. For the moment, though, we need not take s maximal.

Since $f(x) = F\left(x^{d_1}, x^{d_2}, \ldots, x^{d_r}\right)$, the above describes a factorization of $f(x)$, except that we need to take some caution as some v_j may be negative so the expressions $F_i(x^{v_1}, \ldots, x^{v_t})$ may not be polynomials in x. For $1 \leq i \leq s$, define w_i as the integer satisfying

$$
J F_i\left(x^{v_1}, \ldots, x^{v_t}\right) = x^{w_i} F_i\left(x^{v_1}, \ldots, x^{v_t}\right).
\tag{9}
$$

We obtain from (8) that

$$
x^{u_1 v_1 + \cdots + u_t v_t + w_1 + \cdots + w_s} f(x) = \prod_{i=1}^{s} x^{w_i} F_i\left(x^{v_1}, \ldots, x^{v_t}\right).
$$

The definition of w_i implies that this product is over polynomials in $\mathbb{Z}[x]$ that are not divisible by x. The conditions $a_0 \neq 0$ and $d_0 = 0$ imposed on $f(x)$ in the introduction imply that $f(x)$ is not divisible by x. Hence, the exponent of x appearing on the left must be 0, and we obtain the factorization

$$
f(x) = \prod_{i=1}^{s} x^{w_i} F_i\left(x^{v_1}, \ldots, x^{v_t}\right) = \prod_{i=1}^{s} J F_i\left(x^{v_1}, \ldots, x^{v_t}\right).
\tag{10}
$$

The factorization given in (10) is crucial to our algorithm. As we are interested in the case that $f(x)$ has no non-constant reciprocal factor, we restrict our attention to this case. From (10), we see that the polynomials $x^{w_i} F_i(x^{v_1}, \ldots, x^{v_t})$ cannot have a non-constant reciprocal factor. There are, however, still two possibilities that we need to consider for each $i \in \{1, 2, \ldots, s\}$:

(i') $F_i(y_1, \ldots, y_t)$ is reciprocal.
(ii') $J F_i(x^{v_1}, \ldots, x^{v_t}) \in \mathbb{Z}$.

Although we will not need to know a connection between (i') and (ii'), we show here that if (i') holds for some i, then (ii') does as well. We consider then the possibility that

$$J F_i(y_1^{-1}, \ldots, y_t^{-1}) = \pm F_i(y_1, \ldots, y_t). \tag{11}$$

In other words, suppose that

$$y_1^{e_1} \cdots y_t^{e_t} F_i(y_1^{-1}, \ldots, y_t^{-1}) = \pm F_i(y_1, \ldots, y_t), \tag{12}$$

where $e_j = e_j(i)$ is the degree of $F_i(y_1, \ldots, y_t)$ as a polynomial in y_j. Substituting $y_j = x^{v_j}$ into (12), we obtain

$$x^{w_i + e_1 v_1 + \cdots + e_t v_t} F_i(x^{-v_1}, \ldots, x^{-v_t}) = \pm x^{w_i} F_i(x^{v_1}, \ldots, x^{v_t}). \tag{13}$$

By the definition of w_i, the polynomial on the right does not vanish at 0. Assume (ii') does not hold. Let α be a zero of this polynomial. Then substituting $x = 1/\alpha$ into (13) shows that $1/\alpha$ is also a zero. On the other hand, we have already demonstrated in (10) that the right-hand side of (13) is a factor of $f(x)$. This contradicts that $f(x)$ has no non-constant reciprocal factor. Hence, (ii') holds.

We make use of a special case of a result due to the third author in [9]. In particular, the more general result implies that the above idea can in fact always be used to factor $f(x)$ if $f(x)$ has two nonreciprocal irreducible factors. In other words, there exist a matrix M and v_j satisfying (6) and a factorization of the form (7) that leads to a non-trivial factorization of $f(x)$, if it exists, through the substitution $y_j = x^{v_j}$. We are interested in the case that $f(x)$ has no non-constant reciprocal factor. In this case, we can obtain a complete factorization of $f(x)$ into irreducibles.

Theorem 1. *Fix*

$$F = F(x_1, \ldots, x_r) = a_r x_r + \cdots + a_1 x_1 + a_0,$$

where the a_j are nonzero integers. There exists a finite computable set of matrices S with integer entries, depending only on F, with the following property: Suppose the vector $\overrightarrow{d} = \langle d_1, d_2, \ldots, d_r \rangle$ is in $\mathbb{Z}^r$ with $d_r > d_{r-1} > \cdots > d_1 > 0$ and such that $f(x) = F(x^{d_1}, x^{d_2}, \ldots, x^{d_r})$ has no non-constant reciprocal factor. Then there is an $r \times t$ matrix $M = (m_{ij}) \in S$ of rank $t \leq r$ and a vector $\overrightarrow{v} = \langle v_1, v_2, \ldots, v_t \rangle$ in $\mathbb{Z}^t$ such that (6) holds and the factorization given by (7) in $\mathbb{Z}[y_1, \ldots, y_t]$ of a polynomial in t variables $y_1, y_2, \ldots, y_t$ as a product of s irreducible polynomials over $\mathbb{Q}$ implies the factorization of $f(x)$ given by (10) as a product of polynomials in $\mathbb{Z}[x]$ each of which is either irreducible over $\mathbb{Q}$ or a constant.

We are ready now to apply the above to assist us in Algorithm A. As suggested by the statement of Theorem 1, we take the coefficients a_j of $f(x)$ and consider the multi-variable polynomial $F = F(x_1, \ldots, x_r)$. We compute the set S. Since F is a linear polynomial with $r + 1$ terms and height H, the time required to compute S is $O_{r,H}(1)$. Since $f(x) = F(x^{d_1}, \ldots, x^{d_r})$ has no non-constant reciprocal factors, there is a matrix $M = (m_{ij}) \in S$ of rank $t \le r$ and a vector $\overrightarrow{v}$ in $\mathbb{Z}^t$ as in Theorem 1. We go through each of the $O_{r,H}(1)$ matrices M in S and solve for the vectors $\overrightarrow{v} = \langle v_1, v_2, \ldots, v_t \rangle$ in $\mathbb{Z}^t$ satisfying $\overrightarrow{d} = M\overrightarrow{v}$, where t is the number of columns in M and we interpret $\overrightarrow{d}$ and $\overrightarrow{v}$ as column vectors. From the definition of S, we have that the rank of M is t and $t \le r$. Hence, there can be at most one such vector $\overrightarrow{v}$ for each $M \in S$. However, for each $\overrightarrow{d}$, there may be many $M \in S$ and $\overrightarrow{v}$ for which $\overrightarrow{d} = M\overrightarrow{v}$, and we will consider all of them.

We make use of the following simple result in this section and the next.

Theorem D. *There is an algorithm with the following property. Given an $r \times t$ integral matrix $M = (m_{ij})$ of rank $t \le r$ and $\max\{|m_{ij}|\} = O_{r,H}(1)$ and given an integral vector $\overrightarrow{d} = \langle d_1, \ldots, d_r \rangle$ with $\max\{|d_j|\} = O_{r,H}(n)$, the algorithm determines whether there is an integral vector $\overrightarrow{v} = \langle v_1, \ldots, v_t \rangle$ for which (6) holds, and if such a $\overrightarrow{v}$ exists, the algorithm outputs the solution vector $\overrightarrow{v}$. Furthermore, $\max\{|v_j|\} = O_{r,H}(n)$ and the algorithm runs in time $O_{r,H}(\log n)$.*

Proof. There are a variety of ways we can determine if $\overrightarrow{d} = M\overrightarrow{v}$ has a solution and to determine the solution if there is one within the required time $O_{r,H}(\log n)$. We use Gaussian elimination. Performing elementary row operations on M and multiplying by entries from the matrix as one proceeds to use only integer arithmetic allows us to rewrite M in the form of an $r \times t$ matrix $M' = (m'_{ij})$ with each $m'_{ij} \in \mathbb{Z}$ and the first t rows of M' forming a $t \times t$ diagonal matrix with nonzero integers along the diagonal. These computations only depend on the entries of M and, hence, take time $O_{r,H}(1)$. We perform the analogous row operations and integer multiplications on the vector $\overrightarrow{d} = \langle d_1, d_2, \ldots, d_r \rangle$ to solve $\overrightarrow{d} = M\overrightarrow{v}$ for $\overrightarrow{v}$. As the entries of M are integers that are $O_{r,H}(1)$ and each d_j is an integer that is $O_{r,H}(n)$, these operations take time $O_{r,H}(\log n)$. We are thus left with an equation of the form $\overrightarrow{d'} = M'\overrightarrow{v}$ where the entries of M' are integers that are $O_{r,H}(1)$ and the components of $\overrightarrow{d'} = \langle d'_1, d'_2, \ldots, d'_r \rangle$ are integers that are $O_{r,H}(n)$.

For each $j \in \{1, 2, \ldots, t\}$, we check if $d'_j \equiv 0 \pmod{m'_{jj}}$. If for some $j \in \{1, 2, \ldots, t\}$ we have $d'_j \not\equiv 0 \pmod{m'_{jj}}$, then a solution to the original equation $\overrightarrow{d} = M\overrightarrow{v}$, if it exists, must be such that $v_j \notin \mathbb{Z}$. In this case, an integral vector $\overrightarrow{v}$ does not exist. Now, suppose instead that $d'_j \equiv 0 \pmod{m'_{jj}}$ for every $j \in \{1, 2, \ldots, t\}$. Then we divide d'_j by m'_{jj} to determine the vector $\overrightarrow{v}$. This vector may or may not be a solution to the equation

$\vec{d} = M\vec{v}$. We check whether it is by a direct computation. If it is not a solution to the equation $\vec{d} = M\vec{v}$, then there are no solutions to the equation. Otherwise, $\vec{v}$ is an integral vector satisfying $\vec{d} = M\vec{v}$. Checking whether $d'_j \equiv 0 \pmod{m'_{jj}}$ for $1 \leq j \leq t$, solving for $\vec{v}$ if it holds, and checking whether $\vec{d} = M\vec{v}$ all takes time $O_{r,H}(\log n)$. We also have $O_{r,H}(n)$ as a bound for the absolute value of the components v_j of $\vec{v}$. We output $\vec{v}$ if it exists which takes time $O_{r,H}(\log n)$. Combining the running times above, the theorem follows.   □

Algorithm D is performed for each of the $O_{r,H}(1)$ matrices M in S. The running time for each application of Theorem D is $O_{r,H}(\log n)$, so the total running time spent applying Algorithm D for the various $O_{r,H}(1)$ matrices in S is $O_{r,H}(\log n)$. This leads to $O_{r,H}(1)$ factorizations of the form given in (7) into irreducibles, each having a potentially different value for s. For each of these, we compute the values of $F_i\big(x^{v_1}, \ldots, x^{v_t}\big)$ and determine w_i as in (9). We produce then $O_{r,H}(1)$ factorizations of $f(x)$ as in (10). As we obtain these factorizations, we keep track of the number of non-constant polynomials $x^{w_i} F_i\big(x^{v_1}, \ldots, x^{v_t}\big)$ appearing in (10). We choose a factorization for which this number is maximal. Recalling that (10) follows from $\vec{d} = M\vec{v}$ and (7), we deduce from Theorem 1 that the factorization of $f(x)$ we have chosen provides a factorization of $f(x)$ with each $x^{w_i} F_i\big(x^{v_1}, \ldots, x^{v_t}\big)$ either irreducible or constant. Recalling that the polynomials $F_i(y_1, \ldots, y_t)$ in (7) are independent of n and that the components of $\vec{v}$ are bounded in absolute value by $O_{r,H}(n)$, we see that producing the factorization of $f(x)$ into irreducibles and constants as in (10) takes time $O_{r,H}(\log n)$. For a factorization of $f(x)$ into irreducibles over $\mathbb{Q}$, we multiply together the constants appearing on the right of (10) and one of the irreducible polynomials $J F_i\big(x^{v_1}, \ldots, x^{v_t}\big)$. This does not affect the bound given for the running time of Algorithm A.

Thus, we have demonstrated an algorithm for Theorem A as stated in the introduction and justified that the algorithm satisfies the statement of Theorem A as well as (i), (ii) and (iii). Combining the above running time estimates, we deduce that the algorithm also has the stated running time bound given in Theorem A.

3. THE PROOF OF THEOREM B

As mentioned in the introduction, our proof of Theorem B relies heavily on the recent work of Bombieri and Zannier outlined by Zannier in an appendix in [11]. In particular, as a direct consequence of their more general work, we have

Theorem 2. *Let*

$$F(x_1, \ldots, x_k), G(x_1, \ldots, x_k) \in \mathbb{Q}[x_1, \ldots, x_k]$$

*be coprime polynomials. There exists an effectively computable number $B(F,G)$
with the following property. If $\vec{u} = \langle u_1, \ldots, u_k \rangle \in \mathbb{Z}^k$, $\xi \neq 0$ is algebraic and*

$$F\left(\xi^{u_1}, \ldots, \xi^{u_k}\right) = G\left(\xi^{u_1}, \ldots, \xi^{u_k}\right) = 0,$$

*then either ξ is a root of unity or there exists a nonzero vector $\vec{v} \in \mathbb{Z}^k$ having
components bounded in absolute value by $B(F,G)$ and orthogonal to $\vec{u}$.*

It is important for our algorithm that the quantities $B(F,G)$ are effectively
computable. We note that the fact $B(F,G)$ is effectively computable is not
explicitly stated in the appendix of [11], but U. Zannier (private communication) has pointed out that the approach given there does imply that this
is the case. The more recent paper [1] notes explicitly that $B(F,G)$ can be
calculated.

Our description of Algorithm B has similarities to the third author's application of Theorem 2 in [10] and [11]. In particular, we make use of the
following lemma which is Corollary 6 in Appendix E of [11]. A proof is given
there.

Lemma 2. *Let ℓ be a positive integer and $\vec{v} \in \mathbb{Z}^\ell$ with $\vec{v}$ nonzero. The
lattice of vectors $\vec{u} \in \mathbb{Z}^\ell$ orthogonal to $\vec{v}$ has a basis $\vec{v_1}', \vec{v_2}', \ldots, \vec{v_{\ell-1}}'$ such
that the maximum absolute value of a component of any vector $\vec{v_j}'$ is bounded
by $\ell/2$ times the maximum absolute value of a component of $\vec{v}$.*

For our algorithm, we can suppose that $f(x)$ does not have a cyclotomic
factor and do so. We consider only the case that $f(0)g(0) \neq 0$ as computing
$\gcd_{\mathbb{Z}}\left(f(x), g(x)\right)$ can easily be reduced to this case by initially removing an
appropriate power of x from each of $f(x)$ and $g(x)$ (that is, by subtracting the
least degree of a term from each exponent). This would need to be followed
up by possibly multiplying by a power of x after our gcd computation.

We furthermore only consider the case that the content of $f(x)$, that is
the greatest common divisor of its coefficients, and the content of $g(x)$ are
1. Otherwise, we simply divide by the contents before proceeding and then
multiply the final result by the greatest common divisor of the two contents.
We express our two polynomials in the form

$$f(x) = \sum_{j=0}^{k} a_j x^{d_j} \quad \text{and} \quad g(x) = \sum_{j=0}^{k} b_j x^{d_j},$$

where above we have possibly extended the lists of exponents and coefficients
describing $f(x)$ and $g(x)$ so that the exponent lists are identical and the
coefficient lists are allowed to include coefficients which are 0. Also, we take
$d_k > d_{k-1} > \cdots > d_1 > 0$. Thus, $d_0 = 0$, $a_0 b_0 \neq 0$ and $k \leq 2r$. The time
required to modify $f(x)$ and $g(x)$ so that they are not divisible by x and
have content 1 and to adjust the exponent and coefficient lists as above is
$O_{r,H}(\log n)$.

Before continuing with the algorithm, we motivate it with some discussion. Let $w(x)$ denote $\gcd_{\mathbb{Z}}\big(f(x), g(x)\big)$. We will apply Theorem 2 to construct two finite sequences of polynomials in several variables F_u and G_u with integer coefficients and a corresponding finite sequence of vectors $\overrightarrow{d}^{(u)}$ that will enable us to determine a polynomial in $\mathbb{Z}[x]$ that has the common zeros, to the correct multiplicity, of $f(x)$ and $g(x)$. This then will allow us to compute $w(x)$.

Let ξ be a zero of $w(x)$, if it exists. Observe that $\xi \neq 0$, and since ξ is a zero of $f(x)$ which has no cyclotomic factors, we have ξ is not a root of unity. Since ξ is a common zero of $f(x)$ and $g(x)$, we have

$$\sum_{j=0}^{k} a_j \xi^{d_j} = \sum_{j=0}^{k} b_j \xi^{d_j} = 0.$$

We recursively construct F_u, G_u and $\overrightarrow{d}^{(u)}$, for $0 \leq u \leq s$, where s is to be determined, beginning with

$$F_0 = F_0(x_1, \ldots, x_k) = \sum_{j=0}^{k} a_j x_j \quad \text{and} \quad G_0 = G_0(x_1, \ldots, x_k) = \sum_{j=0}^{k} b_j x_j,$$

(14)

and $\overrightarrow{d}^{(0)} = \langle d_1, d_2, \ldots, d_k \rangle$. As u increases, the number of variables defining F_u and G_u will decrease. The value of s then will be $\leq k$. Observe that

$$F_0\big(x^{d_1}, \ldots, x^{d_k}\big) = f(x) \quad \text{and} \quad G_0\big(x^{d_1}, \ldots, x^{d_k}\big) = g(x).$$

We deduce that F_0 and G_0, being linear, are coprime in $\mathbb{Q}[x_1, \ldots, x_k]$ and that

$$F_0\big(\xi^{d_1}, \ldots, \xi^{d_k}\big) = G_0\big(\xi^{d_1}, \ldots, \xi^{d_k}\big) = 0.$$ (15)

Now, suppose for some $u \geq 0$ that nonzero polynomials F_u and G_u in $\mathbb{Z}[x_1, \ldots, x_{k_u}]$ and a vector $\overrightarrow{d}^{(u)} = \langle d_1^{(u)}, \ldots, d_{k_u}^{(u)} \rangle \in \mathbb{Z}^{k_u}$ have been determined, where $k_u < k_{u-1} < \cdots < k_0 = k$. Furthermore, suppose that F_u and G_u are coprime in $\mathbb{Q}[x_1, \ldots, x_{k_u}]$ and that we have at least one zero ξ of $w(x)$ such that

$$F_u\big(\xi^{d_1^{(u)}}, \ldots, \xi^{d_{k_u}^{(u)}}\big) = G_u\big(\xi^{d_1^{(u)}}, \ldots, \xi^{d_{k_u}^{(u)}}\big) = 0.$$ (16)

In particular, $\xi \neq 0$ and ξ is not a root of unity. Note that the $d_j^{(u)}$ may be negative. We will require

$$J F_u\big(x^{d_1^{(u)}}, \ldots, x^{d_{k_u}^{(u)}}\big) \mid f(x) \quad \text{and} \quad J G_u\big(x^{d_1^{(u)}}, \ldots, x^{d_{k_u}^{(u)}}\big) \mid g(x).$$ (17)

Observe that $J F_u\big(x^{d_1^{(u)}}, \ldots, x^{d_{k_u}^{(u)}}\big)$ and $f(x)$ are in $\mathbb{Z}[x]$. We take (17) to mean that there is a polynomial $h(x) \in \mathbb{Z}[x]$ such that

$$f(x) = h(x) \cdot J F_u\big(x^{d_1^{(u)}}, \ldots, x^{d_{k_u}^{(u)}}\big),$$

with an analogous equation holding for $g(x)$ and $JG_u\big(x^{d_1^{(u)}},\ldots,x^{d_{k_u}^{(u)}}\big)$. In particular, we want $JF_u\big(x^{d_1^{(u)}},\ldots,x^{d_{k_u}^{(u)}}\big)$ and $JG_u\big(x^{d_1^{(u)}},\ldots,x^{d_{k_u}^{(u)}}\big)$ to be nonzero. Note that these conditions which are being imposed on F_u and G_u are satisfied for $u = 0$ provided $w(x)$ is not constant. For $0 \le u < s$, we describe next how to recursively construct F_{u+1} and G_{u+1} having analogous properties. The specifics of the algorithm and its running time will be discussed later.

There is a computable bound $B(F_u, G_u)$ as described in Theorem 2. We deduce that there is a nonzero vector $\overrightarrow{v} = \langle v_1, v_2, \ldots, v_{k_u}\rangle \in \mathbb{Z}^{k_u}$ such that each $|v_i| \le B(F_u, G_u)$ and $\overrightarrow{v}$ is orthogonal to $\overrightarrow{d}^{(u)}$. From Lemma 2, there is a $k_u \times (k_u - 1)$ matrix $\mathcal{M}$ with each entry of $\mathcal{M}$ having absolute value $\le k_u B(F_u, G_u)/2$ and such that $\overrightarrow{d}^{(u)} = \mathcal{M}\overrightarrow{v}^{(u)}$ for some $\overrightarrow{v}^{(u)} \in \mathbb{Z}^{k_u-1}$, where we view the vectors as column vectors. We define integers m_{ij} (written also $m_{i,j}$) and $v_j^{(u)}$, depending on u, by the conditions

$$\mathcal{M} = \begin{pmatrix} m_{11} & \cdots & m_{1,k_u-1} \\ \vdots & \ddots & \vdots \\ m_{k_u 1} & \cdots & m_{k_u,k_u-1} \end{pmatrix} \quad \text{and} \quad \overrightarrow{v}^{(u)} = \big\langle v_1^{(u)},\ldots,v_{k_u-1}^{(u)}\big\rangle.$$

The relations

$$x_i = y_1^{m_{i1}}\cdots y_{k_u-1}^{m_{i,k_u-1}} \quad \text{for } 1 \le i \le k_u$$

transform the polynomials $F_u(x_1,\ldots,x_{k_u})$ and $G_u(x_1,\ldots,x_{k_u})$ into polynomials in some, possibly all, of the variables $y_1,\ldots,y_{k_u-1}$. These new polynomials we call $\mathcal{F}_u$ and $\mathcal{G}_u$, respectively. More precisely, we define

$$\mathcal{F}_u(y_1,\ldots,y_{k_u-1}) = JF_u\big(y_1^{m_{11}}\cdots y_{k_u-1}^{m_{1,k_u-1}},\ldots,y_1^{m_{k_u 1}}\cdots y_{k_u-1}^{m_{k_u,k_u-1}}\big) \qquad (18)$$

and

$$\mathcal{G}_u(y_1,\ldots,y_{k_u-1}) = JG_u\big(y_1^{m_{11}}\cdots y_{k_u-1}^{m_{1,k_u-1}},\ldots,y_1^{m_{k_u 1}}\cdots y_{k_u-1}^{m_{k_u,k_u-1}}\big). \qquad (19)$$

The polynomials $\mathcal{F}_u$ and $\mathcal{G}_u$ will depend on the matrix $\mathcal{M}$ so that there may be many choices for $\mathcal{F}_u$ and $\mathcal{G}_u$ for each F_u and G_u. We need only consider one such $\mathcal{F}_u$ and $\mathcal{G}_u$ and do so. Note that this still may require considering various $\mathcal{M}$ until we find one for which $\overrightarrow{d}^{(u)} = \mathcal{M}\overrightarrow{v}^{(u)}$ is satisfied for some $\overrightarrow{v}^{(u)} \in \mathbb{Z}^{k_u-1}$. The equation $\overrightarrow{d}^{(u)} = \mathcal{M}\overrightarrow{v}^{(u)}$ implies that for some integers $e_f(u)$ and $e_g(u)$ we have

$$\mathcal{F}_u\big(x^{v_1^{(u)}},\ldots,x^{v_{k_u-1}^{(u)}}\big) = x^{e_f(u)}F_u\big(x^{d_1^{(u)}},\ldots,x^{d_{k_u}^{(u)}}\big) \qquad (20)$$

and

$$\mathcal{G}_u\big(x^{v_1^{(u)}},\ldots,x^{v_{k_u-1}^{(u)}}\big) = x^{e_g(u)}G_u\big(x^{d_1^{(u)}},\ldots,x^{d_{k_u}^{(u)}}\big). \qquad (21)$$

In particular, $\mathcal{F}_u$ and $\mathcal{G}_u$ are nonzero. Also,

$$J\mathcal{F}_u\big(x^{v_1^{(u)}},\ldots,x^{v_{k_u-1}^{(u)}}\big) \mid f(x) \quad \text{and} \quad J\mathcal{G}_u\big(x^{v_1^{(u)}},\ldots,x^{v_{k_u-1}^{(u)}}\big) \mid g(x). \qquad (22)$$

Furthermore, with ξ as in (16), we have

$$\mathcal{F}_u\big(\xi^{v_1^{(u)}},\ldots,\xi^{v_{k_u-1}^{(u)}}\big) = \mathcal{G}_u\big(\xi^{v_1^{(u)}},\ldots,\xi^{v_{k_u-1}^{(u)}}\big) = 0.$$

The idea is to suppress the variables, if they exist, which do not occur in $\mathcal{F}_u$ and $\mathcal{G}_u$ and the corresponding components of $\overrightarrow{v}^{(u)}$ to obtain the polynomials F_{u+1} and G_{u+1} and the vector $\overrightarrow{d}^{(u+1)}$ for our recursive construction. However, there is one other matter to consider. The polynomials $\mathcal{F}_u$ and $\mathcal{G}_u$ may not be coprime, and we require F_{u+1} and G_{u+1} to be coprime. Hence, we adjust this idea slightly.

Let

$$D_u = D_u(y_1, \ldots, y_{k_u-1}) = \gcd_{\mathbb{Z}}(\mathcal{F}_u, \mathcal{G}_u) \in \mathbb{Z}[y_1, \ldots, y_{k_u-1}]. \qquad (23)$$

Recall that $f(0)g(0) \neq 0$. Hence, (20), (21) and (22) imply that the polynomial $J\,D_u\big(x^{v_1^{(u)}}, \ldots, x^{v_{k_u-1}^{(u)}}\big)$ divides $\gcd_{\mathbb{Z}}(f, g)$ in $\mathbb{Z}[x]$. We define

$$F_{u+1} = \frac{\mathcal{F}_u(y_1, \ldots, y_{k_u-1})}{D_u(y_1, \ldots, y_{k_u-1})} \quad \text{and} \quad G_{u+1} = \frac{\mathcal{G}_u(y_1, \ldots, y_{k_u-1})}{D_u(y_1, \ldots, y_{k_u-1})}, \qquad (24)$$

and set $k_{u+1} \leq k_u - 1$ to be the total number of variables $y_1, \ldots, y_{k_u-1}$ appearing in F_{u+1} and G_{u+1}. Note that F_{u+1} and G_{u+1} are coprime and that (17) holds with u replaced by $u + 1$ and the appropriate change of variables.

We describe next how the recursive construction will end. Suppose we have just constructed F_u, G_u and $\overrightarrow{d}^{(u)}$ and proceed as above to the next step of constructing F_{u+1}, G_{u+1} and $\overrightarrow{d}^{(u+1)}$. At this point, D_{u-1} will have been defined but not D_u. We want to find $\mathcal{M}$ and a $\overrightarrow{v}^{(u)}$ such that $\overrightarrow{d}^{(u)} = \mathcal{M}\overrightarrow{v}^{(u)}$ where $\mathcal{M}$ is a $k_u \times (k_u - 1)$ matrix with entries bounded in absolute value by $k_u B(F_u, G_u)/2$. So we compute $B(F_u, G_u)$ and the bound $k_u B(F_u, G_u)/2$ on the absolute values of the entries of $\mathcal{M}$. We consider such $\mathcal{M}$ and apply Algorithm D to see if there is an integral vector $\overrightarrow{v}^{(u)}$ for which $\overrightarrow{d}^{(u)} = \mathcal{M}\overrightarrow{v}^{(u)}$. Once such an $\mathcal{M}$ and $\overrightarrow{v}^{(u)}$ are found, we can proceed with the construction of F_{u+1} and G_{u+1} given above. On the other hand, it is possible that no such $\mathcal{M}$ and $\overrightarrow{v}^{(u)}$ will be found. Given Theorem 2, this will be the case only if the supposition that (16) holds for some zero ξ of $w(x)$ is incorrect. In particular, (16) does not hold for some zero ξ of $w(x)$ if F_u and G_u are coprime polynomials in < 2 variables (i.e., $k_u \leq 1$), but it is also possible that (16) does not hold for some u with F_u and G_u polynomials in ≥ 2 variables (i.e., $k_u \geq 2$). Given that $\mathcal{M}$ is a $k_u \times (k_u - 1)$ matrix, we consider it to be vacuously true that no $\mathcal{M}$ and $\overrightarrow{v}^{(u)}$ exist satisfying $\overrightarrow{d}^{(u)} = \mathcal{M}\overrightarrow{v}^{(u)}$ in the case that $k_u \leq 1$. If no such $\mathcal{M}$ and $\overrightarrow{v}^{(u)}$ exist, we consider the recursive construction of the polynomials F_u and G_u complete and set $s = u$. We will want the values of D_u for every $1 \leq u \leq s-1$, so we save these as we proceed.

The motivation discussed above can be summarized into a procedure to be used for Algorithm B as follows. Beginning with F_0 and G_0 as in (14) and $\overrightarrow{d}^{(0)} = \langle d_1, \ldots, d_k \rangle$, we construct the multi-variable polynomials F_u and G_u and vectors $\overrightarrow{d}^{(u)} = \langle d_1^{(u)}, \ldots, d_{k_u}^{(u)} \rangle \in \mathbb{Z}^{k_u}$ recursively. Given F_u, G_u and $\overrightarrow{d}^{(u)}$, we compute $B(F_u, G_u)$ and search for a $k_u \times (k_u - 1)$ matrix $\mathcal{M}$ with integer entries having absolute value $\leq k_u B(F_u, G_u)/2$ for which $\overrightarrow{d}^{(u)} = M\overrightarrow{v}^{(u)}$ is solvable with $\overrightarrow{v}^{(u)} = \langle v_1^{(u)}, \ldots, v_{k_u-1}^{(u)} \rangle \in \mathbb{Z}^{k_u-1}$. We check for solvability and determine the solution $\overrightarrow{v}^{(u)}$ if it exists by using Algorithm D. If no such M and $\overrightarrow{v}^{(u)}$ exist, then we set $s = u$ and stop our construction. Otherwise, once such an $M = (m_{ij})$ and $\overrightarrow{v}^{(u)}$ are determined, we define F_{u+1} and G_{u+1} using (18), (19), (23) and (24). After using (24) to construct F_{u+1} and G_{u+1}, we determine the variables $y_1, \ldots, y_{k_u-1}$ which occur in F_{u+1} and G_{u+1} and define $\overrightarrow{d}^{(u+1)}$ as the vector with corresponding components from $v_1^{(u)}, \ldots, v_{k_u-1}^{(u)}$; in other words, if y_j is the ith variable occurring in F_{u+1} and G_{u+1}, then $v_j^{(u)}$ is the ith component of $\overrightarrow{d}^{(u+1)}$.

For the running time for this recursive construction, we use that $B(F_u, G_u)$ is $O_{r,H}(1)$ as u varies and, furthermore, the numbers $B(F_u, G_u)$ can be computed in time $O_{r,H}(1)$. In particular, this implies that for a fixed u, there are $O_{r,H}(1)$ choices for M and, hence, a total of $O_{r,H}(1)$ possible values for F_{u+1} and G_{u+1} independent of the value of $\overrightarrow{d}^{(u)}$. In other words, without even knowing the values of $d_1, \ldots, d_k$, we can use Theorem 2 to deduce that there are at most $O_{r,H}(1)$ possibilities for F_1 and G_1. For each of these possibilities, another application of Theorem 2 implies that there are at most $O_{r,H}(1)$ possibilities for F_2 and G_2. And so on. As $s \leq k \leq 2r$, we deduce that the total number of matrices M that we need to consider during the recursive construction is bounded by $O_{r,H}(1)$. The recursive construction depends on n only when applying Theorem D to see if $\overrightarrow{d}^{(u)} = M\overrightarrow{v}^{(u)}$ holds for some $\overrightarrow{v}^{(u)}$ and to determine $\overrightarrow{v}^{(u)}$ if it exists. For a fixed M, Theorem D implies that these computations can be done in time $O_{r,H}(\log n)$. As the total number of M to consider is bounded by $O_{r,H}(1)$, we deduce that the recursive construction of the F_u, G_u and $\overrightarrow{d}^{(u)}$ takes time $O_{r,H}(\log n)$.

As we proceed in our recursive construction of the F_u and G_u, an important aspect of the construction is that the m_{ij} are bounded in absolute value by $O_{r,H}(1)$ and, hence, the coefficients and exponents appearing in F_u and G_u are bounded by $O_{r,H}(1)$. In other words, F_u and G_u can be written in time $O_{r,H}(1)$. Another important aspect of the construction is to note that as we are dividing by D_u to construct F_{u+1} and G_{u+1}, we obtain not simply that $JD_u\big(x^{v_1^{(u)}}, \ldots, x^{v_{k_u-1}^{(u)}}\big)$ divides $\gcd_{\mathbb{Z}}(f,g)$ in $\mathbb{Z}[x]$ but also

$$\prod_{j=0}^{u} JD_j\big(x^{v_1^{(j)}}, \ldots, x^{v_{k_j-1}^{(j)}}\big) \text{ divides } \gcd_{\mathbb{Z}}(f,g) \text{ in } \mathbb{Z}[x]. \tag{25}$$

This can be seen inductively by observing that

$$J\mathcal{F}_u\big(x^{v_1^{(u)}}, \ldots, x^{v_{k_u-1}^{(u)}}\big) = \frac{f(x)}{\displaystyle\prod_{j=0}^{u-1} JD_j\big(x^{v_1^{(j)}}, \ldots, x^{v_{k_j-1}^{(j)}}\big)} \tag{26}$$

and

$$J\,\mathcal{G}_u\big(x^{v_1^{(u)}},\ldots,x^{v_{k_u-1}^{(u)}}\big) = \frac{g(x)}{\displaystyle\prod_{j=0}^{u-1} J\,D_j\big(x^{v_1^{(j)}},\ldots,x^{v_{k_j-1}^{(j)}}\big)}. \tag{27}$$

Algorithm B ends by making use of the identity

$$\gcd_{\mathbb{Z}}\big(f(x),g(x)\big) = \prod_{u=0}^{s-1} J\,D_u\big(x^{v_1^{(u)}},\ldots,x^{v_{k_u-1}^{(u)}}\big). \tag{28}$$

We justify (28). Recall that we have denoted the left side by $w(x)$. Observe that (25) implies that the expression on the right of (28) divides $w(x)$. By the definition of s, when we arrive at $u = s$ in our recursive construction, (16) fails to hold for every zero ξ of $w(x)$. Therefore, taking $u = s-1$ in (24), (26) and (27) implies that the right-hand side of (28) vanishes at all the zeros of $w(x)$ and to the same multiplicity. As noted earlier, we are considering the case that the contents of $f(x)$ and $g(x)$ are 1. We deduce that (28) holds. Observe that the computation of $\gcd_{\mathbb{Z}}\big(f(x),g(x)\big)$ by expanding the right side of (28) involves $O_{r,H}(1)$ additions of exponents of size $O_{r,H}(n)$. This computation can be done in time $O_{r,H}(\log n)$. Theorem B follows.

References

[1] E. Bombieri, D. Masser, U. Zannier, Anomalous subvarieties - structure theorems and applications, preprint.

[2] J.H. Conway, A.J. Jones, Trigonometric Diophantine equations (On vanishing sums of roots of unity), *Acta Arith.* **30** (1976), 229–240.

[3] R. Crandall, C. Pomerance, *Prime numbers, a computational perspective*, Springer Verlag, New York, 2001.

[4] M. Filaseta, A. Schinzel, On testing the divisibility of lacunary polynomials by cyclotomic polynomials, *Math. Comp.* **73** (2004), 957–965.

[5] G. Hajós, Solution of Problem 41 (Hungarian), *Mat. Lapok* **4** (1953), 40–41.

[6] A.K. Lenstra, Factoring multivariate polynomials over algebraic number fields, *SIAM J. Comput.* **16** (1987), 591–598.

[7] A.K. Lenstra, H.W. Lenstra, Jr., L. Lovász, Factoring polynomials with rational coefficients, *Math. Ann.* **261** (1982), 515–534.

[8] D.A. Plaisted, Sparse complex polynomials and polynomial reducibility, *J. Comput. System Sci.* **14** (1977), 210–221.

[9] A. Schinzel, Reducibility of lacunary polynomials, I, *Acta Arith.* **16** (1969/1970), 123–159.

[10] ———, Reducibility of lacunary polynomials, XII, *Acta Arith.* **90** (1999), 273–289.

[11] ———, *Polynomials with special regard to reducibility*, Encyclopedia of Mathematics and its Applications **77**, Cambridge University Press, 2000.

[12] ———, On the greatest common divisor of two univariate polynomials, I, *A panorama of number theory or the view from Baker's garden (Zürich, 1999)*, 337–352, Cambridge University Press, 2002.

CONSEQUENCES OF THE CONTINUITY OF THE MONIC INTEGER TRANSFINITE DIAMETER

JAN HILMAR

ABSTRACT. We consider the problem of determining the monic integer transfinite diameter $t_M(I)$ for real intervals I of length less than 4. We show that $t_M([0,x])$, as a function in $x > 0$, is continuous, therefore disproving two conjectures due to Hare and Smyth. Consequently, for $n > 2 \in \mathbb{N}$, we define the quantity

$$b_{\max}(n) = \sup_{b > \frac{1}{n}} \left\{ b \,\Big|\, t_M([0,b]) = \tfrac{1}{n} \right\}$$

and give lower and upper bounds of $b_{\max}(n)$. Finally, we improve the lower bound for $b_{\max}(n)$ for $3 \leq n \leq 8$.

1. INTRODUCTION

Let $I \subset \mathbb{R}$ be a closed interval of length less than 4 and $\mathcal{M}_n[x]$ be the set of monic polynomials of degree n with integer coefficients. We define the *monic integer transfinite diameter* $t_M(I)$ of I to be the quantity

$$t_M(I) = \lim_{n \to \infty} \inf_{p_n \in \mathcal{M}_n} \|p_n\|_I^{1/n} . \tag{1}$$

Here $\|p_n\|_I = \sup_{x \in I} |p_n(x)|$ is the supremum norm of the polynomial $p_n(x)$. The problem of determining the monic integer transfinite diameter was first tackled by Borwein, Pinner and Pritsker in [3]. Their techniques were further developed by Hare and Smyth in [4]. The problem is intimately connected to the problem of determining $t_{\mathbb{Z}}(I)$, the *integer transfinite diameter*, where the condition that the polynomials be monic is removed. Interestingly, removing this condition makes the problem much harder, as no exact values of $t_{\mathbb{Z}}(I)$ are known, but $t_M(I)$ can be computed explicitly in some cases. The following lemma is an essential tool in doing so.

Lemma 1 ([3]). *Let $q(x) = a_0 + \cdots + a_d x^d \in \mathbb{Z}_n[x]$ be an irreducible polynomial with $a_d > 1$ and all roots in the closed interval $I \subset \mathbb{R}$ of finite length. Further, assume that $p_n(x) \in \mathcal{M}_n[x]$. Then*

$$a_d^{-\frac{1}{d}} \leq \|p_n\|_I^{\frac{1}{n}} .$$

2000 *Mathematics Subject Classification*. 11C08.

Key words and phrases. Monic integer transfinite diameter, monic integer Chebyshev constant.

The proof of this essentially classical result can be found in [3] or [4] and will be omitted here.

As a consequence, $a_d^{-1/d} \leq t_M(I)$, so that polynomials $q(x)$ as in the lemma are used to determine lower bounds on $t_M(I)$. As a consequence, they are called *obstruction polynomials* for $t_M(I)$ with *obstruction* $a_d^{-1/d}$. Since all obstructions give a lower bound, it is of interest to find the supremum

$$m(I) \;=\; \sup\left\{ a_d^{-1/d}\,\Big|\, q(x) = a_d x^d + \cdots + a_0, a_d > 1 \right\}.$$

Here the supremum is taken over all polynomials with integer coefficients and all roots in the interval I. If the supremum is attained, $m(I)$ is called the *maximal obstruction* for I.

Suppose now we have an interval I with maximal obstruction $m(I)$ and find $p_n(x) \in \mathcal{M}_n$ with $\|p_n\|_I = m(I)^n$. In this case, $m(I) \geq t_M(I) \geq m(I)$, so that we have determined an exact value for $t_M(I)$. Such $p_n(x)$, if it exists, is said to *attain the maximal obstruction*. Some examples of this situation are as follows:

(1) If $I = [0,1]$, then $\frac{1}{2}$ is the maximal obstruction by $q(x) = 2x - 1$. At the same time, $\|x(1-x)\|_I = \frac{1}{4}$, so that $t_M(I) = \frac{1}{2}$.

(2) For an integer $n > 1$, consider $I_n = [0, \frac{1}{n}]$. Then $\frac{1}{n}$ is the maximal obstruction by $q(x) = nx - 1$. At the same time, $\|x\|_{I_n} = \frac{1}{n}$, so that $t_M(I_n) = \frac{1}{n}$.

These are just some examples to illustrate the technique. A more complete list of known values of $t_M(I)$ can be found in [3] and [4].

It was shown in [4] that the maximal obstruction is not always attained by some $p_n(x)$ and explicit conditions for when it cannot be attained were given. The authors conjecture, however, that $t_M(I) = m(I)$ for all I. That this is not the case is a consequence of the continuity of a particular function, proved in Section 2.

2. Continuity of $t_M(x)$, $x \geq 0$

In [4], the authors consider intervals of the form $I = [0, b]$, where, for $1 < n \in \mathbb{N}$, we have $\frac{1}{n} < b < \frac{1}{n-1}$. From $q(x) = nx - 1$, we know that $t_M([0,b]) \geq \frac{1}{n}$ and equality holds in a neighbourhood to the right of $\frac{1}{n}$ (see Theorem 2). Much more interesting is the behaviour of the function

$$t_M(x) \;=\; t_M([0,x]) \tag{2}$$

for $x \geq 0$ to the left of $\frac{1}{n}$, $n > 1$. Hare and Smyth suspected that the function had a discontinuity at $x = \frac{1}{n}$, which would agree with their conjecture that $m(I) = t_M(I)$.

To study the behaviour of $t_M(x)$, it is useful to look back at the classical paper [2] of Borwein and Erdélyi in the theory of the (non-monic) transfinite diameter. In this paper, the authors define the function $t_{\mathbb{Z}}(x)$ in the equivalent way and state that this function is continuous, though without the details of the proof. We will now provide the details for $t_M(x)$.

Let $T_n(x)$ be the nth Chebyshev polynomial on $[-1, 1]$, defined by

$$T_n(x) = \cos(n \arccos x). \tag{3}$$

This can be rewritten as

$$T_n(x) = \tfrac{1}{2}\left[\left(x + \sqrt{x^2 - 1}\right)^n + \left(x - \sqrt{x^2 - 1}\right)^n\right].$$

From this it immediately follows that

$$T_n(x) \leq \left(x + \sqrt{x^2 - 1}\right)^n \text{ for } x \geq 1. \tag{4}$$

We will also need Chebyshev's inequality from [1]:

Lemma 2. *Let $q \in \mathbb{R}[x]$. Then, for $x \in \mathbb{R} \backslash [-1, 1]$,*

$$|q(x)| \leq |T_n(x)| \, \|q\|_{[-1,1]} . \tag{5}$$

We can then prove:

Lemma 3. *Let $b > b_0 > 0$, $p_n \in \mathbb{R}_n[y]$. Then, for every $\delta > 0$, there exists $k_{b,\delta}$, not depending on n, such that*

$$\|p_n\|_{[0,b+\delta]} \leq (1 + k_{b,\delta})^n \, \|p_n\|_{[0,b]}, \tag{6}$$

with $\lim_{\delta \to 0} k_{b,\delta} = 0$ for fixed b.

Proof. Given $p_n \in \mathbb{R}_n[y]$, let $y \in [0, b]$ and $x = \tfrac{2}{b}y - 1$. Then $x \in [-1, 1]$. Put $q_n(x) = p_n(y)$. Then, by Lemma 2, for $x \notin [-1, 1]$, $y \notin [0, b]$, we have

$$|p_n(y)| = |q_n(x)| \leq |T_n(x)| \, \|q_n\|_{[-1,1]}$$
$$= \left|T_n\left(\tfrac{2}{b}y - 1\right)\right| \|p_n\|_{[0,b]} .$$

Note also that

$$\max_{y \in [b, b+\delta]} \left|T_n\left(\tfrac{2}{b}y - 1\right)\right| = \max_{x \in [1, 1+2\frac{\delta}{b}]} |T_n(x)|$$
$$= \|T_n\|_{[1, 1+2\frac{\delta}{b}]} .$$

This clearly implies that

$$\|p_n\|_{[b,b+\delta]} \leq \|T_n\|_{[1, 1+2\frac{\delta}{b}]} \|p_n\|_{[0,b]} .$$

Using inequality (4) above, we see that

$$\|T_n\|_{[1, 1+2\frac{\delta}{b}]} \leq \left(1 + 2\tfrac{\delta}{b}\left(1 + \sqrt{1 + \tfrac{b}{\delta}}\right)\right)^n .$$

The result now follows by letting $k_{b,\delta} = 2\frac{\delta}{b}\left(1 + \sqrt{1 + \frac{b}{\delta}}\right) > 0$ and observing that

$$
\begin{aligned}
\|p_n\|_{[0,b+\delta]} &= \max\left\{\|p_n\|_{[0,b]}, \|p_n\|_{[b,b+\delta]}\right\} \\
&\leq \max\left\{\|p_n\|_{[0,b]}, (1 + k_{b,\delta})^n \|p_n\|_{[0,b]}\right\} \\
&= (1 + k_{b,\delta})^n \|p_n\|_{[0,b]}.
\end{aligned}
$$

□

Using this inequality, we also get that, for $b, \delta > 0$ fixed,

$$
\|p_n\|_{[0,b-\delta]} \geq \|p_n\|_{[0,b]}\left(\frac{1}{1 + k_{b-\delta,\delta}}\right)^n. \tag{7}
$$

Note also that $\lim_{\delta \to 0} k_{b-\delta,\delta} = 0$.

We can now use this to prove

Theorem 1. *The function $t_M(x)$ is continuous on $(0, \infty)$.*

Proof. First, note that $t_M(x)$ is (non-strictly) increasing in x. Let $b \in (0, \infty)$, $\epsilon > 0$ and choose $\delta = \min\{\delta_1, \delta_2\}$, where δ_1 is chosen such that $k_{b,\delta_1} < \frac{\epsilon}{t_M(b)}$ and δ_2 is such that $\frac{k_{b-\delta_2,\delta_2}}{1 + k_{b-\delta_2,\delta_2}} < \frac{\epsilon}{t_M(b)}$.

Let $0 < |b - x| < \delta$. The argument splits into two cases:

(1) Suppose that $0 < b - x < \delta \leq \delta_1$. Since $t_M(x)$ is increasing, we have

$$
\begin{aligned}
0 \leq t_M(x) - t_M(b) &\leq t_M(b + \delta_1) - t_M(b) \\
&= \lim_{n \to \infty}\left(\inf_{p_n \in \mathcal{M}_n[x]} \|p_n\|_{[0,b+\delta_1]}^{1/n} - \inf_{p_n \in \mathcal{M}_n[x]} \|p_n\|_{[0,b]}^{1/n}\right) \\
&\leq \lim_{n \to \infty}\left(\inf_{p_n \in \mathcal{M}_n[x]} k_{b,\delta_1} \|p_n\|_{[0,b]}^{1/n}\right) \\
&= t_M(b) k_{b,\delta_1} < \epsilon.
\end{aligned}
$$

(2) Now assume that $0 < x - b < \delta \leq \delta_2$. Here, we get

$$
\begin{aligned}
0 \leq t_M(b) - t_M(x) &\leq t_M(b) - t_M(b - \delta_2) \\
&= \lim_{n \to \infty}\left(\inf_{p_n \in \mathcal{M}_n[x]} \|p_n\|_{[0,b]}^{1/n} - \inf_{p_n \in \mathcal{M}_n[x]} \|p_n\|_{[0,b-\delta_2]}^{1/n}\right) \\
&\leq \lim_{n \to \infty}\left(\inf_{p_n \in \mathcal{M}_n[x]}\left(\frac{k_{b-\delta_2,\delta_2}}{1 + k_{b-\delta_2,\delta_2}}\right) \|p_n\|_{[0,b]}^{1/n}\right) \\
&= t_M(b)\frac{k_{b-\delta_2,\delta_2}}{1 + k_{b-\delta_2,\delta_2}} < \epsilon.
\end{aligned}
$$

Thus, for $0 < |b - x| < \delta$, we have $|t_M(b) - t_M(x)| < \epsilon$ for any $b \in (0, \infty)$, proving continuity for $x > 0$. □

As mentioned before, Borwein and Erdélyi stated this result for the (non-monic) integer transfinite diameter. In fact, if $\mathcal{A}_n[x] \subseteq \mathbb{R}_n[x]$ and

$$t_{\mathcal{A}}(I) = \lim_{n \to \infty} \inf_{0 \neq p_n \in \mathcal{A}_n[x]} \|p_n\|_I^{\frac{1}{n}}, \tag{8}$$

one can define $t_{\mathcal{A}}(x)$ in the equivalent way and the prove continuity of this function for $x \geq 0$ as in Theorem 1.

The continuity of $t_M(x)$ sheds some light on conjectures made by Hare and Smyth in [4]:

Conjecture 1 (Zero-endpoint interval conjecture). *If $I = [0, b]$ with $b \leq 1$, then $t_M(I) = \frac{1}{n}$, where $n = \max\left(2, \lceil \frac{1}{b} \rceil\right)$ is the smallest integer $n \geq 2$ for which $\frac{1}{n} \leq b$.*

Conjecture 2 (Maximal obstruction implies $t_M(I)$ conjecture). *If an interval I of length less than 4 has a maximal obstruction $m(I)$, then $t_M(I) = m(I)$.*

Theorem 1 clearly shows Conjecture 1 to be false: since $t_M(x)$ is continuous to the left of $\frac{1}{n}, n \in \mathbb{N}$ and $t_M([0, \frac{1}{n}]) = \frac{1}{n}$, we cannot have $t_M([0, b]) = \frac{1}{n+1}$ for all $b < \frac{1}{n}$, as claimed in the conjecture. A further implication is that for these intervals, $t_M(I) \neq m(I)$, contrary to Conjecture 2.

3. The function $b_{\max}(n)$

It turns out that $t_M(x)$ is indeed constant on a large interval to the right of $\frac{1}{n+1}, n \in \mathbb{N}$. We define for $n > 1 \in \mathbb{N}$,

$$b_{\max}(n) = \sup_{b > \frac{1}{n}} \left\{ b \, \big| \, t_M(b) = \frac{1}{n} \right\}. \tag{9}$$

For $n = 1$, this quantity is not finite, as $t_M(I) = 1$ for $|I| \geq 4$ (see [3] for details). For $n = 2$, we can use the results in [4] to obtain $1.26 \leq b_{\max}(2) < 1.328$. For $n > 2$, we have the following:

Theorem 2. *Let $n > 2 \in \mathbb{N}$. Then*

$$\frac{1}{n} + \frac{1}{n^2(n-1)} < b_{\max}(n) \leq \frac{4n}{(2n-1)^2}.$$

Proof. The first inequality follows from the polynomial

$$P_n(x) = x^{n^2-2}(x^2 - nx + 1).$$

This polynomial, first used in [4], was shown to have the following properties:

(1) $P_n(\frac{1}{n}) = \left(\frac{1}{n}\right)^{n^2}$.
(2) $P_n'(\frac{1}{n}) = 0$ and the polynomial has no other extrema in $[0, \frac{1}{n-1}]$.
(3) $P_n(x)$ has a root $\beta_n = \frac{2}{n+\sqrt{n^2-4}} > \frac{1}{n}$, and $|P_n(x)|$ is strictly increasing in $(\beta_n, \frac{1}{n-1})$.

These properties were used in [4] to show that $\|P_n\|_{[0,\frac{1}{n}+\epsilon]} = \left(\frac{1}{n}\right)^{n^2}$ for some $\epsilon > 0$.

Evaluating $P_n(x)$ at $x = \frac{1}{n} + \frac{1}{n^2(n-1)}$ gives

$$\left| P_n\left(\frac{1}{n} + \frac{1}{n^2(n-1)}\right) \right| = \left(\frac{n^2 - n + 1}{n^2 - n}\right)^{n^2} \frac{n^3 - 3n^2 + 2n - 1}{(n^2 - n + 1)^2}.$$

To show that this is indeed less than $\left(\frac{1}{n}\right)^{n^2}$, first note that the sequence

$$\left\{ \left(\frac{n^2 - n}{n^2 - n + 1}\right)^{n^2} \right\}_{n=1}^{\infty}$$

is increasing and tends to e^{-1}. As it is increasing, we clearly have

$$\left(\frac{n^2 - n}{n^2 - n + 1}\right)^{n^2} \geq \left(\frac{2}{3}\right)^4 \quad \text{for } n > 2. \tag{10}$$

Further, note that for all n,

$$\left(\frac{2}{3}\right)^4 > \frac{n^3 - 3n^2 + 2n - 1}{(n^2 - n + 1)^2}. \tag{11}$$

Thus taking (10) and (11) together, we have, for $n > 2$,

$$\left(\frac{n^2 - n}{n^2 - n + 1}\right)^{n^2} > \frac{n^3 - 3n^2 + 2n - 1}{(n^2 - n + 1)^2}.$$

Rearranging now gives the desired result.

For the upper bound, one has to look directly at (6). Suppose we have some $p_d(x) \in \mathcal{M}_d[x]$ such that $\|p_d\|_{I_{\delta_n}}^{\frac{1}{d}} = \frac{1}{n}$ on an interval $I_{\delta_n} = [0, \frac{1}{n-1} - \delta_n]$. Clearly, $\|p_d\|_{[0,\frac{1}{n-1}]}^{\frac{1}{d}} \geq \frac{1}{n-1}$ since $\frac{1}{n-1} \leq t_M([0, \frac{1}{n-1}])$. Thus, using (6), we get

$$\frac{1}{n-1} \leq \frac{1}{n}\left(1 + k_{\frac{1}{n-1} - \delta_n, \delta_n}\right).$$

Using the explicit expression for $k_{\frac{1}{n-1} - \delta_n, \delta_n}$ obtained in the proof of Lemma 3, we see that then $\delta_n \geq \delta_{\min}(n) = \frac{1}{4n^3 - 8n^2 + 5n - 1}$, thus obtaining

$$b_{\max}(n) \leq \frac{1}{n-1} - \delta_{\min}(n) = \frac{4n}{(2n-1)^2}.$$

$\square$

Using the computational methods outlined in Section 5, we get improved lower bounds for $b_{\max}(n)$ for $n = 3, \ldots, 8$. This is done by finding a $b \in (\frac{1}{n}, \frac{1}{n-1})$ as large as possible and a polynomial $P_n(x)$ with $\|P_n\|_{[0,b]} = \frac{1}{n}$, so that then $b_{\max}(n) \geq b$. The polynomials P_n are given in Table 1. The polynomial

P_3 is a corrected version (see Corrigendum to [4]) of one appearing in [4], which did not have the property claimed, while P_4 appears in [4].

$$0.465 \leq b_{\max}(3), \quad 0.303 \leq b_{\max}(4),$$
$$0.230 \leq b_{\max}(5), \quad 0.184 \leq b_{\max}(6),$$
$$0.148 \leq b_{\max}(7), \quad 0.130 \leq b_{\max}(8).$$

As n gets larger, computations become increasingly difficult, as the difference $\frac{1}{n-1} - \frac{1}{n}$ becomes small compared to $\frac{1}{n}$.

Using (6), one can obtain a new lower bound for $t_M([0,b]), b < 1$:

Lemma 4. *Let* $I_b = [0,b]$, $b < 1$ *and let* $n = \min\{m \in \mathbb{N} \mid \frac{1}{m} > b\}$. *Then*

$$t_M(I_b) \geq \max\left\{\frac{1}{n+1}, \frac{b}{2(1 + \sqrt{1 - nb}) - nb}\right\}.$$

Proof. Let $\delta = \frac{1}{n} - b$. As can easily be seen from (6),

$$t_M\left([0, \tfrac{1}{n} - \delta]\right) \geq t_M\left([0, \tfrac{1}{n}]\right) \frac{1}{1 + k_{\frac{1}{n} - \delta, \delta}}$$

$$= \frac{1 - n\delta}{n(1 + n\delta + 2\sqrt{n\delta})}$$

$$= \frac{b}{2 - nb + 2\sqrt{1 - nb}}.$$

Seeing that $\frac{1}{n+1}$ is a larger lower bound for $b \leq b_{\max}(n+1)$, we get the result. $\qquad\square$

4. The Farey Interval Conjecture

Another open conjecture, this one taken from [3], is the following:

Conjecture 3 (Farey Interval Conjecture). *Let* $\frac{p}{q}, \frac{r}{s} \in \mathbb{Q}$ *with* $q, s > 0$ *be such that* $rq - ps = 1$. *Then*

$$t_M\left(\left[\frac{p}{q}, \frac{r}{s}\right]\right) = \max\left\{\frac{1}{q}, \frac{1}{s}\right\}.$$

Computationally, the authors verified the conjecture for denominators up to 21 and it was proved for an infinite family of such intervals in [4]. It is perhaps worth noting that continuity of $t_M(x)$ cannot be used to find a counterexample to this conjecture, as the following argument shows.

Let $n > 1 \in \mathbb{N}$. We will show that we cannot find a Farey Interval of the form $[\frac{k}{n}, \frac{p}{q}]$ with $1 \leq k < n, n = \min\{q, n\}$ and $\frac{p}{q} > b_{\max}^*(n)$, where

$$b_{\max}^*(n) = \sup_{\frac{k}{n} < b}\left\{b \,\big|\, t_M\left(\left[\tfrac{k}{n}, b\right]\right) = \tfrac{1}{n}\right\}.$$

As can easily be derived from the proof of Lemma 3,

$$b^*_{\max}(n) \leq \frac{k(4n^2 + 1)}{n(2n - 1)^2} .$$

If we wanted to use this to derive a counterexample to the Farey Interval Conjecture, we would need $\frac{p}{q} > b^*_{\max}(n)$. Using the Farey property

$$pn - qk = 1,$$ (12)

we can write this as

$$\frac{1 + qk}{qk} > \frac{4n^2 + 1}{(2n - 1)^2} .$$

From this it follows that $1 + qk < 2n + 1$, so that, using (12) again, $1 + qk = 2n$. Now, as $q > n$, it is clear that $k = 1$ for this to hold.

In the case $k = 1$, one can show that the Farey interval is then of the form $\left[\frac{1}{n}, \frac{1+t}{(1+t)n-1} \right]$, $t \in \mathbb{N}$. But no such interval with $\frac{1+t}{(1+t)n-1} > b^*_{\max}(n) = \frac{4n^2+1}{(2n-1)^2 n}$ exists.

The result for the remaining Farey intervals is obtained by using the transformations $x \mapsto m \pm x, m \in \mathbb{Z}$.

5. COMPUTATIONAL METHODS

In order to improve the lower bounds for $b_{\max}(n)$ given in Theorem 2, we need to turn to computational methods to attempt to find a monic polynomial $P(x) \in \mathbb{Z}[x]$ attaining the maximal obstruction on an interval $[0, b)$ with $\frac{1}{n} < b < \frac{1}{n-1}$. These come in two stages:

(1) Using a modification of the LLL algorithm to find factors $f_i(x)$ of $P(x)$.
(2) Using Linear Programming methods first used in [2] in connection with the integer transfinite diameter with additional equality constraints obtained in [4] to determine the exponents α_i.

We will briefly discuss the implementations of both parts of the algorithm.

(1) LLL is an algorithm that, given a basis **b** for a lattice Λ, produces a 'small' basis for Λ with respect to a given inner product $\langle \cdot, \cdot \rangle$. In their modification of the LLL algorithm for monic polynomials introduced in [3], the authors used the Lattice $\mathbb{Z}_n[x]$ with the basis $\mathbf{b} = (1, x, x^2, \cdots x^n)$ and the inner product

$$\langle p_n, q_n \rangle = \int_a^b p_n(x)q_n(x)dx + a_n b_n$$

for $p_n(x) = a_n x^n + \cdots + a_0, q_n(x) = b_n x^n + \cdots + b_0 \in \mathbb{Z}_n[x]$. The additional factor $a_n b_n$ is used to discourage non-monic factors from appearing, and the algorithm usually produces only one monic basis element of degree n.

In practice, we used the following recursive algorithm to identify factors $f_i(x)$ of $P(x)$ for an interval $I = [a, b]$ where the maximal obstruction polynomial $q(x) = a_d x^d + \cdots + a_0$ is known:

(a) Start with $\mathbf{b} = (1, x, x^2, \cdots, x^k)$ for $k = 20$ (in some cases, a larger basis was required initially).

(b) Run LLL, generating a list of factors $l = \{f_i(x)\}$.

(c) Sieve the list by using the condition that if $f_i(x) \mid P(x)$, then the resultant has to satisfy $|\mathrm{Res}(f_i, q)| = 1$ (see [4]).

(d) For every f_i still in l, define

$$\widehat{\mathbf{b}}_i = (1, f_i(x), f_i(x)x, f_i(x)x^2, \cdots f_i(x)x^k)$$

and re-run the LLL Algorithm with this basis, adding new factors to l.

(e) Repeat steps (a)–(d) until no more new factors are found, at which point we return l.

(2) To determine the exponents α_i of $f_i(x), 1 \leq i \leq N$, we use a technique first used by Borwein and Erdélyi in [2]. Given a list of factors $l = \{f_i(x)\}$, one attempts to minimise m subject to

(i) $\sum_{i=1}^{N} \frac{\alpha_i}{\deg f_i} \log |f_i(x)| \leq m - g(x), \quad \text{for } x \in X,$

(ii) $\sum_{i=1}^{N} \alpha_i = 1,$

(iii) $\sum_{i=1}^{N} \frac{\alpha_i}{\deg f_i} \frac{f'(\beta_s)}{f(\beta_s)} = 0, \quad \text{for } 1 \leq s \leq \deg q, \text{where } q(\beta_s) = 0,$

(iv) $\alpha_i \geq 0 \quad \text{for } 1 \leq i \leq N,$

(13)

over a finite set $X \subset I$. Here, $g(x)$ is a function such that

$$g(x) = \begin{cases} 0, & q(x) = 0, \\ \epsilon(x) > 0, & q(x) \neq 0. \end{cases}$$

The use of this function is theoretically not necessary, but is useful when doing computations, as it avoids having to deal with exact values at points where the polynomial does not need to attain the maximal obstruction.

The first two constraints in (13) are taken from [2] with a slight modification to the first, while the third is unique to the monic case and taken from [4]. This is also where we get the final set of constraints:

Let β_s be a root of $q(x)$ and define $\hat{f}_i^{(s)} = \frac{1}{\deg f_i} \log |f_i(\beta_s)|$. If $b_1 = -\frac{1}{d} \log |a_d|, b_2, \ldots, b_l$ is an independent generating set for the $\mathbb{Z}$-lattice generated by $-\frac{1}{d}\log|a_d|$ and the $\hat{f}_i^{(s)}$, let $c_{j,i}^{(s)}$ be such that

$$\sum_{j=1}^{l} c_{j,i}^{(s)} b_j = \hat{f}_i^{(s)}.$$

Then we get the additional conditions, derived in [4]:

$$\sum_{i=1}^{N} c_{j,i}^{(s)} \alpha_i = \begin{cases} -\frac{1}{\deg q}, & j = 1, \\ 0, & j > 1, \end{cases} \qquad \text{for } 1 \le s \le \deg q \qquad (14)$$

Again, we use a recursive algorithm for determining the exponents. Given a set of points X_k, we use (13) and (14) to determine the optimal exponents $\{\alpha_1^{(k)}, \alpha_2^{(k)}, \ldots, \alpha_N^{(k)}\}$ attaining the minimum value m_k. Then, we construct the normalised 'polynomial'

$$P^{(k)}(x) = \prod_{i=1}^{N} f_i(x)^{\frac{\alpha_i^{(k)}}{\deg f_i}},$$

and add its extrema to X_k to obtain X_{k+1}. Starting with a small set of values $X_1 \subset I$, we repeat this procedure until we get $K \in \mathbb{N}$ such that $|m_K - m_{K-1}| < \epsilon$ for required precision $\epsilon > 0$.

Finally, we compute the supremum norm of $P_K(x) \approx e^{m_K}$ on the interval and verify that $\| P_K \|_I = |a_d|^{-\frac{1}{d}}$. One can attempt to find rational approximations of smaller denominator to the exponents, always checking that the obstruction is still attained. The attaining polynomial $P(x)$ is then found by clearing denominators in the exponents of $P_K(x)$.

Acknowledgement

The author would like to thank Kevin Hare for helpful insights and Chris Smyth for guidance and relentless proofreading. Further, I would like to thank the referee for carefully reading the manuscript and catching some mistakes as well as providing useful comments.

TABLE 1. Polynomials used for lower bounds on $b_{\max}(n)$

$$P_3(x) = x^{45944640}(x^{14} - 11406261x^{13} + 47054086x^{12} - 88456310x^{11}$$
$$+100247244x^{10} - 76341256x^9 + 41208853x^8 - 16202606x^7 + 4692047x^6 - 999261x^5$$
$$+154318x^4 - 16766x^3 + 1211x^2 - 52x + 1)^{2450525}$$
$$(x^8 + 14184x^7 - 34944x^6 + 36442x^5 - 20832x^4 + 7041x^3 - 1405x^2 + 153x - 7)^{877415}$$
$$(x^8 + 4842x^7 - 10935x^6 + 10355x^5 - 5317x^4 + 1594x^3 - 278x^2 + 26x - 1)^{2571030}$$
$$(x^8 + 7812x^7 - 18072x^6 + 17561x^5 - 9271x^4 + 2864x^3 - 516x^2 + 50x - 2)^{595980}$$
$$(x^7 - 1233x^6 + 2406x^5 - 1913x^4 + 791x^3 - 179x^2 + 21x - 1)^{1210840}$$
$$(x^5 - 3x^4 + 7x^3 - 11x^2 + 6x - 1)^{1052898}$$

$$P_4(x) = x^{640}(x^5 + 432x^4 - 456x^3 + 179x^2 - 31x + 2)^{47}$$
$$(x^7 + 8760x^6 - 13342x^5 + 8488x^4 - 2784x^3 + 514x^2 - 50x + 2)^{35}$$

$$P_5(x) = x^{1050990}(x^{10} + 5544095x^9 - 9115714x^8 + 6623719x^7 - 2790988x^6$$
$$+751349x^5 - 133974x^4 + 15818x^3 - 1192x^2 + 52x - 1)^{78796}$$
$$(x^6 + 4950x^5 - 4605x^4 + 1698x^3 - 310x^2 + 28x - 1)^{21825}$$

$$P_6(x) = x^{5232473}(x^5 + 1260x^4 - 852x^3 + 215x^2 - 24x + 1)^{118824}$$
$$(x^7 - 140190x^6 + 132517x^5 - 51966x^4 + 10819x^3 - 1261x^2 + 78x - 2)^{200917}$$

$$P_7(x) = x^{44}(x^5 + 3472x^4 - 1826x^3 + 358x^2 - 31x + 1)$$

$$P_8(x) = x^{12288}(x^2 - 8x + 1)^{246}(x^4 - 576x^3 + 208x^2 - 25x + 1)^{741}$$

REFERENCES

[1] P. Borwein, T. Erdélyi, *Polynomials and polynomial inequalities*, Springer, New York, 1995.

[2] ———, ———, The integer Chebyshev problem, *Math. Comp.* **65** (1996), no. 214, 661–681.

[3] P. Borwein, C.G. Pinner, I.E. Pritsker, Monic integer Chebyshev problem, *Math. Comp.* **72** (2003), 1901–1916.

[4] K. Hare, C. Smyth, The monic integer transfinite diameter, *Math. Comp.* **75** (2006), no. 256, 1997–2019; Corrigendum *ibid.* (to appear).

NONLINEAR RECURRENCE SEQUENCES AND LAURENT POLYNOMIALS

ANDREW HONE

ABSTRACT. Rational recurrence relations whose iterates are Laurent polynomials in the initial data with integer coefficients are said to have the Laurent property. Examples of such recurrences have appeared in a wide variety of contexts, including integer sequences in number theory (elliptic divisibility and Somos sequences), integrable systems (discrete soliton equations), and algebraic combinatorics (in particular, Fomin and Zelevinsky's theory of cluster algebras). For suitable initial data, recurrences that have this Laurent property generate integer sequences, and if they also have a rational invariant then their iteration produces infinitely many solutions of an associated Diophantine equation. After presenting a classification of certain second-order nonlinear recurrences with the Laurent property, we discuss which of them are explicitly solvable (or integrable, in a suitable sense). Subsequently the solution of the initial value problem is obtained for a family of third-order nonlinear recurrences that share the Laurent property. For suitable choices of initial data, the iteration of these recurrences furnishes infinitely many integer points on a pencil of cubic surfaces.

1. INTRODUCTION

Integer sequences that are generated by linear recurrence relations have a long history. As well as being the subject of intensive study in number theory, they find modern applications in computer science and cryptography [16]. However, the theory of nonlinear recurrence sequences has only begun to be developed relatively recently. The iteration of a kth-order nonlinear recurrence relation of the form

$$x_{n+k} = F(x_n, x_{n+1}, \ldots, x_{n+k-1}) \equiv F(\mathbf{x}_n), \tag{1.1}$$

where $\mathbf{x}_n = (x_n, x_{n+1}, \ldots, x_{n+k-1})$, is equivalent to iterating the map

$$\varphi : \begin{pmatrix} x_0 \\ x_1 \\ \vdots \\ x_{k-1} \end{pmatrix} \mapsto \begin{pmatrix} x_1 \\ x_2 \\ \vdots \\ F(\mathbf{x}_0) \end{pmatrix}, \tag{1.2}$$

2000 *Mathematics Subject Classification.* 11B37, 11B83, 11D25, 33E05, 37J35.
Key words and phrases. Laurent polynomials, nonlinear recurrence sequences, integrable maps.

which is just a particular sort of discrete dynamical system (in $\mathbb{R}^k$ or $\mathbb{C}^k$, say). If we want (1.1) to generate sequences of integers, then we can certainly choose $F \in \mathbb{Z}[\mathbf{x}]$, and take initial data in $\mathbb{Z}^k$, but in general the corresponding map (1.2) will not have a unique inverse. Furthermore, in that case such sequences will generically exhibit doubly exponential growth, i.e., $\log |x_n|$ will grow exponentially with n.

A simple example in the above class is the quadratic map defined by the recurrence

$$x_{n+1} = x_n^2 + c \qquad (1.3)$$

with a parameter c, which is a prototypical model of chaos [12]. In order to get an integer sequence, it suffices to take $x_0, c \in \mathbb{Z}$. However, note that only the special cases $c = 0, -2$ are exactly solvable [11], and these are also the only values of c for which (1.3) admits another map that commutes with it (see [58] and references). The theory of linear recurrence sequences relies heavily on the fact that they are explicitly solvable. Thus it is natural to look for nonlinear recurrences that are also solvable or 'integrable' in some sense.

In this article we shall be concerned with the case that the map (1.2) is birational and equivalent to a recurrence of the particular form

$$x_{n+k}\, x_n = f(x_{n+1}, \ldots, x_{n+k-1})\,, \qquad (1.4)$$

where f is a polynomial in its arguments. It turns out that among rational recurrences of this kind there is a large class that generate integer sequences from suitable initial data. One of the first known examples of this type was found by Michael Somos from an investigation of the combinatorics of elliptic theta functions, and the sequences that he found are known as Somos sequences. Subsequently, it was realized that the recurrence relations for Somos sequences have a very special property: the iterates are polynomials in the initial data and their inverses (Laurent polynomials) with integer coefficients. Recurrence relations with this property are said to exhibit the *Laurent phenomenon*, and examples of this phenomenon have surfaced in a wide variety of different contexts, ranging from cluster algebras in algebraic combinatorics [17] to integrable partial difference equations in soliton theory [62].

The aim of this article is to describe a fascinating area of overlap between number theory, algebraic combinatorics and integrable systems, with the common theme being recurrences and the sequences of integers or polynomials that they generate. The discussion is led by examples, and technical details are kept to a minimum. The next section provides a brief review of Somos sequences and the Laurent property, in order to put them into a wider context and hopefully whet the reader's appetite. In Section 3 we outline some notions of what it means for a map to be integrable, including an arithmetical criterion for integrability due to Halburd [21], which gives a simple way to detect whether a recurrence should be explicitly solvable or not. We also explain an interesting property of recurrences with the Laurent property:

they have rational invariants (conserved quantities), then they can generate infinitely many solutions of associated Diophantine equations. Section 4 illustrates these ideas by considering second-order recurrences of the form (1.4) for $k = 2$; a classification of those with the Laurent property is presented.

If a conic has an integer point, then there are well known sufficient conditions for it to have infinitely many integer points [38]. In the fifth section we present an analogous result for a family of cubic surfaces, which arises from considering a two-parameter family of third-order nonlinear recurrences. These third-order recurrences both have the Laurent property, and have two rational conserved quantities that define a pencil of cubic surfaces. The explicit solution of the initial value problem for these recurrences is derived, and it is shown that, for suitable parameter values and initial data, the iterates of the nonlinear recurrence produce infinitely many integer points on the associated cubic surfaces. The final section contains some conclusions.

2. SOMOS SEQUENCES AND THE LAURENT PHENOMENON

Somos made the numerical observation that, with the four initial data $x_0 = x_1 = x_2 = x_3 = 1$, the fourth-order recurrence

$$x_{n+4}\, x_n = x_{n+3}\, x_{n+1} + x_{n+2}^2 \tag{2.1}$$

produces a sequence of integers [52], that is

$$1, 1, 1, 1, 2, 3, 7, 23, 59, 314, 1529, 8209, 83313, \dots \; . \tag{2.2}$$

He noticed the same sort of thing for the Somos-k recurrences

$$x_{n+k}\, x_n = \sum_{j=1}^{[k/2]} x_{n+k-j}\, x_{n+j} \tag{2.3}$$

with $k = 5, 6, 7$: if all initial values are 1 then the iterates are all integers; but when $k = 8$ denominators appear.

To begin with, various elementary ways were found to prove that the sequence (2.2) consists of integers, but eventually it was realized that there are deeper reasons behind this fact. If one takes the initial values x_0, x_1, x_2, x_3 as variables, then the next two iterates are

$$x_4 = \frac{x_3 x_1}{x_0} + \frac{x_2^2}{x_0}, \qquad x_5 = \frac{x_3 x_2}{x_0} + \frac{x_2^3}{x_1 x_0} + \frac{x_3^2}{x_1},$$

which are polynomials in these variables and in the inverse powers x_0^{-1}, x_1^{-1}. One can regard such expressions as Laurent series that have been truncated, and hence they are referred to as *Laurent polynomials*. It is clear that the iterates x_6 and x_7 are also Laurent polynomials, and they have integer coefficients, so they lie in the ring $\mathbb{Z}\big[x_0^{\pm 1}, x_1^{\pm 1}, x_2^{\pm 1}, x_3^{\pm 1}\big]$. However, what is surprising is that obtaining x_8 from (2.1) requires division by x_4, but a remarkable cancellation occurs, so that x_8 is also a Laurent polynomial in the variables , x_1, x_2, x_3, and so are all the subsequent iterates. A recurrence whose iterates are Laurent polynomials in the initial data with integer coefficients is said to have the *Laurent property*. By inserting coefficients on the right hand side of (2.1), one obtains the general Somos-4 recurrence, that is

$$x_{n+4}\,x_n = \alpha\,x_{n+3}\,x_{n+1} + \beta\,x_{n+2}^2\,, \qquad (2.4)$$

and this also has the Laurent property; more precisely,

$$x_n \in \mathbb{Z}\big[x_0^{\pm 1}, x_1^{\pm 1}, x_2^{\pm 1}, x_3^{\pm 1}, \alpha, \beta\big]$$

holds for all n. It is an obvious consequence of a recurrence having the Laurent property that if all the initial data have the value 1, and if the coefficients take integer values, then all the iterates must be integers. Subsequently, a variety of other recurrences of the form (1.4) with the Laurent property were discovered [20, 47].

Somewhat earlier, when Mills, Robbins and Rumsey made their study of Charles Dodgson's condensation method for computing determinants [36], which led to the alternating sign matrix conjecture [5], they considered the case $\alpha = 1$ of the recurrence

$$D_{\ell,m,n+1}D_{\ell,m,n-1} = \alpha\,D_{\ell+1,m,n}D_{\ell-1,m,n} + \beta\,D_{\ell,m+1,n}D_{\ell,m-1,n}\,, \qquad (2.5)$$

where the indices ℓ, m, n range over a suitable subset of $\mathbb{Z}^3$, and observed that it generates Laurent polynomials in the initial data. Thus the equation (2.5) became known within the algebraic combinatorics community, where it is usually referred to as the octahedron recurrence [43]; while in the theory of integrable systems it is known as a particular form of the discrete Hirota equation [62], a partial difference equation with soliton solutions. In soliton theory, equations that are quadratic in the dependent variables, such as (2.3) and (2.5), are referred to as being in Hirota bilinear form, and they can by rewritten very elegantly using certain bilinear operators that Hirota introduced [24].

The Laurent property has reappeared most recently in Fomin and Zelevinsky's theory of cluster algebras (see [19] and references). A cluster algebra is a commutative algebra with a distinguished set of generators that live in clusters of k elements attached to the vertices of a k-regular tree. At any two adjacent vertices the clusters $\mathcal{C} = \{x_n, x_{n+1}, \dots, x_{n+k-1}\}$ and $\mathcal{C}' = \{x_{n+1}, x_{n+2}, \dots, x_{n+k}\}$ have all but one element in common, and the exchange relation between such an adjacent pair of clusters has the form

$$x_{n+k}\,x_n = c_1\,M_1(x_{n+1}, \dots, x_{n+k-1}) + c_2\,M_2(x_{n+1}, \dots, x_{n+k-1}) \qquad (2.6)$$

for suitable monomials M_j and coefficients c_j; this is just a special case of (1.4). In [17], the general machinery of cluster algebras was shown to be extremely effective in proving the Laurent property for a wide variety of recurrences, mostly (but not all) of the form (2.6). In particular, Fomin and Zelevinsky there gave the first proof of the Laurent property for the octahedron (discrete Hirota) recurrence (2.5). Subsequently, Speyer has developed a combinatorial model to prove more detailed properties of the Laurent polynomials generated by this recurrence. In particular, he has shown that all the coefficients of these Laurent polynomials are 1, and the analogous property for the Laurent polynomials generated by Somos-4 (that all coefficients are positive) is a corollary of this [53]. Propp directed a group of extremely talented students in the REACH project (Research Experiences in Algebraic Combinatorics at Harvard), who developed a different combinatorial model to show that the integers (2.2) count the number of perfect matchings of a sequence of graphs.

Actually, there is another way to demonstrate that the sequence (2.2) consists of integers, which is to take the sequence of rational numbers

$$0, \ -1, \ 1/4, \ -5/9, \ -20/49, \ 116/529,$$
$$-3741/3481, \ 8385/98596, \ -239785/2337841, \ldots \tag{2.7}$$

that are the X coordinates of the odd multiples of the point $(0, 1)$ on the elliptic curve

$$Y^2 = 4X^3 - 4X + 1.$$

The denominators in (2.7) are always perfect squares, and they are precisely the squares of the integers (2.2). It was understood in unpublished work of several number theorists that the iterates of Somos-4 or Somos-5 recurrences are associated with a sequence of points $P_0 + nP$ on an elliptic curve[1] E— for example, compare the discussion of Zagier [63] with the results of Elkies quoted in [7]. The algebraic part of the construction was detailed in the thesis of Swart [55] (who also mentions unpublished results of Nelson Stephens), while the author has given the explicit analytic solution of the initial value problem for the general Somos-4 [26] and Somos-5 [29] recurrences in terms of the Weierstrass sigma function, and van der Poorten has recently presented another construction based on the continued fraction expansion of the square root of a quartic [40]. For various results on Somos sequences of higher order, see [4, 8, 35, 41, 42].

A particular class of Somos sequences was already studied in detail in the 1940s, namely the *elliptic divisibility sequences* (EDS) considered by Morgan Ward [60, 61]. These sequences are generated by a special case of the recurrence (2.4), namely

$$x_{n+4}\, x_n = x_2^2\, x_{n+3}\, x_{n+1} - x_1\, x_3\, x_{n+2}^2\,, \tag{2.8}$$

with initial data

$$x_1 = 1, x_2, x_3, x_4 \in \mathbb{Z} \qquad \text{with} \qquad x_2 \mid x_4\,.$$

The iterates x_n of an EDS are integers with the divisibility property that $x_m \mid x_n$ whenever $m \mid n$, and they correspond to the multiples nP of a point P on an elliptic curve E, so they are given by values of the division polynomials of E (for a description of these see, e.g., Exercise 3.7 in [50]).

---

[1]Historical notes can be found at `http://www.math.wisc.edu/~propp/somos.html`

The arithmetical properties of EDS and Somos sequences—in particular the distribution of primes therein—are a subject of current interest [10, 15, 49, 51]. Some of these properties are reviewed in the book [16] (see section 1.1.20, for instance), where it is suggested that such bilinear recurrence sequences are natural generalizations of linear ones, with many analogous features. For example, the Fibonacci sequence is a divisibility sequence generated by the linear recurrence $F_{n+1} = F_n + F_{n-1}$ with initial values $F_0 = 0$, $F_1 = 1$, and the even index terms $x_n = F_{2n}$ satisfy the bilinear recurrence

$$F_{2n+4} F_{2n-4} = 9F_{2n+2} F_{2n-2} - 8F_{2n}^2,$$

so this is a degenerate EDS corresponding to the case where the curve E becomes singular.

3. CONSERVED QUANTITIES AND INTEGRABLE MAPS

In order to completely integrate a differential equation of order k (or an equivalent system of k coupled first-order equations), one usually requires sufficiently many conserved quantities. Certainly if there are $k - 1$ independent such quantities, then this will be enough to reduce the problem to a single quadrature, but sometimes one does not need as many as this. In particular, in the traditional setting of classical Hamiltonian mechanics one has an even number $k = 2N$ of first-order equations (Hamilton's equations), and it turns out that only $k/2 = N$ suitable conserved quantities are required.

More precisely, in the setting of classical mechanics one has a real symplectic manifold $\mathcal{M}$ equipped with a closed, nondegenerate two-form ω (the symplectic form). For a real-valued function h on $\mathcal{M}$, its associated Hamiltonian vector field X_h is defined by setting the contraction of ω by X_h to be $i_{X_h}\omega = -dh$. In mechanics, one is usually interested in a particular choice of function h, called the Hamiltonian, which physically corresponds to the total energy, and Hamilton's equations for this h are the k equations

$$\dot{\mathbf{x}} = X_h(\mathbf{x}), \qquad (3.1)$$

where $\mathbf{x} = \mathbf{x}(t)$ denotes a point in $\mathcal{M}$ and the overdot means the time derivative d/dt. The system is said to be *completely integrable* (in the sense of Liouville-Arnold) if there are $k/2 = N$ functionally independent quantities $h_1 = h, h_2, \ldots, h_N$ that are in involution, meaning that

$$\omega(X_{h_j}, X_{h_k}) = 0$$

for all j, k. The latter condition implies that all the corresponding pairs of Hamiltonian vector fields X_{h_j}, X_{h_k} commute with one another, and also that each h_j is a conserved quantity (first integral) for the system, i.e., $\dot{h}_j = 0$ for $j = 1, \ldots, N$. In that case, a theorem of Liouville says that the system (3.1) can be integrated by quadratures.

For a completely integrable system, it is clear that each trajectory $\mathbf{x}(t) \in \mathcal{M}$ generated by the flow (3.1) lies on an N-dimensional level set defined by

$h_1 = \text{constant}, \ldots, h_N = \text{constant}$, where the value of these constants is fixed by the initial data $\mathbf{x}(0)$. Arnold's reformulation of Liouville's theorem includes the statement that each compact, connected level set of the first integrals is diffeomorpic to an N-dimensional torus, and describes the local coordinates (action-angle variables) in the neighbourhood of such a torus [1]. It is also possible to formulate Hamiltonian mechanics and complete integrability in the more general setting of Poisson manifolds, equipped with a (possibly degenerate) Poisson bracket [9]; in particular this allows k, the dimension of $\mathcal{M}$, to be odd. However, the symplectic setting will be sufficient for our purposes here.

After this brief excursion into Hamiltonian systems, we must return to the main topic of interest, namely recurrences. While the complete integrability of differential equations has a long history, the corresponding notions for discrete systems (difference equations, maps and recurrences) have only been introduced quite recently. Indeed, discrete integrable systems are the subject of much ongoing research (see [3, 54] for a range of recent results) and it is fair to say that there is no single definition of a 'discrete integrable system' that applies universally. However, here we shall discuss various definitions which are appropriate in the context of recurrences, as well as some of the relationships between them.

As already mentioned in the introduction, iterating a kth-order recurrence relation is equivalent to iterating a map in k dimensions, φ say. If $k = 2N$ is even and φ preserves a symplectic form ω (so that the pullback $\varphi^*\omega = \omega$), then we say that this is a symplectic map. In this setting, there is a direct analogue of the Liouville theorem [58], and thus it makes sense to define symplectic maps [6] or correspondences [59] to be *completely integrable* (in the Liouville-Arnold sense) when they have N independent conserved quantities (invariants) that are in involution.

Somos-4 recurrences furnish a very simple example of an integrable symplectic map, via the introduction of some new variables. Upon setting

$$u_n = \frac{x_{n-1}x_{n+1}}{x_n^2}, \tag{3.2}$$

in terms of the x_n satisfying (2.4), we get the second-order recurrence

$$u_{n+1}u_n^2 u_{n-1} = \alpha u_n + \beta, \tag{3.3}$$

which is equivalent to a birational map of the plane ($k = 2N = 2$), that is

$$\varphi : \begin{pmatrix} u \\ v \end{pmatrix} \mapsto \begin{pmatrix} v \\ (\alpha v + \beta)/(uv^2) \end{pmatrix} \tag{3.4}$$

which (away from the singularities) is defined on $\mathbb{R}^2$ (or $\mathbb{C}^2$) and preserves the symplectic form

$$\omega = (uv)^{-1}\, du \wedge dv. \tag{3.5}$$

The second-order recurrence (3.3) has the conserved quantity

$$J = u_n u_{n-1} + \alpha \left(\frac{1}{u_n} + \frac{1}{u_{n-1}} \right) + \frac{\beta}{u_n u_{n-1}} \qquad (3.6)$$

(invariant with n), which is equivalent to a conserved quantity for the map (3.4), and so (because we only require one invariant for Liouville's theorem to apply when $N = 1$) it is a completely integrable map.

¿From the expression (3.6) we see that each level set $J = $ constant is a quartic curve of genus one given by

$$u^2 v^2 - Juv + \alpha(u+v) + \beta = 0 \,,$$

and uniformizing this curve leads to the formula $u_n = \wp(z) - \wp(z_0 + nz)$ for the iterates of (3.4) in terms of the Weierstrass $\wp$ function of a birationally equivalent cubic curve $Y^2 = 4X^3 - g_2 X - g_3$, for suitable g_2, g_3, z_0, z (see [26, 29] for details). Figure 1 shows a plot of several different orbits for the map (3.4), obtained by taking various different starting points (u_0, u_1) and plotting successive pairs of iterates (u_{n-1}, u_n) in the (u, v) plane. Each starting point gives a different value of J and hence a different level set defined by (3.6).

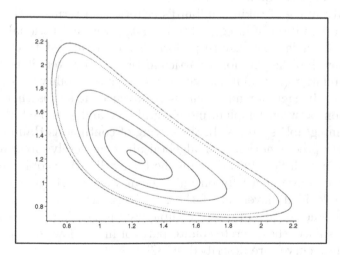

FIGURE 1. *A family of orbits for (3.4), the map of the plane associated with Somos-4.*

Recently, Halburd has proposed an alternative, arithmetical, definition of integrability in the context of rational recurrences (or rational maps). For non-zero $x \in \mathbb{Q}$ written in lowest terms as $x = p/q$ for $p, q \in \mathbb{Z}$, its *height* is $H(x) = \max\{|p|, |q|\}$, while its *logarithmic height* is $h(x) = \log H(x)$. According to [21] a rational recurrence defined over $\mathbb{Q}$ is said to be *Diophantine integrable* if the logarithmic height $h(x_n)$ of the iterates grows no faster than a polynomial in n. According to this definition, the only integrable first-order recurrences with constant coefficients are the Möbius transformations

$$x_{n+1} = \frac{ax_n + b}{cx_n + d}, \qquad ad - bc \neq 0. \tag{3.7}$$

These are the birational automorphisms of the projective plane ($x_n \in \mathbb{P}^1$). The recurrence (3.7) can be exactly solved upon linearization via $x_n = p_n/q_n$ to give the system

$$\begin{pmatrix} p_{n+1} \\ q_{n+1} \end{pmatrix} = \begin{pmatrix} a & b \\ c & d \end{pmatrix} \begin{pmatrix} p_n \\ q_n \end{pmatrix}$$

(which is equivalent to a second-order scalar linear equation by elimination of either p_n or q_n), and for $a, b, c, d \in \mathbb{Q}$ the logarithmic heights $h(x_n)$ of all aperiodic rational orbits grow linearly in n. This should be contrasted with the case $c = 0$ of the quadratic recurrence (1.3): $x_{n+1} = x_n^2$ is explicitly solvable, with the solution being $x_n = x_0^{2^n}$, and it is also linearized by setting $w_n = \log x_n$, but it is not Diophantine integrable because $h(x_n) = 2^n h(x_0)$ for $x_0 \in \mathbb{Q}$. Here we consider only autonomous recurrences (where the coefficients are constant), but the definition of Diophantine integrability applies equally well to non-autonomous recurrences whose coefficients depend on n, such as the discrete Painlevé equations [45].

A huge advantage of taking Halburd's criterion, in terms of the growth of heights, as a definition of integrability is that, given any rational recurrence defined over $\mathbb{Q}$, it is very easy to perform a numerical test to check the asymptotic growth of heights for a particular orbit. However, it is not clear *a priori* what relation (if any) it has with complete integrability in the Liouville-Arnold sense. In fact, for autonomous recurrences it seems that there is a direct relation between Diophantine integrability and so called algebraically completely integrable systems. In the setting of continuous Hamiltonian systems there is a precise notion of an algebraically completely integrable system [57], for which each flow linearizes on the (generalized) Jacobian of an algebraic curve, or more generally on an Abelian variety [31] that corresponds to the (complexified) level set of the first integrals. Quite recently, some discrete analogues of algebraically completely integrable systems have been found; for instance, the discrete counterparts of linear flows on the Jacobians of hyperelliptic curves are described in [32].

Let us consider the simplest example of second-order recurrences, that are equivalent to birational maps of the plane. If such a map is completely integrable *à la* Liouville-Arnold, then it must preserve a symplectic form ω and have a conserved quantity. If the further requirement of algebraic complete integrability is imposed, then the conserved quantity must be a rational function, so that the level sets are algebraic varieties (in this case, plane curves). Now the map induces an automorphism on the plane curve corresponding to each level set, and if it is non-trivial, i.e., infinite order, then by the Hurwitz theorem the genus of the curve must be at most one. In the case where the map is defined over $\mathbb{Q}$, the logarithmic heights of rational points grow linearly in n on a rational curve and quadratically in n on an elliptic one [50], and hence the map is Diophantine integrable. Thus for birational maps of the plane we see that Diophantine integrability is necessary for algebraic integrability, but whether the converse should hold is not so immediately clear. For non-trivial third-order recurrences, similar arguments show that if there are two independent rational conserved quantities then the corresponding map is Diophantine integrable. However, in three dimensions there is the further subtlety that the generic level set need not be irreducible: we shall see an example of this in Section 5. In higher dimensions there are many more possibilities.

We should also mention that there are several other integrability criteria for nonlinear recurrences. For example, Hietarinta and Viallet have considered the degrees d_n of the iterates as rational functions of initial data, and proposed that the algebraic entropy $\lim_{n \to \infty} (\log d_n)/n$ should vanish for integrable maps [23]. Also, Roberts and Vivaldi have studied the distribution of orbit lengths in rational maps of the plane reduced to finite fields $\mathbb{F}_p$ for different primes p, and thence used the Hasse-Weil bound to identify algebraically integrable cases of such maps [46].

3.1. Conserved quantities and Diophantine equations.

There is one additional feature that was not mentioned in our earlier discussion of the Somos-4 recurrence, namely the fact that it generates solutions of a quartic Diophantine equation in four variables. If we rewrite the formula (3.6) for the conserved quantity J in terms of the original variables x_n, we obtain the equation

$$\alpha \left(x_{n-1} x_{n+1}^3 + x_n^3 x_{n+2} \right) + \beta \, x_n^2 x_{n+1}^2 + x_{n-1}^2 x_{n+2}^2 = J \, x_{n-1} x_n x_{n+1} x_{n+2} \,.$$
(3.8)

For fixed J, this defines a quartic threefold which can be viewed as a fibre bundle over the elliptic curve defined by (3.6), with the transformations

$$x_n \mapsto A \, x_n, \qquad x_n \mapsto B^n \, x_n$$

(for arbitrary non-zero A, B) generating gauge symmetries along the fibres. Alternatively, (3.8) can be considered as a quartic Diophantine equation. Given coefficients α, $\beta \in \mathbb{Z}$ (or in $\mathbb{Q}$), if the Somos-4 recurrence (2.4) with a set of integer initial data (x_0, x_1, x_2, x_3) generates a non-periodic sequence of iterates satisfying $x_n \in \mathbb{Z}$ for all n, then there are infinitely many integer quadruples $(x_{n-1}, x_n, x_{n+1}, x_{n+2})$ that satisfy (3.8), with $J \in \mathbb{Q}$ being uniquely determined as long as all the initial data are non-zero. This is a particular instance of a general feature shared by all recurrences that both have the Laurent property and possess a rational invariant: generically, the orbit of suitable initial data will generate infinitely many solutions of an associated Diophantine equation.

Lemma 3.1. *Suppose that a kth-order rational recurrence of the form (1.1) has coefficients in $\mathbb{Q}[\mathbf{c}]$ (for some set of parameters $\mathbf{c}$) and has the Laurent property, i.e., $x_n \in \mathbb{Z}[x_0^{\pm 1}, x_1^{\pm 1}, \ldots, x_{k-1}^{\pm 1}, \mathbf{c}]$ for all n. Suppose further that this recurrence also has a rational invariant given by*

$$K = \frac{f_1(x_n, x_{n+1}, \ldots, x_{n+k-1}, \mathbf{c})}{f_2(x_n, x_{n+1}, \ldots, x_{n+k-1}, \mathbf{c})}$$

for $f_1, f_2 \in \mathbb{Z}[x_n, x_{n+1}, \ldots, x_{n+k-1}, \mathbf{c}]$. If $f_2 \neq 0$ for some fixed integer values of $\mathbf{c}$ and initial data $x_j = 1$ or -1 for $j = 0, \ldots, k-1$, then the value of $K \in \mathbb{Q}$ is fixed, and the recurrence generates infinitely many integer solutions of the Diophantine equation

$$f_1(x_n, x_{n+1}, \ldots, x_{n+k-1}, \mathbf{c}) = K \, f_2(x_n, x_{n+1}, \ldots, x_{n+k-1}, \mathbf{c})$$

as long as the corresponding orbit is aperiodic.

The integer sequence (2.2) provides a particular example of the above result: setting $\alpha = \beta = 1$, the initial data $1, 1, 1, 1$ produce the value $J = 4$ in (3.8), and for $n \geq 0$ any four adjacent terms of this increasing sequence provide a distinct solution of the equation. In [56] it is proved that the iterates of the general Somos-4 recurrence satisfy the stronger property that $x_n \in \mathbb{Z}[x_0^{\pm 1}, x_1, x_2, x_3, \alpha, \beta, (\alpha^2 + \beta J)]$ for $n \geq 0$, which yields a broader set of sufficient criteria for integer sequences. In the following sections we will see analogous results for some other recurrences.

4. SECOND-ORDER RECURRENCES

In this section we consider second-order recurrences of the form (1.4), and present a classification result for these. Note that, in what follows, we exclude those recurrences of the form

$$x_{n+1} \, x_{n-1} = c x_n^d ,$$

which generate Laurent monomials. This trivial case is not Diophantine integrable for $d \geq 3$, but it is explicitly solvable for all d, and can be linearized via the substitution $w_n = \log x_n$ to give $w_{n+1} - d w_n + w_{n-1} = \log c$.

Theorem 4.1. *Let $f(x) \in \mathbb{Z}[x]$, with f being of degree d in x and having the form*

$$f(x) = x^M F(x)$$

for a non-constant polynomial F such that $F(0) \neq 0$. The recurrence

$$x_{n+1} \, x_{n-1} = f(x_n) \qquad (4.1)$$

possesses the Laurent property (that is, $x_n \in \mathcal{R} := \mathbb{Z}[x_0^{\pm 1}, x_1^{\pm 1}, \mathbf{c}]$ for all $n \in \mathbb{Z}$, where $\mathbf{c}$ denotes the coefficients of f) if and only if one of the following three cases holds:
(i) $M = 0$ and f satisfies

$$f(x) = \pm \lambda^{-d/2} x^d f(\lambda/x), \quad \text{with } \lambda = f(0); \qquad (4.2)$$

(ii) $M = 1$ *and the polynomial* F *(of degree* $D = d - 1$*) satisfies*

$$F(x) = \pm \widehat{\lambda}^{-D} x^D F\left(\widehat{\lambda}^2/x\right), \quad \text{with } \widehat{\lambda} = F(0); \qquad (4.3)$$

(iii) $M \geq 2$ *and* F *is an arbitrary non-constant polynomial.*

Remark. The above result was first given (in a slightly different form) in [28]. Case (i) ($f(0) \neq 0$) was also found in 2001 by Speyer, and his proof of the Laurent property is reproduced by Musiker [39]. This case is also covered by the Caterpillar Lemma in [17].

Proof. The proofs for cases (i) and (iii) were described in [28], but most of the details for case (ii) were omitted, so here we mostly concentrate on this case. Doing modular arithmetic in the ring $\mathcal{R}$, which is a unique factorization domain, it is straightforward to show that the stated conditions (4.2) and (4.3) are necessary for the Laurent property in cases (i) and (ii) respectively, while there are no conditions in case (iii). To see why (4.2) is also sufficient in case (i), it suffices to do induction, including as part of the hypothesis that any adjacent pair of iterates x_{j-1}, x_j for $1 \leq j \leq n$ are coprime elements of $\mathcal{R}$. In case (iii) it is even easier: not only $x_n \in \mathcal{R}$ but also x_n/x_{n-1} and $x_{n+1} x_{n-1}/x_n^2 \in \mathcal{R}$ for all n.

For case (ii), first of all note that the recurrence

$$x_{n+1} x_{n-1} = x_n F_n, \qquad F_n := F(x_n) \qquad (4.4)$$

immediately implies the third-order relation

$$x_{n+2}\, x_{n-1} = F_{n+1}\, F_n \qquad (4.5)$$

for all n. Now take as the inductive hypothesis that

$$x_j = G_{j+1}\, G_{j+2}$$

for $0 \leq j \leq n$, where $G_j \in \mathcal{R}$ are given recursively by

$$G_{j+2}\, G_{j-1} = F_{j-1} \qquad (4.6)$$

for the same range of j, starting from $G_1 = 1$, $G_2 = x_0$, $G_3 = x_1 x_0^{-1}$ (which are all units in $\mathcal{R}$). Assume also that any three adjacent terms G_j, G_{j+1}, G_{j+2} are pairwise coprime in this range. By using (4.5), the next iterate of (4.4) is

$$x_{n+1} = F_n F_{n-1} x_{n-2}^{-1} = F_{n-1} F_n/G_{n-1} G_n = G_{n+3} G_{n+2}$$

with $G_{n+2} \in \mathcal{R}$ (by hypothesis) and $G_{n+3} = F_n/G_n$, in accordance with (4.6) for $j = n+1$. So to have $x_{n+1} \in \mathcal{R}$, it suffices to show that $F_n \equiv 0$ (mod G_n) so that $G_{n+3} \in \mathcal{R}$. Now $x_{n-1} = G_n G_{n+1} \equiv 0$ and similarly $x_{n-2} \equiv 0$ (mod G_n), and by (4.5) we have $F_n = F(F_{n-1} F_{n-2}/x_{n-3})$, so that mod G_n we get

$$F_n \equiv F\left(\widehat{\lambda}^2/x_{n-3}\right) \equiv \pm \frac{\widehat{\lambda}^D F_{n-3}}{x_{n-3}^D},$$

using (4.3). The denominator of the last expression has x_{n-3}^D, but $x_{n-3} = G_{n-1} G_{n-2}$ which is coprime to G_n (since G_{n-1} and G_{n-2} are, by assumption while the numerator has $F_{n-3} = G_{n-3}G_n \equiv 0$, so $F_n \equiv 0$ as required. Finally note that

$$G_{n+3}G_n = F_n = F(G_{n+1}G_{n+2}) \equiv \widehat{\lambda}$$

both $\mod G_{n+1}$ and $\mod G_{n+2}$. Thus G_{n+1}, G_{n+2}, G_{n+3} are pairwise coprime, which completes the inductive step.   □

It is worth remarking at this point that the conditions stated in the above theorem essentially require that f and F should be reciprocal polynomials in cases (i) and (ii) respectively. In general, if we work over $\mathbb{C}$, then we can always rescale x_n so that $\lambda = 1$ and $\widehat{\lambda} = 1$ in the respective cases, but since we require $f \in \mathbb{Z}[x]$ we do not have this freedom. In fact, if μ denotes the leading coefficient of f, then in case (i) we have from (4.2) that $\mu = \pm \lambda^{1-d/2}$, so for odd $d \geq 3$ we must have $\lambda = 1$, while for even $d \geq 4$ there are the two possibilities $\lambda = 1$ or -1; the degrees $d = 1, 2$ are special. Similarly in case (ii), from (4.3) we obtain $\mu = \pm \widehat{\lambda}^{1-D}$, so we require that $\widehat{\lambda} = 1$ for $D \geq 2$ (or equivalently, $d \geq 3$); and $d = 2$ is special once more. For case (iii) we always have $d \geq 3$.

Each recurrence of the form (4.1) is equivalent to the iteration of a map of the plane given by

$$\phi : \quad \begin{pmatrix} x \\ y \end{pmatrix} \mapsto \begin{pmatrix} y \\ f(y)/x \end{pmatrix}, \tag{4.7}$$

which preserves the symplectic form

$$\omega = (xy)^{-1} \, \mathrm{d}x \wedge \mathrm{d}y. \tag{4.8}$$

As we shall see, the special cases $d = 1, 2$ are the only ones that are algebraically integrable, in the sense that they admit a rational invariant that defines an algebraic curve.

In the case $d = 1$, if we require the Laurent property then we must have $\mu = \lambda^{1/2}$, so λ is a perfect square and we have the map defined by

$$x_{n+1} x_{n-1} = \mu x_n + \mu^2 \tag{4.9}$$

whose iterates begin

$$x_0, \quad x_1, \quad \frac{\mu(x_1 + \mu)}{x_0}, \quad \frac{\mu^2(x_0 + x_1 + \mu)}{x_0 x_1}, \quad \frac{\mu(x_0 + \mu)}{x_1},$$

and thereafter it repeats wih period five. This is a special case of the recurrence

$$x_{n+1} x_{n-1} = \mu x_n + \lambda,$$

called the Lyness recurrence [2], which is integrable because it has a conserved quantity defining an elliptic curve, i.e.,

$$K = x_{n-1} + x_n + \frac{\mu(x_{n-1}^2 + x_n^2) + (\lambda + \mu^2)(x_{n-1} + x_n) + \mu \lambda}{x_{n-1} x_n}, \tag{4.10}$$

and the iterates correspond to a sequence of points $P_0 + nP$ on the curve. Thus the special case (4.9), whose orbits are all periodic, is known as the Lyness 5-cycle, and corresponds to the case when P is a torsion point of order 5: by varying the initial data x_0, x_1 one gets a family of elliptic curves with such a point. Actually, this example was apparently known to Gauss, as one aspect of his *pentagramma mirificum*, and it has reappeared in several other contexts: it is equivalent to the functional relation (Y-system) that appears in the thermodynamic Bethe ansatz for an A_2 scattering theory [64], and (for $\mu = 1$) it is the recurrence for the rank 2 cluster algebra associated with this root system [18].

For $d = 2$ then $\mu = \pm 1$, and for case (i) with the positive sign we have

$$x_{n+1}\, x_{n-1} = x_n^2 + \nu\, x_n + \lambda\,, \tag{4.11}$$

or alternatively for the negative sign we get $x_{n+1}\, x_{n-1} = -x_n^2 + \lambda$. The second choice can be transformed into the first by taking $x_n = \kappa_n \tilde{x}_n$ where $\kappa_n^2 = 1$ and $\kappa_{n+1}\kappa_{n-1} = -1$ for all n, so it is enough to consider (4.11). This recurrence is integrable, having a conserved quantity that defines a conic, namely

$$L = \frac{x_{n-1}}{x_n} + \frac{x_n}{x_{n-1}} + \nu \left(\frac{1}{x_{n-1}} + \frac{1}{x_n} \right) + \frac{\lambda}{x_{n-1}x_n}\,. \tag{4.12}$$

Furthermore, the iterates also satisfy a linear recurrence, viz.

$$x_{n+1} + x_{n-1} = Lx_n - \nu\,, \tag{4.13}$$

so in terms of the initial data we have $L = L(x_0, x_1) \in \mathcal{R}$ and also $x_n \in \mathbb{Z}[\nu, x_0, x_1, L] \subset \mathcal{R}$ for all n, which is even stronger than the Laurent property. There are two important observations to make here. Firstly, the iterates of (4.13), and hence of (4.11), can be written in terms of Chebyshev polynomials with argument $L/2$. (We shall use this fact explicitly in the next section.) Secondly, if $|L| \geq 2$ then the conic defined by (4.12) has non-negative discriminant, and then iterating the recurrence with integer initial data (x_0, x_1) generates infinitely many integer points (x_{n-1}, x_n) on this conic.

The latter example of conics with infinitely many integer points is a particular case of a result due to Gauss (see chapter 8 in [38], for instance); the iteration of the recurrence (4.11) is equivalent to the usual procedure for generating solutions of Pell's equation. When $\nu = 0$, $\lambda = 1$ the recurrence (4.11) generates the coefficient-free cluster algebra of rank 2 associated with the affine root system $A_1^{(1)}$ [48]. During Graham Everest's talk in Bristol, he discussed the primitive prime divisors in the integer sequence $x_n = n^2 + 1$, which is generated by (4.11) with $\nu = -2$, $\lambda = 5$ and $x_0 = 1$, $x_1 = 2$. See Everest and Harman [13], in this Proceedings.

If $d = 2$ and $f(0) = 0$, so that we are in case (ii), then either $\mu = 1$, ar we have $f(x) = x(x + \widehat{\lambda})$, which is just a special case of (4.11), or $\mu =$

and the recurrence is given by

$$x_{n+1} x_{n-1} = -x_n(x_n + \widehat{\lambda}), \qquad (4.14)$$

which is also integrable, having the conserved quantity

$$\tilde{J} = \frac{x_{n-1}^4 + x_n^4 + \widehat{\lambda}(x_{n-1} - x_n)^2(2x_{n-1} + 2x_n + \widehat{\lambda})}{x_{n-1}^2 x_n^2} \, ;$$

$\tilde{J}$ = constant defines a quartic curve of genus zero. Moreover, for (4.14) the iterates also satisfy the sixth-order linear recurrence

$$x_{n+6} + (\tilde{J} - 1)(x_{n+4} - x_{n+2}) - x_n = 0 \, ,$$

which provides an alternative proof that $x_n \in \mathcal{R}$ based on the the fact that $\tilde{J} = \tilde{J}(x_0, x_1) \in \mathcal{R}$ and $x_j \in \mathcal{R}$ for $j = 0, 1, \ldots, 5$.

We can now assert that, for $d \geq 3$, all of the other recurrences that satisfy the conditions of Theorem 4.1 are not algebraically integrable: they do not have an algebraic invariant. Using standard inequalities for heights, one can show (as in [21]) that for a recurrence of the form (4.1) with f a rational function of degree $d \geq 3$, the logarithmic heights $h(x_n)$ grow exponentially; for generic (aperiodic) orbits the leading order asymptotics are

$$\log h(x_n) \sim \widehat{C} n \, , \qquad \widehat{C} = \log \left(\frac{d + \sqrt{d^2 - 4}}{2} \right).$$

So in particular we see that for integer coefficients **c** the recurrences (4.1) with the Laurent property generate integer sequences (e.g. starting from $x_0 = 1 = x_1$), but for $d \geq 3$ these recurrences are not Diophantine integrable, and hence (by the arguments of section 3) cannot be algebraically integrable. For example, consider the recurrence

$$x_{n+1} x_{n-1} = x_n^3 + 1 \, .$$

The above theorem implies that this produces an integer sequence from the initial data $x_0 = 1 = x_1$, that is

$$1, \, 1, \, 2, \, 9, \, 365, \, 5403014, \, 432130991537958813, \, \ldots \, ,$$

whose asymptotic growth is given by

$$h(x_n) = \log x_n \sim C E^n \, , \qquad E = \frac{3 + \sqrt{5}}{2} \, ,$$

for some $C > 0$. During the meeting in Bristol, Bryan Birch's immediate reaction was that such doubly exponentially growing sequences are 'horrid', which would seem to suggest that aesthetic sensibilities in number theory are very similar to those in integrable systems. Nevertheless, there are many natural questions that one might consider, e.g., regarding the appearance of primitive prime divisors, or about the p-adic properties of such sequences.

5. THIRD-ORDER RECURRENCES AND CUBIC SURFACES

In this section we shall consider a family of third-order recurrences with the Laurent property. However, in contrast to the preceding results on second-order recurrences, we shall not attempt to classify all nonlinear recurrences of the form

$$x_{n+3}\, x_n = f(x_{n+1}, x_{n+2})$$

that have the Laurent property, but rather we present some detailed results for the two-parameter family of recurrences given by

$$x_{n+3}\, x_n = x_{n+2}\, x_{n+1} + b(x_{n+2} + x_{n+1}) + c. \tag{5.1}$$

The particular cases $b = 1, c = 0$ and $b = 0, c = 1$ were considered recently by Heideman and Hogan [22] and by Musiker [39] respectively. In turn, (5.1) is a special case of a more general rational recurrence, of the form

$$x_{n+3} x_n = \frac{a_0 + a_1(x_{n+2} + x_{n+1}) + a_3 x_{n+2} x_{n+1}}{a_3 + b_1(x_{n+2} + x_{n+1}) + b_3 x_{n+2} x_{n+1}}, \tag{5.2}$$

that was found by Hirota *et al.* [25] to have two independent rational conserved quantities. More recently, Matsukaidara and Takahashi have also considered (5.2) and other third-order recurrences in terms of coupled pairs of second-order recurrences [34].

The recurrence (5.1) is equivalent to the map

$$\varphi: \quad \begin{pmatrix} x \\ y \\ z \end{pmatrix} \longmapsto \begin{pmatrix} y \\ z \\ (yz + b(y+z) + c)/x \end{pmatrix} \tag{5.3}$$

in three dimensions. It turns out that for suitable values of K_1 and K_2, each orbit of the map lies on the intersection of the cubic surfaces defined by

$$\begin{aligned} F_1 \equiv\ & (x^2 + z^2)y - K_1\, xyz + b(x^2 + z^2 + 2xy + 2yz + 2zx) \\ & + b^2(x + y + z) + c(x + z) + bc = 0 \end{aligned} \tag{5.4}$$

and

$$F_2 \equiv (y^2 + z^2)x + (x^2 + y^2)z - K_2\, xyz + b(xy + yz + zx + y^2) + cy = 0. \tag{5.5}$$

The key to this is the construction of the conserved quantities, which relies on the fact that the iterates of (5.1) also satisfy a homogeneous linear recurrence of sixth order. The following result is proved by direct calculation.

Theorem 5.1. *The third-order nonlinear recurrence (5.1) has two independent conserved quantities K_1 and K_2, defined by*

$$K_j = \frac{N_j(x_n, x_{n+1}, x_{n+2})}{x_n\, x_{n+1}\, x_{n+2}}, \qquad j = 1, 2, \tag{5.6}$$

where

$$\begin{aligned} N_1 =\ & (x_n^2 + x_{n+2}^2)(x_{n+1} + b) + 2b(x_n\, x_{n+1} + x_{n+1}\, x_{n+2} + x_{n+2}\, x_n) \\ & + b^2(x_n + x_{n+1} + x_{n+2}) + c(x_n + x_{n+2} + b), \end{aligned}$$

and

$$N_2 = (x_{n+2}^2 + x_{n+1}^2) x_n + (x_n^2 + x_{n+1}^2) x_{n+2} \\ + b (x_n x_{n+1} + x_{n+1} x_{n+2} + x_{n+2} x_n + x_{n+1}^2) + c x_{n+1}.$$

If the values of these two quantities are fixed in terms of three initial data, x_0, x_1, x_2 say, then for all n the iterates x_n satisfy the fifth-order inhomogeneous linear recurrence

$$x_{n+5} + x_{n+4} - K_1 (x_{n+3} + x_{n+2}) + x_{n+1} + x_n + b (K_2 + 4) = 0, \qquad (5.7)$$

which has the sixth-order recurrence

$$x_{n+6} - (K_1 + 1)(x_{n+4} - x_{n+2}) - x_n = 0, \qquad (5.8)$$

as a consequence. All adjacent triples of iterates $(x, y, z) = (x_n, x_{n+1}, x_{n+2})$ lie on the pencil of cubic surfaces defined by

$$\tilde{F}(x, y, z) \equiv F_1(x, y, z) + \zeta F_2(x, y, z) = 0, \qquad (5.9)$$

with F_1 as in (5.4) and F_2 as in (5.5).

It is possible to interpret the previous theorem in terms of integrability: there is a natural Poisson bracket associated with each surface in the pencil (5.9), which induces a symplectic structure on the surface, but the technical details of this construction will be presented elsewhere. We proceed to present the explicit solution of the initial value problem for the third-order recurrence (5.1). This is based on the linearization (5.8), which leads to expressions in terms of Chebyshev polynomials.

Proposition 5.2. *The even index terms of a sequence generated by the third-order nonlinear recurrence (5.1), with initial data x_0, x_1, x_2 regarded as variables, are given by the explicit formula*

$$x_{2n} = A_+ T_n(K_1/2) + B_+ U_n(K_1/2) - C_+/(K_1 - 2), \qquad (5.10)$$

where K_1 is the conserved quantity defined by (5.6),

$$C_+ = -\frac{b(x_0 + 2x_1 + x_2 + b)}{x_1}, \qquad (5.11)$$

and T_n and U_n are Chebyshev polynomials of the first and second kind respectively; the coefficients A_+, B_+ are rational functions of K_1, C_+, x_0, x_2 given by

$$A_+ = 2x_0 - \frac{2x_2}{K_1} + \frac{2C_+(K_1 - 1)}{K_1(K_1 - 2)}, \quad B_+ = -x_0 + \frac{2x_2 - C_+}{K_1}. \qquad (5.12)$$

Similarly, the odd index terms are given by the formula

$$x_{2n+1} = A_- T_n(K_1/2) + B_- U_n(K_1/2) - C_-/(K_1 - 2), \qquad (5.13)$$

where

$$C_- = -\frac{b(x_0 x_1 + 2x_0 x_2 + x_1 x_2 + b(x_0 + x_1 + x_2) + c)}{x_0 x_2}, \qquad (5.14)$$

and the coefficients A_-, B_- are rational functions of K_1, C_-, x_1, x_3, given by

$$A_- = 2x_1 - \frac{2x_3}{K_1} + \frac{2C_- + (K_1 - 1)}{K_1(K_1 - 2)}, \qquad B_- = -x_1 + \frac{2x_3 - C_-}{K_1}, \qquad (5.15)$$

with $x_3 = (x_2 x_1 + b(x_2 + x_1) + c)/x_0$. The quantities $C_\pm$ are related to the two conserved quantities K_1, K_2, as defined in (5.6), by

$$C_+ + C_- = -b(K_2 + 4), \qquad C_+ C_- = b^2(K_1 + 2K_2 + 6). \qquad (5.16)$$

The generating function $G(t) := \sum_{n=0}^{\infty} x_n t^n$ for the sequence has the explicit form

$$G(t) = \frac{[(x_0 + x_1 t)(1 - K_1 t^2) + (x_2 + x_3 t)t^2](1 - t^2) + (C_+ + C_- t)t^4}{(1 - K_1 t^2 + t^4)(1 - t^2)}.$$

$$(5.17)$$

Proof. We work in the field of rational functions $\mathbb{Q}(x_0, x_1, x_2, b, c)$ where all the formulae are defined. (This avoids having to exclude the degenerate numerical cases $K_1 = 0$ or 2.) ¿From the sixth-order linear recurrence (5.8) it follows that the iterates satisfy the inhomogeneous recurrence

$$x_{n+4} - K_1 x_{n+2} + x_n = C_\pm \qquad (5.18)$$

where the quantity on the right hand side varies with the parity of n, being given by (5.11)/(5.14) for even/odd n respectively. Symmetric functions of C_+ and C_- must be conserved quantities: adding (5.18) to itself upshifted by $n \to n + 1$ and comparing with (5.7) gives the first relation (5.16), while the second relation comes from a direct calculation.

Recall that the Chebyshev polynomials of the first and second kind are defined by

$$T_n(\cos\theta) = \cos(n\theta) \quad \text{and} \quad U_n(\cos\theta) = \frac{\sin(n\theta)}{\sin\theta}$$

respectively, and hence provide two linearly independent solutions of the homogeneous linear difference equation

$$f_{n+1}(s) - 2s\, f_n(s) + f_{n-1}(s) = 0.$$

With the argument $s = \cos\theta = K_1/2$, it is clear from (5.18) that the subsequence of even index terms x_{2n} satisfies this second-order linear recurrence with the addition of the inhomogeneous term C_+ on the right hand side. Hence, up to the addition of a constant, this subsequence is given by a linear combination of the two Chebyshev polynomials evaluated at $s = K_1/2$. Using the initial data it is straightforward to calculate the formula (5.12) for the coefficients. (This and other expressions must be modified slightly when applying to the numerical cases $K_1 = 0$ or 2.) The formula (5.13) for the odd index terms is found similarly, together with the corresponding expressions (5.15) for A_- and B_- in (5.15); they are obtained immediately by replacing $x_0 \to x_1$, $x_2 \to x_3$, $C_+ \to C_-$ in equation (5.12). The generating function

(5.17) can be found by multiplying (5.18) by t^n and summing over even/odd n separately. □

Remark. ¿From the recurrences (5.18) for even/odd index terms, it is easy to see that these two subsequences each satisfy a recurrence of the form (4.11), and the points $(x, z) = (x_n, x_{n+2})$ alternately lie on each of pair of conics of the form (4.12), given by

$$x^2 + z^2 - K_1 x z - C_\pm (x + z + c/b) + b^2 - 2c = 0 \qquad (5.19)$$

for even/odd n respectively. The intersection of the surfaces $K_1 = \text{constant}$, $K_2 = \text{constant}$ can be understood directly by solving (5.4) for y and substituting in (5.5) to obtain a quartic equation for x, z which factorizes into the two irreducible components given by (5.19). Of course, the solution of the third-order recurrence can also be written down directly in terms of the roots of the sextic polynomial associated with (5.8), but the formulation in Proposition 5.2 gives much more detailed information about the structure of the Laurent polynomials generated by (5.1).

Corollary 5.3. *With K_1 and $C_\pm$ defined as above, the iterates of (5.1) satisfy $x_{2n} \in \mathbb{Z}[x_0, x_2, K_1, C_+]$, $x_{2n+1} \in \mathbb{Z}[x_1, x_3, K_1, C_-]$ for all n.*

The above corollary is a stronger form of the Laurent property, because the quantities $x_3, K_1, C_\pm$ are themselves Laurent polynomials in x_0, x_1, x_2 (and polynomials in b, c). As a consequence, there are many ways to choose the initial data in order to generate integer sequences from the recurrence (5.1), and much stronger results than Lemma 3.1 can be obtained. For instance, if the pencil of surfaces (5.9) contains a pair of integer points (x_0, x_1, x_2) and (x_1, x_2, x_3), and also K_1, C_+, C_- are all integers, then generically it contains infinitely many integer points. For illustration, the following result on solutions of these Diophantine equations is given without proof.

Theorem 5.4. *If arbitrary integers $(k, \ell, m) \in \mathbb{Z}^3$ are given, then for the two sets of parameters (i) $b = k + m$, $c = m^2 - \ell^2 - 2k$, $K_1 = k^2 - \ell^2 - 2$, $K_2 = 2k$, and (ii) $b = k + m - 1$, $c = m(m - 1) - \ell(\ell - 1) - 2k + 1$, $K_1 = k(k - 1) - \ell(\ell - 1) - 2$, $K_2 = 2k - 1$, the cubic pencil (5.9) contains infinitely many triples of integer solutions, except possibly when the point (k, ℓ, m) lies on one of a finite set of special lines in $\mathbb{Z}^3$.*

Remark. The preceding result arises by considering the orbit of the point $(1, 1, \ell - m - 1)$ under (5.1). Generically this orbit consists of infinitely many distinct integer triples, except for certain values of (k, ℓ, m) lying along lines in $\mathbb{Z}^3$ where the corresponding orbit is periodic. From (5.8) it is straightforward to show that the only possible periods are 1, 2, 3, 4, 6, 8, or 12.

6. CONCLUSIONS

The Laurent phenomenon is an unexpected feature of certain rational recurrences, whose iterates are all Laurent polynomials in the initial data. This unexpected property has been observed in many different contexts, ranging from Dodgson condensation to EDS and Somos sequences, to the theory of cluster algebras. Zelevinsky has given the following informal definition: "a cluster algebra is a machine for generating non-trivial Laurent polynomials" [65]. The Laurent polynomials in cluster algebras are very interesting in their own right. The exchange relations in cluster algebras correspond to recurrences of the particular form (2.6), but we have seen various examples that are not of this specific type. In fact, the Laurent phenomenon is not restricted to recurrences of the form (1.4) either: consider the fifth-order recurrence

$$x_{n+3}x_n^3 x_{n-1}^2 = x_{n+2}^3 x_{n-1}^3 - x_{n+2}^2 x_{n+1}^3 x_{n-2} + a(x_{n+1}x_n)^6, \tag{6.20}$$

for example. (It is left to the reader to show that this has the Laurent property.) There seem to be many other examples of a similar kind.

The Laurent property for a recurrence implies that, for suitable initial data (and integer coefficients), it generates integer sequences. However, generically the logarithmic heights $h(x_n)$ of such integers show exponential growth with n. Similarly, the growth rate of other measures of height, such as the total degree of the Laurent polynomials [23], or their Mahler measure [14], gives a measure of entropy for the recurrence, and generically the entropy is non-zero. The condition that $h(x_n)$ has polynomial growth seems to be necessary for the algebraic integrability of the corresponding rational map, but this may not be straightforward to prove in general (i.e. for maps in dimension four or more). It seems that sequences generated by integrable nonlinear recurrences have a great deal in common with linear ones [16], and that many properties of linear recurrence sequences should admit an extension to the setting of integrable recurrences. Above we have discussed only very simple examples of integrable maps or recurrences, which can be understood by elementary means, but the theory of integrable systems has a wealth of techniques to explain the structure of maps and recurrences of higher order [54].

Finally, we would like to comment on the unexpected link with Diophantine equations. All of the examples of recurrences discussed here that both have the Laurent property and have at least one invariant correspond to integrable maps. However, it is possible to have insufficiently many conserved quantities to satisfy the requirements of Liouville's theorem. For example, the recurrence

$$x_{n+3}x_n = x_{n+2}^2 + x_{n+1}^2 + \mathcal{J} \tag{6.21}$$

was considered by Dana Scott in the case $\mathcal{J} = 0$ [20]. It has the conserved quantity

$$\mathcal{N} = (x_{n+2}^2 + x_{n+1}^2 + x_n^2 + \mathcal{J})/(x_{n+2}x_{n+1}x_n), \tag{6.22}$$

and so each orbit lies on a cubic surface $\mathcal{N} = $ constant. Each such surface is defined by an equation of the form

$$x^2 + y^2 + z^2 + \mathcal{J} = \mathcal{N}xyz, \tag{6.2}$$

that was considered (as a Diophantine equation) by Mordell [37]; for $\mathcal{J} = 0$, $\mathcal{N} = 3$ it is Markoff's equation [33]. For $\mathcal{J}, \mathcal{N} \in \mathbb{Z}$ the iteration of the recurrence (6.21) starting from any integer solution triple will generically generate infinitely many integer solutions of (6.23). However, the recurrence cannot have a second rational (or algebraic) invariant: the logarithmic heights grow like $h(x_n) \sim \widehat{C} \left((1 + \sqrt{5})/2 \right)^n$ for some $\widehat{C} > 0$, and the algebraic entropy is also $\log \left((1 + \sqrt{5})/2 \right)$ (see [27] for more details).

REFERENCES

[1] V.I. Arnold, *Mathematical methods of classical mechanics*, 2nd edition, Springer Verlag, Berlin, 1989.

[2] G. Bastien, M. Rogalski, On some algebraic difference equations $u_{n+2}u_n = \psi(u_{n+1})$ in $\mathbb{R}_*^+$, related to families of conics or cubics: generalization of the Lyness' sequences, *J. Math. Anal. Appl.* **300** (2004), 303–333.

[3] A.I. Bobenko, R. Seiler (eds.), *Discrete integrable geometry and physics* Oxford University Press, Oxford, 1999.

[4] H.W. Braden, V.Z. Enolskii, A.N.W. Hone, Bilinear recurrences and addition formulae for hyperelliptic sigma functions, *J. Nonlin. Math. Phys.* **12** (2005), Supplement 2, 46–62.

[5] D.M. Bressoud, *Proofs and confirmations: the story of the alternating sign matrix conjecture*, Cambridge University Press Cambridge, 1999.

[6] M. Bruschi, O. Ragnisco, P.M. Santini, G.-Z. Tu, Integrable symplectic maps, *Physica D* **49** (1991), 273–294.

[7] R.H. Buchholz, R.L. Rathbun, An infinite set of Heron triangles with two rational medians, *Amer. Math. Monthly* **104** (1997), 107–115.

[8] D. Cantor, On the analogue of the division polynomials for hyperelliptic curves, *J. reine angew. Math.* **447** (1994), 91–145.

[9] J.-P. Dufour, N.T. Zung, *Poisson structures and their normal forms*, Birkhäuser, Basel, 2005.

[10] M. Einsiedler, G. Everest, T. Ward, Primes in elliptic divisibility sequences, *LMS Journal of Computation and Mathematics* **4** (2001), 1–13.

[11] S.N. Elaydi, *An introduction to difference equations*, Springer Verlag, New York, 1996.

[12] _____, *Discrete Chaos*, Chapman and Hall/CRC, Boca Raton, 2000.

[13] G. Everest, G. Harman, On primitive divisors of $n^2 + b$, this volume, 142–154.

[14] G. Everest, T. Ward, *Heights of polynomials and entropy in algebraic dynamics*, Springer Verlag, Berlin, 1998.

[15] G. Everest, V. Miller, N. Stephens, Primes generated by elliptic curves, *Proc. Amer. Math. Soc.* **132** (2003), 955–963.

[16] G. Everest, A. van der Poorten, I. Shparlinski, T. Ward, *Recurrence sequences*, AMS Mathematical Surveys and Monographs **104**, Amer. Math. Soc., Providence, 2003.

[17] S. Fomin, A. Zelevinsky, The Laurent phenomenon, *Adv. Appl. Math.* **28** (2002), 119–144.

[18] _____, _____, Y-systems and generalized associahedra, *Ann. of Math.* **158** (2003), 977–1018.

[19] _____, _____, Cluster algebras IV: coefficients, *Compos. Math.* **143** (2007), 112–164.

[20] D. Gale, The strange and surprising saga of the Somos sequences, *Mathematical Intelligencer* **13** (**1**) (1991), 40–42; Somos sequence update, *ibid.* **13** (**4**) (1991), 49–50; reprinted in *Tracking the Automatic Ant*, Springer Verlag, 1998.

[21] R.G. Halburd, Diophantine integrability, *J. Phys. A: Math. Gen.* **38** (2005), L263–L269.

[22] P. Heideman, E. Hogan, New family of Somos-like recurrences, preprint (2004) at http://www.math.wisc.edu/~propp/SSL/heiho.pdf

[23] J. Hietarinta, C. Viallet, Singularity Confinement and Chaos in Discrete Systems, *Phys. Rev. Lett.* **81** (1998), 325–328.

[24] R. Hirota, Exact solution of the Korteweg–deVries equation for multiple collision of solitons, *Phys. Rev. Lett.* **27** (1972), 1192–1194.

[25] R. Hirota, K. Kimura, H. Yahagi, How to find the conserved quantities of nonlinear discrete equations, *J. Phys. A: Math. Gen.* **34** (2001), 10377–10386.

[26] A.N.W. Hone, Elliptic curves and quadratic recurrence sequences, *Bull. London Math. Soc.* **37** (2005), 161–171; Corrigendum, *ibid.* **38** (2006), 741–742.

[27] _____, Diophantine non-integrability of a third-order recurrence with the Laurent property, *J. Phys. A: Math. Gen.* **39** (2006), L171–L177.

[28] _____, Singularity confinement for maps with the Laurent property, *Phys. Lett. A* **361** (2007), 341–345.

[29] _____, Sigma function solution of the initial value problem for Somos 5 sequences, *Trans. Amer. Math. Soc.* (2007) at press; math.NT/0501554.

[30] _____, *Discrete dynamics, integrability and integer sequences*, Imperial College Press (2007), in preparation.

[31] J. Hurtubise, Separation of variables and the geometry of Jacobians, *SIGMA* **3** (2007) 017, 14 pages.

[32] V.B. Kuznetsov, P. Vanhaecke, Bäcklund transformations for finite-dimensional integrable systems: a geometric approach, *J. Geom. Phys.* **44** (2002), 1–40.

[33] A.A. Markoff, Sur les formes quadratiques indéfinies, *Math. Ann.* **15** (1879), 381–406.

[34] J. Matsukaidara, D. Takahashi, Third-order integrable difference equations generated by a pair of second-order equations, *J. Phys. A: Math. Gen.* **39** (2006), 1151–1161.

[35] S. Matsutani, Recursion relation of hyperelliptic PSI-functions of genus two, *Int. Transforms Spec. Func.* **14** (2003), 517–527.

[36] W.H. Mills, D.P. Robbins, H. Rumsey, Alternating-sign matrices and descending plane partitions, *J. Combin. Theory Ser. A* **34** (1983), 340–359.

[37] L.J. Mordell, On the integer solutions of the equation $x^2 + y^2 + z^2 + 2xyz = n$, *J. London Math. Soc.* **28** (1953), 500–510.

[38] _____, *Diophantine equations*, Academic Press, London and New York, 1969.

[39] G. Musiker, *Cluster algebras, Somos sequences and exchange graphs*, Bachelor's thesis, Harvard University, 2002.

[40] A.J. van der Poorten, Elliptic curves and continued fractions, *J. Integer Sequences* **8** (2005), Article 05.2.5.

[41] _____, Curves of genus 2, continued fractions, and Somos sequences, *J. Integer Sequences* **8** (2005), Article 05.3.4.

[42] A.J. van der Poorten, C.S. Swart, Recurrence relations for elliptic sequences: every Somos 4 is a Somos k, *Bull. London Math. Soc.* **38** (2006), 546–554.

[43] J. Propp, The many faces of alternating-sign matrices, *Disc. Math. Theoret. Comp. Sci. Proc.* **AA** (**DM-CCG**) (2001), 43–58.

[44] _____, *The "bilinear" forum*, http://www.math.wisc.edu/~propp/

[45] A. Ramani, B. Grammaticos, J. Satsuma, Bilinear discrete Painlevé equations, *J. Phys. A: Math. Gen.* **28** (1995), 4655–4665.

[46] J.A.G. Roberts, F. Vivaldi, Arithmetical method to detect integrability in maps, *Phys. Rev. Lett.* **90** (2003), 034102, 4 pages.

[47] R. Robinson, Periodicity of Somos sequences, *Proc. Amer. Math. Soc.* **116** (199?) 613–619.

[48] P. Sherman, A. Zelevinsky, Positivity and canonical bases in rank 2 cluster algebras of finite and affine types, *Moscow Math. J.* **4** (2004), 947–974.

[49] R. Shipsey, *Elliptic divisibility sequences*, PhD thesis, Goldsmiths College, University of London, 2000.

[50] J.H. Silverman, *The arithmetic of elliptic curves*, Graduate Texts in Mathematics **106**, Springer Verlag, Berlin, 1986.

[51] ———, *p*-adic properties of division polynomials and elliptic divisibility sequences, *Math. Annal.* **332** (2005), 443–471; Addendum 473–474.

[52] N.J.A. Sloane, *On-line encyclopedia of integer sequences*, http://www.research.att.com/~njas/sequences, sequence A006720.

[53] D. Speyer, Perfect matchings and the octahedron recurrence (2004), math.CO/0402452.

[54] Y.B. Suris, *The problem of integrable discretization: hamiltonian Approach*, Birkhäuser, Basel, 2003.

[55] C.S. Swart, *Elliptic curves and related sequences*, PhD thesis, Royal Holloway, University of London, 2003.

[56] C.S. Swart, A.N.W. Hone, Integrality and the Laurent phenomenon for Somos 4 sequences, Math. Proc. Camb. Phil. Soc. (2007), in press. See also math.NT/0508094.

[57] P. Vanhaecke, *Integrable systems in the realm of algebraic geometry*, 2nd edition, Springer Verlag, Berlin, 2005.

[58] A.P. Veselov, Integrable maps, *Russ. Math. Surveys* **46** (1991), 1–51.

[59] ———, What is an integrable mapping?, *What is integrability?*, 251–272, Springer Verlag, 1991.

[60] M. Ward, Memoir on elliptic divisibility sequences, *Amer. J. Math.* **70** (1948), 31–74.

[61] ———, The law of repetition of primes in an elliptic divisibility sequence, *Duke Math. J.* **15** (1948), 941–946.

[62] A. Zabrodin, A survey of Hirota's difference equations *Teor. Mat. Fiz.* **113** (1997), 179–230 (in Russian); solv-int/9704001.

[63] D. Zagier, 'Problems posed at the St. Andrews Colloquium, 1996,' Solutions, 5th day; available at http://www-groups.dcs.st-and.ac.uk/~john/Zagier/Problems.html

[64] Al.B. Zamolodchikov, On the thermodynamic Bethe ansatz equations for reflectionless ADE scattering theories, *Phys. Lett. B* **253** (1991), 391–394.

[65] A. Zelevinsky, talk at LMS meeting on cluster algebras & Teichmüller Theory, Leicester, May 2006.

CONJUGATE ALGEBRAIC NUMBERS ON CONICS: A SURVEY

JAMES MCKEE

ABSTRACT. A survey of results concerning algebraic numbers α such that all the Galois conjugates of α lie on a conic: a circle, an ellipse, a parabola, a hyperbola, or a pair of straight lines.

1. INTRODUCTION

If α is an algebraic number, with (Galois) conjugates $\alpha = \alpha_1, \ldots, \alpha_d$, then we refer to $S(\alpha) := \{\alpha_1, \ldots, \alpha_d\}$ as a *conjugate set*. Of course

$$S(\alpha) = S(\alpha_2) = \cdots = S(\alpha_d).$$

In 1857, Kronecker [8] published a result that has proved to be as useful as it is beautiful: if α is an algebraic integer and $S(\alpha)$ is a subset of the unit circle $|z| = 1$, then α is a root of unity. Indeed the same conclusion holds if $\alpha \neq 0$ and $S(\alpha)$ is a subset of the closed unit disc $|z| \leq 1$. The map $\theta : z \mapsto z + 1/z$ establishes a correspondence between conjugate sets of algebraic integers on the unit circle, and conjugate sets of totally real algebraic integers lying in the interval $[-2, 2]$. The same map establishes a correspondence between conjugate sets of algebraic numbers. For algebraic numbers generally, rather than integers, there is little more that one can say: by translation and scaling, any interval with rational endpoints would serve in place of $[-2, 2]$.

There are several ways in which one might imagine generalising Kronecker's theorem. One could consider polynomials in several variables (Montgomery and Schinzel [9], Boyd [2], Smyth [15], Dubickas and Smyth [4]); one could extend to algebraic nonintegers (Robinson [13], Dubickas and Smyth [4]); one could allow one or more conjugates to leave the unit disc (Pisot numbers, Salem numbers, and their generalisations); one could consider constraining the conjugates to lie on different curves in the complex plane (Robinson [13], Ennola [5], Ennola and Smyth [6, 7], Smyth [16], Berry [1]). In this article we consider the last of these possibilities, and give a survey of all that is known concerning algebraic numbers whose conjugates lie on a conic. This draws together the work of several authors, over a period of nearly 150 years. Perhaps surprisingly, the final cases (the degenerate conics that give pairs of lines) were not completed until 2003, in the PhD thesis of Neil Berry

---

2000 *Mathematics Subject Classification.* 11R06.
Key words and phrases. Special algebraic numbers, conics.

[1]. Questions of integrality are often more subtle, and for results concerning algebraic integers we shall usually just give pointers to the literature, except for the more fundamental results of Kronecker [8] and Robinson [12].

Starting with Kronecker's theorem, the paper will move in chronological order through circles, parabolas, ellipses, hyperbolas, and finally pairs of lines. The longest part of the paper deals with this last case, giving an exposition of part of Berry's thesis. Most of the earlier results are given simply as statements of fact, but, for the benefit of the newcomer to the area, the foundational results of Kronecker and Robinson are discussed in a more leisurely manner, complete with proofs.

On numerous occasions we shall use the shorthand 'σ is an automorphism' to mean that σ is an element of the Galois group of a finite Galois extension of $\mathbb{Q}$, containing all algebraic numbers under consideration at the time. The map $z \mapsto \sqrt{z}$ will always be the branch that maps $\mathbb{C} = \{0\} \cup \{z \mid -\pi < \arg(z) \leq \pi\}$ to $\{0\} \cup \{z \mid -\pi/2 < \arg(z) \leq \pi/2\}$.

Acknowledgment. I am grateful to the referee for comments that led to a slight strengthening of Theorem 24.

2. KRONECKER'S THEOREM (1857)

Theorem 1 (Kronecker, 1857 [8]). *If α is an algebraic integer such that α and all its conjugates have modulus at most 1, then either $\alpha = 0$ or α is a root of unity ($\alpha^n = 1$ for some positive integer n).*

One simple proof of this fundamental result rests on the following obvious Lemma.

Lemma 2. *For any real numbers B and D, there are only finitely many algebraic integers α of degree at most D such that α and all its conjugates have modulus at most B.*

Proof. The coefficients of the minimal polynomial of an algebraic integer α are symmetric polynomials in its conjugates: if the degree and the size of all the conjugates are bounded, then so are all the coefficients. Hence only finitely many minimal polynomials of such α exist.   □

Proof of Theorem 1. Let α be an algebraic integer such that α and all its conjugates have modulus at most 1, and let D be the degree of α. Then $\alpha^n \in \mathbb{Q}(\alpha)$ has degree at most D for all n, and since α and all its conjugates have modulus at most 1, the same is true for α^n and all its conjugates. Applying Lemma 2 (with $B = 1$), we deduce that the α^n cannot all be distinct: $\alpha^m = \alpha^n$ for some $n \neq m$. Thus either $\alpha = 0$ or α is a root of unity.   □

An immediate consequence of Kronecker's Theorem is that if α is a nonzero algebraic integer with $S(\alpha)$ contained in the unit disc, then α and all its conjugates lie on the unit circle. This more trivial result has a simpler proof:

let c be the constant term in the minimal polynomial of such an α; then c is a nonzero rational integer, so $|c| \geq 1$; yet c is (plus or minus) the product of α and all its conjugates, all with modulus at most 1, and hence all must have modulus exactly 1.

Using the transformation $z \mapsto z + 1/z$, Kronecker deduced that the only totally real algebraic integers with all their conjugates in the interval $[-2, 2]$ are those of the form $2\cos(\pi q)$ with $q \in \mathbb{Q}$. Robinson's later work on circles used an easy generalisation of this transformation, which we record here as a Lemma.

Lemma 3. *Given $R^2 \in \mathbb{N}$, with $R > 0$, the map*

$$\theta : z \mapsto z + R^2/z$$

establishes a one-to-one correspondence between sets of conjugate algebraic integers on $|z| = R$ and sets of conjugate algebraic integers in the real interval $[-2R, 2R]$.

Proof. Suppose that α is an algebraic integer and that $|\alpha| = R$. The key idea is that on the circle $|z| = R$, the map $z \mapsto R^2/z$ is just complex conjugation, so maps algebraic integers to algebraic integers. Hence $\theta(\alpha)$ is the sum of two algebraic integers, so is an algebraic integer. Moreover $\theta(\alpha)$ equals twice the real part of α, so is in the interval $[-2R, 2R]$. Conversely, if β is an algebraic integer in $[-2R, 2R]$, with minimal polynomial $m_\beta(x)$ of degree n, then the preimages of β under θ (the numbers on $|z| = R$ that have real part $\beta/2$) are roots of $z^n m_\beta(z + R^2/z)$, so are algebraic integers.

If $R \in \mathbb{Q}$, then $\{R\}$ corresponds to $\{2R\}$ under θ, and $\{-R\}$ corresponds to $\{-2R\}$. If $R \notin \mathbb{Q}$, then the conjugate set $\{R, -R\}$ corresponds to $\{2R, -2R\}$. Apart from these cases, θ is a two-to-one map: if α has degree n, then $n = 2s$ is even (conjugates occur in complex conjugate pairs), and $S(\alpha)$ corresponds to a set of s algebraic integers in the open interval $(-2, 2)$. These s numbers form a conjugate set $S(\theta(\alpha))$: none can have degree below s else α would satisfy an equation of degree below $2s$. □

Remark 4. *Replacing the restriction $R^2 \in \mathbb{N}$ by $R^2 \in \mathbb{Q}$, the map θ in Lemma 3 establishes a correspondence between conjugate sets of algebraic numbers on $|z| = R$, and sets of conjugate algebraic numbers in the interval $[-2R, 2R]$. The map $\theta/2$ could be used here if the interval $[-R, R]$ is preferred.*

This Remark illustrates the flavour of much of what follows, as we move into the world of algebraic numbers rather than algebraic integers. In Remark 4 we relate the problem of finding algebraic numbers whose conjugates lie on a special circle to that of finding totally real algebraic numbers with all conjugates in a certain interval. Later results for algebraic numbers will similarly relate conjugate sets on conics to conjugate sets contained in certain intervals on the real line, or to conjugate sets that satisfy some other simple description.

3. ROBINSON'S WORK ON INTERVALS (1959–1962)

Although we shall on the whole avoid questions of integrality, such questions usually amount to establishing a map from the conic to the real line, and then determining conditions on the totally real image of an algebraic number α that are sufficient to imply the integrality of α. With this in mind, Robinson's work on algebraic integers whose conjugates lie in a real interval assumes fundamental importance.

The remarks following Kronecker's theorem show in particular that there are infinitely many conjugate sets of algebraic integers contained in the real interval $[-2, 2]$. By translation, the same is true for any real interval of length 4 with integral endpoints. Schur showed that if $|a| < 2$ then the interval $[-a, a]$ can contain only finitely many sets of conjugate algebraic integers, and Pólya pointed out to him that the argument could be extended to any interval of length strictly less than 4 [14]. In 1959 ([10], [11], but see [12] (1962) for the detail) Robinson established the result that if an interval has length strictly greater than 4, then it contains infinitely many conjugate sets of algebraic integers. In this section, we give Robinson's proof of this important result. Apart from the trivial cases mentioned above, nothing is known about the case of a real interval of length exactly 4.

Theorem 5 (Robinson, 1962 [12]). *Let I be a real interval of length strictly greater than 4. Then I contains infinitely many sets of conjugate algebraic integers.*

This theorem underpins all the work on conjugate sets of algebraic integers on conics. In all cases, the technique is to find some analogue of the map θ used in the proof of Lemma 3 that sets up a correspondence between conjugate sets on the conic and conjugate sets in an interval. The detail may be quite subtle, as integrality is not generally preserved by rational maps, but the punchline will always be an invocation of Theorem 5.

Robinson's proof rests on an explicit formula for the coefficients of Chebyshev polynomials, $T_n(x)$, defined by $T_n(2\cos\theta) = 2\cos(n\theta)$ ($n = 0, 1, 2, \ldots$), and satisfying the recurrence $T_{n+1}(x) = xT_n(x) - T_{n-1}(x)$ for $n \geq 1$.

Lemma 6. *The nth Chebyshev polynomial, $T_n(x)$ is given by*

$$T_n(x) = x^n + \sum_{k=1}^{\lceil n/2 \rceil} (-1)^k \frac{n}{k} \binom{n-k-1}{k-1} x^{n-2k}.$$

Proof. One can use induction and the recurrence formula, or see [12] for another inductive proof that makes use of a related family of polynomials satisfying the same recurrence. □

Proof of Theorem 5. Let I be a real interval of length strictly greater than 4. Then we can choose a subinterval $J \subseteq I$ with rational endpoints and length strictly greater than 4, say $J = [c - 2\lambda, c + 2\lambda]$, where c and λ are rational numbers (henceforth fixed), and $\lambda > 1$. For $n = 0, 1, \ldots$, let

$$P_n(x) = \lambda^n T_n\left(\frac{x - c}{\lambda}\right),$$

where $T_n(x)$ is the nth Chebyshev polynomial, defined above. Observe that $|P_n(x)| \leq 2\lambda^n$ for $x \in J$, and that moreover $P_n(x)$ oscillates n times between the bounds $\pm 2\lambda^n$ as x ranges over J.

From the explicit formula in Lemma 6, we can write

$$P_n(x) = x^n + \sum_{k=1}^{n} a_k x^{n-k},$$

where a_k is a polynomial in n of degree k, with rational coefficients. Moreover, crucially, the explicit formula shows that n divides a_k (as a polynomial in n). Thus we have, for any fixed k,

$$a_k = \frac{r_0 n^k + r_1 n^{k-1} + \cdots + r_{k-1} n}{s} \tag{1}$$

for some integers s, $r_0, r_1, \ldots, r_{k-1}$.

Since $\lambda > 1$, we can choose ℓ such that

$$\lambda^\ell(\lambda - 1) \geq 1. \tag{2}$$

Take m to be the least common multiple of all the integers s appearing in (1) for $1 \leq k \leq \ell$. Let n be any multiple of m. Then the coefficients $a_1, \ldots, a_\ell$ are integers, since the numerator in (1) is divisible by m and the denominator divides m.

We now perturb the tail of $P_n(x)$ to give a polynomial with integer coefficients that is close to $P_n(x)$ on the interval J. We choose real numbers $b_{\ell+1}$, $\ldots, b_n$ with $0 \leq b_k < 1$ such that

$$Q_n(x) = P_n(x) + \sum_{k=\ell+1}^{n} b_k P_{n-k}(x)$$

has integer coefficients. Since $|P_{n-k}(x)| \leq 2\lambda^{n-k}$ for $x \in J$, we have (using (2))

$$|Q_n(x) - P_n(x)| < \sum_{k=\ell+1}^{\infty} 2\lambda^{n-k} = \frac{2\lambda^n}{\lambda^\ell(\lambda - 1)} \leq 2\lambda^n$$

for $x \in J$. Since the maxima and minima of $P_n(x)$ in J have modulus $2\lambda^n$, the signs of $Q_n(x)$ and $P_n(x)$ will agree at these critical points. Therefore $Q_n(x)$ has a root between each consecutive pair of turning points of $P_n(x)$ in the interval J, and hence has n distinct roots in that interval.

Since there are infinitely many n that are multiples of m, we can produce infinitely many Q_n of degree n with integer coefficients and distinct roots in J. This is enough to complete the proof. □

Robinson gives a variant of the above argument that produces *irreducible* polynomials $Q_n(x)$, although (as he observes) this is not needed for the main result.

For algebraic numbers, the situation is much simpler:

Remark 7. *Let I be an interval with positive length. Then I contains infinitely many conjugate sets of algebraic numbers.*

Proof. We can translate I by a rational number without affecting either the hypothesis or the conclusion, so may assume that $0 \in I$. Take a positive rational number λ such that the scaled interval λI contains the interval $[-2, 2]$. Then the algebraic numbers $(2/\lambda) \cos(\pi q)$, for $q \in \mathbb{Q}$, lie with all their conjugates in I. $\qquad\square$

4. CIRCLES WITH RATIONAL CENTRE (1969)

In 1969, Robinson [13] posed the question: which circles $|z-a| = R$ contain infinitely many sets of conjugate algebraic integers? It is clear that a must be real, for any conjugate sets must lie on the intersection of the circle and its reflection in the real axis. Robinson gave a complete answer (Theorem 8 below) to his question for rational values of a. We use $C(a, R)$ to denote the circle $|z - a| = R$.

Theorem 8. *Let $a = p/q$, where $p \in \mathbb{Z}$, $q \in \mathbb{N}$, and $\gcd(p, q) = 1$. Let R be a positive real number. Then $C(a, R)$ contains infinitely many conjugate sets of algebraic integers in precisely the following cases:*

- $q = 1$ and $R^n \in \mathbb{Z}$ for some $n \in \mathbb{N}$;
- $q \geq 2$, $R > q$, and $q(R^2 - a^2) \in \mathbb{Z}$;
- $q = 2$, $R > 2$, and $4(R^4 - 1/16) \in \mathbb{Z}$.

Robinson considered also the easier problem of which circles $C(a, R)$ contain infinitely many conjugate sets of algebraic numbers, with $a \in \mathbb{Q}$:

Remark 9. *Let R be a positive real number and let a be a rational number. Then the following statements are equivalent:*

(1) *$C(a, R)$ contains a conjugate set of algebraic numbers;*
(2) *$R^n \in \mathbb{Q}$ for some $n \in \mathbb{N}$;*
(3) *$C(a, R)$ contains infinitely many conjugate sets of algebraic numbers.*

Proof. By translation, the result holds for any rational a if and only if it holds for $a = 0$, which we now suppose.

Clearly (3) implies (1).

Suppose that (1) holds, and that $P(z) = z^n + c_1 z^{n-1} + \cdots + c_n$ is a polynomial with rational coefficients and all its roots on the circle $C(0, R)$. Since roots occur in complex conjugate pairs, except possibly for $\pm R$, the product of all the roots of $P(z)$ is $\pm R^n$. Hence $R^n = \pm c_n \in \mathbb{Q}$, which gives (2).

Suppose that (2) holds. We may suppose that $n = 2m$ is even (if not, then double it). By Remark 7, the interval $[-2R^m, 2R^m]$ contains infinitely many sets of conjugate algebraic numbers, and by Remark 4 the same is true for the circle $C(0, R^{2m})$ (using $R^{2m} \in \mathbb{Q}$). Taking the mth roots of such sets give infinitely many conjugate sets of algebraic numbers on $C(0, R)$, and so (3) holds. $\qquad\square$

Robinson gave more detail about the structure of possible conjugate sets on $C(0, R)$. Suppose that $P(z) = z^n + c_1 z^{n-1} + \cdots + c_n$ is an irreducible polynomial with rational coefficients and all its roots on $C(0, R)$. Since the set of roots of $P(z)$ is closed under complex conjugation, and on $C(0, R)$ complex conjugation is given by $z \mapsto R^2/z$, we must have

$$c_n P(z) = z^n P(R^2/z).$$

Equating coefficients of z^k gives (for $1 \le k \le n$)

$$c_n c_{n-k} = R^{2n-2k} c_k.$$

From the proof of Remark 9 we see that $R^n \in \mathbb{Q}$, and hence R^{2k} is rational whenever $c_k \ne 0$. The set of values of k for which $R^{2k} \in \mathbb{Q}$ is an additive subgroup of $\mathbb{Z}$, and hence equals $\ell\mathbb{Z}$ for some ℓ. Thus $c_k = 0$ unless $\ell \mid k$, and note also that $\ell \mid n$ since $R^{2n} \in \mathbb{Q}$. Hence $P(z) = Q(z^\ell)$ for some polynomial $Q(z)$ with rational coefficients. The roots of Q all lie on $C(0, R^\ell)$. Combining this with Remark 4 we have extracted the following from Robinson's work, as noted in [4]:

Theorem 10. *Let R be a positive real number, some power of which is rational. Let $\ell \in \mathbb{N}$ be minimal such that $R^{2\ell} \in \mathbb{Q}$. Then the minimal polynomial over $\mathbb{Q}$ of an algebraic number lying with all its conjugates on $C(0, R)$ is one of the following:*

- $z^\ell \pm R^\ell$ *(only possible if $R^\ell \in \mathbb{Q}$);*
- $z^{2\ell} - R^{2\ell}$ *(only possible if $R^\ell \notin \mathbb{Q}$);*
- $z^{\ell s} P(z^\ell + R^{2\ell}/z^\ell)$ *for some irreducible monic polynomial $P \in \mathbb{Q}[z]$ of degree s, having all its zeros in the open interval $(-2R^\ell, 2R^\ell)$.*

Conversely, each such polynomial is the minimal polynomial of a conjugate set of algebraic numbers lying on $C(0, R)$.

5. CIRCLES WITH IRRATIONAL CENTRE (1973–1976)

Robinson [13] conjectured that if a is not rational, then no circle $C(a, R)$ can contain infinitely many conjugate sets of algebraic integers. In 1973, Ennola [5] disproved this conjecture. He gave a precise description of the circles $C(a, R)$ that contain infinitely many conjugate sets of algebraic numbers under the restriction that a is totally real, and then dealt with the more difficult problem of integrality in this restricted case.

Ennola observed first that if $C(a, R)$ contains infinitely many conjugate sets of algebraic numbers, then the centre a must be algebraic, and then restricted attention to the special case where a is totally real. In this case, he gave necessary conditions for a circle to contain at least one set of conjugate algebraic numbers with at least three members, and then showed that such circles contain infinitely many conjugate sets. This latter part was achieved with the aid of a 4-to-1 map from the circle to a real interval. For questions of integrality, one invokes Theorem 5, but this is slightly delicate as one needs to study which preimages of the 4-to-1 map are actually integers: see the paper [5] for details. Here we content ourselves with a sketch of the ideas as they apply to algebraic numbers.

The first observation in the previous paragraph is trivial:

Remark 11. *If a and R are not both algebraic numbers, then $C(a, R)$ can contain at most two algebraic numbers.*

Proof. If z is algebraic, then so are its real and imaginary parts. Suppose that $x_1 + iy_1$ and $x_2 + iy_2$ both lie on $C(a, R)$, and that both are algebraic. If $x_1 \neq x_2$, then the equation $(x_1 - a)^2 + y_1^2 = (x_2 - a)^2 + y_2^2$ would give a algebraic, and then the equation $(x_1 - a)^2 + y_1^2 = R^2$ would give R algebraic. So if a and R are not both algebraic then we must have $x_1 = x_2$, and hence at most two algebraic numbers on the circle. □

We can push this a little further, and note that if the centre is not rational, then the number of conjugate sets of degree at most two is tiny.

Remark 12. *If a is not rational, then for any fixed $R > 0$, $C(a, R)$ contains at most two rational points and at most one pair of conjugate quadratic algebraic numbers.*

Proof. The first point is trivial, and the second is almost so. Suppose that $z = x + iy$ is a quadratic number on the circle $C(a, R)$. Then x, $y^2 \in \mathbb{Q}$. If we had two such quadratic numbers $x_1 + iy_1$ and $x_2 + iy_2$ with $x_1 \neq x_2$, then the equation $(x_1 - a)^2 + y_1^2 = (x_2 - a)^2 + y_2^2$ would give $a \in \mathbb{Q}$. □

To make further progress, Ennola exploits the use of field automorphisms. Suppose that $P(z)$ is an irreducible polynomial of degree $n \geq 3$ with rational coefficients, and with all its roots $z_1, \ldots, z_n$ on $C(a, R)$. Ennola establishes the remarkable result that if a is totally real then

$$\left(a - \theta(a)\right)^2 = R^2 + \theta(R^2) \tag{3}$$

for all automorphisms θ such that $\theta(a) \neq a$.

To show this, we start with the equation of our circle

$$(z - a)(\bar{z} - a) = R^2 .$$

For $1 \leq i \leq n$, and any automorphism θ, write

$$\theta(z_i) = a + R\xi_i , \quad \theta(\bar{z}_i) = a + R\eta_i ,$$

and then applying θ to the equation of the circle gives

$$R^2 \xi_i \eta_i + R\left(a - \theta(a)\right)(\xi_i + \eta_i) + \left(a - \theta(a)\right)^2 - \theta(R^2) = 0 .$$

We suppose that a is irrational (else we can apply Robinson's work), and take θ such that $\theta(a) \neq a$. Solving for η_i, gives

$$\eta_i = \frac{A + B\xi_i}{C + D\xi_i}$$

where $A = \theta(R^2) - (a - \theta(a))^2$, $B = (\theta(a) - a)R$, $C = -B$, $D = R^2$. Since we can take at least three different values of i, and $|\eta_i| = |\xi_i| = 1$, the bilinear map $z \mapsto \frac{A+Bz}{C+Dz}$ maps the unit circle to itself. From the theory of bilinear mappings ([3], page 351, or see Ennola's paper for a slightly different but equivalent treatment), $A\bar{C} = B\bar{D}$, giving (and crucially using that a is totally real)

$$\theta(R^2) - (a - \theta(a))^2 = -R^2 \,,$$

which is (3).

Next we show that a is at worst cubic, and that a and R^2 are tied together rather closely.

Theorem 13. *Suppose that $P(z)$ is an irreducible polynomial of degree $n \geq 3$ with rational coefficients, and with all its roots $z_1, \ldots, z_n$ on the circle $C(a, R)$, and that a is irrational and totally real. Then there is a cubic polynomial $g(x) = x^3 + Ax^2 + Bx + C$ with rational coefficients such that both $g(a) = 0$ and $R^2 = g'(a)$.*

Proof. We have equation (3) for all automorphisms θ such that $\theta(a) \neq a$.

If a is quadratic, then taking θ such that $\theta(a) \neq a$ we have $\theta^2(a) = a$, and (3) gives $a' = a - R^2/(a - \theta(a)) \in \mathbb{Q}$. Then $g(z) = (z - a)(z - \theta(a))(z - a')$ works.

If a has degree at least 3, then we can take automorphisms θ_1, θ_2 such that $a, \theta_1(a), \theta_2(a)$ are distinct. From (3) with $\theta = \theta_1$, $\theta = \theta_2$, and $\theta = \theta_1^{-1}\theta_2$ (and then applying θ_1 to this last), we get

$$R^2 + \theta_1(R^2) = (a - \theta_1(a))^2 \,,$$
$$R^2 + \theta_2(R^2) = (a - \theta_2(a))^2 \,,$$
$$\theta_1(R^2) + \theta_2(R^2) = (\theta_1(a) - \theta_2(a))^2 \,.$$

Subtracting the third of these from the sum of the previous two gives (after dividing by 2)

$$R^2 = (a - \theta_1(a))(a - \theta_2(a)) \,. \qquad (4)$$

If a had degree 4 or more, then we could choose θ_3 with $\theta_3(a)$ distinct from $a, \theta_1(a), \theta_2(a)$, and replace $\theta_2(a)$ by $\theta_3(a)$ in (4), giving two distinct values for R^2. We conclude that a has degree 3, with minimal polynomial

$$(z - a)(z - \theta_1(a))(z - \theta_2(a)),$$

and (4) gives $R^2 = g'(a)$. $\qquad \square$

We remark that since $R^2 > 0$, equation (4) shows that for a cubic and totally real, a is either the smallest or the largest of its conjugates. Moreover that same equation gives the pleasing geometric result that the other two conjugates of a are inverse points with respect to our circle.

If a is a quadratic or cubic real number that is a root of the polynomial $g(x) = X^3 + Ax^2 + Bx + C$ (with rational coefficients), and $g'(a) > 0$, then defining $R > 0$ by $R^2 = g'(a)$, Ennola shows that there are infinitely many conjugate sets of algebraic numbers lying on the circle $C(a, R)$. The method of proof is to construct a rational map from the circle to a real interval, and apply Remark 7. We merely sketch the details here.

Writing $g(z) = (z - a)(z - a_2)(z - a_3)$, the equation of our circle is (using $R^2 = g'(a)$)

$$(z - a)(\bar{z} - a) = (a - a_2)(a - a_3).$$

From this, a little algebra gives $g(z) = (z - a)^2(z + \bar{z} - a_2 - a_3)$ whenever z is on the circle. This implies that $g(z)/(z - a)^2$ is real for z on the circle. The same is true for the quartic $f(z) = z^4 - 2Bz^2 - 8Cz + B^2 - 4AC$, since on the circle this equals $(z - a)^2((z + \bar{z})^2 - 4a_2a_3)$. Hence the map $z \mapsto f(z)/g(z)$ sends the circle to the real line, and Ennola notes that the image is the interval between $4a$ and $4a'$, where a' is the nearer of a_2 and a_3 to a. There are infinitely many algebraic numbers in this interval, and their preimages lie with their conjugates on the circle: Ennola's proof of this requires an explicit description of the inverse map.

To disprove Robinson's conjecture, the question of integrality is raised, and Ennola gives conditions for the existence of infinitely many conjugate sets of algebraic integers on a circle with totally real irrational centre, for which we refer the reader to his paper [5].

Moving to centres that are not totally real, Ennola and Smyth cleverly combine field automorphisms with a group of bilinear maps (sometimes called Möbius, or homographic, or linear fractional, or fractional linear, and called linear in [7]; these are maps of the form $z \mapsto (az+b)/(cz+d)$) that permute the conjugates of the centre. The starting point is the observation (generalising Robinson) that complex conjugation on $C(a, R)$ can be realised as a bilinear map:

$$\Gamma : z \mapsto \frac{az + R^2 - a^2}{z - a}.$$

On $C(a, R)$, $\bar{z} = \Gamma(z)$.

Suppose that $S(\beta) \subseteq C(a, R)$, where a is not totally real. After Remark 12 we may restrict attention to $n \geq 3$. If $S(a) = \{a_1, \ldots, a_d\}$, then we can choose automorphisms θ_i ($1 \leq i \leq d$) with $\theta_i(a) = a_i$. For $1 \leq i \leq d$, define the bilinear map Γ_i by applying θ_i to the coefficients of Γ:

$$\Gamma_i : z \mapsto \frac{a_i z + \theta_i(R^2) - a_i^2}{z - a_i}.$$

The Γ_i permute the conjugates of β, for if θ is any automorphism then

$$\Gamma_i\big(\theta(\beta)\big) = \theta_i\Big(\Gamma\big(\theta_i^{-1}\theta(\beta)\big)\Big).$$

As John Conway pointed out (note added in proof in [7]), the group H generated by the Γ_i is a subgroup of the symmetric group on n symbols (permuting the conjugates of β), and each element of H either preserves or reverses the order of the conjugates of β on the circle, hence H is dihedral or cyclic. Ennola and Smyth show that H is in fact dihedral, and that there are real numbers ρ_1 and ρ_2 (satisfying $(x - \rho_1)(x - \rho_2) = x^2 + cx + b$ for some rational c and b, and with ρ_1, ρ_2 inverse with respect to our circle) such that each $\Gamma_i\Gamma$ has ρ_1 and ρ_2 as fixed points. From this, the general shape of the minimal polynomial of the centre a, and of β, can be deduced. Having introduced the key players in their proof, we now merely quote their results.

Let $\mathcal{B}$ be the set of those algebraic numbers β of degree at least 3 such that $S(\beta) \subseteq C\big(a(\beta), R(\beta)\big)$ (for some $a(\beta)$, $R(\beta)$). Divide $\mathcal{B}$ into disjoint subsets:

$$\mathcal{B}_* = \{\beta \in \mathcal{B} \mid \text{some conjugate of } \beta \text{ is real}\},$$

$$\mathcal{B}_{tr} = \{\beta \in \mathcal{B} \mid \beta \text{ totally imaginary, } a(\beta) \text{ totally real}\},$$

$$\mathcal{B}_n = \{\beta \in \mathcal{B} \mid \beta \text{ totally imaginary, } a(\beta) \text{ of degree } n \text{ and not totally real}\}.$$

The set $\mathcal{B}_{tr}$ is treated above in Robinson and Ennola's earlier work.

Given rational numbers s, b, c, with $c^2 > 4b$, let $\rho_1 < \rho_2$ be the (real) roots of $x^2 + cx + b = 0$. Take n to be an integer that is at least 3, and define $\xi_1 = s - n\rho_1$, $\xi_2 = s - n\rho_2$, $d = \xi_1\xi_2 = s^2 + nsc + n^2 b$. Put $K = \mathbb{Q}(\rho_1)$, which either equals $\mathbb{Q}$ or is a quadratic extension of it. If $d > 0$, then we define Δ to be the open interval $(-2\sqrt{d}, 2\sqrt{d})$, and we then let $\mathcal{A}$ denote the set of all totally real algebraic numbers all of whose conjugates lie in Δ. Define

$$g(z) = \big(\xi_2(z - \rho_1)^n - \xi_1(z - \rho_2)^n\big)/(\xi_2 - \xi_1), \qquad (5)$$

noting that this depends on the choice of n, s, b, and c. Put $\eta = \xi_1/\xi_2$, and $\chi = 1$ or 2 according as n is even or odd.

Theorem 14. *Every $\beta \in \mathcal{B}_*$ has minimal polynomial of the form $g(z)$, given by (5), for some $n \geq 3$ and some s, b, $c \in \mathbb{Q}$ satisfying $c^2 > 4b$, $d \neq 0$, and*

$$\eta \notin K^p \text{ for each odd prime } p|n, \text{ and, if } n \text{ is even, } d > 0, \ d \notin \mathbb{Q}^2. \qquad (6)$$

Conversely, given $n \geq 3$ and s, b, $c \in \mathbb{Q}$ satisfying $c^2 > 4b$, $d \neq 0$ and (6), the polynomial $g(z)$ defined by (5) is irreducible over $\mathbb{Q}$ and has all its zeros on a circle, and moreover in $\mathcal{B}_*$, for χ of these zeros are real. The centre of the circle has degree n/χ over $\mathbb{Q}$.

Theorem 15. *Every $\beta \in \mathcal{B}_n$ has minimal polynomial of the form*

$$P(z) = A\prod_j \big(\xi_2(z - \rho_1)^{2n} + \xi_1(z - \rho_2)^{2n} - \alpha_j(z^2 + cz + b)^n\big), \qquad (7)$$

where $\alpha \in \mathcal{A}$ has conjugates α_j, the constant A is rational, and s, b, $c \in \mathbb{Q}$ satisfy $c^2 > 4b$, $d > 0$, and (6).

Conversely, given $\alpha \in \mathcal{A}$, $n \geq 3$, and s, b, $c \in \mathbb{Q}$ such that $c^2 > 4b$, $d > 0$, and (6) holds, the polynomial $P(z)$ given by (7) is irreducible over $\mathbb{Q}$, all its zeros β_i are non-real, and the β_i lie on the circle $C(a, b + ac + a^2)$, where a is the only real zero of $g(z)$ (given by (5)) if n is odd, and a is the real zero of $g(z)$ further from $-c/2$ if n is even. Thus $\beta_i \in \mathcal{B}_n$.

Chris Smyth points out that if an algebraic number is in either $\mathcal{B}_*$ or $\mathcal{B}_n$, with all its conjugates on the circle C, then the centre of C lies in $\mathcal{B}_*$ (unless its degree is below 3). In this way one gets a nest of smaller and smaller circles.

6. PARABOLAS, ELLIPSES AND HYPERBOLAS (1982)

We simply quote the results from Smyth [16].
Define

$$S_P = \{ \text{algebraic } F \in \mathbb{R} \mid F > 0, \text{ all other conjugates of } F \text{ are } < 0 \},$$

$$S_E = \{ \text{algebraic } B \in \mathbb{R} \mid B > 1, \text{ all other conjugates have modulus less than } 1 \},$$

$$S_H = \{ \text{algebraic } B \mid B^2 \neq 1, |B| = 1, \text{ all conjugates other than } B^{\pm 1} \text{ are real} \}.$$

The sets of algebraic numbers whose conjugates all lie on a parabola, ellipse, or hyperbola, will be described in terms of these sets. Note that S_E contains the set of Pisot numbers, and that the image of S_H under $z \mapsto z + 1/z$ is the set of totally real algebraic numbers γ such that $\gamma \in (-2, 2)$ and all other conjugates of γ have modulus greater than 2.

For an algebraic number B, we define $k(B)$ to be the smallest positive integer such that $B^{k(B)}$ has no conjugate of the form $\omega B^{\pm k(B)}$, for a nontrivial root of unity ω. As before, we use T_k for the Chebyshev polynomial of degree k.

For a, $F \in \mathbb{R}$, with $F > 0$, let $P(a, F)$ be the parabola in the complex plane with equation

$$z(t) = a + (t + iF/2)^2 / F, \qquad (t \in \mathbb{R}).$$

For B, a, $R \in \mathbb{R}$, with $R > 0$, $B > 1$, and with $\epsilon = \pm 1$, let $E(a, R, B, \epsilon)$ be the ellipse

$$z(t) = a + R(t\sqrt{B} + \epsilon/(t\sqrt{B})), \qquad (t \in \mathbb{R}).$$

For a, $R \in \mathbb{R}$, with $R > 0$, and with $\epsilon = \pm 1$, and for $B \in \mathbb{C}$ with $|B| = 1$ and $B \neq \pm 1$, let $H(a, R, B, \epsilon)$ be the hyperbola

$$z(t) = a + R(t\sqrt{B} + \epsilon/(t\sqrt{B})), \qquad (t \in \mathbb{R}).$$

Theorem 16. *Suppose that α is an algebraic number of degree at least 9 with $S(\alpha) \subseteq P(a, F)$. Then $a \in \mathbb{Q}$, $F \in S_P$, and α has a conjugate of the form*

$$a + \tfrac{1}{4}\left(\sqrt{\beta} + \sum_{j=1}^{n} \sqrt{-F_j}\right)^2, \tag{8}$$

where $\{F = F_1, \ldots, F_n\}$ is the set of conjugates of F, and β is totally positive. Conversely, given such a, F and β, the number given by (8) lies with all its conjugates on $P(a, F)$.

Theorem 17. *Suppose that α is an algebraic number of degree at least 25 with $S(\alpha) \subseteq E(a, R, B, \epsilon)$. Then a, $R^2 \in \mathbb{Q}$, $B^{k(B)} \in S_E$.*
Put

$$\alpha^* = T_{k(B)}\left(\frac{(\alpha - a)\sqrt{\epsilon}}{R}\right). \tag{9}$$

Then $S(\alpha^*) \subseteq E(0, 1, B^{k(B)}, 1)$.
In view of this, we need only consider $a = 0$, $R = 1$, $k(B) = 1$, $\epsilon = 1$. In this case, α has a conjugate of the form $\nu + 1/\nu$, where

$$\nu = \tfrac{1}{2}\left(\beta + \sqrt{\beta^2 - 4}\right)\left(\prod_{j=1}^{n} B_j\right)^{1/2}, \tag{10}$$

where $B = B_1$ and either $B \in \mathbb{Q}$ and $n = 1$ or $\{B_1^{\pm 1}, \ldots, B_n^{\pm 1}\}$ is the conjugate set of B. Moreover β is totally real, and lies with its conjugates in the interval $[-2, 2]$.
Conversely, suppose that $B^{k(B)} \in S_E$ and β totally real with all conjugates in $[-2, 2]$. Define ν by (10). Then $\alpha^* = \nu + 1/\nu$ lies with all its conjugates on $E(0, 1, B^{k(B)}, 1)$. Suppose further that a, $R^2 \in \mathbb{Q}$, and $\epsilon = \pm 1$. Then if α is a root of (9), $S(\alpha) \subseteq E(a, R, B, \epsilon)$.

Theorem 18. *Suppose that α is an algebraic number of degree at least 25 with $S(\alpha) \subseteq H(a, R, B, \epsilon)$. Then a, $R^2 \in \mathbb{Q}$, $k(B) = 1$ or 2, and $B^{k(B)} \in S_H$.*
If $B \neq \pm i$, and α^* is defined by (9), then $S(\alpha^*) \subseteq H(0, 1, (\epsilon B)^{k(B)}, 1)$. Furthermore if $k(B) = 2$ then t is positive. If $B = \pm i$, then $\alpha^* = (\alpha - a)^2$ lies with all its conjugates on the vertical line $\Re(z) = 2\epsilon R^2$, and hence $\alpha^* = 2\epsilon R^2 + i\beta$ for some totally real β (see Lemma 19 below).
In view of this, we need only consider $B \neq \pm i$, $a = 0$, $R = 1$, $k(B) = 1$, $\epsilon = 1$. Then α has a conjugate of the form $\nu + 1/\nu$, where ν is given by (10), with $\{B_1^{\pm 1}, \ldots, B_n^{\pm 1}\}$ the conjugate set of B, and β totally real, with all conjugates having modulus at least 2.
Conversely, let $B \in S_H$, $B \neq \pm i$, and let β be totally real, with all conjugates of β having modulus at least 2. Let a, $R^2 \in \mathbb{Q}$, $\epsilon = \pm 1$, and $B^{k(B)} \in S_H$. Define $\alpha^* = \nu + 1/\nu$, where ν is given by (10). Then if α is a root of (9), $S(\alpha) \subseteq H(a, R, B, \epsilon)$.

And if $B = \pm i$, β is totally real, $a \in \mathbb{Q}$, and $R^2 \in \mathbb{Q}$, then $\alpha = a + \sqrt{2\epsilon R^2 + i\beta}$ lies with all its conjugates on $H(a, R, B, \epsilon)$.

A non-trivial Corollary in [16] is that if a parabola contains infinitely many conjugate sets of algebraic numbers then its focus is rational, and that if an ellipse or a hyperbola contains infinitely many conjugate sets of algebraic numbers then its foci are either rational or are conjugate quadratic numbers.

7. Pairs of lines (2003)

7.1. A single line. For a real number p, let $L(p)$ denote the straight line $\Re(z) = p$.

Lemma 19. *Let α be an algebraic number that lies with its conjugates on a single straight line in the complex plane. Then either α is totally real, or $\alpha = p + i\beta$, where $p \in \mathbb{Q}$ and β is totally real.*

Proof. We follow the argument in [16]. Clearly if α is not totally real, then the straight line in question is of the form $L(p)$, for some $p \in \mathbb{R}$. Indeed if α' is any non-real conjugate of α, then $p = (\alpha' + \overline{\alpha'})/2$, so that p is algebraic. Let p' be any conjugate of p. Then $p' = (\alpha'' + \overline{\alpha''})/2$ for some α'' that is a conjugate of α, and hence $p' \in L(p)$. Thus

$$p' + \overline{p'} = 2p. \tag{11}$$

Let σ be an automorphism that sends p to one of the conjugates of p that has maximal absolute value. Applying σ to equation (11), we get

$$p_1 + p_2 = 2\sigma(p),$$

for some p_1 and p_2 that are conjugates of p. From our maximality assumption, we must have $p_1 = p_2$, hence $p' = \overline{p'}$, so p' is real, and $p = \Re(p') = p'$. Thus $p \in \mathbb{Q}$.

It is clear now that β is totally real, for if β' is any conjugate of β, then one of $p \pm i\beta'$ is a conjugate of $p + i\beta = \alpha$, and hence β' must be real. $\square$

Remark 20 (Lemma 1(a) in [16]). *The idea used in the above proof (applying an automorphism that sends an algebraic number to a conjugate with maximal absolute value) can be used similarly to show that for any distinct conjugate algebraic numbers α_1, α_2, α_3 one never has*

$$\alpha_1 \pm \alpha_2 = \pm 2\alpha_3,$$

for any choice of signs.

7.2. The four cases: $+$, $=$, $\|$, $\times$. Let α be an algebraic number that lies with its conjugates on a pair of straight lines in the complex plane, but not on a single line. Since we can draw two straight lines to cover any four points, we shall suppose that α has degree at least 5. If all the non-real conjugates of α have the same real part p, then the pair of lines is

$$\mathbb{R} \cup L(p),$$

and for obvious geometric reasons we refer to this as the '+' case. Note that this case covers all α that have only two non-real conjugates.

If α has non-real conjugates α_1, α_2 with different real parts p and q, then there are three possible ways to draw two straight lines through all of α_1, $\overline{\alpha_1}$, α_2, $\overline{\alpha_2}$. If all non-real conjugates have imaginary part $\pm k$ for some k, then a possible pair of lines is

$$\Xi(k) := \{z : \Im(z) = k\} \cup \{z : \Im(z) = -k\},$$

which we call the '=' case. If p and q are the only real parts of any conjugates of α, then a possible pair of lines is

$$L(p) \cup L(q),$$

which we call the '||' case. Finally, there is the '×' case, where each line is neither horizontal nor vertical. The pair of lines is then of the shape:

$$X(a, \theta) := \{z : z = a + te^{\pm i\theta}, \text{ for some } t \in \mathbb{R}\},$$

for some real number a (the point where the two lines intersect) and some angle $\theta \in (0, \pi/2)$.

Our task now is to determine, for each of the four cases $(+, =, ||, \times)$ which algebraic numbers lie with all their conjugates on such a pair of lines. As for non-degenerate conics, we shall give ourselves the flexibility of considering only algebraic numbers of sufficiently large degree, and we shall be content with a description in terms of algebraic numbers of some more simple special form.

7.3. The + case. If α has just two non-real conjugates, then it lies with all its conjugates on the pair of lines $\mathbb{R} \cup L(p)$, where p is the real part of the pair of non-real conjugates, and we can say nothing more about this case. All other possibilities for the + case are covered by the following Theorem.

Theorem 21. *Let p and q be totally real algebraic numbers with $q < 0$, and with $\mathbb{Q}(p) \subseteq \mathbb{Q}(q)$. Let m_q be the minimal polynomial of q, which of course splits over $\mathbb{Q}(p)$ as $m_q = f_q g_q$, where f_q is the minimal polynomial of q over $\mathbb{Q}(p)$. Suppose that all the roots of g_q are positive, and that f_q has at least two negative roots. Then $\alpha = p + \sqrt{q}$ has more than two non-real conjugates, and $S(\alpha) \subseteq \mathbb{R} \cup L(p)$. Moreover all algebraic numbers β with more than two non-real conjugates such that $S(\beta) \subseteq \mathbb{R} \cup L(p)$ for some p arise in this way.*

Proof. Suppose that p, q satisfy all the conditions of the Theorem, with p of degree r and q of degree rs. Then m_q has degree rs, and splits over $\mathbb{Q}(p)$ as a product of s polynomials of degree r, one of which is f_q, and the others have only positive roots. Let G be the Galois group of the normal closure of $\mathbb{Q}(q)$. If $\theta \in G$, then $\theta(p) = p$ if and only if $\theta(q)$ is a root of f_q. If β is any non-real conjugate of $\alpha = p + \sqrt{q}$, then $\beta = \theta(p) \pm \sqrt{\theta(q)}$ for some $\theta \in G$ and being non-real implies that $\theta(q) < 0$, hence $\theta(q)$ is a root of f_q, imply that $\theta(p) = p$, and β lies on $L(p)$. The fact that f_q has at least two negative roots implies that α has at least four non-real conjugates.

The converse requires the work, of course. Suppose that β has more than two non-real conjugates and that $S(\beta) \subseteq \mathbb{R} \cup L(p)$. The proof proceeds in four steps: (i) p is totally real; (ii) letting $p + \sqrt{q}$ be one of the non-real conjugates of β, with $q < 0$, q is totally real; (iii) $p \in \mathbb{Q}(q)$; (iv) m_q has the desired form.

Step (i): p is totally real. Take β_1 and β_2 to be non-real conjugates of β with distinct imaginary parts. Then we have

$$\beta_1 + \overline{\beta_1} = \beta_2 + \overline{\beta_2} = 2p. \tag{12}$$

Applying any automorphism φ gives

$$\varphi(\beta_1) + \varphi(\overline{\beta_1}) = \varphi(\beta_2) + \varphi(\overline{\beta_2}) = 2p', \tag{13}$$

say. Suppose first that $\varphi(\beta_1)$ is real, and $\varphi(\overline{\beta_1}) = p + iq$ is not. Now

- if $\varphi(\beta_2)$ and $\varphi(\overline{\beta_2})$ are both real, then (13) gives a contradiction, as $\varphi(\beta_1) + \varphi(\overline{\beta_1})$ is not real;
- if $\varphi(\beta_2)$ and $\varphi(\overline{\beta_2})$ both have real part p, then equating real parts in (13) gives $\varphi(\beta_1) = p$, and then (12) gives a contradiction with Remark 20;
- if one of $\varphi(\beta_2)$ and $\varphi(\overline{\beta_2})$ is real and the other has real part p, then equating real parts in (13) we see that the real one equals $\varphi(\beta_1)$, contradicting distinctness of $\beta_1, \beta_2, \overline{\beta_2}$.

A similar contradiction occurs if $\varphi(\overline{\beta_1})$ is real and $\varphi(\beta_1)$ is not. Thus $\varphi(\beta_1)$ and $\varphi(\overline{\beta_1})$ are either both real, or both have real part p. Next suppose that both $\varphi(\beta_1)$ and $\varphi(\overline{\beta_1})$ are real. Then p' is real, by (13). Finally, if both $\varphi(\beta_1)$ and $\varphi(\overline{\beta_1})$ have real part p, then adding (13) to its complex conjugate gives $4p = 2p' + 2\overline{p'}$, contradicting Remark 20.

Step (ii): putting $\beta' = p + \sqrt{q}$ for one of the non-real conjugates of β, with q real and negative, we show that q is totally real. Applying automorphisms to

$$(\beta' - \overline{\beta'})^2 = 4q,$$

and gleaning from the proof of Step (i) that any automorphism either maps both β' and $\overline{\beta'}$ to the real line, or both to the line $L(p)$, we see that all conjugates of q are real.

Step (iii): $p \in \mathbb{Q}(q)$. For if not, we could take an automorphism fixing q but not p, and hence mapping $p + \sqrt{q}$ to $p' \pm \sqrt{q}$ with $p' \neq p$, giving a conjugate of β that is not on $\mathbb{R} \cup L(p)$.

Step (iv): m_q has the desired form. We have that $q = (\beta - p)^2 < 0$ is a root of f_q, the minimal polynomial of q over $\mathbb{Q}(p)$, and indeed the other roots of f_q will be of the form $(\beta' - p)^2$, where β' is a conjugate of β. Since is assumed to have more than two non-real conjugates, there will be some $\notin \{\beta, \overline{\beta}\}$ with real part p, giving at least two negative roots for f_q. The roots of $g_q = m_q/f_q$ are of the form $(\beta' - p')^2$, where β' is a conjugate of β and $p' \neq p$ is a conjugate of p. Since q is totally real, β' must be real (the alternative of having real part p leads to $(\beta' - p')^2$ not being real), and hence the roots of g_q are all positive. $\qquad\square$

7.4. The $=$ case.

Theorem 22. *Let $h = h_1$ be a totally real algebraic number, with $h < 0$, and with all other conjugates $h_2, \ldots, h_d$ being positive. Let r be a totally real algebraic number, and let $\epsilon_j \in \{1, -1\}$ for $1 \leq j \leq d$. Then the algebraic number*

$$\alpha = r + \epsilon_1 \sqrt{h} + \epsilon_2 \sqrt{h_2} + \cdots + \epsilon_d \sqrt{h_d} \qquad (14)$$

lies with all its conjugates on $\Xi(\sqrt{-h})$.

Conversely, any algebraic number β with $S(\beta) \subseteq \Xi(k)$ (for some $k > 0$) arises in this way, with $h = -k^2$.

Proof. Applying any automorphism to (14) will permute the h_j, change the sign of some of the ϵ_j, and send r to the real line, hence mapping α to another element of $\Xi(\sqrt{-h})$. Proving the converse requires rather more effort.

Suppose that β lies with all its conjugates on $\Xi(k)$, and put $h_1 = -k^2$. Then certainly $h_1 < 0$. The proof proceeds in three steps: (i) h_1 is totally real; (ii) the other conjugates of h_1 ($h_2, \ldots, h_d$, say) are all positive; (iii) there is a choice of the ϵ_j that makes $r = \beta - \epsilon_1 \sqrt{h_1} - \epsilon_2 \sqrt{h_2} - \cdots - \epsilon_d \sqrt{h_d}$ totally real.

Step (i): h_1 is totally real. Putting $\gamma = \beta - \overline{\beta}$, we have $\gamma = \pm 2ki$. Applying an automorphism φ, we have

$$\gamma' := \varphi(\gamma) = \varphi(\beta) - \varphi(\overline{\beta}) = \beta' - \beta'' = \pm 2\varphi(k)i, \qquad (15)$$

where $\beta' = \varphi(\beta)$ and $\beta'' = \varphi(\overline{\beta})$. Since β' and β'' are on $\Xi(k)$, we have that γ' is either real, or has imaginary part $\pm 2k$. Suppose that $\gamma' = \eta \pm 2ki$ for some non-zero real η. Then

$$\gamma' - \overline{\gamma'} = \pm 4ki = \pm 2\gamma, \qquad (16)$$

contradicting Remark 20. Hence γ' is either real or purely imaginary. Hence $\varphi(h_1) = \varphi(2ki)^2/4 = (\gamma')^2/4$ is real. Thus h_1 is totally real.

Step (ii): $h_2, \ldots, h_d$ are all positive. For suppose that $h_i \neq h_1$ is a negative conjugate of h_1. Taking φ that maps h_1 to h_i, we have (using $\beta - \overline{\beta} = \pm 2ki = \pm 2\sqrt{h_1}$)

$$\varphi(\beta) - \varphi(\overline{\beta}) = \pm 2\sqrt{h_i},$$

and similar reasoning to that in Step (i) shows that this is either real or equal to $\pm 2\sqrt{h_1}$. Since the former case is excluded ($h_i < 0$) we have $\sqrt{h_i} = \pm\sqrt{h_1}$, so that $h_i = h_1$, a contradiction.

Step (iii): there is a choice of the ϵ_j that makes

$$r = \beta - \epsilon_1 \sqrt{h_1} - \epsilon_2 \sqrt{h_2} - \cdots - \epsilon_d \sqrt{h_d}$$

totally real. Choose automorphisms $\varphi_1, \ldots, \varphi_d$ such that $\varphi_i(h_i) = h_1$. The imaginary part of $\varphi_i(\beta)$ is $\pm k$, so we can choose ϵ_i such that $\varphi_i(\beta - \epsilon_i\sqrt{h_i})$ is real. For these ϵ_i, define r by $r = \beta - \epsilon_1\sqrt{h_1} - \epsilon_2\sqrt{h_2} - \cdots - \epsilon_d\sqrt{h_d}$, so that

$$\beta = r + \epsilon_1\sqrt{h_1} + \cdots + \epsilon_d\sqrt{h_d}. \tag{17}$$

Note that since $\varphi_i(\beta - \epsilon_i\sqrt{h_i})$ is real, and φ_i permutes the h_j, and all the h_j for $j > 1$ are positive, we have that $\varphi_i(r)$ is real. Now we claim that for any j between 1 and d, if we define

$$\delta_i = \begin{cases} \epsilon_i & \text{if } i \neq j, \\ -\epsilon_i & \text{if } i = j, \end{cases}$$

then $\beta_j = r + \sum_{i=1}^{d} \delta_i\sqrt{h_i}$ is a conjugate of β. For if we apply φ_j to (17), then apply complex conjugation, then apply φ_j^{-1}, we get β_j. Crucially this uses the fact that $\varphi_j(r)$ is real.

We now have enough information to see that r is totally real. For suppose that $\varphi(r)$ is not real. Then the imaginary part of $\varphi(r)$ must be $\pm 2k$, from (17). Define j by $\varphi(h_j) = h_1$. Then the imaginary parts of $\varphi(\beta) - \varphi(r)$ and $\varphi(\beta_j) - \varphi(r)$ both have magnitude k, but have opposite signs, so that not both $\varphi(\beta)$ and $\varphi(\beta_j)$ lie on $\Xi(k)$, a contradiction. $\qquad\square$

7.5. The $\parallel$ case.

Theorem 23. *(i) Let p and q be distinct real algebraic numbers, neither of them rational, such that $p + q$ is rational, and $h_1 = (p-q)^2/4$ is totally real, with all other conjugates of h_1 ($h_2, \ldots, h_d$, say) being negative. Let r be a totally real algebraic number, and let $\epsilon_j = \pm 1$ for j in the range $1 \leq j \leq d$. Then*

$$\alpha = \frac{p+q}{2} + \epsilon_1\sqrt{h_1} + \cdots + \epsilon_d\sqrt{h_d} + ir$$

lies with all its conjugates on $L(p) \cup L(q)$.

(ii) Let p be a real cubic algebraic number that has two non-real conjugates with real part q, and let r be totally real. Then

$$\alpha = p + ir$$

lies with all its conjugates on $L(p) \cup L(q)$.

(iii) All algebraic numbers β that lie with all their conjugates on some $L(p) \cup L(q)$, but not on a single line, arise as one of (i) or (ii).

A simple idea to transform the $\parallel$ case to the $=$ case is to multiply $\alpha - (p+q)/2$ by i. In some cases, this achieves the desired reduction, but there is more to the story than this, as evinced by part (ii) of the Theorem, which has no analogue in the $=$ case.

Proof. It is clear that for either (i) or (ii) we have $S(\alpha) \subseteq L(p) \cup L(q)$. There remains (iii). We suppose that β lies with all its conjugates on some $L(p) \cup L(q)$, but not on a single line.

We proceed in six steps: (i) the number of conjugates of β on each of the two lines is either 1 or is even; (ii) p and q are not rational; (iii) if φ maps a conjugate of β from one line to the other, then it maps either half or all of the conjugates from that one line to the other; (iv) the numbers of conjugates of β on each of the two lines are either equal, or one is twice the other; (v) the case where the numbers of conjugates on the two lines are equal is covered by the first part of the Theorem; (vi) the case where one line has twice the number of conjugates of the other is covered by the second part of the Theorem.

Step (i): the number of conjugates of β on each of the two lines is either 1 or is even. This is almost immediate from Remark 20, for if say there were an odd number of conjugates on $L(p)$, then one of these would equal p, and if there were any others, then the sum of a complex conjugate pair would give $2p$, contradicting the Remark.

Step (ii): p and q are not rational. Suppose that p, say, were rational. If $L(p)$ contained just one conjugate of β, then this would be p, and since not all conjugates lie on the same line we would have p being conjugate to some other number, contradicting $p \in \mathbb{Q}$. We are reduced to the case where the line $L(p)$ contains a non-real conjugate of β, say β_1. Applying an automorphism φ that sends β_1 to a number on the other line, $L(q)$, and equating real parts in $\varphi(\beta_1) + \varphi(\overline{\beta_1}) = 2p$ gives either $q + p = 2p$ or $2q = 2p$, contradicting $p \neq q$.

Step (iii): if φ maps a conjugate of β from one line to the other, then it maps either half or all of the conjugates from that one line to the other. Let us suppose that $\varphi(\beta_1)$ has real part q, where β_1 is a conjugate of β with real part p. If β_1 is the only conjugate on its line, then the conclusion of Step (iii) is trivial; so we may suppose not, and that by Step (i) there are an even number of conjugates $\beta_1, \overline{\beta_1}, \ldots, \beta_c, \overline{\beta_c}$ with real part p.

Applying φ to $\beta_1 + \overline{\beta_1} = 2p$, and taking real parts, gives

$$q + \Re\big(\varphi(\overline{\beta_1})\big) = 2\Re\big(\varphi(p)\big),$$

and hence $\varphi(p)$ has real part either q or $(p+q)/2$. If the real part of $\varphi(p)$ is q, then all of the conjugates on $L(p)$ will be mapped to $L(q)$; if the real part of $\varphi(p)$ is $(p+q)/2$, then each complex conjugate pair of conjugates of β on $L(p)$ will be mapped one to each line, so that exactly half of the conjugates on $L(p)$ will be mapped to $L(q)$.

Step (iv): the numbers of conjugates of β on the two lines are either equal ($d/2$ on each) or one is twice the other ($d/3$ on one line and $2d/3$ on the other). Let φ map a conjugate with real part p to one with real part q; then it must map some conjugate with real part q to one with real part p. Split the conjugates of β into four sets, $B_{pp}, B_{pq}, B_{qp}, B_{qq}$, where $\gamma \in B_{rs}$ means that γ has real part r and $\varphi(\gamma)$ has real part s. Let $b_{rs} = |B_{rs}|$. We have of course that $b_{pq} = b_{qp}$. By Step (iii), we have either $b_{pq} = b_{pp}$ or $b_{pp} = 0$; and either $b_{qp} = b_{qq}$ or $b_{qq} = 0$. All four cases give $b_{pp} + b_{pq} = r(b_{qp} + b_{qq})$, where $r = 1, 2$, or $1/2$.

Step (v): the case where the numbers of conjugates on the two lines are equal is covered by the first part of the Theorem. First we show that in this case $p+q$ is rational. Note from the proof of Step (iii) that any automorphism either preserves the sets of conjugates on the two lines (Type 1), or sends conjugate pairs to different lines (Type 2). Say β has real part p, and $\varphi(\beta)$ has real part q. Then applying automorphisms to

$$p + q = \left(\beta + \overline{\beta} + \varphi(\beta) + \overline{\varphi(\beta)}\right)/2$$

we see that $p + q$ is fixed by automorphism of both Types. Hence $p + q$ is rational. Now $i\left(\beta - (p+q)/2\right)$ lies with all its conjugates on $\Xi\left((p-q)/2\right)$. We can appeal to the $=$ case to complete Step (v).

Step (vi): the case where one line has twice the number of conjugates of the other is covered by the second part of the Theorem. Suppose that $d/3$ of the conjugates have real part p, and that $2d/3$ have real part q. From Step (iii) of the proof, we note that any automorphism either permutes the conjugates with real part p, or sends them all to the other line. Take β_1 a conjugate with real part p. Applying automorphisms to

$$p = (\beta_1 + \overline{\beta_1})/2$$

we see that $S(p) \subseteq L(p) \cup L(q)$. By Remark 20, no conjugates other than p can have real part p. Note that q itself cannot be a conjugate of p, else applying an automorphism that maps q to p would map all the conjugates of β on $L(q)$ to $L(p)$, which contradicts the assumption that twice as many have real part q. Applying Step (iv) to $S(p)$ (rather than to $S(\beta)$), we conclude that p has exactly two conjugates with real part q, so p is cubic. Replacing β by one of its conjugates on $L(p)$, we have $\beta = p + ir$ for some r. It remains to show that r is totally real. Suppose that $\varphi(r)$ is not real. The real part of $\varphi(p)$ is either p or q: suppose the former (the other case is similar). Then $\varphi(\beta) = \varphi(p + ir) = \varphi(p) \pm i\varphi(r)$ does not have real part p, so must have real part q, giving $q - p$ for the real part of $\pm i\varphi(r)$. But then $\varphi(\overline{\beta})$ has real part $p - (q - p) = 2p - q$, which is neither p nor q, a contradiction. Hence r is totally real. □

7.6. **The $\times$ case.** The most difficult pair-of-lines case is when an algebraic number α and all its conjugates lie on the pair of lines $X(a, \theta)$ for some a and θ. Berry provides an answer that is almost complete, but as with Smyth's work on conics there is a degree restriction. Perhaps surprisingly, a need not be rational; but at worst it is quadratic, provided that the degree of α is at least 10. We note that any cubic α will lie with all its conjugates on some $X(a, \theta)$, where a is a real conjugate.

Theorem 24. *(a) Let a be a real quadratic algebraic number, with conjugate $a' \neq a$, and let r be totally real. Define*

$$\alpha = \frac{a + a'}{2} + i\frac{a' - a}{2} + r\sqrt{i|a' - a|}\,.$$

Then $S(\alpha) \subseteq X(a, \pi/4)$.

(b) *Let a be a real quadratic algebraic number, with conjugate $a' \neq a$, and let r be totally real. Take $n \in \mathbb{N}$ square-free such that $a \in \mathbb{Q}(\sqrt{n})$. Take integers A, B, K with $A \geq 0$, $K > 0$, K square-free, and $A^2 n + B^2 K = E^2$ for some $E \in \mathbb{N}$. Define*

$$u = \frac{A\sqrt{n} + B\sqrt{-K}}{E}$$

and

$$\alpha = \frac{a + a'}{2} + u\frac{a' - a}{2} + r\sqrt{u^2 + 1}.$$

Then $S(\alpha) \subseteq X(a, \theta)$, *where* $u = |u|e^{2i\theta}$.

(c) *Let a be rational, and r totally real. Then*

$$\alpha = a + \sqrt{ir}$$

lies with all its conjugates on $X(a, \pi/4)$.

(d) *Let a be rational, let $u = u_1 = e^{2i\theta}$ be a reciprocal algebraic number with just two conjugates on the unit circle (u_1 and $1/u_1$) and all other conjugates real (u_2, $1/u_2$, $\ldots$, u_g, $1/u_g$), and let r be totally real. Then*

$$\alpha = a + re^{i\theta}\sqrt{|u_2 \cdots u_g|}$$

lies with all its conjugates on $X(a, \theta)$.

(e) *Let a be rational, let $u = u_1 = e^{2i\theta}$ ($0 < \theta < \pi/2$) be an even reciprocal algebraic number with just four conjugates on the unit circle ($\pm u_1$ and $\pm 1/u_1$) and all other conjugates real ($\pm u_2$, $\pm 1/u_2$, $\ldots$, $\pm u_g$, $\pm 1/u_g$), and let r be totally real. Then*

$$\alpha = a + re^{i\theta}\sqrt{2\cos(\theta)(u_2^2 + 1) \cdots (u_g^2 + 1)}$$

lies with all its conjugates on $X(a, \theta)$.

(f) *Any algebraic number that lies with all its conjugates on $X(a, \theta)$ for some a and θ, with a either rational or real quadratic, and $\theta \in (0, 2\pi)$, arises in one of the above five ways, (a), (b), (c), (d), (e).*

(g) *If an algebraic number of degree at least 10 lies with all its conjugates on $X(a, \theta)$ for some a and θ, with $\theta \in (0, 2\pi)$, then a is either rational or real quadratic.*

Proof. It is not hard to see that (a), (b), (c), (d), (e) hold. (For (d), note that the pairs $\{u_i, 1/u_i\}$ are permuted by any automorphism; for (e) note that the quartets $\{u_i, -u_i, 1/u_i, -1/u_i\}$ are permuted.) We shall now prove (g), and then (f).

We assume that α has degree at least 10, and lies with all its conjugates on $X(a, \theta)$. If the degree of α is even, then we shall list its conjugates at α_1, $\overline{\alpha_1}$, $\ldots$, α_d, $\overline{\alpha_d}$, where we suppose that $\alpha = \alpha_1$, $\ldots$, α_d lie on one of our two lines, with their complex conjugates on the other. We shall soon reduce to this case, but before we have done so we allow the possibility that $\alpha = a$, and that the remaining conjugates are $\alpha_1, \overline{\alpha_1}, \ldots, \alpha_d, \overline{\alpha_d}$, where again we have $\alpha_1, \ldots, \alpha_d$ on one of our two lines. Moreover, in both cases, we can order the conjugates such that $\arg(\alpha_i - \alpha_j) = \theta = -\arg(\overline{\alpha_i} - \overline{\alpha_j})$ whenever $i > j$.

The steps in the proof of (g) are as follows: (i) a is algebraic; (ii) a is not one of the conjugates of α (so that the degree of α must be even); (iii) a is totally real; (iv) a is rational or quadratic.

7.6.1. *a is algebraic.* Clearly we have (given that the degree is at least 4)

$$e^{2i\theta} = (\alpha_1 - a)/(\overline{\alpha_1 - a}) = (\alpha_2 - a)/(\overline{\alpha_2 - a}),$$

and rearranging gives

$$a = \frac{\alpha_1 \overline{\alpha_2} - \overline{\alpha_1} \alpha_2}{\alpha_1 - \overline{\alpha_1} - \alpha_2 + \overline{\alpha_2}}, \tag{18}$$

which reveals that a is algebraic.

7.6.2. *α and a are not conjugate (and hence the degree of α is even).* Suppose that $\alpha = a$, so that the degree of α is $2d + 1$. Let σ be an automorphism that sends α to α_1, and suppose that $\sigma^{-1}(\alpha) = \alpha_t$ (the case $\sigma^{-1}(\alpha) = \overline{\alpha_t}$ is entirely similar). We thus have $d - 1$ equations of the shape

$$(\alpha_t - \alpha)(\overline{\alpha_j} - \alpha) = (\overline{\alpha_t} - \alpha)(\alpha_j - \alpha), \qquad 1 \leq j \leq d, j \neq t.$$

Applying σ gives $d - 1$ equations

$$(\alpha - \alpha_1)\big(\sigma(\overline{\alpha_j}) - \alpha_1\big) = (\alpha' - \alpha_1)\big(\sigma(\alpha_j) - \alpha_1\big), \tag{19}$$

where $\alpha' = \sigma(\overline{\alpha_t})$. This implies that

$$\frac{\sigma(\alpha_j) - \alpha_1}{\sigma(\overline{\alpha_j}) - \alpha_1}$$

is constant ($j \neq t$), and in particular it has constant argument. If α' is one of the α_i, then this constant argument is 0; yet taking j such that one of $\alpha_j, \overline{\alpha_j}$ is $\overline{\alpha_1}$, we get that one of the numerator or denominator of (19) has argument $\pm\pi/2$ and the other does not, giving a contradiction. If α' is one of the $\overline{\alpha_j}$, then considering arguments in (19) we see that none of the $\sigma(\alpha_j)$ lie on the same line as α_1, and hence at least two of the $\sigma(\overline{\alpha_j})$ do, giving at least two of the $\alpha_j - \alpha_1$ with the same argument, and hence these α_j on the same line as α_1, a contradiction. This last part requires $d - 1 \geq 2$, and hence α has degree at least 7. Berry [1] gives a further argument to exclude degree 5, but since the main Theorem allows us to assume degree at least 10 we need not pursue this more detailed result.

7.6.3. *a is totally real.* We have that α has even degree $2d$, with $d \geq 5$, and it is in proving that a is totally real that we shall use this lower bound on the degree. Let $\Omega = \{\alpha_1, \dots, \alpha_d\}$, the set of conjugates that lie on the same line as $\alpha = \alpha_1$. Let us call this line L: the other line is $\overline{L}$. Let ω be the mean of the elements of Ω:

$$\omega = \frac{1}{d} \sum_{i=1}^{d} \alpha_i .$$

Note that if ω is real, then $\omega = a$, and then $a = (\omega + \overline{\omega})/2$ is preserved by all automorphisms, so is rational. We are therefore reduced to the case that ω is not real. We suppose that a is not rational, and let τ be any automorphism that moves a.

First we observe that if τ maps three elements α_i, α_j, α_k of Ω to the same line, then $\tau(\overline{\alpha_i})$, $\tau(\overline{\alpha_j})$, $\tau(\overline{\alpha_k})$ are collinear. For

$$\tau(e^{2i\theta}) = \frac{\tau(\alpha_i) - \tau(\alpha_j)}{\tau(\overline{\alpha_i}) - \tau(\overline{\alpha_j})} = \frac{\tau(\alpha_i) - \tau(\alpha_k)}{\tau(\overline{\alpha_i}) - \tau(\overline{\alpha_k})} = \frac{\tau(\alpha_j) - \tau(\alpha_k)}{\tau(\overline{\alpha_j}) - \tau(\overline{\alpha_k})},$$

and since at least two of $\tau(\overline{\alpha_i})$, $\tau(\overline{\alpha_j})$, $\tau(\overline{\alpha_k})$ lie on the same line, consideration of arguments shows that the third does too.

Next we observe that if σ is an automorphism that maps a pair α_i, $\overline{\alpha_i}$ to Ω, or to $\overline{\Omega}$, then $\sigma(a) = a$. We use $d > 4$ to note that at least three elements of $\sigma(\Omega)$ are collinear, and we may suppose (by relabelling if necessary) that they are all in Ω, and that $\sigma(\alpha_i)$ is one of them (if both α_i and $\overline{\alpha_i}$ map to $\overline{\Omega}$ then there must be some other complex conjugate pair that maps to Ω). Let α_j, α_k be two others that are mapped to Ω by σ; by our previous observation the complex conjugates of α_i, α_j, α_k are collinear and hence in Ω, and applying σ to

$$e^{2i\theta} = \frac{\alpha_i - \alpha_j)}{\overline{\alpha_i} - \overline{\alpha_j}} = \frac{\alpha_i - \alpha_k}{\overline{\alpha_i} - \overline{\alpha_k}} = \frac{\alpha_j - \alpha_k}{\overline{\alpha_j} - \overline{\alpha_k}}$$

gives $\sigma(e^{2i\theta}) \in \mathbb{R}$. But now applying σ to

$$e^{2i\theta} = \frac{\alpha_i - a}{\overline{\alpha_i} - a},$$

we see that $\sigma(a)$ lies on L. The hypothesis on σ is symmetric in L and $\overline{L}$, so $\sigma(a)$ must also lie on $\overline{L}$, and hence $\sigma(a) = a$.

It follows that for $j = 1, \dots, d$, our automorphism τ (which moves a) must map one of α_j, $\overline{\alpha_j}$ to Ω and the other to $\overline{\Omega}$. We deduce that

$$\tau(\overline{\Omega}) = \overline{\tau(\Omega)}. \tag{20}$$

Next we observe that $|\tau(e^{2i\theta})| = 1$. For any i and j we have

$$\tau(e^{2i\theta}) = \frac{\tau(\alpha_i) - \tau(\alpha_j)}{\tau(\overline{\alpha_i}) - \tau(\overline{\alpha_j})}. \tag{21}$$

Since $\tau(\overline{\Omega}) = \overline{\tau(\Omega)}$, we have

$$\max_{i,j} |\tau(\alpha_i) - \tau(\alpha_j)| = \max_{k,l} |\tau(\overline{\alpha_k}) - \tau(\overline{\alpha_l})|,$$

and taking i and j to maximise first the numerator and then the denominator in (21) we deduce both $|\tau(e^{2i\theta})| \geq 1$ and $|\tau(e^{2i\theta})| \leq 1$.

Finally we conclude that a is totally real. For averaging (20) gives

$$\tau(\overline{\omega}) = \overline{\tau(\omega)},$$

and applying τ to the average of

$$e^{2i\theta} = \frac{\alpha_j - a}{\overline{\alpha_j} - a}$$

then gives

$$\tau(e^{2i\theta}) = \frac{\tau(\omega) - \tau(a)}{\overline{\tau(\omega)} - \tau(a)},$$

and since $|\tau(e^{2i\theta})| = 1$ we have that $\tau(a)$ is real.

7.6.4. a *is rational or quadratic.*

We preserve all the above notation and conventions. Suppose that φ is an automorphism for which $\varphi(a) = a' \neq a$. We show first that for no i and j (perhaps equal) does $\varphi(\{\alpha_i, \overline{\alpha_i}\}) = \{\alpha_j, \overline{\alpha_j}\}$. For suppose such i and j exist. Replacing φ by $\overline{\varphi}$ if necessary, we can suppose that $\varphi(\alpha_i) = \alpha_j$. Recalling

$$e^{2i\theta} = \frac{\alpha_i - a}{\overline{\alpha_i} - a} = \frac{\alpha_i - \alpha_s}{\overline{\alpha_i} - \alpha_s}$$

(for all $s \neq i$) we have

$$\varphi(e^{2i\theta}) = \frac{\alpha_j - a'}{\overline{\alpha_j} - a'} = \frac{\alpha_j - \varphi(\alpha_s)}{\overline{\alpha_j} - \varphi(\overline{\alpha_s})}$$

for all $s \neq i$.

Next we deduce that $\varphi(\alpha_s) \in \overline{\Omega}$ for all $s \neq i$. For if not, say with $\varphi(\alpha_s) = \alpha_t$, we would have (from an earlier argument) that $\varphi(\overline{\alpha_s}) \in \overline{\Omega}$. Now if $\varphi(\overline{\alpha_s}) \neq \overline{\varphi(\alpha_s)}$, then using $|\varphi(e^{2i\theta})| = 1$ we would get

$$|\alpha_j - \alpha_s| = |\overline{\alpha_j} - \overline{\alpha_m}|$$

for some $m \neq s$, giving $\alpha_s + \alpha_m = 2\alpha_j$, contradicting Remark 20. Thus $\varphi(\overline{\alpha_s}) = \overline{\varphi(\alpha_s)}$, and hence

$$\frac{\alpha_j - a'}{\overline{\alpha_j} - a'} = \frac{\alpha_t - a'}{\overline{\alpha_t} - a'},$$

giving $a' = a$ (compare (18), with 1, 2 replaced by j, t).

Since $\varphi(\alpha_s) \in \overline{\Omega}$ for all $s \neq i$, we must have $\varphi(\overline{\alpha_s}) \in \Omega$ for all $s \neq i$. As above, we can never have $\varphi(\overline{\alpha_s}) = \overline{\varphi(\alpha_s)}$ for $s \neq i$. Hence there is a permutation $r \mapsto r^*$ of $\{1, \ldots, d\} \setminus \{j\}$ that has no fixed points, such that

$$|\alpha_r - a'| = |\overline{\alpha_r} - a'| = |\alpha_{r^*} - a'| = |\overline{\alpha_{r^*}} - a'|,$$

and hence

$$\alpha_r + \alpha_{r*} = 2p, \tag{22}$$

for all $r \neq j$, where p is on L such that the angle apa' is $\pi/2$. Moreover, there is therefore no conjugate α' of α such that $\alpha_j + \alpha' = 2p$, as all conjugates other that α_j and $\overline{\alpha_j}$ have been paired up by (22).

We are trying to show that for no i and j (perhaps equal) does $\varphi(\{\alpha_i, \overline{\alpha_i}\}) = \{\alpha_j, \overline{\alpha_j}\}$, and we are now reduced to the case where at most one such pair (i, j) exists, and with the permutation $r \mapsto r^*$ pairing up all the conjugates of α other than α_j and $\overline{\alpha_j}$ via (22). Choose $r \neq j$, and choose an automorphism τ such that $\tau(\alpha_r) = \alpha_j$. Since $\alpha_j + \alpha'$ never equals $2p$, for any conjugate α' of α, we have

$$2\tau(p) = \tau(\alpha_r) + \tau(\alpha_{r*}) = \alpha_j + \alpha' \neq 2p,$$

(with $\alpha' = \tau(\alpha_{r*})$), and so $\tau(p) \neq p$. We split into two cases, and produce a contradiction in each: (A) $\alpha' \in \Omega$; (B) $\alpha' \in \overline{\Omega}$.

First, then, suppose that $\alpha' \in \Omega$. Then $\tau(p) = (\alpha_j + \alpha')/2 \in L$. Using $\tau(p) = (\tau(\alpha_s) + \tau(\alpha_{s*}))/2$ for $s \neq j$, we have that $\tau(\alpha_s) \in L$ for all $s \neq j$. Take some $s \neq j$, and put $\tau(\alpha_s) = \alpha_a$, $\tau(\alpha_{s*}) = \alpha_b$. Then

$$2\tau(p) = \alpha_j + \alpha' = \alpha_a + \alpha_b.$$

Now take m such that $\tau(\alpha_m)$ is one of α_{a*} or α_{b*}, say α_{a*}. Then

$$2\tau(p) = \tau(\alpha_m) + \tau(\alpha_{m*}) = \alpha_{a*} + \alpha'',$$

say. Thus we have both

$$\alpha_a + \alpha_b = \alpha_{a*} + \alpha''$$

and

$$\alpha_a + \alpha_{a*} = \alpha_b + \alpha_{b*} \qquad (= 2p).$$

Adding these gives $2\alpha_a = \alpha'' + \alpha_{b*}$, contradicting Remark 20.

Next we treat the case $\alpha' \in \overline{\Omega}$. Then $\tau(p) = (\alpha_j + \alpha')/2$ can be on neither L nor $\overline{L}$. Choose $s \neq \{j, r\}$. Since $\alpha_s + \alpha_{s*} = 2p$, and $\tau(p)$ is not on L or $\overline{L}$, we must have one of $\tau(\alpha_s)$, $\tau(\alpha_{s*})$ on L and the other on $\overline{L}$. Suppose the former (the other case goes through in the same way). Then from

$$2\tau(p) = \alpha_j + \alpha' = \tau(\alpha_s) + \tau(\alpha_{s*})$$

we get

$$\alpha_j - \tau(\alpha_s) = \tau(\alpha_{s*}) - \alpha',$$

so the line through α_j and $\tau(\alpha(s))$ (which is L) is parallel to that through α' and $\tau(\alpha_{s*})$ (which is L'), which contradicts their intersection at a.

To sum up, we have shown that for no i and j (perhaps equal) does $\varphi(\{\alpha_i, \overline{\alpha_i}\}) = \{\alpha_j, \overline{\alpha_j}\}$. It follows that there is a permutation $r \mapsto r^*$ of $\{1, \ldots, d\}$ with no fixed points, such that

$$|\alpha_r - a'| = |\alpha_{r*} - a'|$$

for all r. Moreover, the formula $\alpha_r + \alpha_{r*} = 2p$ now reads

$$\alpha_r + \alpha_{r*} = 2\omega, \tag{23}$$

for all r.

We are inching towards the conclusion that a is quadratic (given the existence of φ such that $\varphi(a) = a' \neq a$). We now show that ω, the average of the conjugates on L, is quadratic (or rational). Let τ be any automorphism. Suppose that one of $\tau(\alpha_1)$, $\tau(\alpha_{1*})$ is in Ω and the other is in $\overline{\Omega}$. Then from (23), $\tau(\omega)$ lies neither on L nor on $\overline{L}$, and from (23) again one of $\tau(\alpha_2)$, $\tau(\alpha_{2*})$ lies on L and the other on $\overline{L}$. One could then rearrange $\tau(\alpha_1) + \tau(\alpha_{1*}) = \tau(\alpha_2) + \tau(\alpha_{2*})$ to show that L and $\overline{L}$ are parallel, which is nonsense. We conclude that both $\tau(\alpha_1)$ and $\tau(\alpha_{1*})$ lie on the same line, L or $\overline{L}$, and hence so does ω. Then, using (23) yet again, we conclude that τ either maps Ω to Ω, or maps Ω to $\overline{\Omega}$. Thus $\tau(\omega)$ is one of ω or $\overline{\omega}$, which implies that ω is quadratic (or rational). We also conclude that (with $a' \neq a$) any automorphism maps Ω either to itself or to $\overline{\Omega}$.

We can now show that a is quadratic, and that $e^{2i\theta}$ either equals i, or is quartic, in which case its conjugates are $\pm e^{\pm 2i\theta}$. Moreover, in this quartic case, the automorphisms that send $e^{2i\theta}$ to $-e^{\pm 2i\theta}$ are precisely those that send a to a'. We use the various formulas

$$e^{2i\theta} = \frac{\alpha_r - a}{\overline{\alpha_r} - a} = \frac{\alpha_r - \alpha_s}{\overline{\alpha_r} - \overline{\alpha_s}},$$

$$a = \frac{\alpha_r \overline{\alpha_s} - \overline{\alpha_r} \alpha_s}{\alpha_r - \overline{\alpha_r} - \alpha_s + \overline{\alpha_s}}.$$

Apply an arbitrary automorphism τ to the first of these. Suppose first that $\tau : \Omega \to \Omega$. Then we have (for suitable choices of r and s in the above, and for some a and b)

$$\tau(e^{2i\theta}) = \frac{\alpha_1 - \alpha_d}{\overline{\alpha_a} - \overline{\alpha_b}} = \frac{\alpha_1 - \tau(a)}{\overline{\alpha_a} - \tau(a)} = \frac{\alpha_d - \tau(a)}{\overline{\alpha_b} - \tau(a)} = \frac{\omega - \tau(a)}{\overline{\omega} - \tau(a)}.$$

From the first and the last of the above string of equal numbers, we have $|\tau(e^{2i\theta})| = 1$. Since our ordering of the indices makes $|\alpha_1 - \alpha_d|$ maximal amongst all $|\alpha_r - \alpha_s|$, we must therefore have either $\alpha_1 = \overline{\alpha_a}$, $\alpha_d = \overline{\alpha_b}$, or $\alpha_1 = \overline{\alpha_b}$, $\alpha_d = \overline{\alpha_a}$. The former gives $\tau(a) = a$ and $\tau(e^{2i\theta}) = e^{2i\theta}$. The latter gives

$$\tau(a) = \frac{\alpha_1 \overline{\alpha_1} - \alpha_d \overline{\alpha_d}}{\alpha_1 + \overline{\alpha_1} - \alpha_d - \overline{\alpha_d}}$$

and $\tau(e^{2i\theta}) = -e^{2i\theta}$.

Next suppose that $\tau : \Omega \to \overline{\Omega}$. A similar argument gives the same two possibilities for $\tau(a)$, now with $\tau(e^{2i\theta}) = \pm e^{-2i\theta}$.

With just two possibilities for $\tau(a)$ we conclude that a is quadratic, and we see also that $e^{2i\theta}$ is at worst quartic, and that in the quartic case its conjugates are of the form claimed above.

This concludes the proof of (g), and we now move to (f), dealing first with the case where a is real quadratic, and then rational.

7.6.5. *The quadratic case.* We now suppose that α lies with all its conjugates on $X(a, \theta)$, where a is real quadratic, with conjugate $a' \neq a$. We note that either $\theta = \pi/4$ or both a and ω are in the field $\mathbb{Q}(e^{2i\theta})$. For suppose an automorphism τ maps ω to $\overline{\omega}$ and fixes $e^{2i\theta}$. Choose r and s such that $\tau(\alpha_r) = \overline{\alpha_1}$ and $\tau(\alpha_s) = \overline{\alpha_d}$. Then applying τ to

$$e^{2i\theta} = \frac{\alpha_r - \alpha_s}{\overline{\alpha_r} - \overline{\alpha_s}}$$

gives

$$e^{2i\theta} = \frac{\overline{\alpha_1} - \overline{\alpha_d}}{\alpha_a - \alpha_b},$$

say, and to achieve modulus 1 we must have $\{a, b\} = \{1, d\}$. This gives $e^{2i\theta} = \pm e^{-2i\theta}$, hence $\theta = \pi/2$ or $\theta = \pi/4$. The former is the $+$ case, so we conclude that unless $\theta = \pi/4$ we must have $\omega \in \mathbb{Q}(e^{2i\theta})$. Now, with $\theta \neq \pi/4$, we choose any automorphism σ that fixes $e^{2i\theta}$ (and hence fixes ω). Then

$$e^{2i\theta} = \frac{\omega - a}{\overline{\omega} - a} = \frac{\omega - \sigma(a)}{\overline{\omega} - \sigma(a)},$$

and hence $\sigma(a) = a$. Thus we have $a \in \mathbb{Q}(e^{2i\theta})$ too.

Next we establish the formula

$$\omega = \frac{a + a'}{2} + \frac{a - a'}{2}e^{2i\theta}.$$

From the final part of the proof of (g), which required only that a was totally real, we know that there is an automorphism τ that fixes ω, maps a to a' and maps $e^{2i\theta}$ to $-e^{2i\theta}$. Applying this to $e^{2i\theta} = (\omega - a)/(\overline{\omega} - a)$ and eliminating $\overline{\omega}$ gives the stated formula for ω.

Next we show that for some integers A, B, E, K, n, with $A \geq 0$, $B \neq 0$, $K > 0$, $n > 1$, n squarefree, K squarefree (perhaps equal to 1), and $A^2 n + B^2 K = E^2$ we have

$$e^{2i\theta} = \frac{A\sqrt{n} + B\sqrt{-K}}{E}.$$

Since a is real quadratic, $a \in \mathbb{Q}(\sqrt{n})$ for some squarefree integer $n > 1$. If $\theta = \pi/4$, we can take $A = 0$, $B = E = K = 1$. Otherwise, $e^{2i\theta}$ lies in the quartic extension $\mathbb{Q}(e^{2i\theta}) = \mathbb{Q}(\sqrt{n}, 2i \sin(\theta))$, an imaginary quadratic extension of $\mathbb{Q}(\sqrt{n})$, and hence $e^{2i\theta}$ is of the stated form. Since $e^{2i\theta}$ has modulus 1, we must have $A^2 n + B^2 K = E^2$.

Now suppose that $\theta = \pi/4$ (this leads to case(a)). Since α and ω are both on L, we can write

$$\alpha = \omega + r\sqrt{i|a' - a|}$$

for some real number r, and indeed we see that r is algebraic. Recalling $\omega = (a + a')/2 + i(a - a')/2$ (in this case), we see that the effect of a automorphism τ on ω is determined by its effect on a and i. With a being mapped to a or a', and i being mapped to $\pm i$, there are four cases to consider. Recalling also that $\tau(\omega) = \omega$ if and only if $\tau(\alpha)$ is on L, we check that in all cases $\tau(r)$ is real. Thus we are in case (a) of the Theorem.

We now deal with $\theta \neq \pi/4$. We have seen that $e^{2i\theta}$ is of the form stated in part (b) of the Theorem. Again since α and ω both lie on L we have

$$\alpha = \omega + r\sqrt{2\cos(\theta)}e^{2i\theta}$$

for some real number r, which is algebraic (since $e^{2i\theta}$ is). Again, considering all possibilities for the images of a and $e^{2i\theta}$, and using

$$\omega = (a + a')/2 + e^{2i\theta}(a - a')/2,$$

we check that all conjugates of r are real. Thus we are in case (b) of the Theorem.

7.6.6. *The rational case.* To complete the proof of part (f), we need to treat the case $a \in \mathbb{Q}$. We can simplify matters greatly by translating to achieve $a = 0$. We put $u = e^{2i\theta} = \alpha_r/\overline{\alpha_r}$ (for any r between 1 and d). Observing that u and $1/u = \overline{u}$ are conjugates, the minimal polynomial of u is reciprocal.

First we show that the non-real conjugates of u are among $\pm u^{\pm 1}$. Suppose that $|\tau(u)| > 1$ for some automorphism τ. Take s with $|\alpha_s|$ maximal (then s is either 1 or d, and perhaps $|\alpha_1| = |\alpha_d|$). Then since $\tau(u) = \tau(\alpha_r)/\tau(\overline{\alpha_r})$ for all r, we must have both $\tau(u) = \alpha_s/\alpha'$ and $\tau(u) = \overline{\alpha_s}/\alpha''$ for some α' and α'' conjugates of α. Now $\tau(u)$ has argument $\pm 2\theta$, or $\pi \pm 2\theta$, or 0, or π. If the argument is 2θ or $\pi + 2\theta$, then α'' has argument -3θ or $\pi - 3\theta$, and the only solution with $\theta \in (0, \pi/2)$ is $\theta = \pi/4$, giving $u = i$, contradicting $|\tau(u)| > 1$. Similarly we can exclude arguments -2θ or $\pi - 2\theta$ for $\tau(u)$. Hence $\tau(u)$ is real. Since conjugates of u come in reciprocal pairs, we have also that if $|\tau(u)| < 1$ then $\tau(u)$ is real. There remains the case $|\tau(u)| = 1$, with u not real. Then $\tau(u)$ is one of $\alpha_1/\overline{\alpha_1}$, $\alpha_1/\overline{\alpha_d}$, $\overline{\alpha_1}/\alpha_1$, $\overline{\alpha_1}/\alpha_d$, and the cases that mix α_1 and α_d occur only if $\alpha_1 = -\alpha_d$. We conclude that $\tau(u)$ is one of u, $-u$, $1/u$, $-1/u$, as claimed.

We may suppose that α is on L. We put $u = u_1 = e^{2i\theta}$. From the above discussion, we can split into three cases: (i) $u = i$; (ii) u has conjugates $u_i^{\pm 1}$ for $1 \leq i \leq g$, with $u_2, \ldots, u_g$ all real; (iii) u has conjugates $\pm u_i^{\pm 1}$ for $1 \leq i \leq g$, with $u_2, \ldots, u_g$ all real. In (ii) and (iii) we allow $g = 1$.

First, then, consider the case $u = i$. Let $r = \alpha^2/i$. Then certainly r is real. For any automorphism σ, the argument of $\sigma(\alpha)$ is one of $\pm \pi/4$ or $\pi \pm \pi/4$, and hence $\sigma(r)$ is always real. Thus we are in case (c) of the Theorem.

Next suppose that u has conjugates $u_i^{\pm 1}$ for $1 \leq i \leq g$, with $u_2, \ldots, u_g$ all real. Define $r = r(u_2, \ldots, u_g)$ by

$$\alpha = re^{i\theta}\sqrt{|u_2 \cdots u_g|}. \tag{24}$$

Then certainly r is real, but perhaps not totally real. Let τ_2 be an automorphism satisfying $\tau_2(u_2) = u$. Then $\tau_2(\alpha) = \pm\tau_2(r)e^{i\theta}\sqrt{|u_2\cdots u_g|}$. If $\tau_2(r)$ is not real, then since $\tau_2(\alpha)$ lies on $X(0,\theta)$, the argument of $\pm\tau_2(r)$ must be -2θ: in this case, if we replace u_2 by $1/u_2$, then the new r given by (24) will have $\tau_2(r)$ real (and now we have $\tau_2(u_2) = 1/u$). Similarly we can, after replacing certain of the u_i by $1/u_i$, find automorphisms τ_i ($2 \leq i \leq g$) such that $\tau_i(u_i) = u^{\pm 1}$, and such that each $\tau_i(r)$ is real. Now applying τ_i, then complex conjugation, then τ_i^{-1}, we see that α is conjugate to $\alpha/|u_i|$ for $2 \leq i \leq g$.

Now let σ be any automorphism. We aim to show that $\sigma(r)$ is real. Take t between 1 and g such that $\sigma(u_t) = u^{\pm 1}$. Since $\sigma(r)$ is real if and only if $\overline{\sigma}(r)$ is real, we may suppose (replacing σ by $\overline{\sigma}$ if need be) that $\sigma(u_t) = u$. If $t \geq 2$, then applying σ to both α and its conjugate $\alpha/|u_t|$, we see that both $\sigma(\alpha)$ and $\pm\sigma(\alpha)/u_1$ lie on $X(0,\theta)$. The same conclusion holds if $t = 1$, applying σ to both α and $\overline{\alpha} = \alpha/u_1$. If $\sigma(r)$ were not real, this would be impossible. Hence r is totally real, and we are in case (d) of the Theorem.

The case (iii) is entirely similar, leading to case (e) of the Theorem. $\quad\square$

8. CONCLUDING REMARKS AND QUESTIONS

8.1. **Integrality.**
Moving from algebraic numbers to algebraic integers usually presents a serious challenge, and much less is known in general. Even the case of conjugate sets in real intervals is not completely understood. Is there a real interval of length 4 with non-integral endpoints that contains only finitely many conjugate sets of algebraic integers? Is there such an interval that contains infinitely many?

8.2. **Lowering the degree bounds.**
Can any of the degree bounds be reduced? For parabolas, ellipses, hyperbolas, and pairs of lines, the Theorems include a lower bound on the degree. In all cases, this bound is an artefact of the particular proof, and perhaps an alternative method could lower the bound.

8.3. **Degree greater than 2.**

8.3.1. *More than two lines.*
A natural generalisation of the two-line result is to ask which algebraic numbers lie with all their conjugates on the union of d lines, for some $d > 2$.

8.3.2. *Higher degree curves.*
One might define the *geometric degree* of an algebraic number α to be the minimal degree of a polynomial $f(x, y)$ with integer coefficients such that all the conjugates of α are of the form $x+iy$ with x, y real and $f(x, y) = 0$. The algebraic numbers considered in this survey have geometric degree 1 or 2. What can be said about algebraic numbers of higher degree?

References

[1] N.M. Berry, *On relationships between conjugate algebraic numbers*, PhD thesis, University of Edinburgh, 2003.

[2] D.W. Boyd, Kronecker's theorem and Lehmer's problem for polynomials in several variables, *J. Number Theory* **13** (1981), 116–121.

[3] J.C. Burkill, H. Burkill, *A second course in mathematical analysis*, Cambridge University Press, 1970.

[4] A. Dubickas, C.J. Smyth, Two variations on a theorem of Kronecker, *Expo. Math.* **23** (2005), 289–294.

[5] V. Ennola, Conjugate algebraic integers on a circle with irrational center, *Math. Z.* **34** (1973), 337–350.

[6] V. Ennola, C.J. Smyth, Conjugate algebraic numbers on a circle, *Ann. Acad. Sci. Fenn. Ser. A* **I** (1974), 582.

[7] _____, _____, Conjugate algebraic numbers on circles, *Acta Arith.* **29** (1976), no. 2, 147–157.

[8] L. Kronecker, Zwei sätse über gleichungen mit ganzzahligen coefficienten, *J. Reine Angew. Math.* **53** (1857), 173–175.

[9] H.L. Montgomery, A. Schinzel, Some arithmetic properties of polynomials in several variables, *Transcendence theory: advances and applications (Proceedings of the Conference, University of Cambridge, 1976)*, 195–203, Academic Press, London, 1977.

[10] R.M. Robinson, Intervals containing infinitely many sets of conjugate algebraic integers, abstract in *Not. Amer. Math. Soc.* **6** (1959), 168–169.

[11] _____, Intervals containing infinitely many sets of conjugate algebraic integers, summary in *Report of the Institute of the Theory of Numbers* (University of Colorado, Boulder, Colorado, June 21–July 17, 1959), 200–206.

[12] _____, Intervals containing infinitely many sets of conjugate algebraic integers, *Mathematical analysis and related topics: essays in honor of George Pólya*, 305–315, Stanford, 1962.

[13] _____, Conjugate algebraic integers on a circle, *Math. Z.* **110** (1969), 41–51.

[14] I. Schur, Über die Verteilung der Wurzeln bei gewissen algebraischen Gleichungen mit ganzzahligen Koeffizienten, *Math. Z.* **1** (1918), 377–402.

[15] C.J. Smyth, A Kronecker-type theorem for complex polynomials in several variables, *Canad. Math. Bull.* **24** (1981), 447–452; Addenda and errata, **25** (1982), 504.

[16] _____, Conjugate algebraic numbers on conics, *Acta Arith.* **40** (1981/82), no. 4, 333–346.

ON POLYNOMIAL ERGODIC AVERAGES AND SQUARE FUNCTIONS

RADHAKRISHNAN NAIR

ABSTRACT. Let $\phi_1, \ldots, \phi_d$ be non-constant polynomials mapping the natural numbers to themselves. Let T_1, ..., T_d be commuting measure-preserving transformations of a probability space (X, β, μ). Let

$$A_N f(x) = \frac{1}{N} \sum_{n=1}^{N} f\left(T_1^{\phi_1(n)} \ldots T_d^{\phi_d(n)} x\right), \qquad (N = 1, 2, \ldots)$$

denote the corresponding ergodic averages constructed for an integrable function f defined on (X, β, μ). Also let

$$V_q f(x) = \left(\sum_{N=1}^{\infty} |A_{N+1}f(x) - A_N f(x)|^q\right)^{\frac{1}{q}},$$

for $q \geq 1$. We show that for $p, q > 1$ there exists $C_{p,q} > 0$ such that

$$\|V_q f(x)\|_p \leq C_{p,q} \|f\|_p.$$

We also give an example to show that this consequence is not possible for $q = 1$ even if f is essentially bounded. Finally we show that if the sequence of natural numbers $(N_k)_{k=1}^{\infty}$ satisfies $1 < a \leq \frac{N_{k+1}}{N_k} \leq b < \infty$, for certain a, b and

$$S f(x) = \left(\sum_{k=1}^{\infty} |A_{N_{k+1}}f(x) - A_{N_k}f(x)|^2\right)^{\frac{1}{2}},$$

then there exists $C > 0$ such that

$$\|Sf\|_2 \leq C\|f\|_2.$$

Here, of course, $\|f\|_p$ denotes the $L^p(X, \beta, \mu)$ norm of f.

1. INTRODUCTION

Suppose that (X, β, μ) is a probability space and that, for $i = 1, 2, \ldots, d$, $T_i : X \to X$ are commuting maps which are measure preserving, that is

$$\mu(T^{-1}A) = \mu(A)$$

for each set A in the σ-algebra β. Here for a set A we have used $T^{-1}A$ to denote $\{x : Tx \in A\}$. Suppose also that $\phi_1, \ldots, \phi_d$ are non-constant

2000 *Mathematics Subject Classification.* Primary: 28D05; Secondary: 37A25.
Key words and phrases. p-norm operator, variation function, polynomial ergodic averages, square functions.

polynomials mapping the natural numbers to themselves. For a measurable function f defined on X and for natural numbers $(N = 1, 2, \dots)$ set

$$A_N f(x) = \frac{1}{N} \sum_{n=0}^{N-1} f\left(T_1^{\phi_1(n)} \dots T_d^{\phi_d(n)} x\right). \tag{1.1}$$

Also for $q > 1$ set

$$V_q f(x) = \left(\sum_{N \geq 1} |A_{N+1} f(x) - A_N f(x)|^q\right)^{\frac{1}{q}}. \tag{1.2}$$

Let $L^p = L^p(X, \beta, \mu)$ denote the space of equivalence classes of functions differing only on sets of measure zero, with norm $\|f\|_p = (\int_X |f|^p d\mu)^{\frac{1}{p}}$. We have the following theorems, the first of which answers a question of M. Weber, put to the author personally.

Theorem 1. *Suppose that $f \in L^p$ for $p > 1$, that $\left(A_N f(x)\right)_{N=1}^{\infty}$ is defined as in (1.1) and that $V_q f$ is defined in (1.2). Then for $q > 1$ there exist constants $C_{p,q} > 0$ such that*

$$\|V_q f\|_p \leq C_{p,q} \|f\|_p.$$

What happens when $p = 1$ is unknown to the author. We call the dynamical system $(X, \beta, \mu, T_1, \dots T_d)$ *good* if for every integrable function f on X we have

$$\lim_{N \to \infty} A_N f(x) = \int_X f d\mu \quad \mu \quad a.e.,$$

where the limit exists. There is a spectral characterisation of this in [4] which provides a means of obtaining good dynamical systems, for instance as automorphisms of compact abelian groups. We have the following result.

Theorem 2. *Suppose that $(X, \beta, T_1, \dots T_d)$ is good and that the measure space (X, β, μ) is non-atomic. Then for any non-constant function f on (X, β, μ) we have*

$$V_1 f(x) = +\infty \quad \mu \quad a.e..$$

Theorem 3. *Suppose that $f \in L^2(X, \beta, \mu)$, that the sequence of natural numbers $(N_k)_{k=1}^{\infty}$ satisfies*

$$1 < a \leq \frac{N_{k+1}}{N_k} \leq b < \infty$$

for certain a, b, and that

$$S f(x) = \left(\sum_{k=1}^{\infty} |A_{N_{k+1}} f(x) - A_{N_k} f(x)|^2\right)^{\frac{1}{2}}.$$

Then there exists $C > 0$ such that

$$\|S f\|_2 \leq C \|f\|_2.$$

For similar results when $d = 1$ see [5]. By a simple change of coordinates it is clear that without loss of generality we may, as we do, choose $\phi_i(x) = x^i$ ($i = 1, 2, \ldots$). These results, and indeed some of their proofs, are motivated by the parallels between polynomial ergodic averages and martingales. In particular, if $(Y_i)_{i=1}^{\infty}$ is a real valued integrable martingale, set

$$S_N = \frac{1}{N} \sum_{i=1}^{N} Y_i \quad (N = 1, 2, \ldots).$$

Then there exist $C_1, C_2 > 0$ such that

$$C_1 ||Y_1||_2 \leq \left|\left| \left(\sum_{N=1}^{\infty} |S_{N+1} - S_N|^2 \right)^{\frac{1}{2}} \right|\right|_2 \leq C_2 ||Y_1||_2.$$

This is known as the Burkholder-Gundy-Silverstinov inequality. In the special case where $(Y_i)_{i=1}^{\infty}$ is a sequence of independent identically distributed random variables the right hand inequality is implied by Theorem 1. Henceforth the letter C, possibly with subscripts, refers to a positive constant, not necessarily the same at each occurrence.

2. Proof of Theorem 1

Notice that for $q > 1$,

$$A_{N+1}f(x) - A_N f(x) = \frac{f\left(T_1^{N+1} \ldots T_d^{(N+1)^d} x - A_N f(x)\right)}{N+1}.$$

Using the $\ell^q(\mathbf{Z})$ triangle inequality we have

$$V_q f(x) \leq \left(\sum_{N \geq 1} \left(\left| \frac{f(T_1^{N+1} \ldots T_d^{(N+1)^d} x)}{N+1} \right| \right)^q \right)^{\frac{1}{q}} + \left(\sum_{N \geq 1} \left(\frac{|A_N f(x)|}{N+1} \right)^q \right)^{\frac{1}{q}}$$

$$= A_{(1)} f(x) + A_{(2)} f(x), \quad \text{say}. \tag{2.1}$$

We need the following lemma.

Lemma 4. *Suppose that* $\phi_1, \ldots, \phi_d, T_1, \ldots, T_d$ *and* (X, β, μ) *are as in the statement of Theorem 1. Let*

$$Mf(x) = \sup_{N \geq 1} \left| \frac{1}{N} \sum_{n=1}^{N} f(T_1^{\phi_1(n)} \ldots T_d^{\phi_d(n)} x) \right|,$$

where $f \in L^p(X, \beta, \mu)$ *with* $p > 1$. *Then there exists* $C_p > 0$ *such that*

$$||Mf||_p \leq C_p ||f||_p.$$

In the special case where $\phi(n) = n^d$, $\phi_1 = \phi$ and $\phi_2 \equiv \phi_3 \equiv \cdots \equiv 1$, this is proved in [2]. It is stated in [3], without detailed proof, that Lemma 4 holds in full generality, proved using the same method as in [1]. In the case $p = 2$, this lemma appears in [1]. As a consequence of Lemma 4, because $|A_N f(x)| \leq M f(x)$ for all $N \geq 1$, we see that there exists $C_{p,q} > 0$ such that

$$\|A_{(2)} f\|_p \leq C_{p,q} \|f\|_p,$$

hence Theorem 1 is proved if we can prove the following lemma:

Lemma 5. *For $A_{(1)} f$ defined by (2.1), if $f \in L^p(X, \beta, \mu)$ with $p > 1$, then*

$$\|A_{(1)} f\|_p \leq C_{p,q} \|f\|_p.$$

Proof. Let χ_A denote the characteristic function of the set A, that is

$$\chi_A(x) = \begin{cases} 1 & \text{if } x \in A, \\ 0 & \text{if } x \notin A. \end{cases}$$

Let $\eta > 0$ and set

$$a_n(x) = |f(T_1^n \dots T_d^{n^d} x)| \chi_{[x \in X : f(T_1^n \dots T_d^{n^d} x) \leq \eta(n+1)]}(x)$$

and

$$b_n(x) = |f(T_1^n \dots T_d^{n^d} x)| \chi_{[x \in X : f(T_1^n \dots T_d^{n^d} x) > \eta(n+1)]}(x).$$

Observe that in the notation for $(a_n)_{n \geq 1}$ and $(b_n)_{n \geq 1}$, we suppress mention of η. This is because its specific value plays no role in what follows. Note that

$$|f(T_1^n \dots T_d^{n^d} x)| = a_n(x) + b_n(x).$$

This means that by Minkowski's inequality,

$$A_{(1)} f(x) \leq B_1 f(x) + B_2 f(x),$$

where

$$B_1 f(x) = \left(\sum_{n \geq 0} \left(\frac{a_n(x)}{n+1} \right)^q \right)^{\frac{1}{q}}$$

and

$$B_2 f(x) = \left(\sum_{n \geq 0} \left(\frac{b_n(x)}{n+1} \right)^q \right)^{\frac{1}{q}}.$$

This tells us that

$$\|A_{(1)} f\|_p \leq \|B_1 f\|_p + \|B_2 f\|_p.$$

Hence Lemma 5 is proved if we show that there exists $C_p > 0$ such that

$$\|B_i f\|_p \leq C_{p,q} \|f\|_p, \tag{2.2}$$

for each $i = 1, 2$. We work with weak (1.1) estimates. That is we show that

$$\mu(\{x \in X : B_i f(x) \geq \lambda\}) \leq C_q \frac{\int_X |f| d\mu}{\lambda}, \tag{2.3}$$

and that

$$||B_i f||_\infty \le C_{\infty,q} ||f||_\infty. \tag{2.4}$$

The last inequality (2.4) for $p = \infty$ is self evident from the definition of $\big(a_k(x)\big)_{k=1}^\infty$ and $\big(b_k(x)\big)_{k=1}^\infty$. By the Marcinkiewicz interpolation theorem, the bound (2.2) follows from (2.3). We first prove (2.3) with $i = 1$. We have

$$\mu\left(\{x \in X : B_1 f(x) > \tfrac{\lambda}{2}\}\right) \le \frac{C_q}{\lambda^q} \int_X \sum_{n=0}^\infty \left(\frac{a_n(x)}{n+1}\right)^q d\mu$$

$$= C_q \lambda^{-q} \sum_{n \ge 0} \left(\frac{1}{n+1}\right)^q \int_X a_n(x)^q d\mu.$$

Now

$$\int_X a_n^q(x) d\mu \le C \int_0^\infty y^{q-1} \mu(\{x \in X : a_n(x) > y\}) dy.$$

Hence

$$\mu\left(\{x \in X : B_1 f(x) > \tfrac{\lambda}{2}\}\right)$$

$$\le \frac{C}{\lambda^q} \sum_{n \ge 0} \left(\frac{1}{n+1}\right)^q \int_0^\infty y^{q-1} \mu\left(\{x \in X : a_n(x) > y\}\right) dy.$$

Now from the definition of $a_n(x)$ and the fact that the transformations $T_1, \ldots, T_d$ preserve μ, this is

$$\le \frac{C_q}{\lambda^q} \sum_{n \ge 0} \left(\frac{1}{n+1}\right)^q \int_0^\infty y^{q-1} \mu(\{x \in X : |f(x)| > y\}) dy,$$

which is

$$= \frac{C_q}{\lambda^q} \int_0^\infty \left(\sum_{n \ge [\frac{y}{\lambda}]} \left(\frac{1}{n+1}\right)^q y^{q-1} \mu(\{x \in X : |f(x)| > y\})\right) dy,$$

and hence is

$$\le \frac{C_q}{\lambda^q} \int_0^\infty \left(y^{q-1} \left(\frac{\lambda}{y}\right)^{q-1} \mu(\{x \in X : |f(x)| > y\})\right) dy,$$

which is

$$= \frac{C_q}{\lambda} \int_X |f| d\mu.$$

Because $q > 1$, this is finite and we have shown (2.3) with $i = 1$. We now show (2.3) with $i = 2$. Here

$$\mu(\{B_2 f(x) > 0\}) \leq \sum_{n \geq 0} \mu(\{x : b_n(x) > 0\})$$

$$\leq \sum_{n \geq 0} \mu(\{x : |f(x)| > \lambda(n+1)\})$$

$$\leq \frac{1}{\lambda} \int_X |f| d\mu,$$

and so Theorem 1 is proved.

3. Proof of Theorem 2

For a function f in $L^1(\mu)$ and $\delta > 0$, let

$$E(f, \delta) = \left\{ x \in X : \left| f(x) - \int_X f d\mu \right| > \delta \right\}.$$

Because $(X, \beta, T_1, \ldots, T_d)$ is good, given $\delta_0 > 0$ and $f \in L^1(\mu)$, for almost all x there exists $N_0(x)$ such that if $N > N_0(x)$,

$$\left| A_N f(x) - \int_X f(x) d\mu \right| < \frac{\delta_0}{2}.$$

Pick f to be non-constant and such that there exists $\delta_0 > 0$ with

$$\mu(E(f, \delta_0)) > 0.$$

Then if $x \in E(f, \delta_0)$ and $N > N_0(x)$, using the triangle inequality, we note that

$$|A_{N+1} f(x) - A_N f(x)| = \frac{1}{N+1} \left| A_{N+1} f(x) - f\left(T_1^{N+1} \ldots T_d^{(N+1)^d} x \right) \right|$$

$$\geq \frac{1}{N+1} \left| \int_X f d\mu - f\left(T_1^{N+1} \ldots T_d^{(N+1)^d} x \right) \right|$$

$$- \frac{1}{N+1} \left| A_{N+1} f(x) - \int_X f d\mu \right|.$$

Thus for x in $E(f, \delta_0)$,

$$|A_{N+1} f(x) - A_N f(x)| \geq \frac{\delta_0}{N+1} - \frac{\delta_0}{2(N+1)} = \frac{\delta_0}{2(N+1)}.$$

This means that

$$V_1 f(x) \geq \sum_{N \geq N_0(x)} \frac{\delta_0}{2(N+1)} \chi_{E(f,\delta_0)}(T_1^{N+1} \ldots T_d^{(N+1)^d} x)$$

$$\geq \frac{\delta_0}{2} \left\{ \sum_{l \geq N_0(x)} \frac{1}{l+2} \left\{ \frac{1}{l+1} \sum_{N=N_0(x)}^{l} \chi_{E(f,\delta_0)}(T_1^{N+1} \ldots T_d^{(N+1)^d} x) \right\} \right\},$$

which for suitably large $N_0(x)$ is

$$\geq \frac{\delta}{2}\left(\frac{\mu(E(f,\delta_0))}{2}\right)\sum_{l\geq N_0(x)}\frac{1}{l+1}$$

$$= \infty,$$

as required.

4. PROOF OF THEOREM 3

We begin with some number theoretic preliminaries. Suppose that

$$\psi(x) = a_d x^d + \cdots + a_1 x$$

for integers $a_i (i = 1, 2, \ldots, d)$ and let

$$S(q, a_1, \ldots, a_d) = \frac{1}{q} S(\psi \mid q) \tag{4.1}$$

where

$$S(\psi \mid q) = \sum_{r=0}^{q-1} e^{2\pi i \psi(r) q^{-1}}.$$

For $\beta = (\beta_1, \ldots, \beta_d)$ in $\mathbb{R}^d$ let

$$V_n(\beta) = \frac{1}{n}\int_0^n e^{2\pi i(\beta_d x^d + \cdots + \beta_1 x)}\,dx. \tag{4.2}$$

We have the following lemmas.

Lemma 6 ([6, p. 116]). *If $V_n(\beta)$ is defined by (4.2), then*

$$|1 - V_N(\beta)| \leq C\sum_{j=1}^d |\beta_j|$$

and

$$|V_N(\beta)| \leq C\left[1 + \sum_{j=1}^d |\beta_j| N^j\right]^{-\frac{1}{d}}.$$

Lemma 7. *Let $S(q, a_1, \ldots, a_d)$ be defined by (4.1). Then there exist $\delta_0 > 0$, and $C_{\delta_0} > 0$ such that*

$$|S(q, a_1, \ldots, a_d)| \leq \frac{C_{\delta_0}}{q^{\delta_0}}.$$

In fact Lemma 7 is true for any δ_0 in $(0, \frac{1}{d})$.

Proof. Let $(q, \mathbf{a})$ denote the highest common factor of the integers q and $a_1, a_2, \ldots, a_d$. Then under the assumption that $(q, \mathbf{a}) = 1$, a proof of Lemma 7 appears in [6, p. 112]. To remove this assumption note that

$$S(q, a_1, \ldots, a_d) = \frac{1}{q} S\left(\psi' \,\Big|\, \frac{q}{(q, \mathbf{a})}\right),$$

where
$$\psi(x) = b_d x^d + \cdots + b_1 x,$$
with $b_i = \frac{a_i}{(q,\mathbf{a})}$ $(i = 1, 2, \ldots, d)$. Now because $\left(\frac{q}{(q,\mathbf{a})}, \mathbf{b}\right) = 1$,
$$\left| S\left(\frac{q}{(q,\mathbf{a})}, b_1, \ldots, b_d\right) \right| \leq \frac{C_\delta}{\left(\frac{q}{(q,\mathbf{a})}\right)^{\delta_0}}.$$

Also, using the periodicity properties of the complex exponential function,
$$S(\psi \mid q) = (q, \mathbf{a}) S\left(\psi' \left| \left(\frac{q}{(q,\mathbf{a})}\right)\right.\right).$$

Putting this all together gives
$$|S(q, a_1, \ldots, a_d)| \leq \frac{C_{\delta_0}(q, \mathbf{a})^{1+\delta_0}}{q^{\delta_0}}.$$

Noting that $(q, \mathbf{a}) \leq \min_{i=1,2,\ldots,d}(|a_i|)$ completes the proof. $\qquad\square$

Lemma 8 ([7, p. 198]). *Suppose $F : [X, Y] \to \mathbb{C}$, F is twice differentiable and the derivative of F is monotone on $[X, Y]$ and of constant sign. Suppose that $0 < H < 1$ and that $|F'(x)| \leq H$ in $[X, Y]$. Then*
$$\sum_{X < k \leq Y} e^{2\pi i F(k)} = \int_X^Y e^{2\pi i F(x)} dx + O\left(3 - \frac{2H}{1 - H}\right).$$

For $\delta > 0$ set
$$M_N(\theta) = \{(\alpha_1, \ldots, \alpha_d) \in [0, 1)^d : |\alpha_i - \theta_i| < N^{-i+\delta} (1 \leq i \leq d)\},$$
where $\theta = (\theta_1, \ldots, \theta_d)$ for rationals $\theta_i = \frac{a_i}{q_i}$ $(i = 1, 2, \ldots, d)$. In addition let $q = (q_1, \ldots, q_d)$ denote the highest common factor of the integers $q_1, \ldots, q_d$.

Lemma 9 ([6, p. 116]). *Suppose that $\alpha \in M_N(\theta)$, that $\delta > 0$ and that $q \leq N^\delta$. Also suppose that*
$$\hat{K}_N(\alpha) = \frac{1}{N} \sum_{n=1}^N e^{2\pi i(\alpha_1 n + \cdots + \alpha_d n^d)}.$$

Then
$$\hat{K}_N(\alpha) = \frac{S(q, a_1, \ldots, a_d)}{q} \int_0^N e(\beta_1 x + \cdots + \beta_d x^d) dx$$
$$+ O\left(r(1 + |\beta_1|N + \cdots + |\beta_d|N^d)\right).$$

In particular, if $\alpha = \theta + \beta$ where $\theta_i = \frac{a_i}{q_i}$,
$$\hat{K}_N(\alpha) = S(q, a_1, \ldots, a_d) V_N(\beta) + O(N^{-\frac{1}{2}}).$$

Proof. The first claim of the theorem is quoted from the given reference. For the second let $\theta_j = \frac{a_j}{q_j}$, $\alpha_j = \theta_j + \beta_j$ and $|\beta_j| < N^{-j+\delta}$. Writing $n = qs + r$, where $0 \le s < \frac{N}{q}$ and $r = 1, 2, \ldots, q-1$, for each $j = 1, 2, \ldots, d$, one has

$$\alpha_j n^j = (\theta_j + \beta_j)(qs + r)^j \in \mathbb{Z} + \theta_j r^j + \beta_j q^j s^j + O(N^{-1+2\delta}).$$

Since $q < N^\delta$,

$$\hat{K}_N(\alpha) = \left\{ \frac{1}{q} \sum_{r=1}^{q-1} e^{2\pi i (r\theta_1 + \cdots + r^d \theta_d)} \right\} \left\{ \frac{q}{N} \sum_{s=0}^{\frac{N}{q}} e^{2\pi i (\beta_1 sq + \ldots \beta_d s^d q^d)} \right\} + O(N^{-\frac{1}{2}}).$$

Using Lemma 8 we may replace

$$\frac{q}{N} \sum_{s=0}^{\frac{N}{q}} e^{2\pi i (\beta_1 sq + \ldots \beta_d s^d q^d)}$$

by $V_N(\beta)$ to obtain Lemma 9. $\qquad\square$

We need the following inequality a form of which was stated by J. Bourgain [1, p. 66]. I have not been able to understand his proof, however; so for completeness a different proof is presented here. I thank R.C. Vaughan for assistance with this.

Lemma 10. *There exists $\delta' > 0$ and $C > 0$ such that if α is not in $M_N(\theta)$ for any θ with $q < N^\delta$, we have*

$$\left| \hat{K}_N(\alpha) \right| \le \frac{C}{N^{\delta'}}.$$

Proof. Let $\alpha \in [0,1)^d \backslash M_N$. As a consequence of Dirichlet's theorem on diophantine approximation, for all $i \le d$ we can choose b_i and q_i with $(b_i, q_i) = 1$, $q_i \le N^{i-\frac{1}{3}}$ and $|\alpha_i - \frac{b_i}{q_i}| \le q_i^{-1} N^{\frac{1}{3}-i}$. Suppose first for some i that we have $q_i \ge N^{\frac{1}{3d}}$. Then by [7, Theorems 5.3, p. 68]

$$\hat{K}_N(\alpha) \le N^{1-\frac{\sigma(d)}{d}},$$

where $\sigma(d) \sim \frac{1}{cd^2 \log d}$. Secondly, suppose that $q_i < N^{\frac{1}{3d}}$ for all i. Then $(q_1, \ldots, q_d) < N^{\frac{1}{3}}$. Now, by Lemma 9,

$$\hat{K}_N(\alpha) = \frac{S(q, \mathbf{b})}{q} \int_0^N e(\beta_1 x + \cdots + \beta_d x^d) dx + O\left(q(1 + |\beta_1| N + \cdots + |\beta_d| N^d)\right),$$

where $\beta_i = \alpha_i = \frac{b_i}{q_i}$, $S(q, \mathbf{b}') = \sum_{x=1}^q e\left(\frac{b_1' x + \cdots + b_d' x^d}{q}\right)$ and $\frac{b_i'}{q} = \frac{b_i}{q_i}$. The error term is $\ll N^{\frac{1}{3}} N^{\frac{1}{3}} = N^{\frac{2}{3}} < N^{1-\frac{\sigma(d)}{d}}$. By Lemma 7 and the following remark $S(q, \mathbf{b}) \ll q^{1-\frac{1}{d}+\epsilon}$, for any $\epsilon > 0$. This means that the main term is $\ll N q^{-\frac{1}{d}+\epsilon}$. Since we have $\alpha \notin M_N$ we have $q > N^\delta$. Hence the main term is $\ll N^{1-\delta(\frac{1}{d}+\epsilon)}$. $\qquad\square$

Let ζ be a smooth function defined on $\mathbb{R}$ with support contained in $[-\frac{1}{5}, \frac{1}{5}]$ and equal to 1 on $[-\frac{1}{10}, \frac{1}{10}]$. Let

$$R_s = \left\{ \left(\frac{a_1}{q_1}, \ldots, \frac{a_d}{q_d} \right) \in \mathbb{Q}^d : q = (q_1, \ldots, q_d) \in [2^s, 2^{s+1}) \right\},$$

where we recall $(q_1, \ldots, q_d)$ denotes the highest common factor of the integers $q_1, \ldots, q_d$. Then for $\alpha = (\alpha_1, \ldots, \alpha_d) \in [0,1)^d$ and $\theta = (\frac{a_1}{q_1}, \ldots, \frac{a_d}{q_d})$ we set

$$|\beta| := \max(|\beta_1|, \ldots, |\beta_d|),$$

set

$$\psi_{s,N}(\alpha) = \sum_{\theta \in R_s} S(q, a_1, \ldots, a_d) V_N(\alpha - \theta) \zeta(10^s |\alpha - \theta|),$$

and

$$\hat{L}_N(\alpha) = \sum_{s \geq 0} \psi_{s,N}(\alpha).$$

We have the following lemma.

Lemma 11. *There exist $\sigma > 0$ and $C_\sigma > 0$ such that*

$$\sup_{\alpha \in \mathbb{T}^d} \left| \hat{K}_N(\alpha) - \hat{L}_N(\alpha) \right| \leq \frac{C_\sigma}{N^\sigma}.$$

Proof. Suppose that $\delta > 0$. We should think of δ as very small—of the order of $\frac{1}{100}$, though this is flexible. Suppose α is in $M_N(\theta_0)$ with $q \leq N^\delta$. Also suppose that $\theta_0 \in R_{s_0}$. Note that this means $2^{s_0} < N^\delta$. Observe in addition that for each natural number s the functions $\{\zeta(10^s |\alpha - \theta|)\}_{\theta \in R_s}$ have disjoint supports. The triangle inequality now gives

$$\left| \hat{K}_N(\alpha) - \hat{L}_N(\alpha) \right| \leq \left| \hat{K}_N(\alpha) - S(q, a_1, \ldots, a_d) V_N(\alpha - \theta_0) \zeta(10^s |\alpha - \theta_0|) \right|$$
$$+ \left| \sum_{s \neq s_0} \psi_{s,N}(\alpha) \right|.$$

Lemma 9 and the elementary estimates $|S(q, a_1, \ldots, a_d)| \leq 1$ and $|V_N(\beta)| \leq 1$ give

$$\left| \hat{K}_N(\alpha) - \hat{L}_N(\alpha) \right| \leq \left| 1 - \zeta(10^{s_0}(\alpha - \theta_0)) \right| + C N^{-\frac{1}{2}}$$
$$+ \sum_{s \leq s_1} \sup_{\theta \in R_{s,0}} |V_N(\beta)| + C 2^{-s_1 \delta_0} \qquad (4.3)$$

for an arbitrary natural number s_1, where

$$R_{s,0} = \begin{cases} R_s & \text{if} \quad \theta_0 \notin R_s, \\ R_s \setminus \{\theta_0\} & \text{if} \quad \theta_0 \in R_s. \end{cases}$$

The natural number $s_1 = s_1(N)$ will be chosen optimally later. As $10^{s_0} < N^{4\delta}$, it follows that

$$10^{s_0} |\alpha - \theta_0| \leq N^{4\delta} N^{-1+\delta} = N^{5\delta - 1}$$

and so the first term on the right hand side of (4.3) is zero. Now for two rationals in reduced form $\frac{a}{b}$ and $\frac{a_0}{b_0}$ we have

$$\left| \frac{a}{b} - \frac{a_0}{b_0} \right| \geq \frac{1}{bb_0} .$$

Therefore

$$|\theta_i - \theta_{i,0}| \geq \tfrac{1}{2} q^{-1} 2^{-s_1(N)} \geq \tfrac{1}{4} N^{-2\delta} .$$

Also $|\alpha_i - \theta_i| < N^{-i+\delta}$ so $|\alpha_i - \theta_i|$ is small compared to $|\theta_i - \theta_{i,0}|$. So using the fact that

$$|\alpha_i - \theta_i| \geq |\theta_i - \theta_{i,0}| - |\alpha_i - \theta_{i,0}| \qquad\qquad (i = 1, 2, \dots),$$

we see that $|\alpha_i - \theta_i| > \tfrac{1}{2}|\theta_i - \theta_{0,i}|$. We now set $2^{s_1(N)} \sim N^\delta$. Using Lemma 6,

$$\sum_{s \leq s_1(N)} \sup_{\theta \in R_{s,0}} |V_N(\alpha - \theta)| \leq C \sum_{s \leq s_1(N)} \sup_{\theta \in R_{s,0}} \left(1 + \sum_{j=1}^{d} |\beta_j| N^j \right)^{-\frac{1}{d}}$$

$$\leq C \sum_{s \leq s_1(N)} \sup_{\theta \in R_{s,0}} \left(1 + \sum_{j=1}^{d} |\theta_i - \theta_{i,0}| N^j \right)^{-\frac{1}{d}} ,$$

which, on noting that the number of terms in the first term is $O(\log n)$, is

$$\leq C (\log N) N^{\frac{2\delta-1}{d}} ,$$

thereby proving Lemma 11 if α is in $M_N(\theta)$ for some θ with $q \leq N^\delta$.

Now suppose that α is not in $M_N(\theta)$ for any θ with $q \leq N^\delta$. Evidently

$$\left| \hat{K}_N(\alpha) - \hat{L}_N(\alpha) \right| \leq \left| \hat{K}_N(\alpha) \right| + \left| \hat{L}_N(\alpha) \right| .$$

By Lemma 9 we have

$$\left| \hat{K}_N(\alpha) \right| \leq \frac{C}{N^{\delta'}} .$$

As on the major arcs we have

$$\left| \hat{L}_N(\alpha) \right| \leq \sum_{s \leq s_1(N)} \sup_{\theta \in R_s} |V_N(\alpha - \theta)| + C 2^{-s_1(N)\delta_0}$$

which, using Lemma 6 and the fact that α is not on the major arcs gives

$$|\hat{L}_N(\alpha)| \leq C(\log N) \left(1 + \sum_{j : |\alpha_j - \frac{a_j}{q_j}| N^j \geq N^\delta} \left| \alpha_j - \frac{a_j}{q_j} \right| N^j \right)^{-\frac{1}{d}} + C 2^{-s_1(N)\delta_0}$$

$$\leq C(\log N) \max(N^{-\frac{\delta}{d}}, N^{-\delta\delta_0})$$

as required. $\qquad\qquad\qquad\qquad\qquad\qquad\qquad\qquad\qquad\qquad \square$

Let H denote the Hilbert space $L^2(X, \beta, \mu)$ endowed with the standard inner product

$$\langle f, g \rangle = \int_X f \bar{g} \, d\mu \, .$$

By an abuse of notation let U_i^{-1} denote the adjoint of the unitary U_i defined by

$$U_i f(x) = f(T_i x) \qquad (i = 1, 2, \dots) \, .$$

For each $f \in L^2$ one readily checks that $(\langle f, U_1^{n_1} \dots U_d^{n_d} f \rangle)_{(n_1, \dots, n_d) \in \mathbb{Z}^n}$ defines a positive definite function on $\mathbb{Z}^d$ and so we have the following lemma:

Lemma 12. *Let $\langle ., . \rangle$ denote the standard inner product on L^2. Then if f is in L^2 there exists a measure μ_f on $\mathbb{T}^d$ such that*

$$\langle f, U_1^{n_1} \dots U_d^{n_d} f \rangle = \int_{\mathbb{T}^n} e^{-2\pi i n.t} d\mu_f \, ,$$

where $n = (n_1, \dots, n_d) \in \mathbb{Z}^d$ and $t = (t_1, \dots, t_d) \in \mathbb{T}^d$.

Swapping the order of summation and integration we see that

$$\|S(f)\|_2^2 = \sum_{k \geq 1} \|A_{N_{k+1}} f - A_{N_k} f\|_2^2$$

and so, assuming that $\|f\| = 1$, as we may without loss of generality, in proving Theorem 3 we require to prove that there exists $C > 0$ such that

$$\sum_{k \geq 1} \|A_{N_{k+1}} f - A_{N_k} f\|_2^2 \leq C \, .$$

Using Lemma 12 we see that

$$\sum_{k \geq 1} \|A_{N_{k+1}} f - A_{N_k} f\|_2^2 = \int_{\mathbb{T}^d} \left(\sum_{k \geq 1} \left| \hat{K}_{N_{k+1}}(\alpha) - \hat{K}_{N_k}(\alpha) \right|^2 \right) d\mu_f(\alpha) \, .$$

Thus proving Theorem 3 reduces to showing that there exists $C > 0$ such that

$$\sum_{k \geq 1} \left| \hat{K}_{N_{k+1}}(\alpha) - \hat{K}_{N_k}(\alpha) \right|^2 \leq C \, .$$

Using Lemma 11 and our assumption on $\frac{N_{k+1}}{N_k}$ we see that

$$\|\{\hat{K}_{N_{k+1}}(\alpha) - \hat{K}_{N_k}(\alpha)\}\|_{\ell^2(\mathbb{Z})} \leq C + \|\{\hat{L}_{N_{k+1}}(\alpha) - \hat{L}_{N_k}(\alpha)\}\|_{\ell^2(\mathbb{Z})} \, .$$

Also

$$\|\{\hat{L}_{N_{k+1}}(\alpha) - \hat{L}_{N_k}(\alpha)\}\|_{\ell^2(\mathbb{Z})} \leq \sum_{s \geq 0} \left(\sum_{k \geq 1} |\psi_{s, N_{k+1}}(\alpha) - \psi_{s, N_k}(\alpha)|^2 \right)^{\frac{1}{2}} \, .$$

Hence Theorem 3 is proved if we can show there exists $\delta, C > 0$ such that

$$\sup_{\alpha \in \mathbb{T}^d} \left| \sum_{k \geq 1} \left(\psi_{s,N_{k+1}}(\alpha) - \psi_{s,N_k}(\alpha) \right)^2 \right| \leq \frac{C}{2^{s\delta}}.$$

Let χ denote the characteristic function of the interval $[-1,1]$ in $\mathbb{R}$ and let

$$\psi'_{s,N}(\alpha) = \sum_{\theta \in R_s : q(\theta) = q} S(q, a_1, \ldots, a_d) \chi(N^d |\alpha - \theta|) \zeta(10^s |\alpha - \theta|).$$

Then we have

$$\|(\psi_{s,N_{k+1}} - \psi_{s,N_k})(\alpha)\|_{\ell^2(\mathbb{Z})} \leq \|(\psi'_{s,N_{k+1}} - \psi'_{s,N_k})(\alpha)\|_{\ell^2(\mathbb{Z})}$$

$$+ \|(\psi_{s,N_k} - \psi'_{s,N_k})(\alpha)\|_{\ell^2(\mathbb{Z})} + \|(\psi_{s,N_{k+1}} - \psi'_{s,N_{k+1}})(\alpha)\|_{\ell^2(\mathbb{Z})}.$$

Also

$$\|(\psi_{s,N_{k+1}} - \psi'_{s,N_{k+1}})(\alpha)\|_{\ell^2(\mathbb{Z})} \leq \|(\psi_{s,N_k} - \psi'_{s,N_k})(\alpha)\|_{\ell^2(\mathbb{Z})}.$$

Hence Theorem 3 follows if we can show that

$$\|(\psi'_{s,N_{k+1}} - \psi'_{s,N_k})(\alpha)\|_{\ell^2(\mathbb{Z})} \leq \frac{C}{2^{s\delta}} \tag{4.4}$$

and

$$\|(\psi_{s,N_k} - \psi'_{s,N_k})(\alpha)\|_{\ell^2(\mathbb{Z})} \leq \frac{C}{2^{s\delta}}. \tag{4.5}$$

To show (4.4) note that

$$\|(\psi'_{s,N_{k+1}} - \psi'_{s,N_k})(\alpha)\|^2_{\ell^2(\mathbb{Z})} = \|\Sigma_1\|_{\ell^2(\mathbb{Z})},$$

where

$$\Sigma_1 = \sum_{\theta \in R_s : q(\theta) = q} S(q, a_1, \ldots, a_d) [\chi(N_{k+1}^d |\alpha - \gamma|) - \chi(N_k^d |\alpha - \gamma|)] \zeta(10^s |\alpha - \gamma|).$$

Recall that the functions $(\zeta(10^s |\alpha - \theta|))_{\theta \in R_s}$ are disjointly supported. This means that there is a unique rational $\theta = \gamma$, say, such that this is

$$\leq \sum_{k \geq 1} |s(\gamma)|^2 \left| \chi(N_{k+1}^d |\alpha - \gamma|) - \chi(N_k^d |\alpha - \gamma|) \right|^2,$$

which using Lemma 7 is

$$\leq \frac{C}{2^{s\delta_0}} \sum_{k \geq 1} \left| \chi(N_{k+1}^d |\alpha - \gamma|) - \chi(N_k^d |\alpha - \gamma|) \right|^2$$

$$\leq \frac{C}{2^{s\delta_0}},$$

because at most one term in the above sum is non-zero. We now show (4.5). The left hand side of this equation is, by Lemma 6,

$$\leq \frac{C}{2^{s\delta_0}} \left(\sum_{N_k \leq \frac{1}{|\gamma|^{\frac{1}{d}}}} |V_N(\alpha - \gamma) - 1|^2 + \sum_{N_k > \frac{1}{|\gamma|^{\frac{1}{d}}}} |V_N(\alpha - \gamma)|^2 \right)$$

$$\leq \frac{C}{2^{s\delta_0}} \left(\sum_{N_k \leq \frac{1}{|\gamma|^{\frac{1}{d}}}} |\gamma|^2 a^{2kd} + \sum_{N_k > \frac{1}{|\gamma|^{\frac{1}{d}}}} |\gamma|^{-\frac{2}{d}} (k \log b) \right),$$

$$\leq \frac{C}{2^{s\delta_0}},$$

as required.

References

[1] J. Bourgain, On the maximal ergodic theorem for certain subsets of the integers, *Israel J. Math. Vol.* **61** (1988), 39–72.

[2] ———, Pointwise ergodic theorems for arithemetic sets, *Publ. I.H.E.S. Vol.* **69** (1989), 5–45.

[3] ———, Almost sure convergence in ergodic theory, *Almost Everywhere Convergence*, 145–151, Academic Press, Boston, MA, 1989.

[4] Y. Lacroix, Natural extensions and mixing for semi-group actions, *Séminaires de Probabilités de Rennes*, 1995.

[5] R. Nair and M. Weber, On variation functions for subsequence ergodic averages, *Monatsh. Math.* **128** (1999), 131-150.

[6] R.C. Vaughan, *The Hardy-Littlewood method*, 2nd edition, Cambridge Univerasity Press, 1997.

[7] I.M. Vinogradov, *Selected works*, Springer-Verlag, 1985.

POLYNOMIAL INEQUALITIES, MAHLER'S MEASURE, AND MULTIPLIERS

IGOR E. PRITSKER

ABSTRACT. We survey polynomial inequalities obtained via coefficient multipliers, for norms defined by the contour or the area integrals over the unit disk. Special attention is devoted to the Szegő composition and the inequalities related to Mahler's measure.

We also consider a new height on polynomial spaces defined by the integral over the normalized area measure on the unit disk. This natural analog of Mahler's measure inherits many nice properties such as the multiplicative one. However, this height is a lower bound for Mahler's measure, and it fails an analog of Lehmer's conjecture.

1. THE SZEGŐ COMPOSITION AND POLYNOMIAL INEQUALITIES

This paper is a survey of results on polynomial inequalities obtained via coefficient multipliers, and other topics related to Mahler's measure. Let $\mathbb{C}_n[z]$ and $\mathbb{Z}_n[z]$ be the sets of all polynomials of degree at most n with complex and integer coefficients respectively. *Mahler's measure* of a polynomial $P_n \in \mathbb{C}_n[z]$ is defined by

$$M(P_n) := \exp\left(\frac{1}{2\pi} \int_0^{2\pi} \log |P_n(e^{i\theta})|\, d\theta\right).$$

It is also known as the *contour geometric mean* or as the H^0 *Hardy space norm*. The latter name is explained by the following relation to the Hardy spaces. Defining the Hardy space norm by

$$\|P_n\|_{H^p} := \left(\frac{1}{2\pi} \int_0^{2\pi} |P_n(e^{i\theta})|^p\, d\theta\right)^{1/p}, \quad 0 < p < \infty,$$

we note [18] that $M(P_n) = \lim_{p \to 0+} \|P_n\|_{H^p}$. An application of Jensen's inequality immediately gives that

$$M(P_n) = |a_n| \prod_{|z_j| > 1} |z_j|$$

2000 *Mathematics Subject Classification.* Primary 11C08; Secondary 11G50, 30C10.

Key words and phrases. Polynomials, Mahler's measure, heights, zero distribution, Bergman spaces, inequalities, Szegő composition, approximation by polynomials with integer coefficients.

Research was partially supported by the National Security Agency under grant H98230-06-1-0055, and by the Alexander von Humboldt Foundation.

for $P_n(z) = a_n \prod_{j=1}^n (z - z_j) \in \mathbb{C}_n[z]$.

For a polynomial $\Lambda_n(z) = \sum_{k=0}^n \lambda_k \binom{n}{k} z^k \in \mathbb{C}_n[z]$, define the *Szegő composition* with $P_n(z) = \sum_{k=0}^n a_k z^k \in \mathbb{C}_n[z]$ by

$$
(1.1) \qquad\qquad \Lambda P_n(z) := \sum_{k=0}^n \lambda_k a_k z^k.
$$

If Λ_n is a fixed polynomial, then ΛP_n is a multiplier (or convolution) operator acting on P_n. More information on the history and applications of this composition may be found in [10], [1], [2] and [33]. De Bruijn and Springer [10] proved a very interesting general inequality stated below.

Theorem 1.1. *Suppose that $P_n \in \mathbb{C}_n[z]$. If $\Lambda_n \in \mathbb{C}_n[z]$ and $\Lambda P_n \in \mathbb{C}_n[z]$ are defined by (1.1), then*

$$
(1.2) \qquad\qquad M(\Lambda P_n) \le M(\Lambda_n) M(P_n).
$$

If $\Lambda_n(z) = (1 + z)^n$ then $\Lambda P_n(z) \equiv P_n(z)$ and $M(\Lambda_n) = 1$, so that (1.2) turns into equality, showing sharpness of Theorem 1.1. This result has not received the attention it truly deserves. In particular, it contains the following inequality that is usually attributed to Mahler, who proved it later in [26].

Corollary 1.2. $M(P_n') \le n M(P_n)$

To see this, just note that if $\Lambda_n(z) = nz(1 + z)^{n-1} = \sum_{k=0}^n k \binom{n}{k} z^k$, then $\Lambda P_n(z) = z P_n'(z)$ and $M(\Lambda_n) = n$. Furthermore, (1.2) immediately answers a question about a lower bound for Mahler's measure of derivative raised in [14, pp. 12 and 194]. Following Storozhenko [42], we consider $P_n'(z) = \sum_{k=0}^{n-1} a_k z^k$ and write

$$
\frac{1}{z}\left(P_n(z) - P_n(0)\right) = \sum_{k=0}^{n-1} \frac{a_k}{k+1} z^k = \Lambda P_n'(z),
$$

where

$$
\Lambda_{n-1}(z) = \sum_{k=0}^{n-1} \frac{1}{k+1} \binom{n-1}{k} z^k = \frac{(1+z)^n - 1}{nz}.
$$

The result of de Bruijn and Springer (1.2) gives

Corollary 1.3. [42] *We have $M(P_n(z) - P_n(0)) \le c_n M(P_n')$, where*

$$
c_n := \frac{1}{n} M\left((1+z)^n - 1\right) = \frac{1}{n} \prod_{n/6 < k < 5n/6} 2 \sin \frac{k\pi}{n}.
$$

We note that $c_n \approx (1.4)^n$ as $n \to \infty$. One can produce many other interesting consequences of (1.2), such as the well known estimate for coefficients via Mahler's measure.

Corollary 1.4. *If $P_n(z) = \sum_{k=0}^{n} a_k z^k$ then*

$$|a_k| \leq \binom{n}{k} M(P_n), \quad k = 1, \ldots, n.$$

The above inequality follows at once from (1.2) by letting $\Lambda_n(z) = \binom{n}{k} z^k$, $k = 1, \ldots, n$. Another interesting question is how removing a specific power term from the polynomial affects its Mahler's measure. The answer is below.

Corollary 1.5. *Let $P_n(z) = \sum_{k=0}^{n} a_k z^k$ and $m = 0, \ldots, n$. We have*

$$M\left(\sum_{k \neq m} a_k z^k\right) \leq M\left((1+z)^n - \binom{n}{m} z^m\right) M(P_n).$$

In particular, if $m = 0$ then $M((1+z)^n - 1) = \prod_{n/6 < k < 5n/6} 2 \sin \frac{k\pi}{n} \approx (1.4)^n$ *as $n \to \infty$.*

Again, the proof is a simple application of (1.2) with $\Lambda_n(z) = (1+z)^n - \binom{n}{m} z^m$, so that $\lambda_m = 0$ and $\lambda_k = 1$, $k \neq m$. Finally, we state two variations of (1.2).

Corollary 1.6. *If $P_n(z) = \sum_{k=0}^{n} a_k z^k$ and $m \in \mathbb{N}$, then*

$$(1.3) \qquad M\left(\sum_{k=0}^{n} \lambda_k \binom{n}{k}^{-m} a_k z^k\right) \leq M\left(\sum_{k=0}^{n} \lambda_k z^k\right) M(P_n)$$

and

$$(1.4) \qquad M\left(\sum_{k=0}^{n} \lambda_k^m \binom{n}{k}^{-m} a_k z^k\right) \leq M^m\left(\sum_{k=0}^{n} \lambda_k z^k\right) M(P_n).$$

An important generalization of Theorem 1.1 for the H^p norms was obtained by Arestov [1].

Theorem 1.7. *Suppose that $P_n \in \mathbb{C}_n[z]$. If $\Lambda_n \in \mathbb{C}_n[z]$ and $\Lambda P_n \in \mathbb{C}_n[z]$ are defined by (1.1), then*

$$(1.5) \qquad \|\Lambda P_n\|_{H^p} \leq M(\Lambda_n) \|P_n\|_{H^p}, \quad 0 \leq p \leq \infty.$$

Recall that the Szegő composition can also be viewed as a multiplier or convolution operator in the sense of harmonic analysis. For $P_n(z) = \sum_{k=0}^{n} a_k z^k$, we have

$$\Lambda P_n(z) = \sum_{k=0}^{n} \lambda_k a_k z^k = \left(\sum_{k=0}^{n} \lambda_k z^k\right) * P_n(z).$$

Let $\|\Lambda\|$ be the operator norm of $\Lambda : \mathbb{C}_n[z] \to \mathbb{C}_n[z]$:

$$\|\Lambda\| = \sup_{P_n \in \mathbb{C}_n[z]} \frac{\|\Lambda P_n\|_{H^p}}{\|P_n\|_{H^p}}, \quad 0 \le p \le \infty.$$

A particularly interesting class of multiplier operators is given by the bound (norm) preserving operators. Those are described by the condition $\|\Lambda P_n\|_{H^p} \le \|P_n\|_{H^p}$ for all $P_n \in \mathbb{C}_n[z]$, which holds if and only if $\|\Lambda\| \le 1$. If we choose the multipliers λ_k to satisfy $M(\Lambda_n) \le 1$, then $\|\Lambda\| \le 1$ by Theorem 1.7. Thus we may use the inequality $\|\Lambda P_n\|_{H^p} \le \|P_n\|_{H^p}$ to obtain lower bounds for $\|P_n\|_{H^p}$ via a proper choice of multipliers. There are other interesting norm preserving convolution operators such as the following considered by Sheil-Small [38, pp. 168-171].

Theorem 1.8. *Let* $\limsup\limits_{k \to \infty} |\lambda_k|^{1/k} \le 1$. *If* $\lambda_0 = 1$ *and*

$$\Re\left(\sum_{k=0}^{\infty} \lambda_k z^k\right) > \frac{1}{2}, \quad |z| < 1,$$

then $\|\Lambda P_n\|_{H^p} \le \|P_n\|_{H^p}$ *for any polynomial* P_n, *where* $1 \le p \le \infty$.

In fact, Sheil-Small stated a result for generalized convolution operators that covers more applications. It would be interesting to explore whether Theorem 1.8 remains true for $0 \le p < 1$, i.e., for the range of p including Mahler's measure.

2. AN AREAL ANALOG OF MAHLER'S MEASURE

A natural counterpart of Mahler's measure is obtained by replacing the normalized arclength measure on the unit circle $\mathbb{T}$ by the normalized area measure on the unit disk $\mathbb{D}$. Namely, we define the A^0 *Bergman space norm* by

$$\|P_n\|_0 := \exp\left(\frac{1}{\pi} \iint_{\mathbb{D}} \log |P_n(z)| \, dA\right).$$

This norm is also a multiplicative height of the polynomial P_n, cf. [32]. Furthermore, it has the same relation to Bergman spaces as Mahler's measure to Hardy spaces:

$$\|P_n\|_0 = \lim_{p \to 0+} \|P_n\|_p,$$

see [18], where

$$\|P_n\|_p := \left(\frac{1}{\pi} \iint_{\mathbb{D}} |P_n(z)|^p \, dA\right)^{1/p}, \quad 0 < p < \infty,$$

is the A^p *Bergman space norm*.

In fact, there is a direct relation between Mahler's measure and its areal analog, given below.

Theorem 2.1. *Let* $P_n(z) = a_n \prod_{j=1}^{n}(z - z_j) = \sum_{k=0}^{n} a_k z^k \in \mathbb{C}_n[z]$. *If* P_n *has no roots in* $\mathbb{D}$, *then* $\|P_n\|_0 = M(P_n) = |a_0|$. *Otherwise*

$$(2.1) \qquad \|P_n\|_0 = M(P_n) \exp\left(\frac{1}{2} \sum_{|z_j|<1} (|z_j|^2 - 1)\right).$$

This shows that the value of $\|P_n\|_0$ is influenced by the zeros inside the unit disk more than that of $M(P_n)$. We immediately obtain the following comparison result from Theorem 2.1.

Corollary 2.2. *For any* $P_n \in \mathbb{C}_n[z]$, *we have*

$$(2.2) \qquad e^{-n/2} M(P_n) \leq \|P_n\|_0 \leq M(P_n).$$

Equality holds in the lower estimate if and only if $P_n(z) = a_n z^n$. *The upper estimate turns into equality for any polynomial without zeros in the unit disk.*

If $P_n(z) = \sum_{k=0}^{n} a_k z^k$ then

$$(2.3) \qquad \|P_n\|_0 \geq |a_0|,$$

which follows from the area mean value inequality for the subharmonic function $\log |P_n|$ (cf. [12]). Hence

$$(2.4) \qquad \|P_n\|_0 \geq 1 \quad \text{for all } P_n \in \mathbb{Z}_n[z], \ P_n(0) \neq 0.$$

A well known theorem of Kronecker [21] states that any monic irreducible polynomial $P_n \in \mathbb{Z}_n[z]$, $P_n(0) \neq 0$, with all zeros in the closed unit disk, must be cyclotomic. One can write that statement in the form: $M(P_n) = 1$ for such P_n if and only if P_n is cyclotomic. A direct analog of this result exists for $\|P_n\|_0$.

Theorem 2.3. *Suppose that* $P_n \in \mathbb{Z}_n[z]$, $P_n(0) \neq 0$, *is an irreducible polynomial with all zeros in the closed unit disk. It is cyclotomic if and only if* $\|P_n\|_0 = 1$.

The next natural question is whether one can find a uniform lower bound $\|P_n\|_0 \geq c > 1$ for all non-cyclotomic $P_n \in \mathbb{Z}_n[z]$, $P_n(0) \neq 0$. It is especially interesting in view of Lehmer's conjecture, because $M(P_n) \geq \|P_n\|_0$ by (2.2). However, the answer to the question is negative, as we show with the following example.

Example 2.4. *Consider* $P_n(z) = nz^n - 1$. *It has zeros* z_j, $j = 1, \ldots, n$, *that are equally spaced on the circle* $|z| = n^{-1/n}$. *Note that* $M(P_n) = n$ *and*

$$\|P_n\|_0 = n \exp\left(\frac{n(n^{-2/n} - 1)}{2}\right),$$

by (2.1). *Since*

$$n^{-2/n} = \exp\left(\frac{-2 \log n}{n}\right) = 1 - \frac{2 \log n}{n} + O\left(\frac{\log^2 n}{n^2}\right),$$

we obtain that

$$\|P_n\|_0 = \exp\left(O\left(\frac{\log^2 n}{n}\right)\right) \to 1 \quad as\ n \to \infty.$$

Similarly, we have for the reciprocal polynomial $P_{2n}(z) = z^{2n} + nz^n + 1$ *that*

$$M(P_n) = \frac{n + \sqrt{n^2 - 4}}{2} \sim n \quad as\ n \to \infty,$$

and

$$\|P_n\|_0 = \frac{n + \sqrt{n^2 - 4}}{2} \exp\left(\frac{n}{2}\left(\left(\frac{n + \sqrt{n^2 - 4}}{2}\right)^{2/n} - 1\right)\right) \to 1 \quad as\ n \to \infty.$$

One may notice that for both sequences of polynomials in this example the zeros are asymptotically equidistributed near the unit circle. In fact, this is a part of a more general phenomenon. Consider a polynomial $P_n(z) = a_n \prod_{j=1}^{n}(z - z_j) \in \mathbb{C}_n[z]$, and define its normalized zero counting measure by

$$\nu_n := \frac{1}{n}\sum_{j=1}^{n}\delta_{z_j},$$

where δ_{z_j} is the unit point mass at z_j. Our main result on the asymptotic zero distribution is as follows.

Theorem 2.5. *Suppose that* $P_n \in \mathbb{Z}_n[z]$, $\deg P_n = n$, *is a sequence of polynomials without multiple zeros. If* $\lim_{n\to\infty} \|P_n\|_0^{1/n} = 1$ *then the* ν_n *converge to the normalized arclength measure* $d\theta/(2\pi)$ *on* $\mathbb{T}$ *in the weak* topology, as* $n \to \infty$.

This result extends a theorem of Bilu [4] for Mahler's measure; see also Bombieri [5] and Rumely [36]. From a more general point of view, Theorem 2.5 is a descendant of Jentzsch's result [20] on the asymptotic zero distribution of the partial sums of a power series, and its generalization by Szegő [44]. This area was further developed by Erdős and Turán [13], and by many others.

As an immediate application of Theorem 2.5 we obtain a result on the growth of $\|P_n\|_0$ for polynomials with restricted zeros.

Corollary 2.6. *Suppose that* $P_n \in \mathbb{Z}_n[z]$, $\deg P_n = n$, *is a sequence of polynomials with simple zeros contained in a closed set* $E \subset \mathbb{C}$. *If* $\mathbb{T} \not\subset E$ *then there exists a constant* $C = C(E) > 1$ *such that*

$$\liminf_{n\to\infty} \|P_n\|_0^{1/n} \geq C > 1.$$

This exhibits the geometric growth of $\|P_n\|_0$ for many families of polynomials such as polynomials with real zeros, polynomials with zeros in a sector, etc. Corresponding results with explicit bounds for Mahler's measure were obtained by Schinzel [37], Langevin [22, 23, 24], Mignotte [30], Rhin and

Smyth [35], Dubickas and Smyth [11], and others. A detailed account of these results is contained in Smyth [41].

In a somewhat different direction, we have the following result on the asymptotic behavior of zeros.

Theorem 2.7. *Suppose that* $P_n(z) = a_n z^n + \ldots + a_0 \in \mathbb{C}_n[z]$, $|a_0| \geq 1$, $n \in \mathbb{N}$, *is a sequence of polynomials.*

(a) *If* $\lim_{n \to \infty} \|P_n\|_0 = 1$ *then*

$$(2.5) \qquad \liminf_{n \to \infty} \min_{1 \leq j \leq n} |z_j| \geq 1.$$

(b) *If* $|a_n| \geq 1$ *and* $\lim_{n \to \infty} M(P_n) = 1$, *then*

$$(2.6) \qquad \lim_{n \to \infty} \min_{1 \leq j \leq n} |z_j| = \lim_{n \to \infty} \max_{1 \leq j \leq n} |z_j| = 1.$$

Thus part (a) of Theorem 2.7 indicates that all zeros of P_n are pushed out of $\mathbb{D}$ as $n \to \infty$, while in part (b) they all tend to the unit circle.

3. POLYNOMIAL INEQUALITIES IN BERGMAN SPACES

We obtain the following generalization of Theorems 1.1 and 1.7 for the Bergman space norms.

Theorem 3.1. *Suppose that* $P_n \in \mathbb{C}_n[z]$. *If* $\Lambda_n \in \mathbb{C}_n[z]$ *and* $\Lambda P_n \in \mathbb{C}_n[z]$ *are defined by* (1.1), *then*

$$(3.1) \qquad \|\Lambda P_n\|_p \leq M(\Lambda_n) \|P_n\|_p, \quad 0 \leq p \leq \infty.$$

Note that equality holds in (3.1) for any polynomial $P_n \in \mathbb{C}_n[z]$ when $\Lambda_n(z) = (1+z)^n = \sum_{k=0}^n \binom{n}{k} z^k$, because $\Lambda P_n \equiv P_n$ and $M((1+z)^n) = 1$. This inequality allows to treat many problems in a unified way, and it has numerous interesting consequences. Theorem 3.1 implies that z^n has the smallest Bergman space norm among all monic polynomials.

Corollary 3.2. *If* $P_n \in \mathbb{C}_n[z]$ *is a monic polynomial, then*

$$(3.2) \qquad \|P_n\|_p \geq \|z^n\|_p = \begin{cases} e^{-n/2}, & p = 0, \\ \left(\dfrac{2}{pn+2} \right)^{1/p}, & 0 < p < \infty. \end{cases}$$

It is well known that $\|P_n\|_\infty \geq \|z^n\|_\infty = 1$, see [7, 33].

Another useful estimate compares norms on the concentric disks $D_R := \{z : |z| < R\}$ to that on the unit disk.

Corollary 3.3. *If* $P_n \in \mathbb{C}_n[z]$ *and* $R \geq 1$, *then*

$$(3.3) \qquad \left(\frac{1}{\pi R^2} \iint_{D_R} |P_n(z)|^p \, dA \right)^{1/p} \leq R^n \|P_n\|_p, \qquad p \in (0, \infty),$$

and

(3.4) $$\exp\left(\frac{1}{\pi R^2}\iint_{D_R}\log|P_n(z)|\,dA\right)\le R^n\,\|P_n\|_0,$$

where equality holds for $P_n(z)=z^n$.

Again, in the case $p=\infty$, it is already known that $\max_{z\in D_R}|P_n(z)|\le R^n\|P_n\|_\infty$. See [33].

Another consequence relates $\|P_n\|_p$ to the coefficients of P_n.

Corollary 3.4. *If* $P_n(z)=\sum_{k=0}^n a_k z^k\in\mathbb{C}_n[z]$ *then*

(3.5) $$|a_k|\le\left(\frac{pk+2}{2}\right)^{1/p}\binom{n}{k}\|P_n\|_p,\quad k=0,\dots,n,\ 0<p<\infty,$$

and

(3.6) $$|a_k|\le e^{k/2}\binom{n}{k}\|P_n\|_0,\quad k=0,\dots,n.$$

One can certainly extend the list of corollaries by choosing appropriate polynomials Λ_n.

3.1. Bernstein-type inequalities.

The original Bernstein inequality (cf. [7], [31] and [33]) gives an estimate for the supremum norm of the derivative of a polynomial on D:

$$\|P_n'\|_\infty\le n\|P_n\|_\infty,\qquad P_n\in\mathbb{C}_n[z].$$

Its sharpness is easily seen by considering $P_n(z)=z^n$. Zygmund [47] extended this result to the Hardy spaces by proving that

$$\|P_n'\|_{H^p}\le n\|P_n\|_{H^p},\qquad p\in[1,\infty).$$

As we know from the first section, De Bruijn and Springer [10], and later Mahler [26], showed that

$$M(P_n')\le nM(P_n),$$

thus settling the case $p=0$ for the Hardy space norms. It had been an open question for a long time, whether the above inequality is true for $0<p<1$. After a partial result of Máté and Nevai [29], the question was answered in the affirmative by Arestov [1].

We obtain the following version of the Bernstein inequality for Bergman spaces, as a consequence of Theorem 3.1.

Theorem 3.5. *For any* $P_n\in\mathbb{C}_n[z]$, *we have that*

(3.7) $$\|zP_n'\|_p\le n\|P_n\|_p,\qquad 0\le p<\infty.$$

Note that equality holds here for $P_n(z)=z^n$.

It is also of interest to find the Bernstein inequalities in Bergman spaces exactly matching the classical one in form. For example, when $p=0$, we have

Corollary 3.6. *If $P_n \in \mathbb{C}_n[z]$ then*

$$(3.8) \qquad \|P_n'\|_0 \leq \sqrt{e}\, n\, \|P_n\|_0,$$

where equality holds for $P_n(z) = z^n$.

Furthermore, we obtain by an elementary argument the following

Proposition 3.7. *If $P_n \in \mathbb{C}_n[z]$ then*

$$(3.9) \qquad \|P_n'\|_2 \leq \sqrt{n(n+1)}\, \|P_n\|_2,$$

with equality for $P_n(z) = z^n$.

This suggests that, for arbitrary $p \in (0, \infty)$, one might be able to prove

Conjecture 3.8. *If $P_n \in \mathbb{C}_n[z]$ then*

$$(3.10) \qquad \|P_n'\|_p \leq n \left(1 + \frac{1}{n-1+2/p}\right)^{1/p} \|P_n\|_p, \qquad 0 < p < \infty,$$

with equality for $P_n(z) = z^n$.

Note that Corollary 3.6 may be viewed as the limiting case of this conjecture as $p \to 0$, while the classical Bernstein inequality is obtained by letting p tend to ∞.

3.2. Comparing the Hardy and the Bergman norms.

It is well known [12, 19] that for any function $f \in H^p$ we have

$$\|f\|_p \leq \|f\|_{H^p}, \qquad 0 \leq p \leq \infty.$$

Clearly, we have equality for $p = \infty$. One can prove inequalities for polynomials in the opposite direction, of the form

$$\|P_n\|_{H^p} \leq C(n,p)\, \|P_n\|_p.$$

For example, we have for $p = 0$ that

$$(3.11) \qquad M(P_n) \leq e^{n/2}\, \|P_n\|_0,$$

where equality holds for $P_n(z) = z^n$ (see Corollary 2.2).

The case $p = 2$ is easy to handle, because

$$\|P_n\|_{H^2}^2 = \sum_{k=0}^{n} |a_k|^2 \leq (n+1) \sum_{k=0}^{n} \frac{|a_k|^2}{k+1} = (n+1)\, \|P_n\|_2^2,$$

where $P_n(z) = \sum_{k=0}^{n} a_k z^k$. Hence

Proposition 3.9. *If $P_n \in \mathbb{C}_n[z]$ then*

$$(3.12) \qquad \|P_n\|_{H^2} \leq \sqrt{n+1}\, \|P_n\|_2,$$

with equality for $P_n(z) = z^n$.

It is likely that the following is true.

Conjecture 3.10. *If $P_n \in \mathbb{C}_n[z]$ then*

(3.13) $$\|P_n\|_{H^p} \le (pn/2 + 1)^{1/p} \|P_n\|_p, \qquad 0 < p < \infty,$$

with equality for $P_n(z) = z^n$.

This holds in the limit for $p = \infty$ (trivially) and for $p = 0$ by (3.11).

4. APPROXIMATION BY POLYNOMIALS WITH INTEGER COEFFICIENTS

We consider a related question of approximation by polynomials with integer coefficients on the unit disk. There is a well known condition necessary for approximation by integer polynomials in essentially any norm on $\mathbb{D}$.

Proposition 4.1. *Suppose that $P_n \in \mathbb{Z}_n[z]$, $n \in \mathbb{N}$, converge to f uniformly on compact subsets of $\mathbb{D}$. Then f is analytic in $\mathbb{D}$ and $f^{(k)}(0)/k! \in \mathbb{Z}$ for all $k \ge 0$, $k \in \mathbb{Z}$.*

This necessary condition for the convergence is clearly equivalent to the fact that the power series expansion of f at the origin has integer coefficients.

It is well known that approximation by polynomials with integer coefficients is possible in H^p only in the trivial case. See [16] and [45]. More precisely, we have

Proposition 4.2. *Suppose that $f \in H^p$, $0 < p \le \infty$. If $P_n \in \mathbb{Z}_n[z]$, $n \in \mathbb{N}$, satisfy*

(4.1) $$\lim_{n \to \infty} \|f - P_n\|_{H^p} = 0,$$

then f is a polynomial with integer coefficients.

It remains an open question whether this proposition is true for $p = 0$, i.e. for approximation of functions in Mahler's measure. One can see from the proof of Proposition 4.2, given in Section 6.4, that the main obstacle is that we have no substitute for the triangle inequality in the case of Mahler's measure. Mahler [27] raised an interesting question related to this problem. While it is not possible to have $M(f + g) \le C(M(f) + M(g))$ for a fixed constant C, in general, one can consider a natural analog of the triangle inequality for all polynomials $P_n, Q_n \in \mathbb{C}_n[z]$:

$$M(P_n + Q_n) \le c^n(M(P_n) + M(Q_n)).$$

Mahler [27] showed that $c = 2$ is possible in the above inequality, and asked what is the best (smallest) value of c. He later improved the constant c in [28], and the best currently known range $1.7916 < c < 1.8493$ was obtained by Arestov [2].

Generally, nontrivial approximation by integer polynomials in the supremum norm is valid on sets with transfinite diameter (capacity) less than 1 [16, 45], and it is not possible if the transfinite diameter is greater than or equal to 1. But the transfinite diameter of $\mathbb{D}$ is exactly equal to 1, so that we deal with a borderline case. However, we show that the Bergman space A^p

is different from the Hardy space H^p in this regard, as it does allow approximation by polynomials with integer coefficients.

Theorem 4.3. *Suppose that $f \in A^p$, $1 < p < \infty$. We have*

(4.2)
$$\lim_{n \to \infty} \|f - P_n\|_p = 0,$$

for a sequence of polynomials $P_n \in \mathbb{Z}_n[z]$, $n \in \mathbb{N}$, if and only if f has a power series expansion about $z = 0$ with integer coefficients. Clearly, this is equivalent to $f^{(k)}(0)/k! \in \mathbb{Z}$ for all $k \geq 0$, $k \in \mathbb{Z}$.

Thus there are many functions in A^p that can be approximated by polynomials with integer coefficients. In fact, one can use partial sums of the power series for this purpose. See the proof of Theorem 4.3. However, we do not know whether Theorem 4.3 is valid in the case $0 \leq p \leq 1$. Note that if $f \in A^p$, $p > 1$, has a Taylor expansion with integer coefficients, then $f \in A^q$ for any $q \in [0, p)$ and the partial sums P_n of this expansion satisfy $\|f - P_n\|_q \leq \|f - P_n\|_p \to 0$ as $n \to \infty$.

5. MULTIVARIATE POLYNOMIALS

We believe that many of the results mentioned in this survey are capable of generalization to the multivariate case. However, we do not try to accomplish such an ambitious program here, and restrict ourselves to a few simple remarks. The definition of $\|P_n\|_0$ is easily generalized to the case of multivariate polynomials $P_n(z_1, \ldots, z_d)$ as follows:

$$\|P_n\|_0 := \exp\left(\frac{1}{\pi^d} \int_{\mathbb{D}} \cdots \int_{\mathbb{D}} \log|P_n(z_1, \ldots, z_d)| \, dA(z_1) \ldots dA(z_d)\right).$$

It is also parallel to multivariate Mahler's measure

$$M(P_n) := \exp\left(\frac{1}{(2\pi)^d} \int_{\mathbb{T}} \cdots \int_{\mathbb{T}} \log|P_n(z_1, \ldots, z_d)| \, |dz_1| \ldots |dz_d|\right).$$

We note that many of the properties of $\|P_n\|_0$ are preserved in the multivariate case. Thus it still defines a multiplicative height on the space of polynomials. If P_n is a polynomial with complex coefficients and the constant term a_0, then we can apply the area mean value inequality to the (pluri)subharmonic function $\log|P_n(z_1, \ldots, z_d)|$ in each variable, which gives together with Fubini's theorem that

$$\|P_n\|_0 \geq |a_0|.$$

Furthermore, the above inequality turns into equality if $P_n(z_1, \ldots, z_d) \neq 0$ on $\mathbb{D}^d$, by the area mean value theorem for the (pluri)harmonic function $\log|P_n(z_1, \ldots, z_d)|$. However, it is rather unlikely that some kind of explicit relation such as (2.1) exists for general multivariate polynomials.

We now state an estimate generalizing Corollary 2.2.

Proposition 5.1. *For a polynomial*

$$(5.1) \qquad P_n(z_1, \ldots, z_d) = \sum_{k_1 + \ldots + k_d \leq n} a_{k_1 \ldots k_d} z_1^{k_1} \ldots z_d^{k_d}$$

of degree at most n with complex coefficients, we have

$$(5.2) \qquad e^{-n/2} M(P_n) \leq \|P_n\|_0 \leq M(P_n).$$

Equality holds in the lower estimate for any $P_n(z_1, \ldots, z_d) = a_{k_1 \ldots k_d} z_1^{k_1} \ldots z_d^{k_d}$ with $k_1 + \ldots + k_d = n$. The upper estimate turns into equality for any polynomial not vanishing in $\mathbb{D}^d$.

It is of interest to find explicit values of the multivariate $\|P_n\|_0$. This problem has received a considerable attention in Mahler's measure setting (see [9], [39, 40], [14], [17]), and it remains a very active area of research. In particular, it is of importance to characterize multivariate polynomials with integer coefficients satisfying $\|P_n\|_0 = 1$. Smyth [40] proved a complete Kronecker-type characterization for the multivariate Mahler's measure $M(P_n) = 1$. Thus we expect that one should be able to produce an analog for $\|P_n\|_0$, generalizing Theorem 2.3.

Example 5.2. *The following identities hold for the multivariate $\|P_n\|_0$:*
(a) $\|z_1 + z_2\|_0 = e^{-1/4}$
(b) $\|1 + z_1^{k_1} \ldots z_d^{k_d}\|_0 = 1, \quad k_1, \ldots, k_d \geq 0$
(c) *If the polynomial P_n of the form (5.1) satisfies*

$$|a_{0 \ldots 0}| \geq \sum_{0 < k_1 + \ldots + k_d \leq n} |a_{k_1 \ldots k_d}|,$$

then $\|P_n\|_0 = M(P_n) = |a_{0 \ldots 0}|$.

6. Proofs

6.1. Proofs for Section 1.

Proof of Corollary 1.6. Let $\lambda_k \in \mathbb{C}$, $k = 0, \ldots, n$, be arbitrary fixed numbers, and define the operator

$$A_m P_n(z) := \sum_{k=0}^{n} \lambda_k \binom{n}{k}^{-m} a_k z^k.$$

It is clear from (1.2) that

$$(6.1) \qquad M(A_1 P_n) = M\left(\sum_{k=0}^{n} \lambda_k \binom{n}{k}^{-1} a_k z^k\right) \leq M\left(\sum_{k=0}^{n} \lambda_k z^k\right) M(P_n),$$

which we use as the basis of induction in m. Assuming that

$$M(A_m P_n) \leq M\left(\sum_{k=0}^{n} \lambda_k z^k\right) M(P_n)$$

holds, we obtain that

$$M\left(A_{m+1}P_n\right) = M\left(\sum_{k=0}^{n} \lambda_k \binom{n}{k}^{-m-1} a_k z^k\right)$$

$$\leq M\left(\sum_{k=0}^{n} \lambda_k \binom{n}{k}^{-m} a_k z^k\right) M\left(\sum_{k=0}^{n} z^k\right),$$

where we used (6.1) with λ_k replaced by $\lambda_k \binom{n}{k}^{-m} a_k$, and with $P_n(z) = \sum_{k=0}^{n} z^k$. Since $M\left(\sum_{k=0}^{n} z^k\right) = 1$, it follows that

$$M\left(A_{m+1}P_n\right) \leq M\left(A_m P_n\right), \quad m \in \mathbb{N},$$

and (1.3) is proved by the induction hypothesis.

Let

$$BP_n(z) := \sum_{k=0}^{n} \lambda_k \binom{n}{k}^{-1} a_k z^k = A_1 P_n(z),$$

so that its m-fold composition is

$$B^m P_n(z) = \sum_{k=0}^{n} \lambda_k^m \binom{n}{k}^{-m} a_k z^k.$$

Applying (6.1) m times, we arrive at (1.4). $\qquad \square$

6.2. Proofs for Section 2.

Proof of Theorem 2.1. If P_n does not vanish in $\mathbb{D}$, then $\log|P_n(z)|$ is harmonic in $\mathbb{D}$. Hence $M(P_n) = |a_0|$ and $\|P_n\|_0 = |a_0|$ follow from the contour and area mean value theorems. Assume now that P_n has zeros in $\mathbb{D}$. Applying Jensen's formula, we obtain that

$$\log M(P_n) = \frac{1}{2\pi} \int_0^{2\pi} \log|P_n(e^{i\theta})|\, d\theta = \log|a_n| + \sum_{|z_j|\geq 1} \log|z_j|.$$

Furthermore,

$$\log\|P_n\|_0 = \frac{1}{\pi} \int_0^1 \int_0^{2\pi} \log|P_n(re^{i\theta})|\, r\, dr\, d\theta$$

$$= 2\int_0^1 \left(\frac{1}{2\pi}\int_0^{2\pi} \log|P_n(re^{i\theta})|\, d\theta\right) r\, dr$$

$$= 2\int_0^1 \left(\log|a_n| + \sum_{|z_j|\geq r} \log|z_j| + \sum_{|z_j|<r} \log r\right) r\, dr$$

$$= \log|a_n| + \sum_{|z_j|\geq 1} \log|z_j| + \frac{1}{2}\sum_{|z_j|<1} (|z_j|^2 - 1).$$

Hence

$$\|P_n\|_0 = M(P_n) \exp\left(\frac{1}{2} \sum_{|z_j|<1} (|z_j|^2 - 1)\right).$$

$\square$

Proof of Corollary 2.2. Inequality (2.2) follows from (2.1) after observing that the smallest value of the exponential is achieved when all $z_j = 0$, while the largest value is 1 when all $|z_j| \geq 1$.

$\square$

Proof of Theorem 2.3. If P_n is cyclotomic, then $\|P_n\|_0 = 1$ by Theorem 2.1, because $|z_j| = 1$, $j = 1, \ldots, n$, and $M(P_n) = 1$. Assume now that $\|P_n\|_0 = 1$. Let z_j, $j = 1, \ldots, m$, $m \leq n$, be the zeros of P_n in $\mathbb{D}$. Then we have from (2.1) that

$$(6.2) \qquad \|P_n\|_0 = |a_0| \prod_{j=1}^m \frac{e^{(|z_j|^2-1)/2}}{|z_j|} \geq \prod_{j=1}^m \frac{e^{(|z_j|^2-1)/2}}{|z_j|},$$

where $a_0 \neq 0$ is the constant term of P_n. Define $g(x) := e^{(x^2-1)/2}/x$, $x > 0$, and observe that $g'(x) < 0$ when $x \in (0,1)$, while $g'(x) > 0$ when $x \in (1,\infty)$. Hence

$(6.3) \qquad g(1) = 1$ is the strict global minimum for $g(x)$ on $(0,\infty)$.

It follows from (6.2)-(6.3) that

$$1 < \prod_{j=1}^m g(|z_j|) = \prod_{j=1}^m \frac{e^{(|z_j|^2-1)/2}}{|z_j|} \leq \|P_n\|_0 = 1,$$

which is a contradiction. Hence P_n has no zeros in $\mathbb{D}$, and $M(P_n) = \|P_n\|_0 = 1$ by Theorem 2.1. This implies that P_n is cyclotomic by Kronecker's theorem.

We could also proceed in a different way, by assuming that $\|P_n\|_0 = 1$ and observing from (6.2) that

$$\exp\left(\sum_{j=1}^m \frac{|z_j|^2 - 1}{2}\right) = \frac{1}{|a_0|} \prod_{j=1}^m |z_j|$$

Since the expression on the right is an algebraic number, as well as the sum in the exponent on the left, we obtain that equality is only possible when the latter sum is zero, by the well known result of Lindemann that the exponential of a nonzero algebraic number is transcendental [3]. Hence $|z_j| \geq 1$, $j = 1, \ldots, n$, and $M(P_n) = \|P_n\|_0 = 1$ as before.

$\square$

Proof of Theorem 2.5. We first show that P_n has $o(n)$ zeros in $D_r := \{z : |z| < r\}$ as $n \to \infty$, for any $r < 1$. Assume to the contrary that there is a subsequence of n such that P_n has at least αn zeros, with $\alpha > 0$, in some

D_r, $r < 1$. Suppose that those zeros are $z_j \neq 0$, $j = 1, \ldots, m$, $m \leq n$, and proceed as in the proof of Theorem 2.3 to obtain

(6.4)
$$\prod_{j=1}^{m} g(|z_j|) = \prod_{j=1}^{m} \frac{e^{(|z_j|^2 - 1)/2}}{|z_j|} \leq \|P_n\|_0$$

by (6.2). Since $g(x) = e^{(x^2 - 1)/2}/x$ is strictly decreasing on $(0, 1)$, we have that

$$\prod_{j=1}^{m} g(|z_j|) \geq (g(r))^{\alpha n}.$$

It immediately follows from (6.3) and (6.4) that

$$\limsup_{n \to \infty} \|P_n\|_0^{1/n} \geq (g(r))^{\alpha} > 1,$$

which is in direct conflict with assumptions of this theorem. If P_n has a simple zero at $z = 0$, then $P_n(z) = zQ_{n-1}(z)$ and $\|P_n\|_0 = \|Q_{n-1}\|_0/\sqrt{e}$. Hence we can apply the above argument to Q_{n-1} and come to the same conclusion that P_n has $o(n)$ zeros in $D_r := \{z : |z| < r\}$, $r < 1$, as $n \to \infty$.

The second step is to show that $\lim_{n \to \infty} (M(P_n))^{1/n} = 1$. Note that

(6.5)
$$1 \leq M(P_n) = \|P_n\|_0 \exp\left(\frac{1}{2} \sum_{|z_j| < 1} (1 - |z_j|^2)\right).$$

If P_n has $m = o(n)$ zeros in D_r, $r < 1$, then

$$\exp\left(\frac{1}{2} \sum_{|z_j| < 1} (1 - |z_j|^2)\right) \leq e^{m/2 + n(1 - r^2)/2}.$$

Using this in (6.5), we obtain that

$$1 \leq \liminf_{n \to \infty} (M(P_n))^{1/n} \leq \limsup_{n \to \infty} (M(P_n))^{1/n}$$
$$\leq e^{(1 - r^2)/2} \lim_{n \to \infty} \|P_n\|_0^{1/n} = e^{(1 - r^2)/2}.$$

Hence $\lim_{n \to \infty} (M(P_n))^{1/n} = 1$ follows by letting $r \to 1-$. The proof may now be completed by applying Bilu's result [4] (at least when P_n is irreducible for all $n \in \mathbb{N}$), but we prefer to continue with an independent proof via a standard potential theoretic argument.

Observe that $P_n(z) = a_n \prod_{j=1}^{n}(z - z_j)$ has $o(n)$ zeros in $\mathbb{C} \setminus D_r$, $r > 1$, for otherwise we would have $\liminf_{n \to \infty} (M(P_n))^{1/n} > 1$ as

$$M(P_n) = |a_n| \prod_{|z_j| > 1} |z_j| \geq \prod_{|z_j| > 1} |z_j|.$$

This also implies that

(6.6)
$$\lim_{n \to \infty} |a_n|^{1/n} = 1.$$

Hence any weak* limit ν of the sequence ν_n must satisfy $\operatorname{supp}\nu \subset \mathbb{T}$. Define the logarithmic energy of ν by

$$I(\nu) := \iint \log \frac{1}{|z-t|} d\nu(z)\, d\nu(t).$$

Our goal is to show that $I(\nu) = 0$, which implies that ν has the smallest possible energy among all positive Borel measures of mass 1 supported on $\mathbb{T}$. On the other hand, it is well known in potential theory that the equilibrium measure minimizing the energy integral is unique, and it is equal to the normalized arclength on $\mathbb{T}$ [34, 46]. Thus $\nu = d\theta/(2\pi)$ and the proof would be completed.

Define the discriminant of P_n as $\Delta_n := a_n^{2n-2} \prod_{1\le j<k\le n}(z_j-z_k)^2$. Observe that it is an integer, being a symmetric form with integer coefficients in the roots of $P_n \in \mathbb{Z}_n[z]$. Since P_n has no multiple roots, we have $\Delta_n \ne 0$ and $|\Delta_n| \ge 1$. Therefore

$$(6.7) \qquad \log\frac{1}{|\Delta_n|} = -(2n-2)\log|a_n| + \sum_{j\ne k}\log\frac{1}{|z_j-z_k|} \le 0.$$

Let

$$K_M(z,t) := \min\left(\log\frac{1}{|z-t|}, M\right), \qquad M > 0.$$

It is clear that $K_M(z,t)$ is a continuous function in z and t on $\mathbb{C}\times\mathbb{C}$, and that $K_M(z,t)$ increases to $\log\frac{1}{|z-t|}$ as $M \to \infty$. Using the Monotone Convergence Theorem and the weak* convergence of $\nu_n \times \nu_n$ to $\nu \times \nu$, we obtain that

$$I(\nu) = \lim_{M\to\infty} \iint K_M(z,t)\, d\nu(z)\, d\nu(t)$$

$$= \lim_{M\to\infty}\left(\lim_{n\to\infty} \iint K_M(z,t)\, d\nu_n(z)\, d\nu_n(t)\right)$$

$$= \lim_{M\to\infty}\left(\lim_{n\to\infty}\left(\frac{1}{n^2}\sum_{j\ne k} K_M(z_j,z_k) + \frac{M}{n}\right)\right)$$

$$\le \lim_{M\to\infty}\left(\liminf_{n\to\infty} \frac{1}{n^2}\sum_{j\ne k}\log\frac{1}{|z_j-z_k|}\right)$$

$$= \liminf_{n\to\infty} \frac{1}{n^2}\log\frac{|a_n|^{2n-2}}{\Delta_n}.$$

Hence $I(\nu) \le 0$ follows from (6.6)-(6.7). But $I(\mu) > 0$ for any positive unit Borel measure supported on $\mathbb{T}$, with the only exception for the equilibrium measure $d\mu_{\mathbb{T}} := d\theta/(2\pi)$, $I(\mu_{\mathbb{T}}) = 0$, see [46, pp. 53-89]. $\qquad\square$

Proof of Theorem 2.7. (a) We use the same notation and approach as in the proof of Theorem 2.3. If P_n has no zeros in $\mathbb{D}$, then $\min_{1\le j\le n}|z_j| \ge 1$.

Otherwise, let z_j, $j = 1, \ldots, m$, $m \leq n$, be the zeros of P_n in $\mathbb{D}$. It follows from (6.2)-(6.3) that

$$\|P_n\|_0 = |a_0| \prod_{j=1}^{m} \frac{e^{(|z_j|^2-1)/2}}{|z_j|} \geq g\left(\min_{1 \leq j \leq n} |z_j|\right) > 1.$$

Thus we obtain the result by the continuity of $g(x) = e^{(x^2-1)/2}/x$, $x > 0$, and (6.3).

(b) Note that $\lim_{n \to \infty} \|P_n\|_0 = 1$ in this case too, by (2.2) and (2.3). Hence (2.5) holds true. Furthermore, we have for any zero $z_k \in \mathbb{C} \setminus \mathbb{D}$ that

$$1 \leq |z_k| \leq |a_n| \prod_{|z_j|>1} |z_j| = M(P_n).$$

Thus

$$\lim_{n \to \infty} \max_{1 \leq j \leq n} |z_j| = 1,$$

and (2.6) follows.

$\square$

6.3. Proofs for Section 3.

Proof of Theorem 3.1. Using (1.2) for the polynomial $P_n(rz)$, $r \in [0,1]$, we obtain that

$$\frac{1}{2\pi} \int_0^{2\pi} \log |\Lambda P_n(re^{i\theta})|\, d\theta \leq \log M(\Lambda_n) + \frac{1}{2\pi} \int_0^{2\pi} \log |P_n(re^{i\theta})|\, d\theta.$$

Hence (3.1) follows for $p = 0$, if we multiply this inequality by $r\, dr$ and integrate from 0 to 1. Similarly, we obtain from (1.5) that

$$\frac{1}{2\pi} \int_0^{2\pi} |\Lambda P_n(re^{i\theta})|^p\, d\theta \leq \frac{M^p(\Lambda_n)}{2\pi} \int_0^{2\pi} |P_n(re^{i\theta})|^p\, d\theta, \quad 0 < p < \infty,$$

which gives (3.1) for this range of p after integration with respect to $r\, dr$. When $p = \infty$, (3.1) is identical to (1.5).

$\square$

Proof of Corollary 3.2. Consider a monic polynomial $P_n(z) = z^n + \ldots$ and $\Lambda_n(z) = z^n$. Then $\Lambda P_n(z) = z^n$, so that (3.2) follows from (3.1) and an elementary computation.

$\square$

Proof of Corollary 3.3. Let $\Lambda_n(z) = (1 + Rz)^n = \sum_{k=0}^{n} \binom{n}{k} R^k z^k$. Then $\Lambda P_n(z) = P_n(Rz)$ and $M(\Lambda_n) = R^n$. Hence (3.1) gives that

$$\|P_n(Rz)\|_p \leq R^n \|P_n\|_p, \quad 0 \leq p < \infty \text{ and } R \geq 1.$$

Changing variable and passing to the integrals over D_R, we obtain (3.3) and (3.4). The case of equality for $P_n(z) = z^n$ is verified by a routine calculation.

$\square$

Proof of Corollary 3.4. Let $\Lambda_n(z) = \binom{n}{k} z^k, 0 \le k \le n$. Then $\Lambda P_n(z) = a_k z^k$ and $M(\Lambda_n) = \binom{n}{k}$. It follows from (3.1) that

$$|a_k| \|z^k\|_p = \|a_k z^k\|_p \le \binom{n}{k} \|P_n\|_p, \quad 0 \le p \le \infty.$$

One only needs now to find $\|z^k\|_p$, to show that (3.5) and (3.6) hold true. □

Proof of Theorem 3.5. We use the approach of de Bruijn and Springer [10, 1, 2, 33] by setting $\Lambda_n(z) = nz(1+z)^{n-1} = \sum_{k=0}^n k\binom{n}{k} z^k$. This gives $\Lambda P_n(z) = zP_n'(z)$ and $M(\Lambda_n) = n$. Hence (3.7) is a direct consequence of (3.1). □

Proof of Corollary 3.6. In order to deduce Corollary 3.6 from Theorem 3.5, we only need to observe that $\|zP_n'\|_0 = \|z\|_0 \|P_n'\|_0 = \|P_n'\|_0 / \sqrt{e}$. □

Proof of Proposition 3.7. For $P_n(z) = \sum_{k=0}^n a_k z^k$ we have $P_n'(z) = \sum_{k=0}^n k a_k z^{k-1}$, so that

$$\|P_n\|_2^2 = \sum_{k=0}^n \frac{|a_k|^2}{k+1} \quad \text{and} \quad \|P_n'\|_2^2 = \sum_{k=0}^n k|a_k|^2.$$

It follows that

$$\|P_n'\|_2^2 = \sum_{k=0}^n k(k+1)\frac{|a_k|^2}{k+1} \le n(n+1)\sum_{k=0}^n \frac{|a_k|^2}{k+1} = n(n+1)\|P_n\|_2^2.$$

The case of equality is verified directly. □

6.4. Proofs for Section 4.

Proof of Proposition 4.1. Recall that the uniform convergence of P_n to f on compact subsets of $\mathbb{D}$ implies that f is analytic in $\mathbb{D}$, and that $P_n^{(k)}$ converge to $f^{(k)}$ on compact subsets of $\mathbb{D}$ for any $k \in \mathbb{N}$. In particular,

$$\lim_{n \to \infty} P_n^{(k)}(0) = f^{(k)}(0) \qquad \forall\, k \ge 0,\ k \in \mathbb{Z}.$$

But $P_n^{(k)}(0) = k! a_k$, where $a_k \in \mathbb{Z}$ is a corresponding coefficient of P_n. Hence the result follows. □

Proof of Proposition 4.2. We have that

$$\|P_n - P_{n-1}\|_{H^p} \le \|f - P_n\|_{H^p} + \|f - P_{n-1}\|_{H^p}$$

by the triangle inequality for $p \ge 1$, and

$$\|P_n - P_{n-1}\|_{H^p}^p \le \|f - P_n\|_{H^p}^p + \|f - P_{n-1}\|_{H^p}^p$$

for $0 < p < 1$. In both cases, (4.1) implies that

$$\lim_{n\to\infty} \|P_n - P_{n-1}\|_{H^p} = 0, \qquad 0 < p \le \infty.$$

If $P_n \not\equiv P_{n-1}$ then we let $a_k z^k$ be the lowest nonzero term of $P_n - P_{n-1}$, where $|a_k| \in \mathbb{N}$. Using the mean value inequality [12], we obtain

$$\|P_n - P_{n-1}\|_{H^p} \ge |a_k| \ge 1, \qquad 0 < p \le \infty.$$

This is obviously impossible as $n \to \infty$, so that we have $P_n \equiv P_{n-1}$ for all sufficiently large $n \in \mathbb{N}$. Hence the limit function f is also a polynomial with integer coefficients.

$\square$

Proof of Theorem 4.3. If (4.2) holds then the P_n converge to f on compact subsets of $\mathbb{D}$ by the area mean value inequality:

$$|f(z) - P_n(z)|^p \le \frac{1}{\pi(1-|z|)^2} \iint_{|t-z|<1-|z|} |f(t) - P_n(t)|^p \, dA$$
$$\le \frac{\|f(t) - P_n(t)\|_p^p}{(1-|z|)^2} \to 0, \qquad n \to \infty, \ z \in \mathbb{D}.$$

Hence f has a power series expansion at $z = 0$ with integer coefficients by Proposition 4.1.

Conversely, suppose that $f \in A^p$ is represented by a power series with integer coefficients. Since the partial sums of this series converge to f in A^p norm for $1 < p < \infty$ by Theorem 4 [12, p. 31], we can select the sequence P_n be the sequence of the partial sums.

$\square$

6.5. **Proofs for Section 5.**

Proof of Proposition 5.1. We apply (2.2) in each variable z_j, $j = 1, \ldots, d$, and use Fubini's theorem to prove (5.2). Indeed, (2.2) gives that

$$\frac{1}{2\pi} \int_{\mathbb{T}} \log|P_n(z_1, \ldots, z_d)| \, |dz_1| - \frac{k_1}{2} \le \frac{1}{\pi} \int_{\mathbb{D}} \log|P_n(z_1, \ldots, z_d)| \, dA(z_1)$$
$$\le \frac{1}{2\pi} \int_{\mathbb{T}} \log|P_n(z_1, \ldots, z_d)| \, |dz_1|$$

is true for all $z_2, \ldots, z_d \in \mathbb{C}$. Integrating the above inequality with respect to $dA(z_2)/\pi$, interchanging the order of integration in the lower and upper bounds, and applying (2.2) in the variable z_2, we obtain

$$\frac{1}{(2\pi)^2} \int_\mathbb{T} \int_\mathbb{T} \log |P_n(z_1, \ldots, z_d)| \, |dz_1||dz_2| - \frac{k_1 + k_2}{2}$$

$$\leq \frac{1}{\pi^2} \int_\mathbb{D} \int_\mathbb{D} \log |P_n(z_1, \ldots, z_d)| \, dA(z_1) dA(z_2)$$

$$\leq \frac{1}{(2\pi)^2} \int_\mathbb{T} \int_\mathbb{T} \log |P_n(z_1, \ldots, z_d)| \, |dz_1||dz_2|$$

is true for all $z_3, \ldots, z_d \in \mathbb{C}$. After carrying out this argument for each variable z_j, we arrive at (5.2) in d steps. When $P_n(z_1, \ldots, z_d) \neq 0$ in $\mathbb{D}^d$, we have that $\|P_n\|_0 = M(P_n) = |a_{0\ldots0}|$ by the iterative application of Theorem 2.1. If $P_n(z_1, \ldots, z_d) = a_{k_1 \ldots k_d} z_1^{k_1} \ldots z_d^{k_d}$, where $k_1 + \ldots + k_d = n$, then we evaluate directly that $M(P_n) = |a_{k_1 \ldots k_d}|$ and $\|P_n\|_0 = |a_{k_1 \ldots k_d}| e^{-n/2}$, because $\|z_j\|_0 = e^{-1/2}$, $j = 1, \ldots, n$.   □

Proof of Example 5.2. (a) Applying (2.1), we have that

$$\frac{1}{\pi^2} \int_\mathbb{D} \int_\mathbb{D} \log |z_1 + z_2| \, dA(z_1) dA(z_2) = \frac{1}{\pi} \int_\mathbb{D} \frac{|z_2|^2 - 1}{2} \, dA(z_2) = -\frac{1}{4}.$$

(b) is an immediate consequence of (c).

(c) Let $a_{0\ldots0} = |a_{0\ldots0}| e^{i\phi}$. Observe that $P_n(z_1, \ldots, z_d) + \varepsilon e^{i\phi} \neq 0$ in $\mathbb{D}^d$ for any $\varepsilon > 0$, because

$$|P_n(z_1, \ldots, z_d) + \varepsilon e^{i\phi}| \geq |a_{0\ldots0}| + \varepsilon - \sum_{0 < k_1 + \ldots + k_d \leq n} |a_{k_1 \ldots k_d}| > 0$$

by the triangle inequality. We obtain that $\|P_n + \varepsilon e^{i\phi}\|_0 = M(P_n + \varepsilon e^{i\phi}) = |a_{0\ldots0}| + \varepsilon$ by the area and contour mean value properties of the (pluri)harmonic function $\log |P_n(z_1, \ldots, z_d) + \varepsilon e^{i\phi}|$ in $\mathbb{D}^d$, and the result follows by letting $\varepsilon \to 0$.   □

REFERENCES

[1] V. V. Arestov, On integral inequalities for trigonometric polynomials and their derivatives, *Math. USSR-Izv.* **18** (1982), 1-17.

[2] ———, Integral inequalities for algebraic polynomials on the unit circle, *Math. Notes* **48** (1990), 977–984.

[3] A. Baker, *Transcendental number theory*, Cambridge Univ. Press, New York, 1975.

[4] Y. Bilu, Limit distribution of small points on algebraic tori, *Duke Math. J.* **89** (1997), 465-476.

[5] E. Bombieri, Subvarieties of linear tori and the unit equation: A survey, in *Analytic number theory*, ed. by Y. Motohashi, LMS Lecture Notes **247** (1997), Cambridge Univ. Press, Cambridge, pp. 1-20.

[6] P. Borwein, *Computational excursions in analysis and number theory*, Springer-Verlag, New York, 2002.

[7] P. Borwein and T. Erdélyi, *Polynomials and polynomial inequalities*, Springer-Verlag, New York, 1995.

[8] D. W. Boyd, Variations on a theme of Kronecker, *Canad. Math. Bull.* **21** (1978), 1244-1260.

[9] _____, Speculations concerning the range of Mahler measure, *Canad. Math. Bull.* **24** (1981), 453-469.

[10] N. G. de Bruijn and T. A. Springer, On the zeros of composition-polynomials, *Indag. Math.* **9** (1947), 406-414.

[11] A. Dubickas and C. J. Smyth, The Lehmer constants of an annulus, *J. Théor. Nombres Bordeaux* **13** (2001), 413–420.

[12] P. L. Duren and A. Schuster, *Bergman spaces*, American Mathematical Society, Providence, 2004.

[13] P. Erdős and P. Turán, On the distribution of roots of polynomials, *Ann. of Math.* **51** (1950), 105-119.

[14] G. Everest and T. Ward, *Heights of polynomials and entropy in algebraic dynamics*, Springer-Verlag, London, 1999.

[15] M. Fekete, Über die Verteilung der Wurzeln bei gewissen algebraischen Gleichungen mit ganzzahligen Koeffizienten, *Math. Zeit.* **17** (1923), 228-249.

[16] Le Baron O. Ferguson, *Approximation by polynomials with integral coefficients*, Amer. Math. Soc., Providence, R.I., 1980.

[17] E. Ghate and E. Hironaka, The arithmetic and geometry of Salem numbers, *Bull. Amer. Math. Soc.* **38** (2001), 293-314.

[18] G. H. Hardy, J. E. Littlewood and G. Pólya, *Inequalities*, Cambridge Univ. Press, London, 1952.

[19] H. Hedenmalm, B. Korenblum and K. Zhu, *Theory of Bergman spaces*, Springer-Verlag, New York, 2000.

[20] R. Jentzsch, Untersuchungen zur Theorie der Folgen analytischer Funktionen, *Acta Math.* **41** (1917), 219-270.

[21] L. Kronecker, Zwei Sätze über Gleichungen mit ganzzahligen Koeffizienten, *J. Reine Angew. Math.* **53** (1857), 173-175.

[22] M. Langevin, Méthode de Fekete-Szegö et problème de Lehmer, *C. R. Acad. Sci. Paris Sér. I Math.* **301** (1985), 463–466.

[23] _____, Minorations de la maison et de la mesure de Mahler de certains entiers algébriques, *C. R. Acad. Sci. Paris Sér. I Math.* **303** (1986), 523–526.

[24] _____, Calculs explicites de constantes de Lehmer, Groupe de travail en théorie analytique et élémentaire des nombres, 1986–1987, 52–68, Publ. Math. Orsay, 88-01, Univ. Paris XI, Orsay, 1988.

[25] D. H. Lehmer, Factorization of certain cyclotomic functions, *Ann. of Math.* **34** (1933), 461-479.

[26] K. Mahler, On the zeros of the derivative of a polynomial, *Proc. Roy. Soc. London Ser. A* **264** (1961), 145-154.

[27] _____, On two extremum properties of polynomials, *Illinois J. Math.* **7** (1963), 681–701.

[28] _____, A remark on a paper of mine on polynomials, *Illinois J. Math.* **8** (1964), 1–4.

[29] A. Máté and P. Nevai, Bernstein inequality in L^p for $0 < p < 1$ and $(C, 1)$ bounds of orthogonal polynomials, *Ann. of Math.* **111** (1980), 145-154.

[30] M. Mignotte, Sur un théorème de M. Langevin, *Acta Arith.* **54** (1989), 81–86.

[31] G. V. Milovanović, D. S. Mitrinović and Th. M. Rassias, *Topics in polynomials: extremal problems, inequalities, zeros*, World Scientific, Singapore, 1994.

[32] I. E. Pritsker, An areal analog of Mahler's measure, *Illinois J. Math.* (to appear)

[33] Q. I. Rahman and G. Schmeisser, *Analytic theory of polynomials*, Clarendon Press, Oxford, 2002.

[34] T. Ransford, *Potential theory in the complex plane*, Cambridge University Press, Cambridge, 1995.

[35] G. Rhin and C. J. Smyth, On the absolute Mahler measure of polynomials having all zeros in a sector, *Math. Comp.* **65** (1995), 295–304.

[36] R. Rumely, On Bilu's equidistribution theorem, in *Spectral problems in geometry and arithmetic (Iowa City, IA, 1997)*, Contemp. Math. **237**, Amer. Math. Soc., Providence, RI, 1999, pp. 159-166.

[37] A. Schinzel, On the product of the conjugates outside the unit circle of an algebraic number, *Acta Arith.* **24** (1973), 385-399. (Addendum: *Acta Arith.* **26** (1974/75), 329–331.)

[38] T. Sheil-Small, *Complex polynomials*, Cambridge University Press, Cambridge, 2002.

[39] C. J. Smyth, On measures of polynomials in several variables, *Bull. Australian Math. Soc.* **23** (1981), 49-63. (Corrigendum: G. Myerson and C. J. Smyth, *Bull. Austral. Math. Soc.* **26** (1982), 317–319.)

[40] _____, A Kronecker-type theorem for complex polynomials in several variables, *Canad. Math. Bull.* **24** (1981), 447-452. (Addenda and errata: *Canad. Math. Bull.* **25** (1982), 504.)

[41] _____, The Mahler measure of algebraic numbers: A survey, *this volume*, 322–349.

[42] E. A. Storozhenko, A problem of Mahler on the zeros of a polynomial and its derivative, *Sb. Math.* **187** (1996), 735-744.

[43] G. Szegő, Beiträge zur Theorie der Toeplitzschen Formen, I, *Math. Zeit.* **6** (1920), 167-202.

[44] _____, Über die Nullstellen von Polynomen, die in einem Kreis gleichmässig konvergieren, *Sitzungsber. Ber. Math. Ges.* **21** (1922), 59-64.

[45] R. M. Trigub, Approximation of functions with Diophantine conditions by polynomials with integral coefficients, in *Metric Questions of the Theory of Functions and Mappings*, No. 2, Naukova Dumka, Kiev, 1971, pp. 267-333. (Russian)

[46] M. Tsuji, *Potential theory in modern function theory*, Chelsea Publ. Co., New York, 1975.

[47] A. Zygmund, A remark on conjugate series, *Proc. London Math. Soc.* **34** (1932), 392-400.

INTEGER TRANSFINITE DIAMETER AND COMPUTATION OF POLYNOMIALS

GEORGES RHIN AND QIANG WU

ABSTRACT. In this paper we explain how explicit auxiliary functions are used to compute some interesting families of polynomials. These functions depend on generalisations of the integer transfinite diameter of some compact sets in $\mathbb{C}$. They give better bounds than the classical ones for the coefficients of the minimal polynomial of an algebraic integer α.

1. INTRODUCTION

1.1. Definitions and notation.

Let α be an algebraic integer of degree $d \geq 2$, whose conjugates are $\alpha_1 = \alpha, \alpha_2, \ldots, \alpha_d$, and let

$$P = b_0 X^d + b_1 X^{d-1} + \cdots + b_{d-1} X + b_d,$$

with $b_0 = 1$, be its minimal polynomial. We say that α is *totally positive* if all its conjugates are positive real numbers and α is *reciprocal* if α^{-1} is a conjugate of α. A *Salem number* is a real algebraic number $\alpha > 1$ all of whose conjugates lie in the unit disc $|z| \leq 1$ with at least one (and so all the others but $1/\alpha$) of modulus 1. A *Perron number* is a positive real algebraic integer α such that $\max_{2 \leq i \leq d} |\alpha_i| < \alpha$. For $k \geq 1$ we write $s_k = \sum_{1 \leq i \leq d} \alpha_i^k$. Then s_1 is the *trace* of α and s_1/d is the *absolute trace* of α.

We define the Mahler measure of α (and of P) by

$$M(\alpha) = |b_0| \prod_{i=1}^{d} \max(1, |\alpha_i|)$$

and the *house* of α by

$$\boxed{\alpha} = \max_{1 \leq i \leq d} |\alpha_i|.$$

2000 *Mathematics Subject Classification.* Primary:12Y05; Secondary 11C08, 11R06, 11Y40.

Key words and phrases. Algebraic integer, maximal modulus, Schinzel-Zassenhaus conjecture, Mahler measure, Breusch-Smyth theorem, Perron numbers, explicit auxiliary functions, integer transfinite diameter.

Qiang Wu was supported in part by the Natural Science Foundation of Chongqing grant CSTC n° 2005BB8024.

1.2. The explicit auxiliary functions.

Throughout this paper we will use functions of the following type

$$f(z) = g(z) - \sum_{1 \le j \le J} e_j \log |Q_j(z)|, \qquad (1)$$

where the e_j are positive real numbers and the integer polynomials Q_j belong to a fixed set S. Also g is a real valued function such that f is harmonic outside a finite set containing the roots of the polynomials Q_j. The numbers e_j are always chosen to get the best auxiliary function, i.e. the one with the largest minimum. Firstly we explain briefly how we use such an auxiliary function in the simplest case of the Schur Trace Problem [2]. We want to get a lower bound of the absolute trace of all totally positive algebraic integers, apart from a finite number of exceptions. Here we take $g(x) = x$ and m is the minimum of $f(x)$ for $x \ge 0$. Then

$$\sum_{1 \le i \le d} f(\alpha_i) \ge md \qquad (2)$$

and

$$s_1 \ge dm + \sum_{1 \le j \le J} e_j \log \left| \prod_{1 \le i \le d} Q_j(\alpha_i) \right|.$$

$\prod_{1 \le i \le d} Q_j(\alpha_i)$ is equal to the resultant of P and Q_j and, if we assume that P does not divide any Q_j, this is a nonzero integer. Therefore

$$s_1/d \ge m$$

if α is not a root of any polynomial Q_j. So, the greater m is, the 'better' f is. The scheme of the computations is the following. With the functions f we get bounds of s_k for k less than $K \ge d$ which is often much larger than d (like $2d$ or $3d$). Then we use Newton's formula which gives by induction the coefficients b_k for $1 \le k \le d$, in order to get a large set of polynomials. We compute s_k for $d < k \le K$ and eliminate the polynomial when s_k is not within its bounds. The last computation is done with Pari [14]. The construction of these auxiliary functions is explained in Section 6. Since the function f is harmonic, it takes its minimal value on the boundary $\mathcal{B}$ of the domain which contains all the conjugates of α (which is always $\mathbb{R}$ or an ellipse invariant under complex conjugation). Generally, even in the worst case, we can reduce the number of non-real conjugates that are on the boundary to 1 or 2 (except in the case of Section 5). Hence, we will replace md in (2) by $(d-1)m_0 + m_1$ (respectively by $(d-2)m_0 + 2m_1$) where m_0 is the minimum of $f(x)$ for a real x inside $\mathcal{B}$ and m_1 ($\le m_0$) is the minimum of $f(z)$ for z on $\mathcal{B}$. Therefore the 'efficiency' of the auxiliary functions (1) decreases when the boundary $\mathcal{B}$ becomes less 'flat' from Sections 2 to 5. Now we describe what kind of polynomials we are interested in.

1.3. Some interesting families of polynomials. In Section 2, after a brief summary giving the best values of m that have been recently obtained for the absolute trace of totally positive algebraic integers, we explain the use of this function to compute small Salem numbers. It is not known whether the Mahler measure of α, when it is greater than 1, is greater than the Mahler measure of Lehmer's polynomial. The computation of small Mahler measures is investigated in Section 3. In Section 4 we explain how to solve a problem of Colin Maclachlan. Section 5 is devoted to the exploration of the conjecture of Schinzel and Zassenhaus for the numbers α with small house up to degree 28 and we prove that the smallest Perron numbers satisfy the conjecture of Lind-Boyd up to degree 23. In Section 6 we explain how to get the auxiliary functions used in the previous sections.

This paper is an extended version of the lectures given by the authors in Bristol in April 2006. We take the opportunity to thank the organisers of the workshop 'Number Theory and Polynomials', held at the University of Bristol, for their warm hospitality.

2. THE TOTALLY POSITIVE ALGEBRAIC INTEGERS

In 1984 C. Smyth [18] gave the good lower bound $(1.771\ldots)$ with a suitable function f of the type (1). The best value for m given by McKee and Smyth [12] is equal to 1.77838 and the best known value in April 2006 was 1.782061, recently improved to 1.784109. For more details we refer to the survey by J. Aguirre and J.C. Peral in these proceedings [1]. In these cases the polynomials Q_j were found by an extended heuristic search.

Let α be a Salem number. Then $\beta = \alpha + 1/\alpha + 2$ is a totally positive algebraic integer of degree $\deg(\alpha)/2$. Since all positive powers β^k are totally positive, it is possible to use Smyth's function f to get good lower bounds for s_k equal to the sum of the k-th powers of β. V. Flammang, M. Grandcolas and G. Rhin [7] used this method to verify that the list of Salem numbers less than 1.3 of degree not greater than 40 given by M. Mossinghoff [13] is complete. More recently J. McKee and C. Smyth [12] have found the two Salem numbers of smallest degree (i.e., 20) with trace -2.

3. ALGEBRAIC INTEGERS WITH SMALL MAHLER MEASURE

For an algebraic integer α, we have $M(P) \geq 1$ and Kronecker's theorem implies that $M(P) = 1$ if and only if P is a product of cyclotomic polynomials and a power of X. Lehmer's question is the following:

Lehmer's question. Does there exist a positive constant ε such that, if $M(\alpha) > 1$, then $M(\alpha) \geq 1 + \varepsilon$?

The smallest known value for $M(\alpha) > 1$ has been obtained by D.H. Lehmer [10] and is $M(\alpha) = 1.1762808\ldots$ given by the polynomial of degree 10:

$$L(X) = X^{10} + X^9 - X^7 - X^6 - X^5 - X^4 - X^3 + X + 1.$$

C. Smyth [17] has proved that, if the algebraic number $\alpha \neq 0, 1$ is not reciprocal, then $M(\alpha) \geq \theta_0$, where $\theta_0 = 1.324717\ldots$ is the smallest Pisot number which is the real root of $X^3 - X - 1$. This kind of result had been proved earlier by Breusch [6] with the smaller constant $1.1796\ldots$. P. Voutier [20] has proved that, if an algebraic number α of degree $d > 2$ is not a root of unity, then

$$M(\alpha) > 1 + \frac{1}{4}\left(\frac{\log\log d}{\log d}\right)^3. \tag{3}$$

Exhaustive searches have been made by D. Boyd [3] and [5] and by M. Mossinghoff [13] up to degree 20 and to degree 24 respectively to find all the algebraic numbers α with $M(\alpha) < 1.3$.

Other extensive searches have been made by M. Mossinghoff [13] with heuristic methods up to degree 180. G. Rhin and J.-M. Sac-Épée [15] used both a statistical method and a minimisation method to get polynomials of large degree and small Mahler measure. V. Flammang, G. Rhin and J.-M. Sac-Épée [9] have computed all the numbers α with $M(\alpha) < \theta_0$, $d \leq 36$ and $M(\alpha) < 1.31$ for $d = 38, 40$ and then proved that Mossinghoff's table is complete up to degree 40.

Because of Smyth's theorem, an algebraic integer with Mahler measure less than θ_0 is reciprocal. Its minimal polynomial P is of degree $2d$:

$$P = X^{2d} + c_1 X^{2d-1} + \ldots + c_1 X + 1 = \prod_{i=1}^{2d}(X - \alpha_i).$$

We may suppose that $|\alpha_i| \geq 1$ and $\alpha_{d+i} = 1/\alpha_i$ for $1 \leq i \leq d$. We define the polynomial Q, associated to P by the formula $X^d Q(X + 1/X) = P(X)$. Therefore

$$Q = X^d + b_1 X^{d-1} + \ldots + b_{d-1} X + b_d$$

is a monic polynomial of degree d with integer coefficients whose roots are $\gamma_i = \alpha_i + 1/\alpha_i$ for $1 \leq i \leq d$.

For Q we have $s_1 = \sum_{1 \leq i \leq d} \gamma_i = \sum_{1 \leq i \leq 2d} \alpha_i$. More generally for $k > 1$ we define $\gamma_{i,k} = \alpha_i^k + 1/\alpha_i^k$ and we have $s_k = \sum_{1 \leq i \leq d} \gamma_{i,k} = \sum_{1 \leq i \leq 2d} \alpha_i^k$. Let $1 < M \leq \theta_0$ be a fixed bound. If we suppose that $1 \leq |\alpha_i| \leq M^a$ with $0 < a \leq 1$ then $\gamma_{i,k}$ lies inside the ellipse

$$\mathcal{E}_{ka} = \left\{z = x + iy \text{ such that } \left(\frac{x}{A}\right)^2 + \left(\frac{y}{B}\right)^2 \leq 1\right\},$$

where $A = M^{ka} + M^{-ka}$ and $B = M^{ka} - M^{-ka}$. If $M = \theta_0$ the ellipse $\mathcal{E}_{ka}$ is close to the real axis when k is not too large (for $k = 1$, $B < 0.57$). For degree 34, the classical bound for the trace is $|s_1| \leq 34$ and we get $|s_1| \leq 4$.

4. A PROBLEM OF COLIN MACLACHLAN

Here we explain how to solve the following problem of Colin Maclachlan from the University of Aberdeen. The discrete subgroups of $SL(2, \mathbb{C})$ with two generators, one of order 2 and one of order 3, are related to the algebraic integers α of degree $d \geq 2$ whose conjugates all lie inside the ellipse:

$$\mathcal{E} = \left\{ z = x + iy \text{ such that } \left(\frac{x - 1/2}{5/2} \right)^2 + \left(\frac{y}{2} \right)^2 \leq 1 \right\},$$

with two of them conjugate non-real numbers, while all the others are real in the interval $[-1, 2]$. We denote by $\mathcal{C}$ the set of the minimal polynomials P of the algebraic integers defined above. It is clear that, if $P \in \mathcal{C}$ is of degree d, then $(-1)^d P(1 - X)$ is also in $\mathcal{C}$. So we call $\mathcal{C}_1$ the set of polynomials $P \in \mathcal{C}$ such that if $P(z) = 0$ and z is complex then $\operatorname{Re} z \leq 1/2$. We denote by $\mathcal{E}_1$ the set of the elements z of the ellipse $\mathcal{E}$ with $\operatorname{Re} z \leq 1/2$.

To show that $\mathcal{C}$ is finite, it suffices to show that $\mathcal{C}_1$ is finite. For this, we use the fact that all the roots of P, but 2, lie in the interval $[-1, 2]$ whose transfinite diameter is less than 1 (in fact $3/4$). Maclachlan proved that $d \leq 18$. If we use the classical bounds for the coefficients of the polynomials in $\mathcal{C}_1$ this would give a set of 10^{63} elements. The first step is to prove, with suitable auxiliary functions, that $d \leq 12$. For this we use the function $g(z) = -\log|z|$ inside the auxiliary function (1). This gives an upper bound for $|P(0)|$ which is less than 1 for $d > 12$. The same kind of functions give bounds for $|P(z_i)|$ where the numbers z_i are real or complex algebraic integers which lie in the real interval $[-1, 2]$ or not too far from it. This gives a rather 'small set' of 1000 4-tuples $(b_{d-3}, b_{d-2} \, b_{d-1} \, b_d)$ for $d = 8$ (respectively 332 for $d = 9$). More relations on the last coefficients are given for $d = 10$, 11 and 12.

Next, functions with $g(z) = \pm \operatorname{Re}(z^k)$ give good bounds for the numbers s_k. V. Flammang and G. Rhin [8] found the 15909 polynomials satisfying the initial conditions. All these polynomials are of degree at most 10.

5. ALGEBRAIC INTEGERS OF SMALL HOUSE AND SMALL PERRON NUMBERS

5.1. Algebraic integers of smallest house.

Kronecker's theorem asserts that $\boxed{|\alpha|} = \max_{1 \leq i \leq d} |\alpha_i| = 1$ if and only if α is a root of unity. We define $\mathrm{m}(d)$ to be the minimum of the houses of the nonzero algebraic integers α of degree d that are not roots of unity.

A classical problem, see P. Borwein [2], is to study the behaviour of $\mathrm{m}(d)$ when d varies. On the one hand it is clear that $\mathrm{m}(d) \leq 2^{1/d}$ since the polynomial $X^d - 2$ is irreducible of degree d. On the other hand there is a conjecture of A. Schinzel and H. Zassenhaus [19] which asserts that $\mathrm{m}(d) \geq 1 + c_1/d$, where c_1 is a positive constant. Moreover D. Boyd [4] suggests that the best c_1 should be equal to $\frac{3}{2} \log(\theta_0)$ where $\theta_0 = 1.3247 \ldots$ is the smallest Pisot number which is the real root of $X^3 - X - 1$. This is based on the fact, pointed out by C. Smyth, that for $d = 3k$ the number α with minimal polynomial $X^{3k} + X^{2k} - 1$ has $\boxed{\alpha} = \theta_0^{1/(2k)} = \theta_0^{3/(2d)}$ and it is expected that this is equal to m(d) for this degree. We say that an α which gives m(d) is *extremal*. Boyd has computed m(d) for $d \leq 12$.

The first step of our computations is devoted to the search of the best bound for m(d) up to $d = 28$. We first take Boyd's bound $B = (2 + 1/d)^{1/d}$ and compute all the numbers α of height 1 with house less than B. Now, we take for our further computations the bound B as the minimum of the houses > 1 that we found for a fixed degree d. They are all less than $2^{1/d}$, $b_d = \pm 1$. Here we are in the worst situation since the conjugates of the extremal α belong to a disk $|z| \leq B$. The classical bound gives $|s_1| \leq dB$ but, for $d = 28$ we get $|s_1| \leq 5$. We find [16] all extremal numbers α up to degree 28. They satisfy the conjecture of Schinzel and Zassenhaus with Boyd's constant. The search for degree 28 took 6800 hours on a 2.8Ghz PC. Using Smyth's theorem, we can deduce the following result:

Let α be a nonzero algebraic integer, not a root of unity, and $d = \deg(\alpha)$ at least 13. Then

$$\boxed{\alpha} \geq \exp(3\log(d/2)/d^2). \tag{4}$$

This gives an improvement of a theorem of E.M. Matveev [11] who proved the relation (4) only for reciprocal α. We remark that (4) is better than the inequality that we can deduce from Voutier's inequality (3) for $d \leq 6380$. Another remark is the following: for $1 \leq d \leq 28$, m(d) is given by a polynomial which is a factor of a polynomial with at most four monomials and of length at most 4.

5.2. The smallest Perron numbers.

Lind and Boyd [4] also made the following conjecture:

Conjecture(Lind-Boyd). The smallest Perron number of degree $d \geq 2$ has minimal polynomial

$$
\begin{array}{ll}
X^d - X - 1 & \text{if } d \not\equiv 3, 5 \pmod 6, \\
(X^{d+2} - X^4 - 1)/(X^2 - X + 1) & \text{if } d \equiv 3 \pmod 6, \\
(X^{d+2} - X^2 - 1)/(X^2 - X + 1) & \text{if } d \equiv 5 \pmod 6.
\end{array}
$$

For $1 \leq d \leq 12$ the smallest Perron numbers were computed by Boyd. Using the same method as in 5.1 the second author [22] computed all Perron numbers less than the Perron number which is the real root of $X^d - X - 1$ for $12 < d \leq 22$. They all satisfy the Lind-Boyd conjecture. The smallest house for $d = 23$ is given by a Perron number, which also satisfies this conjecture. If the bound B used for the computation is Boyd's bound then the computation gives all the numbers α with $\boxed{\alpha} < (2 + 1/d)^{1/d}$. We give in Table 1 all the numbers α of degree at most 22 whose house is greater than m(d), less than $2^{1/d}$, which are not a Perron number and whose minimal polynomial is primitive. They all are also factors of polynomials of length at most equal to 4.

6. Construction of explicit auxiliary functions

6.1. Relations between explicit auxiliary functions and integer transfinite diameter.
For simplicity we examine the case of small houses which we dealt with in Section 5. If inside the function (1), with $g(z) = \mathrm{Re}(z)$, we replace the real numbers e_j by rational numbers, we may write

$$f(z) = \mathrm{Re}(z) - \frac{t}{h} \log |H(z)| \geq m,$$

where H is in $\mathbb{Z}[X]$ of degree h and t is a positive real number. We want to get a function f whose minimum m in $|z| \leq B$ is as large as possible. That is to say, we seek a polynomial $H \in \mathbb{Z}[X]$ such that

$$\sup_{|z| \leq B} |H(z)|^{t/h} e^{-\mathrm{Re}(z)} \leq e^{-m}.$$

Now, if we suppose that t is fixed, say $t = 1$, it is clear that we need an effective upper bound for the quantity

$$t_{\mathbb{Z},\phi}(|z| \leq B) = \liminf_{h \geq 1, h \to \infty} \inf_{H \in \mathbb{Z}[X],\, \deg H = h} \sup_{|z| \leq B} |H(z)|^{t/h} \phi(z),$$

where we use the weight $\phi(z) = e^{-\mathrm{Re}(z)}$. To get an upper bound for this quantity, it suffices to get an explicit polynomial $H \in \mathbb{Z}[X]$ and then to use the sequence of successive powers of H.

This is a generalisation of the integer transfinite diameter. For any $h \geq 1$ we say that a polynomial H (not always unique) is an *Integer Chebyshev Polynomial* if the quantity $\sup_{|z| \leq B} |H(z)|^{t/h} \phi(z)$ is minimum. With Wu's algorithm [21], we compute polynomials H of degree less than 30 or 40 and take their irreducible factors as polynomials Q_j. We start with the polynomial $X - 1$, get the best e_1 and take $t = e_1$. After computing J polynomials, we optimise the numbers e_j as explained in the next subsection. This gives us a new number t, and we continue by induction to get $J + 1$ polynomials.

6.2. Optimization of the numbers e_j.
We give a brief outline of the semi-infinite linear programming method introduced into number theory by C.J. Smyth. More details can be found in [9]. To optimize the numbers e_j, we first put the coefficient of $\mathrm{Re}(z)$ equal to $e_0 = 1$. We take a set X_1 of 'well distributed' points of modulus equal to B. By linear programming, we get the maximum μ of the minimum of a finite set of linear forms whose coefficients are $\mathrm{Re}(z_i)$ and $-\log |Q_j(z_i)|$ for $1 \leq j \leq J$ for any z_i in X_1. This gives an auxiliary function f which has a minimum $m > \mu$. We add to X_1 a selection of the points of $|z| = B$ where f has a local minimum. With this new set X we get another value for m and μ. We stop the process when the integer pa of m and μ coincide.

d	polynomial $P(X)$	house
3	$X^3 + X^2 - 1$	$1.15096392\ldots$
5	$(X^5 + X^3 + X - 1)$	$1.13925029\ldots$
5	$(X^6 + X^2 - X - 1)/\Phi_1$	$1.14150997\ldots$
6	$(X^8 - X + 1)/\Phi_6$	$1.09373169\ldots$
7	$(X^{11} - X^3 + X^2 - 1)/(\Phi_1\Phi_2\Phi_6)$	$1.10154059\ldots$
7	$(X^8 - X^4 - X + 1)/\Phi_1$	$1.10335536\ldots$
8	$(X^{10} + X^5 - X + 1)/\Phi_4$	$1.08370432\ldots$
9	$(X^{11} + X - 1)/\Phi_6$	$1.06715088\ldots$
12	$(X^{14} - X + 1)/\Phi_6$	$1.05218083\ldots$
14	$(X^{17} - X^{10} + X^3 - 1)/(\Phi_1\Phi_4)$	$1.05050388\ldots$
15	$(X^{17} + X - 1)/\Phi_6$	$1.04263049\ldots$
16	$(X^{19} - X^{10} + X - 1)/(\Phi_1\Phi_4)$	$1.04163090\ldots$
18	$(X^{20} - X + 1)/\Phi_6$	$1.03602095\ldots$
21	$(X^{23} + X - 1)/\Phi_6$	$1.03712124\ldots$

TABLE 1. List of α of degree d and minimal polynomial $P(X)$ whose house is greater than m(d), less than $2^{1/d}$, which is not a Perron number and whose minimal polynomial is primitive. P is written as a quotient with numerator a trinomial or a quadrinomial. The denominator is a product of at most three cyclotomic polynomials: $\Phi_1 = X - 1$, $\Phi_2 = X + 1$, $\Phi_4 = X^2 + 1$ and $\Phi_6 = X^2 - X + 1$.

6.3. Further refinements.
Refinements of the method are needed when we want to decrease the computing time. Very often, when the bounds for the numbers s_k do not grow too quickly with k, we get, with more sophisticated functions g, relations between s_k and s_{2k}. We generally use some thousands of distinct functions. Most of them are optimized almost automatically when we are given the polynomials Q_j.

REFERENCES

[1] J. Aguirre, J.C. Peral, The trace problem for totally positive algebraic integers, this volume, 1–19.

[2] P. Borwein, *Computational excursions in analysis and number theory*, CMS Books in Mathematics, Springer 2002.

[3] D.W. Boyd, Reciprocal polynomials having small measure, *Math. Comp.* **35** (1980), 1361–1377.

[ ] _____, The maximal modulus of an algebraic integer, *Math. Comp.* **45** (1985), 243–249.

[5] ———, Reciprocal polynomials having small measure II, *Math. Comp.* **53** (1989), 355–357; S1–S5.

[6] R. Breusch, On the distribution of the roots of a polynomial with integral coefficients, *Proc. Amer. Math. Soc.* **2** (1951), 939–941.

[7] V. Flammang, M. Grandcolas, G. Rhin, Small Salem numbers, *Number Theory in Progress, Proceedings of the International Conference on Number Theory, Zakopane, 1997*, Walter de Gruyter, 1999, 165–168.

[8] V. Flammang, G. Rhin, Algebraic integers whose conjugates all lie in an ellipse, *Math. Comp.* **74** (2005), 2007–2015, and `http://www.math.univ-metz.fr/` ~`rhin`.

[9] V. Flammang, G. Rhin, J.M. Sac-Épée, Integer transfinite diameter and polynomials of small Mahler measure, *Math. Comp.* **75** (2006), 1527–1540.

[10] D.H. Lehmer, Factorizations of certain cyclotomic functions, *Ann. of Math.* **34** (1933), 461–479.

[11] E.M. Matveev, On the size of algebraic integers, *Math. Notes* **49** (1991), 437–438.

[12] J. McKee, C.J. Smyth, Salem numbers of trace -2 and traces of totally positive algebraic integers, *Algorithmic number theory*, Lecture Notes in Computer Science **3076**, Springer, Berlin, 2004, 327–337.

[13] M.J. Mossinghoff, Polynomials with small Mahler measure, *Math. Comp.* **67** (1998), 1697–1705; `http://www.cecm.sfu.ca/~mjm/Lehmer/`.

[14] C. Batut, K. Belabas, D. Bernardi, H. Cohen, M. Olivier, *GP-Pari version 2.0.12*, 1998.

[15] G. Rhin, J.M. Sac-Épée, New methods providing high degree polynomials with small Mahler measure, *Experiment. Math.* **12** (2003), 457–461.

[16] G. Rhin, Q. Wu, On the smallest value of the maximal modulus of an algebraic integer, *Math. Comp.* **76** (2007), 1025–1038.

[17] C.J. Smyth, On the product of the conjugates outside the unit circle of an algebraic integer, *Bull. London Math. Soc.* **3** (1971), 169–175.

[18] ———, The mean values of totally real algebraic integers, *Math. Comp.* **42** (1984), 663–681.

[19] A. Schinzel, H. Zassenhaus, A refinement of two theorems of Kronecker, *Michigan Math. J.* **12** (1965), 81–85.

[20] P. Voutier, An effective lower bound for the height of algebraic numbers, *Acta Arith.* **74** (1996), 81–95.

[21] Q. Wu, On the linear independence measure of logarithms of rational numbers, *Math. Comp.* **72** (2002), 901–911.

[22] ———, *The smallest Perron numbers*, preprint, Université de Metz, 2006.

SMOOTH DIVISORS OF POLYNOMIALS

EIRA J. SCOURFIELD

ABSTRACT. Let f be a monic polynomial over the integers that is not necessarily irreducible but has no repeated factor, and let $\omega(m)$ denote the number of solutions of the congruence $f(n) \equiv 0 \pmod{m}$. We establish an asymptotic formula with a good error term for $\sum_{m \leq x} \omega(m)$, and derive asymptotic formulae for the number of positive divisors $m \leq x$ of $f(n)$ summed over $n \leq x$ and, using a result of Hanrot, Tenenbaum and Wu, for the number of these divisors m with no large prime factors.

1. INTRODUCTION

Let $\{f_i : i = 1, ..., l\}$ be a set of l distinct monic irreducible polynomials over $\mathbb{Z}$, where f_i has degree $k_i \geq 2$ $(i = 1, ..., l)$. Let

$$f = \prod_{i=1}^{l} f_i, \tag{1.1}$$

so f is also monic and has degree $k = \sum_{i=1}^{l} k_i$. We remark that f being monic is not crucial for our results, but is assumed for convenience.

The motivation for writing this paper goes back to the old problem of estimating, for f irreducible over $\mathbb{Z}$ and of degree at least 2, the number $d(f(n))$ of divisors of $f(n)$ summed over $n \leq x$. A key result is due to Paul Erdős [2] who, by a complicated elementary method, showed that

$$x \log x \ll \sum_{n \leq x} d(f(n)) \ll x \log x.$$

When f is an irreducible quadratic polynomial, an asymptotic formula for $\sum_{n \leq x} d(f(n))$ was established and studied further in [6], [11], [12], [13] and [17], but, to the author's knowledge, no corresponding asymptotic formula has so far been derived in the case when f has degree at least 3. However, if we restrict ourselves to divisors $m \leq x$ of $f(n)$ with no prime factor exceeding y where y satisfies (1.5) below, we can obtain, for $l \geq 1$, the asymptotic formulae given in Theorem 4 below. To establish this result, we need estimates for

2000 *Mathematics Subject Classification*. Primary: 11N37; Secondary: 11N64, 11N25.
Key words and phrases. Polynomial congruences, divisors of polynomials, smooth divisors, generalized Dickman function.

some sums involving the number $w(m)$ of solutions n $(1 \le n \le m)$ of the congruence

$$f(n) \equiv 0 \pmod{m},$$

given in Theorems 1, 2, and 3. These are of interest in their own right, and are considerably stronger than those required to derive Theorem 4. The results in Theorem 2 are special cases of very general results in [3].

Our first objective is to establish a good asymptotic formula for

$$A(x) := \sum_{m \le x} w(m). \tag{1.2}$$

When $l = 1$, we see from Theorem 1 below that $A(x)/x$ tends to a positive limit as $x \to \infty$, and so $w(m)$ has a mean value in this case. Several authors have written papers giving a very good insight into the problem of estimating the mean value (when this exists) of a multiplicative function, but the conditions imposed are not always satisfied by $w(m)$, even when $l = 1$. E. Wirsing's important memoir [26] on multiplicative functions can be applied to obtain the best estimate for $A(x)$, for any fixed $l \ge 1$, available at that time. From Satz 1.1 of that paper it follows that as $x \to \infty$

$$A(x) \sim \frac{e^{-\gamma l}}{(l-1)!} \frac{x}{\log x} \prod_{p \le x} \left(1 + \sum_{\alpha=1}^{\infty} w(p^\alpha) p^{-\alpha} \right),$$

where p denotes a rational prime. The main term of this asymptotic formula can then be determined by evaluating the product using the estimate

$$\sum_{p \le x} w(p) = l \ \mathrm{li}(x) + O\left(x \exp\left(-(\log x)^{\frac{3}{5} - \varepsilon} \right) \right)$$

(a consequence of the prime ideal theorem and the sentence after (2.4) below). An asymptotic formula for $A(x)$ with an explicit error term follows in the case $l = 1$ from Lemme 2.1 of [22], and for $l \ge 1$ from Lemme 3.9 of [21]; see also (1.10) and (1.12) of [3], where the error term in a more general situation is of the form $O(x^{1-\theta})$ with $\theta < 1/2$ described explicitly. Theorem 1 below improves this for $A(x)$ by giving a precise value for θ that to the author's knowledge is the best available up to now.

Theorem 1. *As $x \to \infty$,*

$$A(x) = x P_{l-1}(\log x) + O(x^{1-\theta}) \tag{1.3}$$

where P_{l-1} is a polynomial of degree $l-1$ and with positive leading coefficient $B/(l-1)!$ with B given by (3.13), and θ is any fixed number satisfying $0 < \theta < \min(\frac{1}{2}, \frac{1}{\Delta_l})$, where Δ_l is given by (3.1) below.

We prove this result in section 3 by induction on the number of factors in (1.1) using the generating function associated with $A(x)$ which is investigated in section 2. The definition for Δ_l that we use requires some notation. If $k_i \ge 3$ for all i, we could use the slightly larger value $\Delta_l = \frac{1}{2}(k_1 + \ldots + k_l + l)$.

Alternatively we could obtain a result of the form (1.3) by using analytic methods and the estimates for the Dedekind zeta-function on the line $\sigma = \frac{1}{2}$ given in [4] and credited in [5] to Kaufman, see [8]. However the value of θ that we were able to derive by this approach is smaller than that described in Theorem 1.

From Theorem 1 we deduce

Corollary 1. *As $x \to \infty$,*

$$C(x) := \sum_{m \le x} \frac{\omega(m)}{m} = P_l(\log x) + O(x^{-\theta}), \tag{1.4}$$

where P_l is a polynomial of degree l and with positive leading coefficient $B/l!$.

Denote by $P(m)$ the greatest prime factor of an integer $m \ge 2$ and let $P(1) = 1$. We require information on the quantities $A(x,y)$ and $C(x,y)$ obtained by imposing the additional condition $P(m) \le y$ on each of the terms in the sums in (1.2) and (1.4). In order to state the next two theorems, we need some more notation. For any $\varepsilon > 0$, define the region H_ε by

$$H_\varepsilon : \quad (\log_2 x)^{\frac{5}{3}+\varepsilon} \le \log y \le \log x, \quad x \ge x_0(\varepsilon), \tag{1.5}$$

where, throughout this paper, $\log_2 x = \log(\log x)$ for $x > e$. Let

$$L_\varepsilon(y) := \exp\left((\log y)^{\frac{3}{5}-\varepsilon}\right). \tag{1.6}$$

Define the function $\rho_l(u)$, a generalization of the Dickman function $\rho(u)$, by the differential-difference equation

$$\left.\begin{array}{l} \rho_l(u) = 0 \quad (u \le 0), \quad \rho_l(u) = \frac{u^{l-1}}{(l-1)!} \quad (0 < u \le 1), \\[2mm] u\rho_l'(u) = (l-1)\rho_l(u) - l\rho_l(u-1) \quad (u > 1). \end{array}\right\} \tag{1.7}$$

When $l = 1$ this reduces to $\rho(u)$ (except that $\rho(0)$ is usually defined to be 1). Define $z_l(u)$ by the differential - difference equation

$$\left.\begin{array}{l} z_l(u) = 0 \quad (u < 0), \quad z_l(u) = 1 \quad (0 \le u \le 1), \\[2mm] u z_l'(u) = -l z_l(u-1) \quad (u > 1). \end{array}\right\} \tag{1.8}$$

We observe (see the sentence after (4.10)) that when $u \ne 0$

$$z_l(u) = \rho_l^{(l-1)}(u). \tag{1.9}$$

Throughout this paper we put $u = \frac{\log x}{\log y} > 1$. Given $n \in \mathbb{N}$, let I_n denote the following union of intervals:

$$I_n = \bigcup_{r=1}^{n} [r + \varepsilon_{n,r}(y), r+1] \cup [n+1, \infty) \tag{1.10}$$

where

$$\varepsilon_{n,r}(y) = \theta^{-1}(n+1-r)\frac{\log_2 y}{\log y} \quad (1 \le r \le n). \qquad (1.11)$$

The results of the following Theorem are special cases of the main results in [3]; see Theorem 1.2 for (i) and Theorem 2.3 for (ii).

Theorem 2. *(i) In the region H_ε*

$$A(x,y) := \sum_{\substack{m \le x \\ P(m) \le y}} \omega(m) = x\left(1 + O\left(\frac{1}{L_\varepsilon(y)}\right)\right) \int_{0-}^u z_l(u-v)d\left(\frac{A(y^v)}{y^v}\right).$$

$$(1.12)$$

(ii) Let $n \ge 1$ be a fixed integer and assume that $u \in I_{n+1-l}$ when $n \ge l$. Then in the region H_ε

$$\begin{aligned} A(x,y) &= x(\log y)^{l-1}\left\{\sum_{r=0}^n a_r \rho_l^{(r)}(u-0)(\log y)^{-r}\right. \\ &\quad \left. +O\left(\rho_l(u)\left(\frac{\log(u+1)}{\log y}\right)^{n+1}\right)\right\}, \end{aligned} \qquad (1.13)$$

where the a_r are given by (2.28) and (2.31).

For sufficiently large u, we see from (4.4) and (4.20) that the terms in the sum of (1.13) decrease as the power of $\log y$ decreases. Thus as $u \to \infty$

$$A(x,y) \sim a_0 x(\log y)^{l-1}\rho_l(u) \qquad (1.14)$$

where $a_0 = B$ given by (3.13). For sufficiently large u the O-term in (1.13) is $O(|\rho_l^{(n+1)}(u)|(\log y)^{-n-1})$ by (4.2).

Next we state the corresponding results for $C(x,y)$. Let G_ε denote the region given by

$$G_\varepsilon : \quad (\log x)^{\frac{5}{8}+\varepsilon} \le \log y \le \log x, \quad x \ge x_0(\varepsilon). \qquad (1.15)$$

Theorem 3. *(i) In the region H_ε*

$$C(x,y) := \sum_{\substack{m \le x \\ P(m) \le y}} \frac{\omega(m)}{m} = \left(1 + O\left(\frac{1}{L_\varepsilon(y)}\right)\right) \int_{0-}^u z_l(u-v)d(C(y^v)). \quad (1.16)$$

(ii) Let $n \ge 1$ be a fixed integer and assume that $u \in I_{n-l}$ when $n > l$. Then in the region H_ε

$$\begin{aligned} C(x,y) &= (\log y)^l\left\{b_0\int_0^u \rho_l(v)dv + \sum_{r=1}^n b_r\rho_l^{(r-1)}(u-0)(\log y)^{-r}\right. \\ &\quad \left. +O\left(\rho_l(u)\frac{(\log(u+1))^n}{(\log y)^{n+1}}\right)\right\} + O\left(\frac{1}{L_\varepsilon(y)}\right) \end{aligned} \qquad (1.17)$$

where the b_r are given by (2.29) and (2.31) and satisfy $b_0 = a_0$, $b_r = a_r + a_{r-1}$ $(r \ge 1)$. In the region G_ε the error term $O(1/L_\varepsilon(y))$ can be removed. As $u \to \infty$, $\int_0^u \rho_l(v)dv = e^{\gamma l} + O(\rho_l(u)/\log u)$ where γ is Euler's constant.

We observe that as $u \to \infty$ in the region H_ε

$$C(x,y) \sim e^{\gamma l} b_0 (\log y)^l$$

where $b_0 = B$ given by (3.13). For sufficiently large u the first O-term in (1.17) is $O(|\rho_l^{(n)}(u)|(\log y)^{-n-1})$ by (4.2).

On receiving a draft of this paper, Professor G. Tenenbaum e-mailed [24] the author to describe how partial summation techniques can be applied to deduce from the main theorems in [3] results of the same type as those in Theorem 3 above, but valid for a general class of non-negative multiplicative functions f studied in [3]. Moreover he is able to eliminate the error term $O(1/L_\varepsilon(y))$ in the asymptotic expansion analogous to (1.17) above in the region H_ε by introducing the function $F(y) = \sum_{P(n) \le y} f(n)/n$ into the argument.

To illustrate an alternative approach, we give in section 5 a direct proof of Theorem 3 that is independent of Theorem 2. We follow the same broad strategy that is standard in this type of problem, but there are differences and simplifications in its implementation. These arise when we express $C(x,y)$ as an integral of a function with a simple pole at $s = 0$, and take the line of integration to be of the form $\sigma = \kappa > 0$; taking the line of integration to the left of $\sigma = 0$ does not seem to improve the result, unlike the corresponding situation for $A(x,y)$ and the more general sums studied in [3]. Factors of the integrand include the generating function $G(s+1) = \sum_{m=1}^\infty \omega(m)m^{-1-s}$ for $C(x)$ and the Laplace transform $\widehat{\rho_l}(s)$ of $\rho_l(u)$ defined in (1.7), and we find that we only need an estimate for $\widehat{\rho_l}(s)$ for $s = \kappa + it$ for $|t|$ large. In contrast, for $A(x,y)$ the analogous integrand is analytic at the critical point $s = 1$ but the corresponding integral is taken along a line of integration to the left of $\sigma = 1$, and rather precise information on $\rho_l(u)$ and on $\widehat{\rho_l}(s)$ when $\sigma = -\xi < 0$ is required, where ξ has a specific standard value; see [3] for the details when $A(x,y)$ is replaced by a general sum of this type.

We remark that our proofs of Theorems 1 and 3 are valid when $\omega(m)$ is replaced by any multiplicative function with an associated generating function of the form

$$G(s) = H(s) \prod_{i=1}^{l} \zeta_{K_i}(s)^{\nu_i}, \tag{1.18}$$

where $H(s)$ is analytic and bounded in $\sigma > \frac{1}{2}$, the K_i are distinct number fields, $\zeta_{K_i}(s)$ is the associated Dedekind zeta-function, and each $\nu_i \in \mathbb{N}$ (whereas in (2.20) each $\nu_i = 1$). In [3] the authors consider a more general situation for a multiplicative function with generating function of the form (1.18) but with the ν_i real numbers such that $\sum_{i=1}^l \nu_i > 0$, where $H(s)$ satisfies certain conditions. As mentioned above, in his e-mails [24] Tenenbaum described how to obtain a generalisation of Theorem 3 from these results and partial summation. Furthermore in section 1.4 of [3] the authors describe a generalization of their work that includes the problem of investigating $A(x,y)$

when the polynomial f in (1.1) is not squarefree, and they give explicitly the corresponding generating function and the generalizations of the Dickman function required.

We turn now to the problem discussed at the beginning of this paper. Let

$$D_l(x,y) = \sum_{n \leq x} \# \{m : m \leq x, P(m) \leq y, m \mid f(n)\} \qquad (1.19)$$

and $D_l(x) = D_l(x,x)$. Since $m \mid f(n)$ if and only if n lies in one of $w(m)$ residue classes modulo m,

$$D_l(x,y) = xC(x,y) + O\big(A(x,y)\big), \qquad (1.20)$$

with a corresponding expression for $D_l(x)$. Hence from Theorems 1, 2 and 3 and equation (1.4) we deduce

Theorem 4. *(i) We have*

$$D_l(x) = \frac{B}{l!}x(\log x)^l + O\big(x(\log x)^{l-1}\big) \qquad (1.21)$$

where B is given by (3.13).
(ii) In the region H_ε

$$
\begin{aligned}
D_l(x,y) &= Bx(\log y)^l \int_0^u \rho_l(v)dv + O\left(x\rho_l(u)(\log y)^{l-1} + \tfrac{x}{L_\varepsilon(y)}\right) \\
&\sim Be^{\gamma l}x(\log y)^l
\end{aligned} \qquad (1.22)
$$

as $u \to \infty$. In the region G_ε the error term in (1.22) is $O(x\rho_l(u)(\log y)^{l-1})$.

The related problems that one would really like to be able to solve are those of obtaining an unconditional asymptotic formula for

$$\Psi(f;x,y) = \#\{1 \leq n \leq x : P(f(n)) \leq y\}$$

for (x,y) in a suitable region, as well as for $\sum_{n \leq x} d(f(n))$ when f is irreducible over $\mathbb{Z}$ and has degree at least 3, but these seem unattainable at present. However, assuming a certain hypothesis that is a quantitive version of Schinzel and Sierpinski's Hypothesis H, G. Martin in a very long technical paper [10] has derived an asymptotic formula for $\Psi(f;x,x^{\frac{1}{u}})$ that is uniform for u in a given range and with some uniformity in the coefficients of f.

The plan of the current paper is as follows. In Section 2 we consider the properties of $w(m)$ and of the Dedekind zeta-function that we need and set up the generating function that is used in the proof of Theorem 1 in Section 3. In Section 4 we look at properties of the generalized Dickman function $\rho_l(u)$ and its Laplace transform. Theorem 3 is proved in section 5.

Acknowledgements. The author is grateful to the referee for a careful reading of this paper and for suggestions that clarified its exposition. She thanks Professor Gérald Tenenbaum for informing her that the results in [3], then being prepared, were relevant to the work in her early draft of this paper, for his e-mails [24] concerning deriving a generalization of Theorem 3 from the results of [3], and for his encouragement. She also thanks Professor Ram Murty for a helpful conversation that led to a better error term in Theorem 1 in the case $l = 1$.

2. PRELIMINARIES

2.1. **Properties of** $\omega(m)$. For $i = 1, ..., l$, recall that f_i is an irreducible monic polynomial over $\mathbb{Z}$ of degree $k_i \geq 2$ and that $f_i \neq f_j$ if $i \neq j$. As in (1.1), $f = \prod_{i=1}^{l} f_i$ so f is also monic and has degree

$$k = \sum_{i=1}^{l} k_i. \tag{2.1}$$

Without loss of generality, we assume

$$k_1 \geq k_2 \geq ... \geq k_l \geq 2. \tag{2.2}$$

Denote the discriminant of f_i by D_i and put

$$D = \prod_{i=1}^{l} D_i. \tag{2.3}$$

Since the f_i are distinct, f_i and f_j are coprime if $i \neq j$. Hence there exist polynomials $u_i, u_j \in \mathbb{Z}[t]$ and $e_{i,j} \in \mathbb{Z} \setminus \{0\}$ such that

$$u_i f_i + u_j f_j = e_{i,j}. \tag{2.4}$$

It follows that if p is a prime with $p \nmid e_{i,j}$ then the congruences

$$f_i(n) \equiv 0 \pmod{p}, \quad f_j(n) \equiv 0 \pmod{p}$$

have no common solution. Let

$$e = \prod_{1 \leq i < j \leq l} e_{i,j}. \tag{2.5}$$

Define $\omega_i(m), \omega(m)$ by

$$\omega_i(m) = \#\{n \in \mathbb{Z}_m : f_i(n) \equiv 0 \pmod{m}\}, \quad i = 1, ..., l, \tag{2.6}$$

$$\omega(m) = \#\{n \in \mathbb{Z}_m : f(n) \equiv 0 \pmod{m}\}. \tag{2.7}$$

It follows from above that if $(m, e) = 1$ then

$$\omega(m) = \sum_{i=1}^{l} \omega_i(m).$$

In particular, if $p \nmid e$, for any $\alpha \geq 1$

$$\omega(p^\alpha) = \sum_{i=1}^{l} \omega_i(p^\alpha).$$

The restriction that f_i is monic does not affect the value of $\omega_i(p^\alpha)$, except for a finite number of primes p. For if f_i^* has leading coefficient $a > 0$ and degree k_i, then

$$a^{k_i-1} f_i^*(n) = f_i(an) \tag{2.8}$$

for some monic polynomial f_i of degree k_i, so if $p \nmid a$, then $\omega_i^*(p^\alpha) = \omega_i(p^\alpha)$.

The results in the following lemma are well known:

Lemma 1. *(i) The functions $\omega(m)$ and $\omega_i(m)$ are multiplicative.*
(ii) If $p^{\beta_i} \| D_i$, then for all $\alpha \geq 2\beta_i + 1$

$$\omega_i(p^\alpha) = \omega_i(p^{2\beta_i+1}). \tag{2.9}$$

In particular if $p \nmid D_i$, then $\omega_i(p^\alpha) = \omega_i(p)$ for all $\alpha \geq 1$.
(iii) We have $\omega_i(p) \leq \min(k_i, p)$ for all primes p, and
$\omega_i(p^\alpha) \ll 1$ for all primes p and $\alpha \geq 1$.

For (ii), see for example Theorem 53, p. 89, in [14], (i) follows from the Chinese Remainder Theorem, and (iii) is a consequence of (ii). From Theorem 54 of [14], $\omega_i(p^\alpha) \leq k_i D_i^2$.

Corollary 2. *If $p \nmid De$, for all $\alpha \geq 1$*

$$\omega(p^\alpha) = \sum_{i=1}^{l} \omega_i(p) = \omega(p), \tag{2.10}$$

so if $p \geq k_1$, $\omega(p^\alpha) \leq k \leq lk_1$.

We remark that $\omega_i(p)$ is the number of linear factors in the factorization of f_i over $\mathbb{Z}_p$.

2.2. **Related algebraic number fields.** With f_i as above, let θ_i be an algebraic integer satisfying $f_i(\theta_i) = 0$, and let K_i be the algebraic number field given by $K_i = \mathbb{Q}(\theta_i)$, and $\mathfrak{O}_i$ be the ring of integers of K_i $(i = 1, ..., l)$. Denote the discriminant of K_i by d_i, and let $o_i = [\mathfrak{O}_i : \mathbb{Z}[\theta_i]]$, the index of the additive group $\mathbb{Z}[\theta_i]$ in $\mathfrak{O}_i$. Then $D_i = d_i o_i^2$; see Proposition 4.4.4 of [1].

If p is a rational prime with $p \nmid d_i$, then p is unramified in K_i, and its factorization into prime ideals of $\mathfrak{O}_i$ takes the form

$$p = \prod_{j=1}^{r} \mathfrak{p}_{i,j}$$

where $\mathfrak{p}_{i,j}$, $j = 1, ..., r$, are distinct prime ideals of $\mathfrak{O}_i$. Moreover

$$N(\mathfrak{p}_{i,j}) = p^{g_{i,j}} \text{ where } \sum_{j=1}^{r} g_{i,j} = k_i;$$

see Theorems 4.8.3, 4.8.5 and 4.8.8 of [1]. We can now connect this factorization of p in $\mathfrak{O}_i$ with $\omega_i(p)$ when $p \nmid D_i$ using Theorems 4.8.13 and 4.8.5 in [1]:

Lemma 2. *Suppose* $p \nmid D_i$. *Write*

$$f_i(n) \equiv \prod_{j=1}^{r} \overline{f_{i,j}}(n) \pmod{p}, \qquad (2.11)$$

where the $\overline{f_{i,j}}$ *are irreducible and monic over* $\mathbb{Z}_p$. *Then, with appropriate ordering, the degree of* $\overline{f_{i,j}}$ *is* $g_{i,j}$. *Hence* $\omega_i(p)$ *is the number of exponents* $g_{i,j}$ *equal to 1, i.e. the number of* $\mathfrak{p}_{i,j}$ *dividing* p *with norm equal to* p.

2.3. **The generating function.** Throughout this paper, we write $s = \sigma + it$. Let

$$G(s) = \sum_{m=1}^{\infty} \omega(m) m^{-s} \quad (\sigma > 1) \qquad (2.12)$$

and p_0 be a fixed integer satisfying

$$p_0 > k2^{k+2}, \quad p \mid De \Rightarrow p \leq p_0.$$

Lemma 3. *For* $\sigma > 1$

$$G(s) = \prod_{p > p_0} (1 - p^{-s})^{-\omega(p)} H_0(s), \qquad (2.13)$$

where $H_0(s)$ *is analytic and can be written as an absolutely convergent product over primes in* $\sigma > \frac{1}{2}$, *and* $H_0(1) \neq 0$.

Proof. By Lemma 1(i), in $\sigma > 1$

$$G(s) = \prod_{p} \left(1 + \sum_{\alpha=1}^{\infty} \omega(p^{\alpha}) p^{-\alpha s} \right).$$

When $p > p_0$, so $p \nmid De$, we have by Lemma 1(ii) that $\omega(p^{\alpha}) = \omega(p)$ for all $\alpha \geq 1$. Hence for $p > p_0$ and $\omega(p) > 0$

$$\left(1 + \sum_{\alpha=1}^{\infty} \omega(p^{\alpha}) p^{-\alpha s} \right) (1 - p^{-s})^{\omega(p)}$$

$$= \left(1 + \omega(p) \sum_{\alpha=1}^{\infty} p^{-\alpha s} \right) \left(\sum_{\beta=0}^{\omega(p)} \binom{\omega(p)}{\beta} (-p^{-s})^{\beta} \right)$$

$$= 1 + \sum_{\alpha=2}^{\infty} a(p^{\alpha}) p^{-\alpha s}$$

where, for $\alpha > \omega(p)$,

$$a(p^{\alpha}) = \omega(p) \sum_{\beta=0}^{\omega(p)} \binom{\omega(p)}{\beta} (-1)^{\beta} = \omega(p)(1 - 1)^{\omega(p)} = 0,$$

and, for $2 \leq \alpha \leq \omega(p)$,

$$a(p^{\alpha}) = \binom{\omega(p)}{\alpha} (-1)^{\alpha} + \omega(p) \sum_{\beta=0}^{\alpha-1} \binom{\omega(p)}{\beta} (-1)^{\beta}.$$

In the latter case, when $\omega(p) > 0$,

$$|a(p^\alpha)| \leq \omega(p) \sum_{\beta=0}^{\omega(p)} \binom{\omega(p)}{\beta} = \omega(p)2^{\omega(p)} \leq k2^k, \qquad (2.14)$$

since $\omega(p) \leq k$. Moreover, since $p > p_0 > 2$, for $\sigma > \frac{1}{2}$ we have

$$\left| \sum_{\alpha=2}^{\omega(p)} a(p^\alpha)p^{-\alpha s} \right| < k2^k \sum_{\alpha=2}^{\infty} p^{-\alpha\sigma} < k2^k \frac{p^{-2\sigma}}{1-2^{-\sigma}} < 4k2^k p^{-1} < 1,$$

so for $p > p_0$, $\sigma > \frac{1}{2}$

$$1 + \sum_{\alpha=2}^{\omega(p)} a(p^\alpha)p^{-\alpha s} \neq 0. \qquad (2.15)$$

It follows that for $\sigma > 1$

$$G(s) = \prod_{p>p_0} \left\{ \left(1 + \sum_{\alpha=2}^{\omega(p)} a(p^\alpha)p^{-\alpha s} \right) \left(1 - p^{-s} \right)^{-\omega(p)} \right\} \prod_{p \leq p_0} \left(1 + \sum_{\alpha=1}^{\infty} \omega(p^\alpha)p^{-\alpha s} \right).$$

When $p \leq p_0$, $\omega(p^\alpha) \leq \prod_{j=1}^{l} \omega_j(p^\alpha) \ll 1$, so for $\sigma \geq \frac{1}{2}$

$$\left| \prod_{p \leq p_0} \left(1 + \sum_{\alpha=1}^{\infty} \omega(p^\alpha)p^{-\alpha s} \right) \right| \leq \prod_{p \leq p_0} \left(1 + \sum_{\alpha=1}^{\infty} \omega(p^\alpha)p^{-\alpha/2} \right) \ll 1.$$

Hence

$$G(s) = \prod_{p>p_0} \left(1 - p^{-s} \right)^{-\omega(p)} H_0(s), \qquad (2.16)$$

where

$$H_0(s) = \prod_{p>p_0} \left(1 + \sum_{\alpha=2}^{\omega(p)} a(p^\alpha)p^{-\alpha s} \right) \prod_{p \leq p_0} \left(1 + \sum_{\alpha=1}^{\infty} \omega(p^\alpha)p^{-\alpha s} \right), \qquad (2.17)$$

which is analytic in $\sigma > \frac{1}{2}$ and is an absolutely convergent product there. Since for $\sigma = 1$ a typical term with $p > p_0 > k2^{k+2}$ in $H_0(s)$ is

$$\gg 1 - k2^{k+2}p^{-2} > 1 - p_0^{-1} > 0,$$

and the terms with $p \leq p_0$ and $s = 1$ are clearly positive, we have $H_0(1) \neq 0$ as required. $\square$

We now use the information in Lemma 2 to express $G(s)$ in terms of Dedekind zeta-functions. For $\sigma > 1$, define

$$\zeta_{K_i}(s) = \sum_{\mathfrak{a}} (N(\mathfrak{a}))^{-s} = \prod_{\mathfrak{p}} \left(1 - (N(\mathfrak{p}))^{-s} \right)^{-1}, \qquad (2.18)$$

where $\mathfrak{a}$ runs over the ideals and $\mathfrak{p}$ the prime ideals of $\mathfrak{O}_i$. By Lemma 2, $N(\mathfrak{p})$ is a rational prime p for exactly $\omega_i(p)$ prime ideals and otherwise $N(\mathfrak{p})$ is a higher power of a prime p. Hence for $\sigma > 1$ we can write

$$\zeta_{K_i}(s) = \prod_{p > p_0} (1-p^{-s})^{-\omega_i(p)} \prod_{\substack{\mathfrak{p} \\ N(\mathfrak{p})=p^g, g \geq 2, p > p_0}} (1-p^{-gs})^{-1} \prod_{\substack{\mathfrak{p} \\ N(\mathfrak{p})=p^g, g \geq 1, p \leq p_0}} (1-p^{-gs})^{-1}$$

where we know that $g \leq k_i$ holds. The second product is absolutely convergent and analytic in $\sigma > \frac{1}{2}$, the third product is finite, bounded and non-zero in $\sigma \geq \delta > 0$, and hence

$$\zeta_{K_i}(s) = \prod_{p > p_0} (1 - p^{-s})^{-\omega_i(p)} H_i(s) \qquad (2.19)$$

where $H_i(s)$ is an absolutely convergent product, analytic and non-zero in $\sigma > \frac{1}{2}$. Hence we have:

Corollary 3. *For $\sigma > 1$*

$$G(s) = \prod_{i=1}^{l} \left(\zeta_{K_i}(s) \big(H_i(s) \big)^{-1} \right) H_0(s) = H(s) \prod_{i=1}^{l} \zeta_{K_i}(s) , \qquad (2.20)$$

where, for $\sigma > \frac{1}{2}$, $H(s)$ is an absolutely convergent product over primes and is analytic, and where $H(1) \neq 0$.

A more general formula of this type was established in the proof of Lemme 3.9 in [21].

We remark that if we had started with polynomials f_i^* that were not all monic, then from (2.8) it would follow that the only change would be in the finite product $H_0(s)$, assuming that p_0 exceeds the magnitude of the leading coefficient of f^*, and hence in $H(s)$ and its value at $s = 1$. This just changes the constants in our theorems.

Next we state some properties of $\zeta_K(s)$ for K an algebraic number field of degree k that we use later. First we need some definitions. Let

$$\lambda = 2^{q+r} \pi^r R h m^{-1} |\Delta|^{-\frac{1}{2}} ,$$

where q is the number of real and r is the number of complex conjugate pairs of monomorphisms $K \to \mathbb{C}$, m is the number of roots of unity in K and R, h, Δ denote the regulator, class number, discriminant of K, respectively. Define

α, β according to the following table:

k	α	β
$k = 2$	$\frac{50}{73}$	$\frac{315}{146}$
$3 \le k \le 6$	$\frac{2}{k} - \frac{8}{k(5k+2)}$	$\frac{10}{5k+2}$
$k \ge 7$	$\frac{2}{k} - \frac{3}{2k^2}$	$\frac{2}{k}$

(2.21)

Lemma 4. *(i) The function $\zeta_K(s)$ has a simple pole at $s = 1$ with residue λ and is analytic on $\mathbb{C}$ otherwise.*
  (ii) Let $j(m) = \#\{\mathfrak{a} \in \mathfrak{D} : N(\mathfrak{a}) = m\}$; for $\sigma > 1$

$$\zeta_K(s) = \sum_{m=1}^{\infty} j(m)m^{-s}. \tag{2.22}$$

Also

$$S(x) := \sum_{m \le x} j(m) = \lambda x + O\big(x^{1-\alpha}(\log x)^{\beta}\big). \tag{2.23}$$

(iii) There exist positive constants c, t_0 such that $\zeta_K(s) \ne 0$ for

$$\sigma \ge 1 - c(\log |t|)^{-2/3}(\log_2 |t|)^{-1/3} \quad (|t| \ge t_0), \tag{2.24}$$

$$\sigma \ge 1 - c(\log t_0)^{-2/3}(\log_2 t_0)^{-1/3} \quad (|t| \le t_0). \tag{2.25}$$

(iv) In the region (2.24)

$$\zeta_K(s) \ll \big(\log |t|\big)^{2/3}\big(\log_2 |t|\big) \tag{2.26}$$

and

$$\frac{\zeta_K'(s)}{\zeta_K(s)} \ll \big(\log |t|\big)^{2/3}\big(\log_2 |t|\big)^{4/3}. \tag{2.27}$$

Part (i) is well known. For part (ii), see [7] when $k = 2$ and [15] for $k \ge 3$, and for part (iii) see [20]. Part (iv) is given by Lemmas 2.6 and 2.8 of [18]. In standard text books (2.23) is established with $\alpha = \frac{1}{k}$, $\beta = 0$, and for $k \ge 3$ E. Landau obtained in [9] $\alpha = \frac{2}{k+1}$, $\beta = 0$.

Corollary 4. *(i) The function $sG(s+1)/(s+1)$ has a pole of order $l-1$ at $s = 0$.*
  (ii) The function $G(s+1)$ has a pole of order l at $s = 0$.

This follows from (2.20) and Lemma 4(i). Hence we can expand both functions in a Laurent series about $s = 0$; for $0 < |s| < \delta$ for a suitable $\delta > 0$ we have

$$\frac{s}{s+1} G(s+1) = \sum_{r=0}^{\infty} a_r s^{r-(l-1)}, \tag{2.28}$$

$$G(s+1) = \sum_{r=0}^{\infty} b_r s^{r-l}. \tag{2.29}$$

We deduce that $b_0 = a_0$, $b_r = a_r + a_{r-1}$ for $r \geq 1$. By using (2.12) and the definitions of $A(x)$ and $C(x)$ we find that for $\sigma > 0$

$$G(s+1) = \frac{s+1}{s} \int_{1-}^{\infty} v^{-s} d\left(\frac{A(v)}{v}\right) = \int_{1-}^{\infty} v^{-s} d\big(C(v)\big). \tag{2.30}$$

It follows from (1.3) and (1.4) that

$$a_r = P_{l-1}^{(l-1-r)}(0) \ (0 \leq r \leq l-1), \qquad b_r = P_l^{(l-r)}(0) \ (0 \leq r \leq l), \tag{2.31}$$

and in particular $a_0 = B = b_0$, where B is given by (3.13). Hence

$$P_{l-1}(\log x) = \sum_{r=0}^{l-1} a_r \frac{(\log x)^{l-1-r}}{(l-1-r)!},$$

$$P_l(\log x) = \sum_{r=0}^{l} b_r \frac{(\log x)^{l-r}}{(l-r)!} = a_0 \frac{(\log x)^l}{l!} + \sum_{r=1}^{l} (a_r + a_{r-1}) \frac{(\log x)^{l-r}}{(l-r)!}.$$

Moreover, by (2.30), (1.3) and (1.4), for $r \geq 1$

$$a_{r+l-1} = \frac{(-1)^{r-1}}{(r-1)!} \int_{1-}^{\infty} \big(A(v) - v P_{l-1}(\log v)\big) (\log v)^{r-1} v^{-2} dv,$$

$$b_{r+l} = \frac{(-1)^{r-1}}{(r-1)!} \int_{1-}^{\infty} \big(C(v) - P_l(\log v)\big) (\log v)^{r-1} v^{-1} dv. \tag{2.32}$$

From Perron's formula, for $x \notin \mathbb{N}$ and $\kappa > 0$ we see that

$$A(x) = \frac{x}{2\pi i} \int_{\kappa-i\infty}^{\kappa+i\infty} \frac{s G(s+1)}{s+1} \frac{x^s}{s} ds \, ;$$

the residue of the integrand at its pole of order l at $s = 0$ is $P_{l-1}(\log x)$. Similarly

$$C(x) = \frac{1}{2\pi i} \int_{\kappa-i\infty}^{\kappa+i\infty} G(s+1) \frac{x^s}{s} ds \, ,$$

and $P_l(\log x)$ is the residue of this integrand at its pole of order $l+1$ at $s = 0$.

3. PROOF OF THEOREM 1

As in section 2.2, for each $i \geq 1$ the algebraic number field K_i is associated with the polynomial f_i and has degree k_i. The corresponding Dedekind zeta-function $\zeta_{K_i}(s)$ has residue λ_i at its simple pole at $s = 1$. Define the quantities $j_i(m)$, α_i, β_i analogously to those defined for K in (2.22) and (2.21). For $i \geq 1$ let

$$\Delta_i = \frac{1}{\alpha_1} + \ldots + \frac{1}{\alpha_i}, \qquad \delta_i = \beta_1 + \ldots + \beta_i + \max(0, i-2). \qquad (3.1)$$

We observe that $\Delta_i > 2$ except when $i = 1$ and $k_1 = 2$ or 3.

Our proof of Theorem 1 depends on the following lemma that we establish by induction on i.

Lemma 5. *To each $i \geq 1$ there corresponds a polynomial P_{i-1} of degree $i-1$ and with leading coefficient*

$$\frac{\lambda_1 \ldots \lambda_i}{(i-1)!} \qquad (3.2)$$

such that

$$A_i(x) := \sum_{m_1 \ldots m_i \leq x} \prod_{r=1}^{i} j_r(m_r) = x P_{i-1}(\log x) + O\left(x^{1-\frac{1}{\Delta_i}}(\log x)^{\delta_i}\right). \qquad (3.3)$$

Proof (by induction on i). The case $i = 1$ follows from (2.23). Assume the result holds for some $i \geq 1$; for convenience write

$$m = m_1 \ldots m_i, \qquad a(m) = \prod_{r=1}^{i} j_r(m_r) \geq 0.$$

Consider $A_{i+1}(x)$. Suppose X, Y (to be chosen later) satisfy $XY = x$, X, $Y \to \infty$ as $x \to \infty$. By a standard hyperbolic argument

$$\begin{aligned}
A_{i+1}(x) &= \sum_{mn \leq x} a(m) j_{i+1}(n) \\
&= \sum_{m \leq X} a(m) S_{i+1}\left(\frac{x}{m}\right) + \sum_{n \leq Y} j_{i+1}(n) A_i\left(\frac{x}{n}\right) \\
&\quad - A_i(X) S_{i+1}(Y) \\
&= \sum\nolimits_1 + \sum\nolimits_2 - A_i(X) S_{i+1}(Y),
\end{aligned} \qquad (3.4)$$

say. By (2.23)

$$\sum\nolimits_1 = \lambda_{i+1} x \sum_{m \leq X} \frac{a(m)}{m} + O\left(x^{1-\alpha_{i+1}}(\log x)^{\beta_{i+1}} \sum_{m \leq X} \frac{a(m)}{m^{1-\alpha_{i+1}}}\right). \qquad (3.5)$$

Let

$$\overline{P}_i(\log X) = \int_{1-}^{X} P_{i-1}(\log v)v^{-1}dv, \tag{3.6}$$

$$C_i = \int_{1-}^{\infty} \left(A_i(v) - w_{i-1}^{\frown}(\log v) \right)v^{-2}dv \, ; \tag{3.7}$$

then $\overline{P}_i$ is a polynomial of degree i and with leading coefficient $\lambda_1 \ldots \lambda_i/i!$ on using (3.2), and C_i is an absolutely convergent integral by (3.3). Using partial summation, we find that

$$\sum_{m \leq X} \frac{a(m)}{m} = \overline{P}_i(\log X) + P_{i-1}(\log X) + C_i + O\left(X^{-\frac{1}{\Delta_i}}(\log X)^{\delta_i}\right),$$

$$\sum_{m \leq X} \frac{a(m)}{m^{1-\alpha_{i+1}}} \ll X^{\alpha_{i+1}}(\log X)^{i-1}.$$

Substituting these estimates in (3.5), we obtain

$$\sum_1 = \lambda_{i+1}x\left(\overline{P}_i(\log X) + P_{i-1}(\log X) + C_i\right)$$
$$+ O\left(xX^{-\frac{1}{\Delta_i}}(\log X)^{\delta_i} + xY^{-\alpha_{i+1}}(\log x)^{\beta_{i+1}+(i-1)}\right) \tag{3.8}$$

since $XY = x$, $X < x$.

Similarly by our hypothesis (3.3)

$$\sum_2 = x\sum_{n \leq Y} \frac{j_{i+1}(n)}{n}P_{i-1}(\log \frac{x}{n}) + O\left(x^{1-\frac{1}{\Delta_i}}(\log x)^{\delta_i}\sum_{n \leq Y} j_{i+1}(n)n^{-1+\frac{1}{\Delta_i}}\right). \tag{3.9}$$

Let

$$Q_{i-1}(\log x) = \int_{1-}^{\infty} \left(S_{i+1}(v) - \lambda_{i+1}v\right)\left(P_{i-1}(\log \frac{x}{v}) + P'_{i-1}(\log \frac{x}{v})\right)v^{-2}dv,$$

which by (2.23) is a polynomial in $\log x$ of degree $i-1$ since the coefficients are absolutely convergent integrals. By partial summation and (2.23) and (3.6)

$$\sum_{n \leq Y} \frac{j_{i+1}(n)}{n}P_{i-1}(\log \frac{x}{n}) = \lambda_{i+1}\left(\overline{P}_i(\log x) - \overline{P}_i(\log X) + P_{i-1}(\log x)\right)$$
$$+ Q_{i-1}(\log x) + O\left(Y^{-\alpha_{i+1}}(\log x)^{\beta_{i+1}+(i-1)}\right),$$

$$\sum_{n \leq Y} j_{i+1}(n)n^{-1+\frac{1}{\Delta_i}} \ll Y^{\frac{1}{\Delta_i}}.$$

Hence substituting in (3.9) and using $XY = x$ again

$$\sum_2 = \lambda_{i+1}x\left(\overline{P}_i(\log x) - \overline{P}_i(\log X) + P_{i-1}(\log x)\right) + xQ_{i-1}(\log x)$$
$$+ O\left(xY^{-\alpha_{i+1}}(\log x)^{\beta_{i+1}+(i-1)} + xX^{-\frac{1}{\Delta_i}}(\log x)^{\delta_{i+1}}\right). \tag{3.10}$$

By (2.23) and (3.3), since $XY = x$,

$$A_i(X)S_{i+1}(Y) = \lambda_{i+1}xP_{i-1}(\log X)$$

$$+ O\left(xY^{-\alpha_{i+1}}(\log x)^{\beta_{i+1}+(i-1)} + xX^{-\frac{1}{\Delta_i}}(\log x)^{\delta_i}\right).$$

$$(3.11)$$

Substituting (3.8), (3.10) and (3.11) in (3.4) we obtain after some cancellations that

$$A_{i+1}(x) = \lambda_{i+1}x\left(\overline{P_i}(\log x) + P_{i-1}(\log x) + C_i\right) + xQ_{i-1}(\log x)$$

$$(3.12)$$

$$+ O\left(x(Y^{-\alpha_{i+1}} + X^{-\frac{1}{\Delta_i}})(\log x)^{\delta_{i+1}}\right)$$

since $\beta_{i+1} + (i-1) < \delta_{i+1}$, $\delta_i < \delta_{i+1}$. We now choose X, Y so that

$$X^{\frac{1}{\Delta_i}} = Y^{\alpha_{i+1}} = \left(\frac{x}{X}\right)^{\alpha_{i+1}}$$

which gives

$$X^{\frac{1}{\Delta_i}} = x^{\frac{1}{\Delta_i + (1/\alpha_{i+1})}} = x^{\frac{1}{\Delta_{i+1}}};$$

then the error term in (3.12) becomes

$$O\left(x^{1-\frac{1}{\Delta_{i+1}}}(\log x)^{\delta_{i+1}}\right).$$

Taking

$$P_i(t) = \lambda_{i+1}(\overline{P_i}(t) + P_{i-1}(t) + C_i) + Q_{i-1}(t),$$

a polynomial of degree i and with leading coefficient $\lambda_1 \ldots \lambda_{i+1}/i!$, we obtain the required result for $A_{i+1}(x)$ in the form of that in (3.3). This completes our proof by induction. $\square$

We apply this lemma with $i = l$. Recall that by Corollary 3

$$G(s) = H(s)\prod_{r=1}^{l} \zeta_{K_r}(s),$$

where $H(s) = \sum_{m=1}^{\infty} h(m)m^{-s}$ is absolutely convergent for $\sigma > \frac{1}{2}$ and $H(1) \neq 0$. Hence

$$A(x) = \sum_{m \leq x} \omega(m) = \sum_{m_0 m_1 \ldots m_l \leq x} h(m_0)\prod_{r=1}^{l} j_r(m_r).$$

We use Lemma 5 with $i = l$ to estimate this sum provided $\Delta_l > 2$, which holds when $l > 1$ or $l = 1$ and $k_1 > 3$. In the exceptional cases we do not have enough information about $H(s)$ when $\sigma = 1 - \frac{1}{\Delta_l}$ so we modify α_1 to ensure that $1 - \alpha_1 > \frac{1}{2}$; for example, when $l = 1$, take $\alpha_1 = \frac{36}{73}$ if $k_1 = 2$ and $\alpha_1 = \frac{26}{53}$ if $k_1 = 3$. Then let $\Delta = \Delta_l$, except in the exceptional cases whe with the modified α_1 we put $\Delta = \frac{1}{\alpha_1}$.

Proof of Theorem 1. Let

$$m = m_1 \ldots m_l, \qquad a(m) = \prod_{r=1}^{l} j_r(m_r) \geq 0.$$

For any $X < x$ with $X \to \infty$ as $x \to \infty$, we have by Lemma 5 that

$$A(x) = \sum_{mn \leq x} a(m)h(n) = \sum_{n \leq X} h(n)A_l\left(\frac{x}{n}\right) + O\left(x(\log x)^{l-1} \sum_{X < n \leq x} |h(n)| n^{-1}\right).$$

The first sum on the right equals

$$x \sum_{n \leq X} h(n)n^{-1}P_{l-1}(\log \tfrac{x}{n}) + O\left(x^{1-\frac{1}{\Delta}}(\log x)^{\delta_l} \sum_{n \leq X} |h(n)| \, n^{-1+\frac{1}{\Delta}}\right)$$

$$= x \sum_{n=1}^{\infty} h(n)n^{-1}P_{l-1}(\log \tfrac{x}{n})$$

$$+ O\left(x \sum_{n > X} |h(n)| \, n^{-1} \left|P_{l-1}(\log \tfrac{x}{n})\right| + x^{1-\frac{1}{\Delta}}(\log x)^{\delta_l} \sum_{n \leq X} |h(n)| \, n^{-1+\frac{1}{\Delta}}\right).$$

The first sum here converges absolutely to a polynomial $P_{l-1}^*(\log x)$ of degree $l - 1$ and with leading coefficient $B/(l - 1)!$ where

$$B = \lambda_1 \ldots \lambda_l H(1). \tag{3.13}$$

Since $1 - \frac{1}{\Delta} > \frac{1}{2}$,

$$\sum_{n \leq X} |h(n)| \, n^{-1+\frac{1}{\Delta}} < \sum_{n=1}^{\infty} |h(n)| \, n^{-1+\frac{1}{\Delta}} \ll 1.$$

For $0 < \eta < \frac{1}{2}$,

$$\sum_{n > X} |h(n)| \, n^{-1} \left|P_{l-1}(\log \frac{x}{n})\right| < X^{\eta - \frac{1}{2}} \sum_{n=1}^{\infty} |h(n)| \, n^{-\frac{1}{2}-\eta} \left|P_{l-1}(\log \frac{x}{n})\right|$$

$$\ll_\eta X^{\eta - \frac{1}{2}}(\log x)^{l-1}.$$

Choose X, η so that $X^{\frac{1}{2}-\eta} = x^{\frac{1}{\Delta}}$; for example let $\eta = \frac{1}{2}(\frac{1}{2} - \frac{1}{\Delta})$,

$$X = x^{\frac{4}{2+\Delta}} = o(x)$$

since $\Delta > 2$. Then we see that all the error terms are $O(x^{1-\frac{1}{\Delta}}(\log x)^\delta)$ where $\delta = \max(\delta_l, l - 1)$. Hence

$$A(x) = \sum_{m \leq x} w(m) = xP_{l-1}^*(\log x) + O(x^{1-\frac{1}{\Delta}}(\log x)^\delta),$$

ich proves Theorem 1. $\square$

Proof of Corollary 1. We use partial summation to deduce this from Theorem 1. We have

$$C(x) = \int_{1-}^{x} v^{-1} d(A(v)) = \overline{P_l}(\log x) + P_{l-1}(\log x) + C + O(x^{-\theta})$$

where $\overline{P_l}$ is given by (3.6) and $C = \int_{1-}^{\infty}(A(v) - vP_{l-1}(\log v))v^{-2}dv = a_l$. Hence

$$C(x) = P_l(\log x) + O(x^{-\theta})$$

where $P_l(t) = \overline{P_l}(t) + P_{l-1}(t) + C$, which is a polynomial of degree l and with leading coefficient $B/l!$ where B is given by (3.13). $\qquad\square$

4. THE FUNCTIONS $\rho_l(u)$ AND $\mu_l(u)$

4.1. The generalized Dickman function.
We defined the generalized Dickman function $\rho_l(u)$ in (1.7) by

$$\left.\begin{array}{l} \rho_l(u) = \frac{u^{l-1}}{(l-1)!} \quad (0 < u \le 1), \qquad \rho_l(u) = 0 \quad (u \le 0), \\[2mm] u\rho_l'(u) = (l-1)\rho_l(u) - l\rho_l(u-1) \quad (u > 1). \end{array}\right\} \qquad (4.1)$$

When $l = 1$, this reduces to the usual Dickman function $\rho(u)$, except that $\rho(0)$ is usually defined to be 1, not 0.

We need the following properties of $\rho_l(u)$ and its derivatives:

Lemma 6. *(i) For $l \ge 2$, $\rho_l(u)$ and its first $(l-2)$ derivatives are continuous for all real u.*

(ii) For $l \ge 1$, $\rho_l^{(l-1)}(u)$ is continuous except at $u = 0$, and

$$\rho_l^{(l-1)}(0-) = 0, \qquad \rho_l^{(l-1)}(0+) = 1.$$

For $n \ge l$, $\rho_l^{(n)}(u)$ is continuous except at $u = 1, 2, \ldots, n+1-l$ where it has a finite jump.

(iii) For each $n \ge 0$, there exists u_0 such that $(-1)^n \rho_l^{(n)}(u) > 0$ for $u \ge u_0$. In particular $\rho_l(u) > 0$ for $u > 0$. Also as $u \to \infty$

$$\left| \rho_l^{(n)}(u) \right| \sim \rho_l(u)(\log u)^n, \qquad (4.2)$$

$$\left| \rho_l^{(n)}(u) \right| = \exp\left(-u\left(\log u + \log_2 u + O(1)\right) \right), \qquad (4.3)$$

$$\rho_l^{(n+1)}(u) = -\log(u \log u)\rho_l^{(n)}(u)\left(1 + o(1)\right). \qquad (4.4)$$

(iv) For $v > 0$ and $u - v > \max(u_0, n+1)$

$$\rho_l^{(n)}(u-v)/\rho_l^{(n)}(u) \le \exp\left((1 + o(1))v \log(u \log u) \right). \qquad (4.5)$$

The function $\rho_l(u)$ belongs to a family of functions studied in [25] using a different notation, and the results in (i), (ii), and (iii) follow from Theorems 1(ii), 3(iii) and equation (3.4) there. Part (iv) is obtained from (4.4) by integrating $\rho_l^{(n+1)}(w)/\rho_l^{(n)}(w)$ over $u - v \le w \le u$.

We turn next to the Laplace transform of $\rho_l^{(n)}(u)$, when defined. Recall that a function $f(u)$ and its Laplace transform $\widehat{f}(s)$, where $u \in \mathbb{R}$ and $s \in \mathbb{C}$, satisfy

$$\widehat{f}(s) = \int_0^\infty e^{-sv} f(v)dv, \quad f(u) = \frac{1}{2\pi i} \int_{c-i\infty}^{c+i\infty} e^{us} \widehat{f}(s)ds \qquad (4.6)$$

when these integrals converge.

Lemma 7. *(i) For $0 \le n < l$ and all $s \in \mathbb{C}$, the Laplace transform $\widehat{\rho_l^{(n)}}(s)$ of $\rho_l^{(n)}(u)$ exists. Moreover*

$$\widehat{\rho_l^{(n)}}(s) = s^n \widehat{\rho_l}(s) \quad (s \in \mathbb{C}), \qquad (4.7)$$

and

$$\widehat{\rho_l}(s) = \widehat{\rho}(s)^l. \qquad (4.8)$$

(ii) We have

$$s^l \widehat{\rho_l}(s) = 1 + O\left(\frac{e^{-\sigma} + |\sigma|}{|t|}\right) \quad \text{if } |t| > \max(e^{-\sigma}, |\sigma|). \qquad (4.9)$$

Part (i) follows from Theorem 2(ii) of [25]; for (ii) see equation (3.6) of [19].

We also use another function closely related to $\rho_l^{(l-1)}(u)$, and its Laplace transform. We defined $z_l(u)$ in (1.8) by

$$z_l(u) = 1 \ (0 \le u \le 1), \quad z_l(u) = 0 \ (u < 0), \quad uz_l'(u) = -lz_l(u-1) \ (u > 1). \qquad (4.10)$$

From (4.1) it follows that, for $u > 1$,

$$u\rho_l^{(l)}(u) = -l\rho_l^{(l-1)}(u-1),$$

and hence $\rho_l^{(l-1)}(u)$ and $z_l(u)$ are equal, except at $u = 0$ where $z_l(0) = 1$ but $\rho_l^{(l-1)}(0)$ has not been defined. By Lemma 6(ii), $z_l^{(r)}(u)$ is continuous for $r \ge 1$, except at $u = 1, \dots, r$ where it has a finite jump.

Lemma 8. *The Laplace transform $\widehat{z_l}(s)$ of $z_l(u)$ is defined for all $s \in \mathbb{C}$ and satisfies*

$$s^l \widehat{\rho_l}(s) = s\widehat{z_l}(s). \qquad (4.11)$$

Proof. See Lemma 7 (i). □

4.2. Expansion of a related function.

Our next aim is to investigate another function $\mu_l(u)$ that, together with its Laplace transform, will be used to establish Theorem 3.

Definition For $u > 0$, let

$$\mu_l(u) = \int_{0-}^u z_l(u-v)d\,(C(y^v)). \qquad (4.12)$$

Lemma 9. *Let $n \in \mathbb{N}$ be fixed and assume that $(x, y) \in H_\varepsilon$. If $n > l$, assume also that $u \in I_{n-l}$, defined in (1.10). Then with the b_r given by (2.29)*

$$\mu_l(u) \;=\; (\log y)^l \Big\{ b_0 \int_0^u \rho_l(v) dv + \sum_{r=1}^n b_r \rho_l^{(r-1)}(u-0)(\log y)^{-r}$$

$$+ O\Big(\rho_l(u) \frac{(\log(u+1))^n}{(\log y)^{n+1}}\Big)\Big\}. \tag{4.13}$$

Proof. By (1.4)

$$C(y^v) = P_l(v \log y) + E(y^v) \qquad \text{where} \quad E(y^v) \ll y^{-\theta v}. \tag{4.14}$$

Since $b_r = P_l^{(l-r)}(0)$ for $0 \leq r \leq l$ by (2.31) and $z_l(u-v) = \rho_l^{(l-1)}(u-v)$, we find on integrating by parts that

$$\mu_l(u) \;=\; (\log y)^l \Big\{ b_0 \int_0^u \rho_l(v) dv + \sum_{r=1}^l b_r \rho_l^{(r-1)}(u-0)(\log y)^{-r} \Big\}$$

$$+ E(y^u) + \int_{0-}^{u-1-0} z_l'(u-v) E(y^v) dv, \tag{4.15}$$

since $z_l'(w) = \rho_l^{(l)}(w) = 0$ for $w < 1$. For $n \leq l$, the result now follows on using (4.14), (4.2) and (4.19) with $n = l$. When $n > l$, we expand the last integral using the method of proof of Lemme 4 of [16].

For $t \geq 1$, let $F_0(t) = -E(t)$ and for $r \geq 1$ define

$$F_r(t) = \frac{(-1)^{r-1}}{(r-1)!} \int_t^\infty E(w) \Big(\log \frac{w}{t}\Big)^{r-1} w^{-1} dw. \tag{4.16}$$

Then for $r \geq 1$ we see that $F_r'(t) = F_{r-1}(t) t^{-1}$ so $\log y F_r(y^v) = \frac{d}{dv}(F_{r+1}(y^v))$; also $F_r(1) = b_{r+l}$ by (2.32), and $F_r(t) \ll t^{-\theta}$ for all $r \geq 0$. For $r \geq 0$, let

$$J_r(u, y) = \int_{1+}^\infty z_l^{(r+1)}(u - \frac{\log t}{\log y}) F_r(t) t^{-1} dt = \log y \int_{0+}^u z_l^{(r+1)}(u-v) F_r(y^v) dv. \tag{4.17}$$

For $n > l$, by integrating $J_0(u, y)$ by parts $(n-l)$ times and recalling that $z_l^{(r+1)}(v)$ has finite jump discontinuities at $v = 1, \ldots, r+1$ and that

$$z_l(0+) - z_l(0-) = 1,$$

we obtain

$$E(y^u) + \int_{0-}^{u-1-0} z_l'(u-v) E(y^v) dv = E(y^u) - J_0(u, y)(\log y)^{-1}$$

$$= \sum_{r=l+1}^n b_r \rho_l^{(r-1)}(u-0)(\log y)^{l-r} - J_{n-l}(u, y)(\log y)^{l-n-1}$$

$$- \sum_{i=0}^{n-l} (\log y)^{-i} \sum_{0 \leq j < \min(i+1, u)} \Big(z_l^{(i)}(j+0) - z_l^{(i)}(j-0)\Big) F_i(y^{u-j}). \tag{4.18}$$

From the properties of $\rho_l^{(n)}(v) = z_l^{(n-l+1)}(v)$ in Lemma 6 we now deduce that

$$J_{n-l}(u,y) \ll \log y \int_{0-}^u \left| \rho_l^{(n)}(u-v) \right| y^{-v\theta} dv \ll \rho_l(u)(\log(u+1))^n . \quad (4.19)$$

We recall that $\rho_l(u) > 0$ for $u > 1$. When $u \le 2u_0$, $J_{n-l}(u,y) \ll 1$ from which (4.19) follows. If $u > 2u_0$, so $u - u_0 > u_0$,

$$\log y \int_{u-u_0}^u \left| \rho_l^{(n)}(u-v) \right| y^{-v\theta} dv \ll y^{-(u-u_0)\theta} \ll \left| \rho_l^{(n)}(u-0) \right|$$

by (4.3) and since in the region H_ε we have

$$\log u \le \log_2 x \le (\log y)^{\frac{3}{5}-\varepsilon_1}, \qquad \varepsilon_1 = \frac{9\varepsilon}{5(5+3\varepsilon)} . \quad (4.20)$$

When $0 \le v \le u - u_0$, so $u - v \ge u_0$, we use (4.5) to obtain

$$\log y \int_0^{u-u_0} \left| \rho_l^{(n)}(u-v) \right| y^{-v\theta} dv$$

$$\le \left| \rho_l^{(n)}(u-0) \right| \log y \int_0^{u-u_0} y^{-v\theta} e^{(1+o(1))v \log u \log_2 u} dv$$

$$\ll \left| \rho_l^{(n)}(u-0) \right| \log y \int_0^{u-u_0} y^{-v\theta/2} dv \ll \left| \rho_l^{(n)}(u-0) \right|$$

on using (4.20). Then (4.19) follows when $u > 2u_0$ on using (4.2).

Since the discontinuities of $z_l^{(i)}(v)$ are bounded, the double sum on the right of (4.18) is

$$\ll \sum_{i=0}^{n-l} (\log y)^{-i} \sum_{0 \le j < \min(i+1,u)} y^{-(u-j)\theta} . \quad (4.21)$$

If $u \ge n - l + 1$, this double sum is

$$\ll y^{-(u-n+l)\theta}(\log y)^{l-n} \ll \rho_l(u)(\log u)^n(\log y)^{l-n-1} \quad (4.22)$$

by (4.2), (4.3) and (4.20). If $\lfloor u \rfloor = h$, $1 \le h \le n-l$, we have $u-h \ge \varepsilon_{n-l,h}(y)$, so by (1.11) $y^{-(u-h)\theta} \le (\log y)^{-(n-l+1-h)}$ and it follows that the terms with $j = h$ in the double sum contribute

$$\ll \sum_{i=h}^{n-l} (\log y)^{-i}(\log y)^{-(n-l+1-h)} \ll (\log y)^{l-n-1}$$

$$\ll \rho_l(u)(\log(u+1))^n(\log y)^{l-n-1} , \quad (4.23)$$

since u is bounded. The remaining terms in (4.21) contribute when $\lfloor u \rfloor = h$

$$\ll y^{-\theta} \ll \rho_l(u)(\log(u+1))^n(\log y)^{l-n-1} . \quad (4.24)$$

From equations (4.15) to (4.19) and (4.21) to (4.24), we deduce (4.13) under the stated conditions. By (4.4) and (4.20), the terms in the sum decrease in size for large u. As $u \to \infty$, $\int_0^u \rho_l(v)dv = e^{\gamma l} - \int_u^\infty \rho_l(v)dv = e^{\gamma l} + O\left(\frac{\rho_l(u)}{\log u}\right)$.

For u large enough we see by (4.2) that the O-term is $O(|\rho_l^{(n)}(u)|(\log y)^{-n-1})$.

$\square$

Lemma 10. *The Laplace transform $\widehat{\mu}_l(s)$ of $\mu_l(u)$ is defined in $\sigma > 0$ and is given by*

$$\widehat{\mu}_l(s) = \widehat{z}_l(s)G\left(1 + \frac{s}{\log y}\right). \qquad (4.25)$$

Proof. This follows from (4.12), (2.30) and the convolution Theorem for Laplace transforms and holds in $\sigma > 0$. We have that

$$\widehat{\mu}_l(s) = \int_0^\infty \mu_l(v)e^{-vs}dv.$$

$\square$

5. Proof of Theorem 3

Next we give a direct proof of this theorem that does not depend on Theorem 2 or [24]. For $\zeta_K(s)$ a Dedekind zeta-function given by (2.18), define $\zeta_K(s, y)$ for $\sigma > 0$ by

$$\zeta_K(s, y) = \sum_{\substack{\mathfrak{a} \\ P(\mathfrak{a}) \leq y}} (N(\mathfrak{a}))^{-s} = \prod_{\substack{\mathfrak{p} \\ N(\mathfrak{p}) \leq y}} \left(1 - (N(\mathfrak{p}))^{-s}\right)^{-1}, \qquad (5.1)$$

where $P(\mathfrak{a}) = \max\{N(\mathfrak{p}) : \mathfrak{p} \mid \mathfrak{a}\}$ for $\mathfrak{a} \neq \mathfrak{O}$. Let

$$G(s, y) = \sum_{\substack{m=1 \\ P(m) \leq y}}^\infty w(m)m^{-s} \qquad (\sigma > 0). \qquad (5.2)$$

Adapting the proof of (2.20) we find that

$$G(s, y) = H(s, y) \prod_{i=1}^l \zeta_{K_i}(s, y), \qquad (5.3)$$

where for $\sigma \geq \frac{1}{2} + \delta$ (for any $\delta > 0$)

$$H(s, y) = H(s)\left(1 + O(y^{\frac{1}{2}-\sigma})\right), \qquad (5.4)$$

since $H(s)/H(s, y)$ consists of a finite number of products of the form $\prod_{p>y}(1 + O(p^{-2\sigma}))$.

Lemma 11. *(i) To each $\varepsilon > 0$, there exists $y_0(\varepsilon)$ such that*

$$\zeta_K(s, y) = \zeta_K(s)(s-1)\log y\, \hat{\rho}\big((s-1)\log y\big)\left(1 + O\left(\frac{1}{L_\varepsilon(y)}\right)\right) \qquad (5.5)$$

uniformly for

$$y \geq y_0(\varepsilon), \quad \sigma \geq 1 - (\log y)^{-\frac{2}{5}-\varepsilon}, \quad |t| \leq L_\varepsilon(y). \qquad (5.6)$$

(ii) To each $\varepsilon > 0$, there exists $y_0(\varepsilon)$ such that

$$G(s, y) = G(s)(s-1)\log y\, \hat{z}_l\big((s-1)\log y\big)\left(1 + O\left(\frac{1}{L_\varepsilon(y)}\right)\right) \qquad (5.7)$$

uniformly when (5.6) holds.

Part (i) is proved in the same way as Lemma 3.5.9.1 of [23], using (2.27); see Lemma 4.1 of [18]. Part (ii) then follows from (5.3), (5.4), (4.8) and (4.11).

Let

$$T = L_{\varepsilon/3}(y) \qquad (5.8)$$

using the definition (1.6).

Lemma 12. *In the region H_ε*

$$C(x, y) = \frac{1}{2\pi i}\int_{\frac{1}{u}-iT\log y}^{\frac{1}{u}+iT\log y} G\left(1 + \frac{s}{\log y}\right)\hat{z}_l(s)e^{us}ds + O\left(\frac{(\log x)^l}{\sqrt{T}}\right). \qquad (5.9)$$

Proof. Let $\kappa = \frac{1}{\log x}$. Then by Perron's formula

$$C(x, y) = \frac{1}{2\pi i}\int_{\kappa-iT}^{\kappa+iT} G(s+1, y)s^{-1}x^s ds + E, \qquad (5.10)$$

where

$$E \ll \sum_{\substack{m=1\\ P(m)\leq y}}^{\infty} \frac{\omega(m)m^{-1-\kappa}}{1 + T\left|\log\frac{x}{m}\right|} \ll \frac{(\log x)^l}{\sqrt{T}}, \qquad (5.11)$$

on considering separately the cases $|x - m| > \frac{x}{\sqrt{T}}$, when $\left|\log\frac{x}{m}\right| \gg \frac{1}{\sqrt{T}}$ and we use that $G(1+\kappa) \ll (\log x)^l$, and $|x - m| \leq \frac{x}{\sqrt{T}}$ when we apply Corollary 1.

We observe that the integral in (5.10) is

$$\ll G(1+\kappa)\int_0^T \frac{dt}{\kappa + t} \ll (\log x)^l \log T, \qquad (5.12)$$

since $-\log \kappa = \log_2 x = o(\log T)$ by (4.20). We now apply Lemma 11(ii) with ε replaced by $\varepsilon/3$ to the integral in (5.10) to obtain

$$C(x,y) = \left(1 + O\left(\tfrac{1}{T}\right)\right) \tfrac{1}{2\pi i} \int_{\kappa-iT}^{\kappa+iT} G(s+1)s \log y \, \widehat{z_l}(s \log y) s^{-1} x^s ds$$

$$+ O\left(\tfrac{(\log x)^l}{\sqrt{T}}\right)$$

$$= \tfrac{1}{2\pi i} \int_{\frac{1}{u}-iT \log y}^{\frac{1}{u}+iT \log y} G\left(1 + \tfrac{s}{\log y}\right) \widehat{z_l}(s) e^{us} ds + O\left(\tfrac{(\log x)^l}{\sqrt{T}}\right),$$

by the change of variable $s \to \tfrac{s}{\log y}$ and using (5.12); this gives (5.9). □

Since $G\left(1 + \tfrac{s}{\log y}\right)$ has a pole of order l at $s = 0$ and

$$\lim_{s \to 0} s^{1-l} \widehat{z_l}(s) = \widehat{\rho_l}(0) = e^{\gamma l},$$

we see that the integrand in (5.9) has a simple pole at $s = 0$. This requires us to use a different strategy from one that would be used in a direct proof of Theorem 2, when the corresponding integrand has no poles, and in that case the integral is estimated by moving the line of integration to the left of $s = 0$.

Lemma 13.

$$J := \tfrac{1}{2\pi i} \int_{\frac{1}{u}+iT \log y}^{\frac{1}{u}+i\infty} G\left(1 + \tfrac{s}{\log y}\right) \widehat{z_l}(s) e^{us} ds \ll \tfrac{(\log x)^l}{\sqrt{T}}. \tag{5.13}$$

Proof. By (4.9) and (4.11), when $\sigma = \tfrac{1}{u}$, $|t| \geq e^{-\frac{1}{u}}$

$$s \widehat{z_l}(s) = 1 + O\left(\tfrac{1}{|t|}\right). \tag{5.14}$$

Hence

$$J = \tfrac{1}{2\pi i} \int_{\frac{1}{u}+iT \log y}^{\frac{1}{u}+i\infty} G(1 + \tfrac{s}{\log y}) s^{-1} e^{us} ds + O\left(\int_{T \log y}^{\infty} (\log \tfrac{t}{\log y})^l t^{-2} dt\right), \tag{5.15}$$

by (2.20) and (2.26). The error term is $O\left(\tfrac{(\log T)^l}{T \log y}\right)$. If $x \notin \mathbb{N}$,

$$C(x) = \tfrac{1}{2\pi i} \int_{\kappa-i\infty}^{\kappa+i\infty} G(1 + s) s^{-1} x^s ds \, ;$$

we see that the main integral in J is just the error term obtained by applying Perron's formula to this integral for $C(x)$ and so is

$$\ll \sum_{m=1}^{\infty} \tfrac{w(m)m^{-1-\kappa}}{1 + T \left|\log \tfrac{x}{m}\right|} \ll \tfrac{(\log x)^l}{\sqrt{T}} \tag{5.16}$$

as in (5.11). Hence (5.13) follows from (5.15) and (5.16). □

Proof of Theorem 3. From Lemmas 10, 12 and 13 and equation (4.20) we deduce that in the region H_ε

$$C(x,y) = \mu_l(u) + O\left(\frac{(\log x)^l}{\sqrt{T}}\right) = \mu_l(u) + O\left(\frac{1}{L_\varepsilon(y)}\right). \qquad (5.17)$$

Also since $\mu_l(u) \gg 1$,

$$C(x,y) = \left(1 + O\left(\frac{1}{L_\varepsilon(y)}\right)\right)\mu_l(u), \qquad (5.18)$$

which gives Theorem 3(i). Part (ii) is a consequence of (5.17) and Lemma 9. We observe that, by (4.3), for large u

$$\left|\rho_l^{(n)}(u)\right|(\log y)^{l-n-1} > \frac{1}{L_\varepsilon(y)}$$

when

$$u(\log u + \log_2 u + O(1)) < (\log y)^{\frac{3}{5}-\varepsilon} - (n+1-l)\log_2 y;$$

this is valid in the region G_ε given by (1.15) since $u = \frac{\log x}{\log y}$. $\square$

REFERENCES

[1] H. Cohen, *A course in computational algebraic number theory*, Springer-Verlag, Berlin Heidelberg New York, 1993.

[2] P. Erdős, On the sum $\sum_{k=1}^{x} d(f(k))$, *J. London Math. Soc.* **27** (1952), 7–15.

[3] G. Hanrot, G. Tenenbaum, J. Wu, Moyennes de certaines fonctions multiplicatives sur les entiers friables 2, *Proc. London Math. Soc.*, to appear.

[4] D.R. Heath-Brown, The growth rate of the Dedekind Zeta-function on the critical line, *Acta Arith.* **49** (1988), 323–339.

[5] _____, Acknowledgement of priority [The growth rate of the Dedekind zeta-function on the critical line, *Acta Arith.* **49** (1988), 323–339], *ibid.* **77**.(1996), 405.

[6] C. Hooley, On the number of divisors of quadratic polynomials, *Acta Math.* **110** (1963), 97–114.

[7] M.N. Huxley, N. Watt, The number of ideals in a quadratic field II, *Israel J. Math.* **120** (2000), Part A, 125–153.

[8] R.M. Kaufman, Estimate of the Hecke L-function on the half line, (Russian), *Zap. Nauchn. Sem. Leningrad. Otdel. Mat. Inst. Steklov. (LOMI)* **91** (1979), 40–51, 180–181.

[9] E. Landau, *Einführung in die elementare und analytische Theorie der algebraische Zahlen und Ideale*, Teubner, Leipzig, 1927; reprint Chelsea, New York, 1949.

[10] G. Martin, An asymptotic formula for the number of smooth values of a polynomial, *J. Number Theory* **93** (2002), 108–182.

[11] J. McKee, On the average number of divisors of quadratic polynomials, *Math. Proc. Camb. Phil. Soc.* **117** (1995), 389–392.

[12] _____, A note on the number of divisors of quadratic polynomials, In: *Sieve methods, exponential sums, and their application in number theory*, London Math. Soc. Lecture Note Series **237**, CUP, Cambridge, 1997, 275–281.

[13] _____, The average number of divisors of an irreducible quadratic polynomial, *Math. Proc. Camb. Phil. Soc.* **126** (1999), 17–22.

[14] T. Nagell, *Introduction to number theory*, Chelsea, New York, 1964.

[15] W.G. Nowak, On the distribution of integer ideals in algebraic number fields, *Math. Nachr.* **161** (1993). 59–74.

[16] E. Saias, Sur le nombre des entier sans grand facteur premier, *J. Number Theory* **32** (1989), 78–99.

[17] E.J. Scourfield, The divisors of a quadratic polynomial, *Proc. Glasgow Math. Assoc.* **5** (1961), 8–20.

[18] _____, On ideals free of large prime factors, *J. de Théorie des Nombres de Bordeaux* **16** (2004), 733–772.

[19] H. Smida, Valeur moyenne des fonctions de Piltz sur les entier sans grand facteur premier, *Acta Arith.* **63** (1993), 21–50.

[20] A.V. Sokolovskii, A Theorem on the zeros of Dedekind's zeta-function and the distance between 'neighbouring' prime ideals, (Russian), *ibid.* **13** (1968), 321–334.

[21] G. Tenenbaum, Sur une question d'Erdős et Schinzel, *A tribute to Paul Erdős*, 405–443, CUP, Cambridge, 1990.

[22] _____, Sur une question d'Erdős et Schinzel II, *Inventiones Math.* **99** (1990), 215–224.

[23] _____, *Introduction to analytic and probabilistic number theory*, CUP, Cambridge, 1995.

[24] _____, private communication (e-mails), received 29.12.06 and 13.02.07.

[25] F.S. Wheeler, Two differential-difference equations arising in number theory, *Trans. Amer. Math. Soc.* **318** (1990), 491–523.

[26] E. Wirsing, Das asymptotische Verhalten von Summen über multiplikativ Funktionen II, *Acta Math. Acad. Sci. Hungar.* **18** (1967), 411–467.

SELF-INVERSIVE POLYNOMIALS WITH ALL ZEROS ON THE UNIT CIRCLE

CHRISTOPHER D. SINCLAIR AND JEFFREY D. VAALER

ABSTRACT. We give a number of sufficient conditions for a self-inversive polynomial to have all zeros on the unit circle.

1. INTRODUCTION

Given a polynomial $g(z) \in \mathbb{C}[z]$ we may create a new polynomial, $g^*(z)$ whose coefficients are obtained from the coefficients of g by reversing their order followed by complex conjugation. That is, if

$$g(z) = c_0 z^N + c_1 z^{N-1} + \cdots + c_{N-1} z + c_N, \qquad (1.1)$$

then

$$g^*(z) = \overline{c_N} z^N + \overline{c_{N-1}} z^{N-1} + \cdots + \overline{c_1} z + \overline{c_0}.$$

Or, more succinctly, $g^*(z) = z^N \overline{g(1/\overline{z})}$. If α is a zero of g then $1/\overline{\alpha}$ is a zero of g^*, and thus the zeros of g^* are determined by 'inverting' the zeros of g over the unit circle.

We define the set of *conjugate reciprocal* polynomials to be the set of $f \in \mathbb{C}[z]$ such that $f = f^*$. And, if there exists some ω on the unit circle such that $f = \omega f^*$, then we will call f an ω-conjugate reciprocal polynomial. The union of all ω-conjugate reciprocal polynomials over all ω on the unit circle is the set of *self-inversive* polynomials. Thus conjugate reciprocal polynomials are simply self-inversive polynomials corresponding to $\omega = 1$. Moreover, there is an isometric bijection between the coefficient space of conjugate reciprocal polynomials and that of ω-conjugate reciprocal polynomials. This correspondence is given by fixing a branch of the N-th root, and associating the conjugate reciprocal polynomial $f(z)$ to the ω-conjugate reciprocal polynomial $\omega f(\omega^{-1/N} z)$.

The zeros of a self-inversive polynomial are either on the unit circle, or come in pairs symmetric with respect to the unit circle. This explains the nomenclature, since the zeros are invariant under 'inversion' with respect to the unit circle. Self-inversive polynomials were first introduced in [1], the original interest being the determination of the number of zeros on the unit circle. Here our goal is similar; we report on conditions for a self-inversive

---

2000 *Mathematics Subject Classification.* Primary: 11C08; Secondary: 51M16, 52A27.
Key words and phrases. Self-inversive polynomials, zeros, unit circle .
312

polynomial to have all zeros on the unit circle. Our results are similar in spirit to recent results of Schinzel [5] and Lakatos and Losonczi [3].

For $p \geq 1$ we define $|g|_p$ to be the p-norm on the coefficients of g. That is, if g is given as in (1.1), then $|g|_p^p = |c_0|^p + |c_1|^p + \cdots + |c_N|^p$.

Theorem 1.1. *If f is a monic self-inversive polynomial of degree N such that*

$$|f|_p^p \leq 2 + \frac{2^p}{(N-1)^{p-1}},$$

then f has all its zeros on the unit circle.

In fact, this result can be strengthened to the following:

Theorem 1.2. *If f is a monic self-inversive polynomial with L non-zero coefficients such that*

$$|f|_p^p \leq 2 + \frac{2^p}{(L-2)^{p-1}},$$

then f has all its zeros on the unit circle.

Theorems 1.1 and 1.2 are sharp in the sense that their right hand sides cannot be unconditionally improved.

It is easily seen that

$$|f|_p^p = \left| \omega f(\omega^{-1/N} z) \right|_p^p,$$

and consequently it suffices to prove Theorems 1.1 and 1.2 for monic conjugate reciprocal polynomials.

2. The Geometry of Conjugate Reciprocal Polynomials

Theorems 1.1 and 1.2 are presented (and proved) in an analytic way. However, they also shed light onto the interesting geometric properties of the set of monic conjugate reciprocal polynomials with all zeros on the unit circle. To see the geometric picture, notice that if $f(z) = z^N + c_1 z^{N-1} + \cdots + c_{N-1} z + 1$ is conjugate reciprocal, then $c_{N-n} = \overline{c_n}$. If N is even, this implies that the middle coefficient, $c_{N/2}$, is real. It follows that a monic conjugate reciprocal polynomial of degree N can be described by $N-1$ real numbers, and we may identify the set of monic conjugate reciprocal polynomials of degree N with $\mathbb{R}^{N-1}$. Perhaps the best way to do this is to introduce the $N-1$ by $N-1$ matrix X_N, whose j, k entry is given by

$$X_N[j,k] = \begin{cases} \frac{\sqrt{2}}{2}\left(\delta_{j,k} + \delta_{N-j,k}\right) & \text{if } 1 \leq j < N/2, \\ \delta_{j,k} & \text{if } j = N/2, \\ \frac{\sqrt{2}}{2}\left(i\delta_{N-j,k} - i\delta_{j,k}\right) & \text{if } N/2 < j < N, \end{cases} \tag{2.1}$$

where $\delta_{j,k} = 1$ if $j = k$ and is zero otherwise. For instance,

$$X_5 = \frac{\sqrt{2}}{2} \begin{bmatrix} 1 & 0 & 0 & i \\ 0 & 1 & i & 0 \\ 0 & 1 & -i & 0 \\ 1 & 0 & 0 & -i \end{bmatrix} \quad \text{and} \quad X_6 = \frac{\sqrt{2}}{2} \begin{bmatrix} 1 & 0 & 0 & 0 & i \\ 0 & 1 & 0 & i & 0 \\ 0 & 0 & \sqrt{2} & 0 & 0 \\ 0 & 1 & 0 & -i & 0 \\ 1 & 0 & 0 & 0 & -i \end{bmatrix}.$$

The $\sqrt{2}/2$ factor is a normalization so that $|\det X_N| = 1$. Given $\mathbf{a} \in \mathbb{R}^{N-1}$, $X_N \mathbf{a}$ is a vector in $\mathbb{C}^{N-1}$. Moreover if $\mathbf{c} = X_N \mathbf{a}$ then $c_{N-n} = \overline{c_n}$, and we may associate a conjugate reciprocal polynomial to $\mathbf{a}$ by specifying that

$$f_{\mathbf{a}}(z) = (z^N + 1) + \sum_{n=1}^{N-1} c_n z^{N-n}, \qquad \mathbf{c} = X_N \mathbf{a}.$$

Finally, we set

$$W_N = \{\mathbf{w} \in \mathbb{R}^{N-1} : f_{\mathbf{w}} \text{ has all zeros on the unit circle}\}. \qquad (2.2)$$

The set W_N was first studied by Petersen and Sinclair [4]. Their primary goal was the determination of the volume of W_N (they show it is the same as the volume of the $N - 1$ dimensional ball of radius 2). A related calculation was done by DiPippo and Howe who determine the volume of the set of real polynomials with all roots on the unit circle [2].

Finding sufficient conditions for a conjugate reciprocal polynomial to have all its zeros on the unit circle corresponds to describing subsets of W_N. When $p = 2$, Theorem 1.1 gives the radius of the largest $N - 1$-sphere (centered at the origin) which is inscribed in W_N.

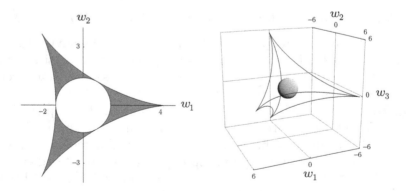

FIGURE 1. W_3 and W_4 with their largest inscribed spheres.

Theorem 1.2 gives us the largest inscribed spheres for *slices* of W_N corresponding to setting a fixed number of coefficients equal to 0. For instance, W_5 is a subset of $\mathbb{R}^4$. The slice of W_5 corresponding to setting coefficients of z^3 and z^2 to zero is given by the first graphic in Figure 2. The second graphic in Figure 2 corresponds to the slice formed by setting the coefficients of z^4 and z to be zero.

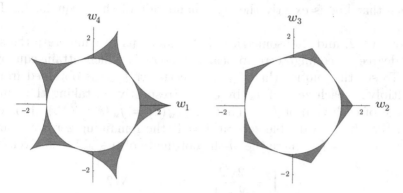

FIGURE 2. Slices of W_5 and their largest inscribed disks

W_N has geometric properties which will allow us to describe two key subsets. The properties that we need are summarized here.

Theorem 2.1 (K. Petersen, C. Sinclair).

(1) W_N is homeomorphic to a closed $(N-1)$-ball,
(2) $\partial W_N = \{\mathbf{w} \in W_N : f_\mathbf{w} \text{ has at least one multiple zero}\}$.
(3) The group of isometries of W_N is isomorphic to the dihedral group of order $2N$, and generated by the isometries given by the $N-1$ by $N-1$ matrices R and C, whose j,k entries are

$$R[j,k] = \delta_{j,k}\cos\left(\frac{2\pi k}{N}\right) - \delta_{j,N-k}\sin\left(\frac{2\pi k}{N}\right),$$

$$C[j,k] = \begin{cases} \delta_{j,k} & \text{if } j \leq N/2, \\ -\delta_{j,k} & \text{if } j > N/2. \end{cases}$$

Theorem 2.1, shows us the way to find the radius of the largest sphere inscribed in W_N: Find the minimal 2-norm of a degree N monic conjugate reciprocal polynomial with all zeros on the unit circle and at least one multiple zero.

Theorem 2.2. Let

$$h(z) = z^N - \frac{2}{N-1}z^{N-1} - \frac{2}{N-1}z^{N-2} - \cdots - \frac{2}{N-1}z + 1.$$

If $f(z)$ is a monic polynomial of degree N, with all zeros on the unit circle and at least one multiple zero, then

$$|f|_2 \geq |h|_2. \tag{2.3}$$

Moreover, there is equality in (2.3) if and only if there exists ξ on the unit circle such that $f(z) = \xi^N h(\xi^{-1} z)$.

Notice that $|h|_p^p$ is exactly the right hand side of the inequality in Theorem 1.1.

Theorem 2.2, and the isometries of W_N allow us to enumerate the set of monic degree N conjugate reciprocal polynomials which attain equality in (2.3). To see this, notice that $f_{R\mathbf{w}}$ (respectively, $f_{C\mathbf{w}}$) is obtained from $f_{\mathbf{w}}$ by multiplying each zero of $f_{\mathbf{w}}$ by $e^{2\pi i/N}$ (respectively, taking the complex conjugate of each zero of $f_{\mathbf{w}}$). That is, $f_{R\mathbf{w}}(z) = f_{\mathbf{w}}(e^{-2\pi i/N} z)$. It is easily verified that h has a double zero at $z = 1$; the remaining zeros are simple. Furthermore, $h = f_{\mathbf{u}}$, where the m-th coordinate of $\mathbf{u} \in \mathbb{R}^{N-1}$ is given by

$$
u_m = \begin{cases} -\dfrac{2\sqrt{2}}{N-1} & \text{if } m < N/2, \\[2mm] -\dfrac{2}{N-1} & \text{if } m = N/2, \\[2mm] 0 & \text{otherwise.} \end{cases} \tag{2.4}
$$

Thus, $f_{R^n \mathbf{u}}$ is a polynomial in ∂W_N with a double zero at $z = e^{2\pi i n/N}$ and the set $\{R^n \mathbf{u} : 0 \le n < N\}$ lies in the intersection of ∂W_N with the sphere of radius $2/\sqrt{N-1}$ centered at the origin.

FIGURE 3. A plot of the zeros of $f_{R^n \mathbf{u}}$ for $0 \le n < N$ when $N = 5$.

For each $0 \le n < N$, the surface of the sphere of radius $2/\sqrt{N-1}$ is tangent to the boundary of W_N at $R^n \mathbf{u}$. The tangent plane of ∂W_N at $R^n \mathbf{u}$ separates $\mathbb{R}^{N-1}$ into two half spaces, one of which contains the origin. The intersection of the half spaces formed in this manner from all of the $R^n \mathbf{u}$ forms a generalized tetrahedron — a simplex. It turns out that this simplex is contained completely in W_N, which gives us another sufficient condition for a self-inversive polynomial to have all zeros on the unit circle.

Theorem 2.3. *Let $\mathbf{u} \in \mathbb{R}^{N-1}$ be defined by (2.4). If*

$$
R^n \mathbf{u} \cdot \mathbf{w} \le \frac{4}{N-1} \qquad \text{for all } 0 \le n < N, \tag{2.5}
$$

then $\mathbf{w} \in W_N$ and $f_{\mathbf{w}}$ has all zeros on the unit circle.

This simplex can also be described as the convex hull of a set of N points in W_N.

Corollary 2.4. *Let* $\mathbf{t} \in \mathbb{R}^{N-1}$ *be given by*

$$
t_m = \begin{cases} 2\sqrt{2} & \text{if } m < N/2, \\ 2 & \text{if } m = N/2, \\ 0 & \text{otherwise.} \end{cases}
$$

That is,

$$
f_{\mathbf{t}}(z) = z^N + 2z^{N-1} + 2z^{N-2} + \cdots + 2z + 1.
$$

If $\mathbf{w}$ *is a convex linear combination of* $\{R^n \mathbf{t} : 0 \leq n < N\}$, *then* $f_{\mathbf{w}}(z)$ *has all zeros on the unit circle.*

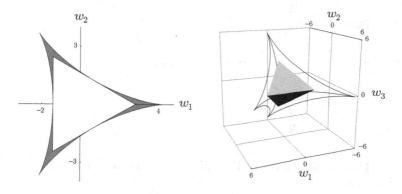

FIGURE 4. W_3 and W_4 with their inscribed simplices.

The right hand side of (2.5) is simply $|\mathbf{u}|_2^2$, and thus the inscribed simplex is described by the equations $R^n \mathbf{u} \cdot \mathbf{w} \leq |\mathbf{u}|_2^2$. Restricting ourselves to slices of W_N formed by setting all but L coefficients of the corresponding conjugate reciprocal polynomials to zero, we may improve Theorem 2.3 by replacing $\mathbf{u}$ with a vector of length $4/\sqrt{L-2}$ (the radius of the largest sphere inscribed in the slice of W_N).

Theorem 2.5. *Let* $\mathbf{w} \in \mathbb{R}^{N-1}$ *be such that* $f_{\mathbf{w}}$ *has exactly L non-zero coefficients, and let* $\mathbf{u}' = (N-1)/(L-2)\mathbf{u}$. *If*

$$
R^n \mathbf{u}' \cdot \mathbf{w} \leq \frac{4}{L-2} \qquad \text{for all } 0 \leq n < N, \tag{2.6}
$$

then $\mathbf{w} \in W_N$ *and* $f_{\mathbf{w}}$ *has all zeros on the unit circle.*

Geometrically, Theorem 2.5 describes a number of convex polytopes which are inscribed into the intersection of W_N with linear subspaces of dimension $L-2$. Fixing one such linear subspace $\mathscr{L}$, and identifying it with $\mathbb{R}^{L-2}$ we may project $\mathbf{u}$ onto a vector $\mathbf{u}^* \in \mathbb{R}^{L-2}$. The isometries of W_N fix $\mathscr{L}$, and thus each isometry induces an action on $\mathbb{R}^{L-2}$. The orbit of $\mathbf{u}^*$ under the induced isometries yields a set of N vectors. It is easy to verify that the projection of $R^n\mathbf{u}'$ onto $\mathscr{L}$ corresponds exactly to the image of $\mathbf{u}^*$ under the action on $\mathbb{R}^{L-1}$ induced by the isometry R^n. The N conditions in (2.6) thus yield at most N distinct half spaces in $\mathbb{R}^{L-2}$. In this manner Theorem 2.5 cuts out a convex polytope inside with at most N codimension one faces lying in the intersection of W_N and $\mathscr{L}$.

Figure 5 shows the slices of W_5 from Figure 2 with the inscribed polytopes guaranteed by Theorem 2.5. Notice in the slice of W_5 determined by setting the coefficients of z^4 and z equal to zero, the inscribed polytope seems to describe the entire slice.

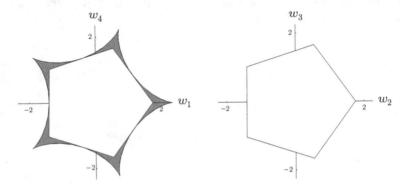

FIGURE 5. Slices of W_5 and their inscribed polytopes

3. Proofs

Viewed through the prism of trigonometric polynomials, the proofs of Theorems 1.1, 1.2, 2.2, 2.3 and 2.5 are elementary.

First notice that if

$$f(z) = z^N + 1 + \sum_{n=1}^{N-1} c_n z^{N-n} \tag{3.1}$$

is conjugate reciprocal, and $M = N/2$, then

$$V(\theta) := e^{Mi\theta} f(e^{i\theta})$$

$$= e^{Mi\theta} + e^{-Mi\theta} + \left\{ \sum_{m=1}^{M-1} c_m e^{(M-m)i\theta} + \overline{c_m} e^{(-M+m)i\theta} \right\} + c_M^*,$$

$$= 2\cos(M\theta) + c_M^*$$

$$+ 2 \left\{ \sum_{m=1}^{M-1} \Re(c_m) \cos\left((M-m)\theta\right) + \Im(c_m) \sin\left((M-m)\theta\right) \right\},$$

where the asterisk on the coefficient c_M indicates that this (middle) coefficient is non-existent when N is odd. From the right hand side of this expression, we see that V is real valued, and has exactly as many zeros in the interval $[0, 2\pi)$ as f has on the unit circle. The number of real zeros of V is also equal to the number of intersections of the graph of $-2\cos(M\theta)$ with that of

$$U(\theta) = \left\{ \sum_{m=1}^{M-1} \Re(c_m)\cos\left((M-m)\theta\right) + \Im(c_m)\sin\left((M-m)\theta\right) \right\} + c_M^*, \quad (3.2)$$

(counting points of tangency of the graphs with multiplicities). If $f = f_{\mathbf{w}}$, then $\mathbf{c} = X_N\mathbf{w}$ and $2\Re(c_m) = \sqrt{2}\,w_m$, $2\Im(c_m) = \sqrt{2}\,w_{N-m} = \sqrt{2}\,w_{2M-m}$ and $c_M = w_M$ (should it exist). In this situation (3.2) reads as

$$U(\theta) = w_M^* + \sqrt{2}\left\{ \sum_{m=1}^{M-1} w_m \cos\left((M-m)\theta\right) + w_{2M-m}\sin\left((M-m)\theta\right) \right\}.$$
$$(3.3)$$

The following lemma is an easy consequence of the Intermediate Value Theorem.

Lemma 3.1. *Let f and U be given as in (3.1) and (3.2). If*

$$(-1)^n U\left(\frac{2\pi n}{N}\right) + 2 \geq 0, \qquad \text{for each} \quad 0 \leq n < N,$$

then f has all zeros on the unit circle.

Proof. Since W_N is closed, it suffices to prove the lemma for the case of strict inequality. That is, we may assume

$$(-1)^n U\left(\frac{2\pi n}{N}\right) + 2 > 0, \qquad \text{for each} \quad 0 \leq n < N.$$

Replacing $(-1)^n$ with $1/\cos(\pi n)$ and clearing denominators, we find that $V(2\pi n/N) = U(2\pi n/N) + 2\cos(\pi n)$ is positive when n is even, and negative when n is odd. Thus, by the Intermediate Value Theorem, V must have at least one zero in each interval $(2\pi n/N, 2\pi(n+1)/N)$, and since there are N such subintervals of $[0, 2\pi)$, V must have at least N zeros in $[0, 2\pi)$. It follows that f has at least (and hence exactly) N zeros on the unit circle. $\quad\square$

Since there is an isometric bijection between conjugate reciprocal polynomials and ω-conjugate reciprocal polynomials, it suffices to prove Theorems 1.1, 1.2 and 2.2 in the case where f is conjugate reciprocal. Theorems 2.3 and 2.5 are stated only for conjugate reciprocal polynomials, but can be used for more general self-inversive polynomials by exploiting the isometric bijection.

3.1. The Proof of Theorems 1.1 and 1.2. Perhaps the simplest restriction on U which satisfies Lemma 3.1 is that $|U(\theta)| \le 2$ for all $\theta \in [0, 2\pi)$. Theorems 1.1 and 1.2 follow from this condition by Cauchy's Inequality, since if exactly $L - 2$ of the c_m are nonzero, then

$$|U(\theta)| \le \left\{ |c_M^*|^p + 2 \sum_{m=1}^{M-1} |c_m|^p \right\}^{1/p} (L-2)^{1/q}, \quad \text{where} \quad \frac{1}{p} + \frac{1}{q} = 1.$$

Solving q in terms of p, and noticing that the expression in braces is nothing more than $|f|_p^p - 2$, we have

$$|U(\theta)| \le (|f|_p^p - 2)^{1/p} (L-2)^{(p-1)/p}.$$

It follows that if

$$|f|_p^p \le 2 + \frac{2^p}{(L-2)^{p-1}},$$

then $|U(\theta)| \le 2$ for all θ, and f has N zeros on the unit circle.

3.2. The Proof of Theorem 2.2. As mentioned previously, Theorem 2.2 comes from finding the intersections of ∂W_N with the closed ball of radius $2/\sqrt{N-1}$ centered at the origin. Call this ball B, and suppose $\mathbf{v} \in B \cap \partial W_N$. Let $\{\mathbf{v}_\ell\} \subseteq \text{interior}(B)$ be a sequence having $\mathbf{v}$ as a limit point. From Theorem 1.1 and the proof of Lemma 3.1, each of the polynomials $f_{\mathbf{v}_\ell}$ has all simple zeros lying on the unit circle. Moreover, the zeros of $f_{\mathbf{v}_\ell}$ are equidistributed in the sense that each arc $\{e^{i\theta} : 2\pi n/N < \theta < 2\pi(n+1)/N\}$, contains exactly one zero.

By Theorem 2.1, $f_{\mathbf{v}}$ must have at least one multiple zero, and since it is a limit point of polynomials $f_{\mathbf{v}_\ell}$ with equidistributed zeros, we may conclude that $f_{\mathbf{v}}$ has a double zero at $e^{2\pi i n/N}$ for some $0 \le n < N$. From our knowledge of the isometries of W_N, if $\mathbf{w} = R^{-n}\mathbf{v}$, then $f_{\mathbf{w}}$ has a double zero at $z = 1$. Forming U as in (3.3), we have

$$U(0) = w_M^* + \sqrt{2} \sum_{m=1}^{M-1} w_m = -2, \tag{3.4}$$

and by Cauchy's inequality,

$$2 \le \left\{ (w_M^*)^2 + \sum_{m=1}^{M-1} w_m^2 \right\}^{1/2} (N-1)^{1/2} = 2,$$

where the equality on the right follows since $\mathbf{w}$ is on the sphere of radius $2/\sqrt{N-1}$. We have equality in Cauchy's inequality, and thus when N is even, $w_1 = w_2 = \cdots = w_{M-1} = w_M/\sqrt{2}$, and when N is odd,

$$w_1 = w_2 = \cdots = w_{M-1/2}.$$

This together with (3.4) implies that $\mathbf{w} = \mathbf{u}$ given as in (2.4), and that $f_{\mathbf{w}}(z) = h(z)$ given as in the statement of the theorem. Moreover, $\mathbf{v} = R^n\mathbf{u}$, and thus $B \cap \partial W_N$ consists entirely of $\{R^n\mathbf{u} : 0 \le n < N\}$, as claimed.

3.3. The Proof of Theorems 2.3 and 2.5.

Using an easy induction argument based on the definition of R, or the fact that $f_{R\mathbf{w}}(z) = f_{\mathbf{w}}(e^{-2\pi i/N}z)$, we can write R^n as the matrix given by

$$R^n[j,k] = \delta_{j,k}\cos\left(\frac{2\pi nk}{N}\right) - \delta_{j,N-k}\sin\left(\frac{2\pi nk}{N}\right).$$

Thus, the m-th coordinate of $R^n\mathbf{u}$ is given by

$$(R^n\mathbf{u})_m = \begin{cases} -\dfrac{2\sqrt{2}}{N-1}\cos\left(\dfrac{2\pi mn}{N}\right) & \text{if } m < N/2\,, \\[2ex] -\dfrac{2}{N-1}(-1)^n & \text{if } m = N/2\,, \\[2ex] \dfrac{2\sqrt{2}}{N-1}\sin\left(\dfrac{2\pi mn}{N}\right) & \text{if } m > N/2\,, \end{cases}$$

and, writing $c(m)$ and $s(m)$ for $\cos\big((M-m)2\pi n/N\big)$ and $\sin\big((M-m)2\pi n/N\big)$ respectively, we have

$$R^n\mathbf{u}\cdot\mathbf{w} = \frac{2}{N-1}(-1)^{n+1}\left(w_M^* + \sqrt{2}\left\{\sum_{m=1}^{M-1} w_m c(m) + w_{2M-m}s(m)\right\}\right)$$

$$= \frac{2}{N-1}(-1)^{n+1}U\left(\frac{2\pi n}{N}\right).$$

By assumption, $R^n\mathbf{u}\cdot\mathbf{w} < 4/(N-1)$, and thus, $(-1)^{n+1}U\left(2\pi n/N\right) \le 2$ for all $0 \le n < N$. We conclude from Lemma 3.1 that $f_{\mathbf{w}}$ has all zeros on the unit circle.

The proof of Theorem 2.5 is essentially the same as that of Theorem 2.3 after replacing $\mathbf{u}$ with $\mathbf{u}'$.

REFERENCES

[1] F.F. Bonsall, M. Marden, Zeros of self-inversive polynomials, *Proc. Amer. Math. Soc.* **3** (1952), 471–475.

[2] S.A. DiPippo, E.W. Howe, Real polynomials with all roots on the unit circle and abelian varieties over finite fields, *J. Number Theory* **73** (1998), 426–450; Corrigendum **83** (2000), 182.

[3] P. Lakatos, L. Losonczi, On zeros of reciprocal polynomials of odd degree, *JI-PAM. J. Inequal. Pure Appl. Math.* **4**, no. 3, article 60, 8 pp. (electronic), 2003.

[4] K.L. Petersen, C.D. Sinclair, Conjugate reciprocal polynomials with all roots on the unit circle. Preprint, 2005.

[5] A. Schinzel, Self-inversive polynomials with all zeros on the unit circle, *Ramanujan J.*, **9** (2005), no. 1–2, 19–23.

THE MAHLER MEASURE OF ALGEBRAIC NUMBERS: A SURVEY

CHRIS SMYTH

ABSTRACT. A survey of results for Mahler measure of algebraic numbers, and one-variable polynomials with integer coefficients is presented. Related results on the maximum modulus of the conjugates ('house') of an algebraic integer are also discussed. Some generalisations are given too, though not to Mahler measure of polynomials in more than one variable.

1. INTRODUCTION

Let $P(x) = a_0 z^d + \cdots + a_d = a_0 \prod_{i=1}^d (z - \alpha_i)$ be a nonconstant polynomial with (at first) complex coefficients. Then, following Mahler [101] its *Mahler measure* is defined to be

$$M(P) := \exp \left(\int_0^1 \log |P(e^{2\pi i t})| dt \right), \qquad (1)$$

the geometric mean of $|P(z)|$ for z on the unit circle. However $M(P)$ had appeared earlier in a paper of Lehmer [94], in an alternative form

$$M(P) = |a_0| \prod_{|\alpha_i| \geq 1} |\alpha_i|. \qquad (2)$$

The equivalence of the two definitions follows immediately from Jensen's formula [88]

$$\int_0^1 \log |e^{2\pi i t} - \alpha| dt = \log_+ |\alpha|.$$

Here $\log_+ x$ denotes $\max(0, \log x)$. If $|a_0| \geq 1$, then clearly $M(P) \geq 1$. This is the case when P has integer coefficients; we assume henceforth that P is of this form. Then, from a result of Kronecker [90], $M(P) = 1$ occurs only if $\pm P$ is a power of z times a cyclotomic polynomial.

In [101] Mahler called $M(P)$ the measure of the polynomial P, apparently to distinguish it from its (naïve) height. This was first referred to as Mahler's measure by Waldschmidt [165, p.21] in 1979 ('mesure de Mahler'), and soon afterwards by Boyd [33] and Durand [75], in the sense of "the function that Mahler called 'measure'", rather than as a name. But it soon *became* a name. In 1983 Louboutin [98] used the term to apply to an algebraic number. We shall follow this convention too — $M(\alpha)$ for an algebraic number α will mean

2000 *Mathematics Subject Classification*. Primary 11R06.
Key words and phrases. Mahler measure, Lehmer problem, house of algebraic integer .

the Mahler measure of the minimal polynomial P_α of α, with d the degree of α, having conjugates $\alpha = \alpha_1, \alpha_2, \ldots, \alpha_d$. The Mahler measure is actually a height function on polynomials with integer coefficients, as there are only a finite number of such polynomials of bounded degree and bounded Mahler measure. Indeed, in the MR review of [98], it is called the Mahler height; but 'Mahler measure' has stuck.

For the Mahler measure in the form $M(\alpha)$, there is a third representation to add to (1) and (2). We consider a complete set of inequivalent valuations $|.|_\nu$ of the field $\mathbb{Q}(\alpha)$, normalised so that, for $\nu|p$, $|.|_\nu = |.|_p$ on $\mathbb{Q}_p$. Here $\mathbb{Q}_p$ is the field of p-adic numbers, with the usual valuation $|.|_p$. Then for a_0 as in (2),

$$|a_0| = \prod_{p<\infty} |a_0|_p^{-1} = \prod_{p<\infty} \prod_{\nu|p} \max(1, |\alpha|_\nu^{d_\nu}), \qquad (3)$$

coming from the product formula, and from considering the Newton polygons of the irreducible factors (of degree d_ν) of P_α over $\mathbb{Q}_p$ (see e.g. [170, p. 73]).

Then [169, pp. 74–79], [23] from (2) and (3)

$$M(\alpha) = \prod_{\text{all } \nu} \max(1, |\alpha|_\nu^{d_\nu}), \qquad (4)$$

and so also

$$h(\alpha) := \frac{\log M}{d} = \sum_{\text{all } \nu} \log_+ |\alpha|_\nu^{d_\nu/d}. \qquad (5)$$

Here $h(\alpha)$ is called the *Weil*, or *absolute* height of α.

2. LEHMER'S PROBLEM

While Mahler presumably had applications of his measure to transcendence in mind, Lehmer's interest was in finding large primes. He sought them amongst the Pierce numbers $\prod_{i=1}^{d}(1 \pm \alpha_i^m)$, where the α_i are the roots of a monic polynomial P having integer coefficients. Lehmer showed that for P with no roots on the unit circle these numbers grew with m like $M(P)^m$. Pierce [120] had earlier considered the factorization of these numbers. Lehmer posed the problem of whether, among those monic integer polynomials with $M(P) > 1$, polynomials could be chosen with $M(P)$ arbitrarily close to 1. This has become known as 'Lehmer's problem', or 'Lehmer's conjecture', the 'conjecture' being that they could not, although Lehmer did not in fact make this conjecture.[1] The smallest value of $M(P) > 1$ he could find was

$$M(L) = 1.176280818\ldots,$$

---

[1] 'Lehmer's conjecture' is also used to refer to a conjecture on the non-vanishing of Ramanujan's τ-function. But I do not know that Lehmer actually made that conjecture either: in [95, p. 429] he wrote "... and it is natural to ask whether $\tau(n) = 0$ for any $n > 0$."

where $L(z) = z^{10} + z^9 - z^7 - z^6 - z^5 - z^4 - z^3 + z + 1$ is now called 'Lehmer's polynomial'. To this day no-one has found a smaller value of $M(P) > 1$ for $P(z) \in \mathbb{Z}[z]$.

Lehmer's problem is central to this survey. We concentrate on results for $M(P)$ with P having integer coefficients. We do not attempt to survey results for $M(P)$ for P a polynomial in several variables. For this we refer the reader to [18], [33], [163], [39], [43], [40], [79, Chapter 3]. However, the one-variable case should not really be separated from the general case, because of the fact that for every P with integer coefficients, irreducible and in genuinely more than one variable (i.e., its Newton polytope is not one-dimensional) $M(P)$ is known [33, Theorem 1] to be the limit of $\{M(P_n)\}$ for some sequence $\{P_n\}$ of one-variable integer polynomials. This is part of a far-reaching conjecture of Boyd [33] to the effect that the set of all $M(P)$ for P an integer polynomial in any number of variables is a closed subset of the real line.

Our survey of results related to Lehmer's problem falls into three categories. We report lower bounds, or sometimes exact infima, for $M(P)$ as P ranges over certain sets of integer polynomials. Depending on this set, such lower bounds can either tend to 1 as the degree d of P tends to infinity (Section 4), be constant and greater than 1 (Section 5), or increase exponentially with d (Section 6). We also report on computational work on the problem (Section 8).

In Sections 3 and 7 we discuss the closely-related function $\overline{|\alpha|}$ and the Schinzel-Zassenhaus conjecture. In Section 9 connections between Mahler measure and the discriminant are covered. In Section 10 the known properties of $M(\alpha)$ as an algebraic number are outlined. Section 11 is concerned with counting integer polynomials of given Mahler measure, while in Section 12 a dynamical systems version of Lehmer's problem is presented. In Section 13 variants of Mahler measure are discussed, and finally in Section 14 some applications of Mahler measure are given.

3. THE HOUSE $\overline{|\alpha|}$ OF α AND THE CONJECTURE OF SCHINZEL AND ZASSENHAUS

Related to the Mahler measure of an algebraic integer α is $\overline{|\alpha|}$, the *house* of α, defined as the maximum modulus of its conjugates (including α itself). For α with $r > 0$ roots of modulus greater than 1 we have the obvious inequality

$$M(\alpha)^{1/d} \leq M(\alpha)^{1/r} \leq \overline{|\alpha|} \leq M(\alpha) \qquad (6)$$

(see e.g. [34]). If α is in fact a unit (which is certainly the case if $M(\alpha) < 2$) then $M(\alpha) = M(\alpha^{-1})$ so that

$$M(\alpha) \leq (\max(\overline{|\alpha|}, \overline{|1/\alpha|}))^{d/2} .$$

In 1965 Schinzel and Zassenhaus [144] proved that if $\alpha \neq 0$ an algebraic integer that is not a root of unity and if $2s$ of its conjugates are nonreal, then

$$\boxed{\alpha} > 1 + 4^{-s-2}. \tag{7}$$

This was the first unconditional result towards solving Lehmer's problem, since by (6) it implies the same lower bound for $M(\alpha)$ for such α. They conjectured, however, that a much stronger bound should hold: that under these conditions in fact

$$\boxed{\alpha} \geq 1 + c/d \tag{8}$$

for some absolute constant $c > 0$. Its truth is implied by a positive answer to Lehmer's 'conjecture'. Indeed, because $\boxed{\alpha} \geq M(\alpha)^{1/d}$ where $d = \deg \alpha$, we have

$$\boxed{\alpha} \geq 1 + \frac{\log M(\alpha)}{d} = 1 + h(\alpha), \tag{9}$$

so that if $M(\alpha) \geq c_0 > 1$ then $\boxed{\alpha} > 1 + \frac{\log(c_0)}{d}$.

Likewise, from this inequality any results in the direction of solving Lehmer's problem will have a corresponding 'Schinzel-Zassenhaus conjecture' version. In particular, this applies to the results of Section 5.1 below, including that of Breusch. His inequality appears to be the first, albeit conditional, result in the direction of the Schinzel-Zassenhaus conjecture or the Lehmer problem.

4. Unconditional lower bounds for $M(\alpha)$ that tend to 1 as $d \to \infty$

4.1. The bounds of Blanksby and Montgomery, and Stewart.

The lower bound for $M(\alpha)$ coming from (7) was dramatically improved in 1971 by Blanksby and Montgomery [22], who showed, again for α of degree $d > 1$ and not a root of unity, that

$$M(\alpha) > 1 + \frac{1}{52d \log(6d)}.$$

Their methods were based on Fourier series in several variables, making use of the nonnegativity of Fejér's kernel

$$\tfrac{1}{2} + \sum_{k=1}^{K} \left(1 - \tfrac{k}{K+1}\right) \cos(kx) = \tfrac{1}{2(K+1)} \left(\sum_{j=0}^{K} e^{ix\left(\frac{K}{2}-j\right)}\right)^2.$$

They also employed a neat geometric lemma for bounding the modulus of complex numbers near the unit circle: if $0 < \rho \leq 1$ and $\rho \leq |z| \leq \rho^{-1}$ then

$$|z - 1| \leq \rho^{-1} \left|\rho\tfrac{z}{|z|} - 1\right|. \tag{10}$$

In 1978 Stewart [158] caused some surprise by obtaining a lower bound of the same strength $1 + \frac{C}{d \log d}$ by the use of a completely different argument. He based his proof on the construction of an auxiliary function of the type used in transcendence proofs.

In such arguments it is of course necessary to make use of some arithmetic information, because of the fact that the polynomials one is dealing with, here the minimal polynomials of algebraic integers, are monic, have integer coefficients, and no root is a root of unity. In the three proofs of the results given above, this is done by making use of the fact that, for α not a root of unity, the Pierce numbers $\prod_{i=1}^{d}(1 - \alpha_i^m)$ are then nonzero integers for all $m \in \mathbb{N}$. Hence they are at least 1 in modulus.

4.2. Dobrowolski's lower bound. In 1979 a breakthrough was achieved by Dobrowolski, who, like Stewart, used an argument based on an auxiliary function to get a lower bound for $M(\alpha)$. However, he also employed more powerful arithmetic information: the fact that for any prime p the resultant of the minimal polynomials of α and of α^p is an integer multiple of p^d. Since this can be shown to be nonzero for α not a root of unity, it is at least p^d in modulus. Dobrowolski [54] was able to apply this fact to obtain for $d \geq 2$ the much improved lower bound

$$M(\alpha) > 1 + \frac{1}{1200}\left(\frac{\log\log d}{\log d}\right)^3. \tag{11}$$

He also has an asymptotic version of his result, where the constant $1/1200$ can be increased to $1 - \varepsilon$ for α of degree $d \geq d_0(\varepsilon)$. Improvements in the constant in Dobrowolski's Theorem have been made since that time. Cantor and Straus [49] proved the asymptotic version of his result with the larger constant $2 - \varepsilon$, by a different method: the auxiliary function was replaced by the use of generalised Vandermonde determinants. See also [125] for a similar argument (plus some early references to these determinants). As with Dobrowolski's argument, the large size of the resultant of α and α^p was an essential ingredient. Louboutin [98] improved the constant further, to $9/4-\varepsilon$, using the Cantor-Straus method. A different proof of Louboutin's result was given by Meyer [108]. Later Voutier [164], by a very careful argument based on Cantor-Straus, has obtained the constant $1/4$ valid for all α of degree $d \geq 2$. However, no-one has been able to improve the dependence on the degree d in (11), so that Lehmer's problem remains unsolved!

4.3. Generalisations of Dobrowolski's Theorem. Amoroso and David [3, 4] have generalised Dobrowolski's result in the following way. Let $\alpha_1, \ldots, \alpha_n$ be n multiplicatively independent algebraic numbers in a number field of degree d. Then for some constant $c(n)$ depending only on n

$$h(\alpha_1)\ldots h(\alpha_n) \geq \frac{1}{d\log(3d)^{c(n)}}. \tag{12}$$

Matveev [107] also has a result of this type, but using instead the modified Weil height $h_*(\alpha) := \max(h(\alpha), d^{-1}|\log\alpha|)$.

Amoroso and Zannier [11] have given a version of Dobrowolski's result for α, not 0 or a root of unity, of degree D over an finite abelian extension of a number field. Then

$$h(\alpha) \geq \frac{c}{D}\left(\frac{\log\log 5D}{\log 2D}\right)^{13}, \tag{13}$$

where the constant c depends only on the number field, not on its abelian extension. Amoroso and Delsinne [9] have recently improved this result, for instance essentially reducing the exponent 13 to 4.

Analogues of Dobrowolski's Theorem have been proved for elliptic curves by Anderson and Masser [12], Hindry and Silverman [84], Laurent [93] and Masser [104]. In particular Masser proved that for an elliptic curve E defined over a number field K and a nontorsion point P defined over a degree $\leq d$ extension F of K that the canonical height $\hat{h}(P)$ satisfies

$$\hat{h}(P) \geq \frac{C}{d^3(\log d)^2}.$$

Here C depends only on E and K. When E has non-integral j-invariant Hindry and Silverman improved this bound to $\hat{h}(P) \geq \frac{C}{d^2(\log d)^2}$. In the case where E has complex multiplication, however, Laurent obtained the stronger bound

$$\hat{h}(P) \geq \frac{C}{d}(\log\log d/\log d)^3.$$

This is completely analogous to the formulation of Dobrowolski's result (11) in terms of the Weil height $h(\alpha) = \log M(\alpha)/d$.

5. RESTRICTED RESULTS OF LEHMER STRENGTH: $M(\alpha) > c > 1$.

5.1. Results for nonreciprocal algebraic numbers and polynomials.
Recall that a polynomial $P(z)$ of degree d is said to be *reciprocal* if it satisfies $z^d P(1/z) = \pm P(z)$. (With the negative sign, clearly $P(z)$ is divisible by $z-1$.) Furthermore an algebraic number α is *reciprocal* if it is conjugate to α^{-1} (as then P_α is a reciprocal polynomial). One might at first think that it should be possible to prove stronger results on Lehmer's problem if we restrict our attention to reciprocal polynomials. However, this is far from being the case: reciprocal polynomials seem to be the most difficult to work with, perhaps because cyclotomic polynomials are reciprocal; we can prove stronger results on Lehmer's problem if we restrict our attention to nonreciprocal polynomials!

The first result in this direction was due to Breusch [44]. Strangely, this paper was unknown to number theorists until it was recently unearthed by Narkiewicz. Breusch proved that for α a nonreciprocal algebraic integer

$$M(\alpha) \geq M(z^3 - z^2 - \tfrac{1}{4}) = 1.1796\ldots . \tag{14}$$

Breusch's argument is based on the study of the resultant of α and α^{-1}, for α a root of P. On the one hand, this resultant must be at least 1 in modulus. But, on the other hand, this is not possible if $M(P)$ is too close to 1, because then all the distances $|\alpha_i - \alpha_i^{-1}|$ are too small. (Note that $\alpha_i = \alpha_i^{-1}$ implies that P is reciprocal.)

In 1971 Smyth [155] independently improved the constant in (14), showing for α a nonreciprocal algebraic integer

$$M(\alpha) \geq M(z^3 - z - 1) = \theta_0 = 1.3247\ldots, \tag{15}$$

the real root of $z^3 - z - 1 = 0$. This constant is best possible here, $z^3 - z - 1$ being nonreciprocal. Equality $M(\alpha) = \theta_0$ occurs only for α conjugate to $(\pm\theta_0)^{\pm 1/k}$ for k some positive integer.[2] Otherwise in fact $M(\alpha) > \theta_0 + 10^{-4}$ ([156]), so that θ_0 is an isolated point in the spectrum of Mahler measures of nonreciprocal algebraic integers. The lower bound 10^{-4} for this gap in the spectrum was increased to $0.000260\ldots$ by Dixon and Dubickas [52, Th. 15]. It would be interesting to know more about this spectrum. All of its known small points come from trinomials, or their irreducible factors:

$1.324717959\cdots = M(z^3 - z - 1) = M(\frac{z^5 - z^4 - 1}{z^2 - z + 1})$;

$1.349716105\cdots = M(z^5 - z^4 + z^2 - z + 1) = M(\frac{z^7 + z^2 + 1}{z^2 + z + 1})$;

$1.359914149\cdots = M(z^6 - z^5 + z^3 - z^2 + 1) = M(\frac{z^8 + z + 1}{z^2 + z + 1})$;

$1.364199545\cdots = M(z^5 - z^2 + 1)$;

$1.367854634\cdots = M(z^9 - z^8 + z^6 - z^5 + z^3 - z + 1) = M(\frac{z^{11} + z^4 + 1}{z^2 + z + 1})$.

The smallest known limit point of nonreciprocal measures is

$$\lim_{n \to \infty} M(z^n + z + 1) = 1.38135\ldots$$

([31]). The spectrum clearly contains the set of all Pisot numbers, except perhaps the reciprocal ones. But in fact it does contain those too, a result due to Boyd [36, Proposition 2]. There are however smaller limit points of reciprocal measures (see [33], [42]).

The method of proof of (15) was based on the Maclaurin expansion of the rational function $F(z) = P(0)P(z)/z^d P(1/z)$, which has integer coefficients and is nonconstant for P nonreciprocal. This idea had been used in 1944 by Salem [133] in his proof that the set of Pisot numbers is closed, and in the same year by Siegel [147] in his proof that θ_0 is the smallest Pisot number. One can write $F(z)$ as a quotient $f(z)/g(z)$ where f and g are both holomorphic and bounded above by 1 in modulus in the disc $|z| < 1$. Furthermore, $f(0) = g(0) = M(P)^{-1}$. These functions were first studied by Schur [146], who completely specified the conditions on the coefficients of a power series $\sum_{n=0}^{\infty} c_n z^n$ for it to belong to this class. Then study of functions of this type, combined with the fact that the series of their quotient has integer coefficients, enables one to get the required lower bound for $M(P)$. To prove that θ_0 is an isolated point of the nonreciprocal spectrum, it was necessary to consider the quotient $F(z)/F_1(z)$, where $F_1(z) = P_1(0)P_1(z)/z^d P_1(1/z)$. Here P_1 is chosen as the minimal polynomial of some $(\pm\theta_0)^{\pm 1/k}$ so that, if $F(z) = 1 + a_k z^k + \ldots$, where $a_k \neq 0$ then also $F_1(z) \equiv 1 + a_k z^k \pmod{z^{k+1}}$.

---

[2]As Boyd [36] pointed out, however, this does not preclude the possibility of equality for some *reciprocal* α. But it was proved by Dixon and Dubickas [52, Cor. 14] that this could not happen.

Thus this quotient, assumed nonconstant, had a first nonzero term of higher order, enabling one to show that $M(P) > \theta_0 + 10^{-4}$.

5.2. Nonreciprocal case: generalizations.

Soon afterwards Schinzel [137] and then Bazylewicz [15] generalised Smyth's result to polynomials over Kroneckerian fields. (These are fields that are either totally real extensions of the rationals, or totally nonreal quadratic extensions of such fields.) For a further generalisation to polynomials in several variables see [142, Theorem 70]. In these generalisations the optimal constant is obtained. If the field does not contain a primitive cube root of unity ω_3 then the best constant is again θ_0, while if it does contain ω_3 then the best constant is the maximum modulus of the roots θ of $\theta^2 - \omega_3\theta - 1 = 0$.

Generalisations to algebraic numbers were proved by Notari [116] and Lloyd-Smith [97]. See also Skoruppa's Heights notes [154] and Schinzel [142].

5.3. The case where $\mathbb{Q}(\alpha)/\mathbb{Q}$ is Galois.

In 1999 Amoroso and David [4], as a Corollary of a far more general result concerning heights of points on subvarieties of $\mathbb{G}_m^n$, solved Lehmer's problem for $\mathbb{Q}(\alpha)/\mathbb{Q}$ a Galois extension: they proved that there is a constant $c > 1$ such that if α is not zero or a root of unity and $\mathbb{Q}(\alpha)$ is Galois of degree d then $M(\alpha) \geq c$.

5.4. Other restricted results of Lehmer strength.

Mignotte [109, Cor. 2] proved that if α is an algebraic number of degree d such that there is a prime less than $d \log d$ that is unramified in the field $\mathbb{Q}(\alpha)$ then $M(\alpha) \geq 1.2$.

Mignotte [109, Prop. 5] gave a very short proof, based on an idea of Dobrowolski, of the fact that for an irreducible noncyclotomic polynomial P of length $L = ||P||_1$ that $M(P) \geq 2^{1/2L}$. For a similar result (where $2^{1/2L}$ is replaced by $1 + 1/(6L)$), see Stewart [159].

In 2004 P. Borwein, Mossinghoff and Hare [28] generalised the argument in [155] to nonreciprocal polynomials P all of whose coefficients are odd, proving that in this case

$$M(P) \geq M(z^2 - z - 1) = \phi.$$

Here $\phi = (1+\sqrt{5})/2$. This lower bound is clearly best possible. Recently Borwein, Dobrowolski and Mossinghoff have been able to drop the requirement of nonreciprocality: they proved in [27] that for a noncyclotomic irreducible polynomial with all odd coefficients then

$$M(P) \geq 5^{1/4} = 1.495348\ldots \quad . \tag{16}$$

In the other direction, in a search [28] of polynomials up to degree 72 with coefficients ± 1 and no cyclotomic factor the smallest Mahler measure found was $M(z^6 + z^5 - z^4 - z^3 - z^2 + z + 1) = 1.556030\ldots$.

Dobrowolski, Lawton and Schinzel [59] first gave a bound for the Mahler measure of an noncyclotomic integer polynomial P in terms of the number k

of its nonzero coefficients:

$$M(P) \geq 1 + \frac{1}{\exp_{k+1} 2k^2}.$$ (17)

Here $\exp_{k+1}$ is the $(k+1)$-fold exponential. This was later improved by Dobrowolski [56] to $1 + \frac{1}{13911}\exp(-2.27k^k)$, and lately [57] to

$$M(P) \geq 1 + \frac{1}{\exp(a3^{\lfloor(k-2)/4\rfloor}k^2 \log k)},$$ (18)

where $a < 0.785$. Furthermore, in the same paper he proves that if P has no cyclotomic factors then

$$M(P) \geq 1 + \frac{0.31}{k!}.$$ (19)

With the additional restriction that P is irreducible, Dobrowolski [55] gave the lower bound

$$M(P) \geq 1 + \frac{\log(2e)}{2e}(k+1)^{-k}.$$ (20)

In [57] he strengthened this to

$$M(P) \geq 1 + \frac{0.17}{2^m m!},$$ (21)

where $m = \lceil k/2 \rceil$.

Recently Dobrowolski [58] has proved that for an integer symmetric $n \times n$ matrix A with characteristic polynomial $\chi_A(x)$, the reciprocal polynomial $z^n \chi_A(z + 1/z)$ is either cyclotomic or has Mahler measure at least 1.043. The Mahler measure of A can then be defined to be the Mahler measure of this polynomial. McKee and Smyth [100] have just improved the lower bound in Dobrowolski's result to the best possible value $\tau_0 = 1.176\ldots$ coming from Lehmer's polynomial. The adjacency matrix of the graph below is an example of a matrix where this value is attained.

The Mahler measure of a graph, defined as the Mahler measure of its adjacency matrix, has been studied by McKee and Smyth [99]. They showed that its Mahler measure was either 1 or at least τ_0, the Mahler measure of the graph $\cdots\vdots\cdots$. They further found all numbers in the interval $[1, \phi]$ that were Mahler measures of graphs. All but one of these numbers is a Salem number.

6. Restricted results where $M(\alpha) > C^d$.

6.1. **Totally real** α. Suppose that α is a totally real algebraic integer of degree d, $\alpha \neq 0$ or ± 1. Then Schinzel [137] proved that

$$M(\alpha) \geq \phi^{d/2}.$$ (22)

A one-page proof of this result was later provided by Höhn and Skoruppa [86]. The result also holds for any nonzero algebraic number α in a Kroneckerian field, provided $|\alpha| \neq 1$. Amoroso and Dvornicich [10, p. 261] gave the interesting example of $\alpha = \frac{1}{2}\sqrt{3 + \sqrt{-7}}$, not an algebraic integer, where $|\alpha| = 1$, $\mathbb{Q}(\alpha)$ is Kroneckerian, but $M(\alpha) = 2 < \phi^2$.

Smyth [157] studied the spectrum of values $M(\alpha)^{1/d}$ in $(1, \infty)$. He showed that this spectrum was discrete at first, and found its smallest four points. The method used is semi-infinite linear programming (continuous real variables and a finite number of constraints), combined with resultant information. One takes a list of judiciously chosen polynomials $P_i(x)$, and then finds the largest c such that for some $c_i \geq 0$

$$\log_+ |x| \geq c - \sum_i c_i \log |P_i(x)| \tag{23}$$

for all real x. Then, averaging this inequality over the conjugates of α, one gets that $M(\alpha) \geq e^c$, unless some $P_i(\alpha) = 0$.

Two further isolated points were later found by Flammang [80], giving the six points comprising the whole of the spectrum in $(1, 1.3117)$. On the other hand Smyth also showed that this spectrum was dense in (ℓ, ∞), where $\ell = 1.31427\ldots$. The number ℓ is $\lim_{n \to \infty} M(\alpha_n)$, where $\beta_0 = 1$ and β_n, of degree 2^n, is defined by $\beta_n - \beta_n^{-1} = \beta_{n-1}(n \geq 1)$. The limiting distribution of the conjugates of β_n was studied in detail by Davie and Smyth [51]. It is highly irregular: indeed, the Hausdorff dimension of the associated probability measure is $0.800611138269168784\ldots$. It is the invariant measure of the map $\mathbb{C} \to \mathbb{C}$ taking $t \mapsto t - 1/t$, whose Julia set (and thus the support of the measure) is $\mathbb{R}$.

Bertin [17] pointed out that from a result of Matveev (22) could be strengthened when α was a nonunit.

6.2. Langevin's Theorem.

In 1988 Langevin [92] proved the following general result, which included Schinzel's result (22) as a special case (though not with the explicit and best constant given by Schinzel). Suppose that V is an open subset of $\mathbb{C}$ that has nonempty intersection with the unit circle $|z| = 1$, and is stable under complex conjugation. Then there is a constant $C(V) > 1$ such that for every irreducible monic integer polynomial P of degree d having all its roots outside V one has $M(P) > C(V)^d$. The proof is based on the beautiful result of Kakeya to the effect that, for a compact subset of $\mathbb{C}$ stable under complex conjugation and of transfinite diameter less than 1 there is a nonzero polynomial with integer coefficients whose maximum modulus on this set is less than 1. (Kakeya's result is applied to the unit disc with V removed.) For Schinzel's result take $V = \mathbb{C}\backslash\mathbb{R}$, $C(\mathbb{R}) = \phi^{1/2}$, where the value of $C(\mathbb{R})$ given here is best possible. It is of course of interest to find such best possible constants for other sets V.

Stimulated by Langevin's Theorem, Rhin and Smyth [129] studied the case where the subset of $\mathbb{C}$ was the sector $V_\theta = \{z \in \mathbb{C} : |\arg z| > \theta\}$. They found a value $C(V_\theta) > 1$ for $0 \leq \theta \leq 2\pi/3$, including 9 subintervals of this range for which the constants found were best possible. In particular, the best constant $C(V_{\pi/2})$ was evaluated. This implied that for $P(z)$ irreducible, of degree d, having all its roots with positive real part and not equal to $z-1$ or z^2-z+1 we have

$$M(P)^{1/d} \geq M(z^6 - 2z^5 + 4z^4 - 5z^3 + 4z^2 - 2z + 1)^{1/6} = 1.12933793\ldots, \quad (24)$$

all roots of $z^6 - 2z^5 + 4z^4 - 5z^3 + 4z^2 - 2z + 1$ having positive real part. Curiously, for some root α of this polynomial, $\alpha + 1/\alpha = \theta_0^2$, where as above θ_0 is the smallest Pisot number.

Recently Rhin and Wu [131] extended these results, so that there are now 13 known subintervals of $[0, \pi]$ where the best constant $C(V_\theta)$ is known. It is of interest to see what happens as θ tends to π; maybe one could obtain a bound connected to Lehmer's original problem. Mignotte [112] has looked at this, and has shown that for $\theta = \pi - \varepsilon$ the smallest limit point of the set $M(P)^{1/d}$ for P having all its roots outside V_θ is at least $1 + c\varepsilon^3$ for some positive constant c.

Dubickas and Smyth [73] applied Langevin's Theorem to the annulus

$$V(R^{-\gamma}, R) = \{z \in \mathbb{C} \mid R^{-\gamma} < |z| < R\},$$

where $R > 1$ and $\gamma > 0$, proving that the best constant $C(V(R^{-\gamma}, R))$ is $R^{\gamma/(1+\gamma)}$.

6.3. Abelian number fields.
In 2000 Amoroso and Dvornicich [10] showed that when α is a nonzero algebraic number, not a root of unity, and $\mathbb{Q}(\alpha)$ is an abelian extension of $\mathbb{Q}$ then $M(\alpha) \geq 5^{d/12}$. They also give an example with $M(\alpha) = 7^{d/12}$. It would be interesting to find the best constant $c > 1$ such that $M(\alpha) \geq c^d$ for these numbers. Baker and Silverman [13], [151], [14] generalised this lower bound first to elliptic curves, and then to abelian varieties of arbitrary dimension.

6.4. Totally p-adic fields.
Bombieri and Zannier [25] proved an analogue of Schinzel's result (22) for 'totally p-adic' numbers: that is, for algebraic numbers α of degree d all of whose conjugates lie in $\mathbb{Q}_p$. They showed that then $M(\alpha) \geq c_p^d$, for some constant $c_p > 1$.

6.5. The heights of Zagier and Zhang and generalisations.
Zagier [171] gave a result that can be formulated as proving that the Mahler measure of any irreducible nonconstant polynomial in $\mathbb{Z}[(x(x-1)]$ has Mahler measure at least $\phi^{d/2}$, apart from $\pm(x(x-1)+1)$. Doche [60, 61] studied the spectrum resulting from the measures of such polynomials, giving a gap to the right of the smallest point $\phi^{1/2}$, and finding a short interval where the smallest limit point lies. He used the semi-infinite linear programming method outlined above. For this problem, however, finding the second point of the spectrum seems to be difficult. Zagier's work was motivated by a far-reaching result of Zhang [173] (see also [169, p. 103]) for curves on a linear torus. He proved

that for all such curves, apart from those of the type $x^i y^j = \omega$, where $i, j \in \mathbb{Z}$ and ω is a root of unity, there is a constant $c > 0$ such that the curve has only finitely many algebraic points (x, y) with $h(x) + h(y) \le c$. Zagier's result was for the curve $x + y = 1$.

Following on from Zhang, there have been recent deep and diverse generalisations in the area of small points on subvarieties of $\mathbb{G}_m^n$. In particular see Bombieri and Zannier [24], Schmidt [145] and Amoroso and David [5, 6, 7, 8].

Rhin and Smyth [130] generalised Zagier's result by replacing polynomials in $\mathbb{Z}(x(x-1))$ by polynomials in $\mathbb{Z}[Q(x)]$, where $Q(x) \in \mathbb{Z}[x]$ is not $\pm$ a power of x. Their proof used a very general result of Beukers and Zagier [21] on heights of points on projective hypersurfaces. Noticing that Zagier's result has the same lower bound as Schinzel's result above for totally real α, Samuels [135] has recently shown that the same lower bound holds for a more general height function. His result includes those of both Zagier and Schinzel. The proof is also based on [21].

7. LOWER BOUNDS FOR $\overline{|\alpha|}$

7.1. General lower bounds. We know that any lower bound for $M(\alpha)$ immediately gives a corresponding lower bound for $\overline{|\alpha|}$, using (9). For instance, from [164] it follows that for α of degree $d > 2$ and not a root of unity

$$\overline{|\alpha|} \ge 1 + \frac{1}{4d} \left(\frac{\log \log d}{\log d} \right)^3. \tag{25}$$

Some lower bounds, though asymptotically weaker, are better for small degrees. For example Matveev [105] has shown that for such α

$$\overline{|\alpha|} \ge \exp \frac{\log(d + 0.5)}{d^2}, \tag{26}$$

which is better than (25) for $d \le 1434$ (see [132]). Recently Rhin and Wu have improved (26) for $d \ge 13$ to

$$\overline{|\alpha|} \ge \exp \frac{3 \log(d/2)}{d^2}, \tag{27}$$

which is better than (25) for $d \le 6380$. See also the paper of Rhin and Wu in this volume.

Matveev [105] also proves that if α is a reciprocal (conjugate to α^{-1}) algebraic integer, not a root of unity, then $\overline{|\alpha|} \ge (p-1)^{1/(pm)}$, where p is the least prime greater than $m = n/2 \ge 3$.

Indeed, Dobrowolski's first result in this area [53] was for $\overline{|\alpha|}$ rather than $M(\alpha)$: he proved that

$$\overline{|\alpha|} > 1 + \frac{\log d}{6d^2}.$$

His argument is a beautifully simple one, based on the use of the power sums $s_k = \sum_{i=1}^{d} \alpha_i^k$, the Newton identities, and the arithmetic fact that, for any prime p, $s_{kp} \equiv s_k \pmod{p}$.

The strongest asymptotic result to date in the direction of the Schinzel-Zassenhaus conjecture is due to Dubickas [62]: that given $\varepsilon > 0$ there is a constant $d(\varepsilon)$ such than any nonzero algebraic integer α of degree $d > d(\varepsilon)$ not a root of unity satisfies

$$\boxed{\alpha} > 1 + \left(\frac{64}{\pi^2} - \varepsilon\right)\left(\frac{\log\log d}{\log d}\right)^3 \frac{1}{d}. \tag{28}$$

Cassels [46] proved that if an algebraic number α of degree d has the property $\boxed{\alpha} \le 1 + \frac{1}{10d^2}$ then at least one of the conjugates of α has modulus 1. Although this result has been superseded by Dobrowolski's work, Dubickas [66] applied the inequality

$$\prod_{k<j} |z_k \overline{z_j} - 1| \le n^{n/2} \left(\prod_{m=1}^{n} \max(1, |z_m|)\right)^{n-1} \tag{29}$$

for complex numbers $z_1, \ldots, z_n$, a variant of one in [46], to prove that

$$M(\alpha)^2 \left|\prod \log|\alpha_i|\right|^{1/d} \ge 1/(2d)$$

for a nonreciprocal algebraic number α of degree d with conjugates α_i.

7.2. The house $\boxed{\alpha}$ for α nonreciprocal.

The Schinzel-Zassenhaus conjecture (8) restricted to nonreciprocal polynomials follows from Breusch's result above, with $c = \log 1.1796 \cdots = 0.165\ldots$, using (9). Independently Cassels [46] obtained this result with $c = 0.1$, improved by Schinzel to 0.2 ([136]), and by Smyth [155] to $\log \theta_0 = 0.2811\ldots$. He also showed that c could not exceed $\frac{3}{2}\log \theta_0 = 0.4217\ldots$. In 1985 Lind and Boyd (see [34]), as a result of extensive computation (see Section 8), conjectured that, for degree d, the extremal α are nonreciprocal and have $\sim \frac{2}{3}d$ roots outside the unit circle. What a contrast with Mahler measure, where all small $M(\alpha)$ are reciprocal! This would imply that the best constant c is $\frac{3}{2}\log \theta_0$. In 1997 Dubickas [64] proved that $c > 0.3096$ in this nonreciprocal case.

7.3. The house of totally real α.

Suppose that α is a totally real algebraic integer. If $\boxed{\alpha} \le 2$ then by [90, Theorem 2] α is of the form $\omega + 1/\omega$, where ω is a root of unity. If for some $\delta > 0$ we have $2 < \boxed{\alpha} \le 2 + \delta^2/(1+\delta)$, then, on defining γ by $\gamma + 1/\gamma = \alpha$, we see that γ and its conjugates are either real or lie on the unit circle, and $1 < \boxed{\gamma} \le 1 + \delta$. This fact readily enables us to deduce a lower bound greater than 2 for $\boxed{\alpha}$ whenever we have a lower bound greater than 1 for $\boxed{\gamma}$. Thus from (7) [144] it follows that for α not of the form $2\cos \pi r$ for any $r \in \mathbb{Q}$

$$\boxed{\alpha} \ge 2 + 4^{-2d-3} \tag{30}$$

[144]. In a similar way (28) above implies that for such α, and $d > d(\varepsilon)$ that

$$\overline{|\alpha|} > 2 + \left(\frac{4096}{\pi^4} - \varepsilon \right) \left(\frac{\log \log d}{\log d} \right)^6 \frac{1}{d^2} \tag{31}$$

[62]. However Dubickas [63] managed to improve this lower bound to

$$\overline{|\alpha|} > 2 + 3.8 \frac{(\log \log d)^3}{d(\log d)^4}. \tag{32}$$

He improved the constant 3.8 to 4.6 in [64].

7.4. The Kronecker constant. Callahan, Newman and Sheingorn [48] define the *Kronecker constant* of a number field K to be the least $\varepsilon > 0$ such that $\overline{|\alpha|} \geq 1 + \varepsilon$ for every algebraic integer $\alpha \in K$. The truth of the Schinzel-Zassenhaus conjecture (8) would imply that the Kronecker constant of K is at least $c/[K : \mathbb{Q}]$. They give [48, Theorem 2] a sufficient condition on K for this to be the case. They also point out, from considering $\alpha\bar{\alpha} - 1$, that if α is a nonzero algebraic integer not a root of unity in a Kroneckerian field then $\overline{|\alpha|} \geq \sqrt{2}$ (See also [111]), so that the Kronecker constant of a Kroneckerian field is at least $\sqrt{2} - 1$.

8. SMALL VALUES OF $M(\alpha)$ AND $\overline{|\alpha|}$

8.1. Small values of $M(\alpha)$. The first recorded computations on Mahler measure were performed by Lehmer in his 1933 paper [94]. He found the smallest values of $M(\alpha)$ for α of degrees 2, 3 and 4, and the smallest $M(\alpha)$ for α reciprocal of degrees 2, 4, 6 and 8. Lehmer records the fact that Poulet (?unpublished) "...has made a similar investigation of symmetric polynomials with practically the same results". Boyd has done extensive computations, searching for 'small' algebraic integers of various kinds. His first major published table was of Salem numbers less than 1.3 [29], with four more found in [30]. Recall that these are positive reciprocal algebraic integers of degree at least 4 having only one conjugate (the number itself) outside the unit circle. These numbers give many examples of small Mahler measures, most notably (from (2)) $M(L) = 1.176\ldots$ from the Lehmer polynomial itself, which is the minimal polynomial of a Salem number. In later computations [32], [38], he finds all reciprocal α with $M(\alpha) \leq 1.3$ and degree up to 20, and those with $M(\alpha) \leq 1.3$ and degree up to 32 having coefficients in $\{-1, 0, 1\}$ ('height 1').

Mossinghoff [114] extended Boyd's tables from degree 20 to degree 24 for $M(\alpha) < 1.3$, and to degree 40 for height 1 polynomials, finding four more Salem numbers less than 1.3. He also has a website [115] where up-to-date tables of small Salem numbers and Mahler measures are conveniently displayed (though unfortunately without their provenance). Flammang, Grandcolas and Rhin [82] proved that Boyd's table, with the additions by Mossinghoff, of the 47 known Salem numbers less than 1.3 is complete up to degree 40. Recently Flammang, Rhin and Sac-Épée [83] have extended these tables, finding all $M(\alpha) < \theta_0$ for α of degree up to 36, and all $M(\alpha) < 1.31$ for α of degree up to 40. This latter computation showed that the earlier tables of Boyd and Mossinghoff for α of degree up to 40 with $M(\alpha) < 1.3$ are complete.

8.2. Small values of $\boxed{\alpha}$. Concerning $\boxed{\alpha}$, Boyd [34] gives tables of the smallest values of $\boxed{\alpha}$ for α of degree d up to 12, and for α reciprocal of degree up to 16. Further computation has recently been done on this problem by Rhin and Wu [132]. They computed the smallest house of algebraic numbers of degree up to 28. All are nonreciprocal, as predicted by Boyd's conjecture (see Section 7.2). Their data led the authors to conjecture that, for a given degree, an algebraic number of that degree with minimal house was a root of a polynomial consisting of at most four monomials.

9. Mahler measure and the discriminant

9.1. Mahler [103] showed that for a complex polynomial

$$P(z) = a_0 z^d + \cdots + a_d = a_0 (z - \alpha_1) \ldots (z - \alpha_d)$$

its discriminant $\operatorname{disc}(P) = a_0^{2d-2} \prod_{i<j} (\alpha_i - \alpha_j)^2$ satisfies

$$|\operatorname{disc}(P)| \le d^d M(P)^{2d-2}. \tag{33}$$

From this it follows immediately that if there is an absolute constant $c > 1$ such that $|\operatorname{disc}(P)| \ge (cd)^d$ for all irreducible $P(z) \in \mathbb{Z}[z]$, then $M(P) \ge c^{d/(2d-2)}$, which would solve Lehmer's problem. This consequence of Mahler's inequality has been noticed in various variants by several people, including Mignotte [109] and Bertrand [16].

In 1996 Matveev [106] showed that in Dobrowolski's inequality, the degree $d \ge 2$ of α could be replaced by a much smaller (for large d) quantity

$$\delta = \max(d/\operatorname{disc}(\alpha)^{1/d}, \delta_0(\varepsilon))$$

for those α for which α^p had degree d for all primes p. (Such α do not include any roots of unity.) Specifically, he obtained for given $\varepsilon > 0$

$$M(\alpha) \ge \exp\left((2 - \varepsilon) \left(\frac{\log \log \delta}{\log \delta} \right)^3 \right) \tag{34}$$

for these α.

Mahler [103] also gives the lower bound

$$\delta(P) > \sqrt{3} |\operatorname{disc}(P)|^{1/2} d^{-(d+2)/2} M(P)^{-(m-1)} \tag{35}$$

for the minimum distance $\delta(P) = \min_{i<j} |\alpha_i - \alpha_j|$ between the roots of P.

9.2. Generalisation involving the discriminant of Schinzel's lower bound.

Rhin [127] generalised Schinzel's result (22) by proving, for α a totally positive algebraic integer of degree d at least 2 that

$$M(\alpha) \geq \left(\frac{\delta_1 + \sqrt{\delta_1^2 + 4}}{2} \right)^{d/2}. \tag{36}$$

Here $\delta_1 = |\operatorname{disc}(\alpha)|^{1/d(d-1)}$. This result apparently also follows from an earlier result of Zaïmi [172] concerning a lower bound for a weighted product of the moduli of the conjugates of an algebraic integer — see the Math Review of Rhin's paper.

10. PROPERTIES OF $M(\alpha)$ AS AN ALGEBRAIC NUMBER

A *Perron number* is an algebraic integer with exactly one conjugate of maximum modulus. It is clear from (2) that $M(\alpha)$ is a Perron number for any algebraic integer α; this seems to have been first observed by Adler and Marcus [1] (see [36]). In the other direction: is the Perron number $1 + \sqrt{17}$ a Mahler measure? See Schinzel [143], Dubickas [71]. Dubickas [70] proves that for any Perron number β some integer multiple of β is a Mahler measure. (These papers also contains other interesting properties of the set of Mahler measures.) Boyd [35] proves that if $\beta = M(\alpha)$ for some algebraic integer α, then all conjugates of β other than β itself either lie in the annulus $\beta^{-1} < |z| < \beta$ or are equal to $\pm\beta^{-1}$.

If α were reciprocal, it might be expected that $M(\alpha)$ would be reciprocal too, while if α were nonreciprocal, then $M(\alpha)$ would be nonreciprocal. However neither of these need be the case: in [36, Proposition 6] Boyd exhibits a family of degree 4 Pisot numbers that are the Mahler measures of reciprocal algebraic integers of degree 6, and in [36, Proposition 2] he notes that for $q \geq 3$ a root α_q of the irreducible nonreciprocal polynomial $z^4 - qz^3 + (q + 1)z^2 - 2z + 1$ then $M(\alpha_q) = \frac{1}{2}(q + \sqrt{q^2 - 4})$ is reciprocal. In fact, since $M(\frac{1}{2}(q + \sqrt{q^2 - 4})) = \frac{1}{2}(q + \sqrt{q^2 - 4})$, this also shows that a number can be both a reciprocal and a nonreciprocal measure. See also [37]. Dixon and Dubickas [52] prove that the set of all $M(\alpha)$ does not form a semigroup, as for instance $\sqrt{2} + 1$ and $\sqrt{3} + 2$ are Mahler measures, while their product is not. (In terms of polynomials, this set is of course equal to the set of all $M(P)$ for P irreducible. If instead we take the set of all (reducible and irreducible) polynomials, then, because of $M(PQ) = M(P)M(Q)$ this larger set *does* form a semigroup.)

In [69] Dubickas proves that the additive group generated by all Mahler measures is the group of all real algebraic numbers, while the multiplicative group generated by all Mahler measures is the group of all positive real algebraic numbers.

We know that $M(P(z)) = M(P(\pm z^k))$ for either choice of sign, and any $k \in \mathbb{N}$. Is this the only way that Mahler measures of irreducible polynomials can be equal? Boyd [32] gives some illuminating examples to show that there can be other reasons that make this happen. The examples were discovered during his computation of reciprocal polynomials of small Mahler measure (see Section 8). For example, for $P_6 = z^6 + 2z^5 + 2z^4 + z^3 + 2z^2 + 2z + 1$ and $P_8 = z^8 + z^7 - z^6 - z^5 + z^4 - z^3 - z^2 + z + 1$ we have

$$M(P_6) = M(P_8) = 1.746793\cdots = M,$$

say, where both polynomials are irreducible. Boyd explains how such examples arise. If $\alpha_i(i = 1,\ldots,8)$ are the roots of P_8, then for different i $M(\alpha_1\alpha_i)$ can equal M, M^2 or M^3. The roots of P_6 are the three $\alpha_1\alpha_i$ with $M(\alpha_1\alpha_i) = M$ and their reciprocals. Clearly $M(\alpha_1^2) = M^2$, while for three other α_i the product $\alpha_1\alpha_i$ is of degree 12 and has $M(\alpha_1\alpha_i) = M^3$. ($P_8$ has the special property that it has roots $\alpha_1, \alpha_2, \alpha_3, \alpha_4$ with $\alpha_1\alpha_2 = \alpha_3\alpha_4 \neq 1$.)

Dubickas [67] gives a lower bound for the distance of an algebraic number γ of degree n and leading coefficient c, not a Mahler measure, from a Mahler measure $M(\alpha)$ of degree D:

$$|M(\alpha) - \gamma| > c^{-D}(2\overline{|\gamma|})^{-nD}. \tag{37}$$

11. Counting polynomials with given Mahler measure

Let $\#(d, T)$ denote the number of integer polynomials of degree d and Mahler measure at most T. This function has been studied by several authors. Boyd and Montgomery [41] give the asymptotic formula

$$c(\log d)^{-1/2} d^{-1} \exp\left(\frac{1}{\pi}\sqrt{105\zeta(3)d}\right)(1 + o(1)), \tag{38}$$

where $c = \frac{1}{4\pi^2}\sqrt{105\zeta(3)e^{-\gamma}}$, for the number $\#(d, 1)$ of cyclotomic polynomials of degree d, as $d \to \infty$.

Dubickas and Konyagin [72] obtain by simple arguments the lower bound $\#(d, T) > \frac{1}{2}T^{d+1}(d+1)^{-(d+1)/2}$, and upper bound $\#(d, T) < T^{d+1}\exp(d^2/2)$, the latter being valid for d sufficiently large. For $T \geq \theta_0$ they derived the upper bound $\#(d, T) < T^{d(1+16\log\log d/\log d)}$. Chern and Vaaler [50] obtained the asymptotic formula $V_{d+1}T^{d+1} + O_d(T^d)$ for $\#(d, T)$ for fixed d, as $T \to \infty$. Here V_{d+1} is an explicit constant (the volume of a certain star body). Recently Sinclair [153] has produced corresponding estimates for counting functions of reciprocal polynomials.

12. A dynamical Lehmer's problem

Given a rational map $f(\alpha)$ of degree $d \geq 2$ defined over a number field K, one can define for α in some extension field of K a canonical height

$$h_f(\alpha) = \lim_{n \to \infty} d^{-n}h(f^n(\alpha)),$$

where f^n is the nth iterate of f, and h is, as before, the Weil height of α. Then $h_f(\alpha) = 0$ if and only if the iterates $f^n(\alpha)$ form a finite set, and an analogue of Lehmer's problem would be to decide whether or not

$$h_f(\alpha) \geq \frac{C}{\deg(\alpha)}$$

for some constant C depending only on f and K. Taking $f(\alpha) = \alpha^d$ we retrieve the Weil height and the original Lehmer problem. There seem to be no good estimates, not even of polynomial decay, for any f not associated to an endomorphism of an algebraic group. See [152, Section 3.4] for more details.

13. VARIANTS OF MAHLER MEASURE

Everest and ní Fhlathúin [77] and Everest and Pinner [78] (see also [79, Chapter 6]) have defined the *elliptic Mahler measure*, based on a given elliptic curve $E = \mathbb{C}/L$ over $\mathbb{C}$, where $L = \langle \omega_1, \omega_2 \rangle \subset \mathbb{C}$ is a lattice, with $\wp_L$ its associated Weierstrass $\wp$-function. Then for $F \in \mathbb{C}[z]$ the (logarithmic) elliptic Mahler measure $m_E(F)$ is defined as

$$\int_0^1 \int_0^1 \log|F(\wp_L(t_1\omega_1 + t_2\omega_2))| dt_1 dt_2. \tag{39}$$

If E is in fact defined over $\mathbb{Q}$ and has a rational point Q with x-coordinate M/N then often $m_E(Nz - M) = 2\hat{h}(Q)$, showing that m_E is connected with the canonical height on E.

Kurokawa [91] and Oyanagi [118] have defined a q-analogue of Mahler measure, for a real parameter q. As $q \to 1$ the classical Mahler measure is recovered.

Dubickas and Smyth [74] defined the *metric Mahler measure* $\mathcal{M}(\alpha)$ as the infimum of $\prod_i M(\beta_i)$, where $\prod_i \beta_i = \alpha$. They used this to define a metric on the group of nonzero algebraic numbers modulo torsion points, the metric giving the discrete topology on this group if and only if Lehmer's 'conjecture' is true (i.e., $\inf_{\alpha:M(\alpha)>1} M(\alpha) > 1$).

Very recently Pritsker [122, 123] has studied an areal analogue of Mahler measure, defined by replacing the normalised arclength measure on the unit circle by the normalised area measure on the unit disc.

14. APPLICATIONS

14.1. Polynomial factorization. I first met Andrzej Schinzel at the ICM in Nice in 1970. There he mentioned to me an application of Mahler measure to irreducibility of polynomials. (After this we had some correspondence about the work leading to [155], which was very helpful to me.) If a class of irreducible polynomials had Mahler measure at least B, then any polynomial of Mahler measure less than B^2 can have at most one factor from that class. For instance, a trinomial $z^d \pm z^m \pm 1$ has, by Vicente Gonçalves' inequality

[162], [117], [124, Th. 9.1.1] $M(P)^2 + M(P)^{-2} \leq ||P||_2^2$, Mahler measure at most ϕ. Since $\phi < \theta_0^2$, by (15) such trinomials can have at most one irreducible noncyclotomic factor. Here $||P||_2$ is the 2-norm of P (the square root of the sum of the squares of its coefficients).

More generally Schinzel (see [54]) pointed out the following consequence of (11): that for any fixed $\varepsilon > 0$ and polynomial P of degree d with integer coefficients, the number of its noncyclotomic irreducible factors counted with multiplicities is $O(d^\varepsilon ||P||_2^{1-\varepsilon})$. See also [138], [140], [121].

14.2. Ergodic theory.
One-variable Mahler measures have applications in ergodic theory. Consider an automorphism of the torus $\mathbb{R}^d/\mathbb{Z}^d$ defined by a $d \times d$ integer matrix of determinant ± 1, with characteristic polynomial $P(z)$. Then the topological entropy of this map is $\log M(P)$ (Lind [96] — see also [33], [79, Theorem 2.6]).

14.3. Transcendence and diophantine approximation.
Mahler measure, or rather the Weil height $h(\alpha) = \log M(\alpha)/d$, plays an important technical rôle in modern transcendence theory, in particular for bounding the coefficients of a linear form in logarithms known to be dependent.

As remarked by Waldschmidt [169, p65], the fact that this height has three equivalent representations, coming from (1), (2) and (4) makes it a very versatile height function for these applications.

If $\alpha_1, \ldots, \alpha_n$ are algebraic numbers such that their logarithms are $\mathbb{Q}$-linearly dependent, then it is of importance in Baker's transcendence method to get small upper estimates for the size of integers $m_1, \ldots, m_n$ needed so that $m_1 \log \alpha_1 + \cdots + m_n \log \alpha_n = 0$. Such estimates can be given using Weil heights of the α_i. See [169, Lemma 7.19] and the remark after it.

Chapter 3 ('Heights of Algebraic Numbers') of [169] contains a wealth of interesting material on the Weil height and other height functions, connections between them, and applications. For instance, for a polynomial $f \in \mathbb{Z}[z]$ of degree at most N for which the algebraic number α is not a root one has

$$|f(\alpha)| \geq \frac{1}{M(\alpha)^N ||f||_1^{d-1}},$$

where $||f||_1$ is the length of f, the sum of the absolute values of its coefficients, and $d = \deg \alpha$ ([169, p83]).

In particular, for a rational number $p/q \neq \alpha$ with $q > 0$, and $f(x) = qx - p$ we obtain

$$\left| \alpha - \frac{p}{q} \right| \geq \frac{1}{M(\alpha)q(\max(|p|+q))^{d-1}}. \tag{40}$$

14.4. Distance of α from 1.
From (40) we immediately get for $\alpha \neq 1$

$$|\alpha - 1| \geq \frac{1}{2^{d-1}M(\alpha)}. \tag{41}$$

Better lower bounds for $|\alpha - 1|$ in terms of its Mahler measure have been given by Mignotte [110], Mignotte and Waldschmidt [113], Bugeaud, Mignotte and Normandin [45], Amoroso [2], Dubickas [63], and [65]. For instance Mignotte and Waldschmidt prove that

$$|\alpha - 1| > \exp\{-(1 + \varepsilon)(d(\log d)(\log M(\alpha)))^{1/2}\} \tag{42}$$

for $\varepsilon > 0$ and α of degree $d \geq d(\varepsilon)$. Dubickas [63] improves the constant 1 in this result to $\pi/4$, and in the other direction [65] proves that for given $\varepsilon > 0$ there is an infinite sequences of degrees d for which an α of degree d satisfies

$$|\alpha - 1| < \exp\left\{-(c - \varepsilon)\left(\frac{d \log M(\alpha)}{\log d}\right)^{1/2}\right\}. \tag{43}$$

Here Dubickas uses the following simple result: if $F \in \mathbb{C}[z]$ has degree t and $F'(1) \neq 0$ then there is a root a of F such that $|a - 1| \leq t|F(1)/F'(1)|$.

14.5. Shortest unit lattice vector. Let K be a number field with unit lattice of rank r, and $M = \min M(\alpha)$, the minimum being taken over all units $\alpha \in K$, α not a root of unity. Kessler [89] showed that then the shortest vector λ in the unit lattice has length $||\lambda||_2$ at least $\sqrt{\frac{2}{r+1}} \log M$.

14.6. Knot theory. Mahler measure of one-variable polynomials arises in knot theory in connection with Alexander polynomials of knots and reduced Alexander polynomials of links — see Silver and Williams [148]. Indeed, in Reidemeister's classic book on the subject [126], the polynomial $L(-z)$ appears as the Alexander polynomial of the $(-2, 3, 7)$-pretzel knot. Hironaka [85] has shown that among a wide class of Alexander polynomials of pretzel links, this one has the smallest Mahler measure. Champanerkar and Kofman [47] study a sequence of Mahler measures of Jones polynomials of hyperbolic links L_m obtained using $(-1/m)$-Dehn surgery, starting with a fixed link. They show that it converges to the Mahler measure of a 2-variable polynomial. (The many more applications of Mahler measures of several-variable polynomials to geometry and topology are outside the scope of this survey.)

15. FINAL REMARKS

15.1. Other sources on Mahler measure. Books covering various aspects of Mahler measure include the following: Bertin and Pathiaux-Delefosse [19], Bertin *et al* [20], Bombieri and Gubler [23], Borwein [26], Schinzel [139], Schinzel [142], Waldschmidt [169].

Survey articles and lecture notes on Mahler measure include: Boyd [31], Boyd [33], Everest [76], Hunter [87], Schinzel [141], Skoruppa [154], Stewart [159], Vaaler [160], Waldschmidt [167].

15.2. Memories of Mahler. As one of a small group of undergraduates in ANU, Canberra in the mid-1960s, we were encouraged to attend graduate courses at the university's Institute of Advanced Studies, where Mahler had a research chair. I well remember his lectures on transcendence with his blackboard copperplate handwriting, all the technical details being carefully spelt out.

15.3. Acknowledgements. I thank Matt Baker, David Boyd, Artūras Dubickas, James McKee, Alf van der Poorten, Georges Rhin, Andrzej Schinzel, Joe Silverman, Michel Waldschmidt, Susan Williams, Umberto Zannier and the referee for some helpful remarks concerning an earlier draft of this survey, which have been incorporated into the final version. This article arose from a talk I gave in January 2006 at the Mahler Measure in Mobile Meeting, held in Mobile, Alabama. I would like to thank the organisers Abhijit Champanerkar, Eriko Hironaka, Mike Mossinghoff, Dan Silver and Susan Williams for a stimulating meeting.

REFERENCES

[1] R.L. Adler, B. Marcus, Topological entropy and equivalence of dynamical systems, *Mem. Amer. Math. Soc.* **20** (1979), no. 219.

[2] F. Amoroso, Algebraic numbers close to 1 and variants of Mahler's measure, *J. Number Theory* **60** (1996), 80–96.

[3] F. Amoroso, S. David, Le théorème de Dobrowolski en dimension supérieure, *C. R. Acad. Sci. Paris Sér. I Math.* **326** (1998), 1163–1166.

[4] _____, _____, Le problème de Lehmer en dimension supérieure. *J. Reine Angew. Math.* **513** (1999), 145–179.

[5] _____, _____, Minoration de la hauteur normalisée des hypersurfaces. *Acta Arith.* **92** (2000), 339–366.

[6] _____, _____, Minoration de la hauteur normalisée dans un tore. *J. Inst. Math. Jussieu* **2** (2003), 335–381.

[7] _____, _____, Distribution des points de petite hauteur dans les groupes multiplicatifs. *Ann. Sc. Norm. Super. Pisa Cl. Sci.* (5) **3** (2004), 325–348.

[8] _____, _____, Points de petite hauteur sur une sous-variété d'un tore. *Compos. Math.* **142** (2006), 551–562.

[9] F. Amoroso, E. Delsinne, Une minoration relative explicite pour la hauteur dans une extension d'une extension abélienne, in *Diophantine geometry, April 12–July 22 2005, proceedings*, Pisa 2007, 24pp.

[10] F. Amoroso, R.A. Dvornicich, A lower bound for the height in abelian extensions, *J. Number Theory* **80** (2000), 260–272.

[11] F. Amoroso, U. Zannier, A relative Dobrowolski lower bound over abelian extensions, *Ann. Scuola Norm. Sup. Pisa Cl. Sci.* (4),1 **29** (2000), 711–727.

[12] M. Anderson, D. Masser, Lower bounds for heights on elliptic curves, *Math. Z.* **174** (1980), 23–34.

[13] M.H. Baker, Lower bounds for the canonical height on elliptic curves over abelian extensions, *Int. Math. Res. Not.* 2003, no. 29, 1571–1589.

[14] M.H. Baker, J.H. Silverman, A lower bound for the canonical height on abelian varieties over abelian extensions, *Math. Res. Lett.* **11** (2004), 377–396.

[15] A. Bazylewicz, On the product of the conjugates outside the unit circle of an algebraic integer, *Acta Arith.* **30** (1976/77), 43–61.

[16] D. Bertrand, Problème de Lehmer et petits discriminants, *Arithmétix* **6** (1982), 14–15.

[17] M.J. Bertin, The operator $x + (1/x) - 2$ and the reciprocal integers, *Number theory (Ottawa, ON, 1996)*, 17–23, CRM Proc. Lecture Notes **19**, Amer. Math. Soc., Providence, RI, 1999.

[18] _____, Mahler's measure: from number theory to geometry, *this volume*, 20–31.

[19] M.J. Bertin, M. Pathiaux-Delefosse, Conjecture de Lehmer et petits nombres de Salem, *Queen's Papers in Pure and Applied Mathematics* **81**, Queen's University, Kingston, ON, 1989.

[20] M.J. Bertin, A. Decomps-Guilloux, M. Grandet-Hugot, M. Pathiaux-Delefosse, J.-P. Schreiber, *Pisot and Salem numbers*, Birkhäuser Verlag, Basel, 1992.

[21] F. Beukers, D. Zagier, Lower bounds of heights of points on hypersurfaces, *Acta Arith.* **79** (1997), 103–111.

[22] P.E. Blanksby, H.L. Montgomery, Algebraic integers near the unit circle, *Acta Arith.* **18** (1971), 355–369.

[23] E. Bombieri, W. Gubler, *Heights in Diophantine geometry,* New Mathematical Monographs **4**, Cambridge University Press, Cambridge, 2006.

[24] E. Bombieri, U. Zannier, Algebraic points on subvarieties of $\mathbf{G}_m^n$. *Internat. Math. Res. Notices* 1995, no. 7, 333–347.

[25] E. Bombieri, U. Zannier, A note on heights in certain infinite extensions of $\mathbb{Q}$, *Atti Accad. Naz. Lincei Cl. Sci. Fis. Mat. Natur. Rend. Lincei (9) Mat. Appl.* **12** (2001), 5–14 (2002).

[26] P. Borwein, *Computational excursions in analysis and number theory,* CMS Books in Mathematics, New York, NY: Springer 2002.

[27] P. Borwein, E. Dobrowolski, M. Mossinghoff, Lehmer's problem for polynomials with odd coefficients, *Ann. of Math.* (to appear).

[28] P. Borwein, K.G. Hare, M.J. Mossinghoff, The Mahler measure of polynomials with odd coefficients, *Bull. London Math. Soc.* **36** (2004), 332–338.

[29] D.W. Boyd, Small Salem numbers, *Duke Math. J.* **44** (1977), 315–328.

[30] _____, Pisot and Salem numbers in intervals of the real line, *Math. Comp.* **32** (1978), 1244–1260.

[31] _____, Variations on a theme of Kronecker, *Canad. Math. Bull.* **21** (1978), 129–133.

[32] _____, Reciprocal polynomials having small measure, *Math. Comp.* **35** (1980), 1361–1377.

[33] _____, Speculations concerning the range of Mahler's measure, *Canad. Math. Bull.* **24** (1981), 453–469.

[34] _____, The maximal modulus of an algebraic integer, *Math. Comp.* **45** (1985), 243–249, S17–S20.

[35] _____, Perron units which are not Mahler measures, *Ergodic Theory Dynam. Systems* **6** (1986), 485–488.

[36] _____, Inverse problems for Mahler's measure, *Diophantine analysis* (Kensington, 1985), 147–158, London Math. Soc. Lecture Note Ser. **109**, Cambridge Univ. Press, Cambridge, 1986.

[37] _____, Reciprocal algebraic integers whose Mahler measures are nonreciprocal, *Canad. Math. Bull.* **30** (1987), 3–8.

[38] _____, Reciprocal polynomials having small measure. II, *Math. Comp.* **53** (1989), 355–357, S1–S5.

[39] _____, Mahler's measure and special values of L-functions, *Experiment. Math.* **7** (1998), 37–82.

[40] _____, Mahler's measure and invariants of hyperbolic manifolds, *Number theory for the millennium*, I (Urbana, IL, 2000), 127–143, A K Peters, Natick, MA, 2002.

[41] D.W. Boyd, H.L. Montgomery, Cyclotomic partitions, *Number theory* (Banff, AB, 1988), 7–25, de Gruyter, Berlin, 1990.

[42] D.W. Boyd, M.J. Mossinghoff, Small limit points of Mahler's measure, *Experiment. Math.* **14** (2005), 403–414.

[43] D.W. Boyd, F. Rodriguez Villegas, Mahler's measure and the dilogarithm. I, *Canad. J. Math.* **54** (2002), 468–492.

[44] R. Breusch, On the distribution of the roots of a polynomial with integral coefficients, *Proc. Amer. Math. Soc.* **2** (1951), 939–941.

[45] Y. Bugeaud, M. Mignotte, F. Normandin, Nombres algébriques de petite mesure et formes linéaires en un logarithme, *C. R. Acad. Sci. Paris Sér. I Math.* **321** (1995), 517–522.

[46] J.W.S. Cassels, On a problem of Schinzel and Zassenhaus. *J. Math. Sci.* **1** (1966), 1–8.

[47] A. Champanerkar, I. Kofman, On the Mahler measure of Jones polynomials under twisting, *Algebr. Geom. Topol.* **5** (2005), 1–22.

[48] T. Callahan, M. Newman, M. Sheingorn, Fields with large Kronecker constants, *J. Number Theory* **9** (1977), 182–186.

[49] D.C. Cantor, E.G. Straus, On a conjecture of D.H. Lehmer, *Acta Arith.* **42** (1982/83), 97–100. Correction: **42** (1983), 327.

[50] S.-J. Chern, J.D. Vaaler, The distribution of values of Mahler's measure, *J. Reine Angew. Math.* **540** (2001), 1–47.

[51] A.M. Davie, C.J. Smyth, On a limiting fractal measure defined by conjugate algebraic integers, *Groupe de Travail en Théorie Analytique et Élémentaire des Nombres*, 1987–1988, 93–103, Publ. Math. Orsay, **89-01**, Univ. Paris XI, Orsay, 1989.

[52] J.D. Dixon, A. Dubickas, The values of Mahler measures, *Mathematika* **51** (2004), 131–148 (2005).

[53] E. Dobrowolski, On the maximal modulus of conjugates of an algebraic integer, *Bull. Acad. Polon. Sci.*, Sér. Sci. Math. Astronom. Phys. **26** (1978), 291–292.

[54] _____, On a question of Lehmer and the number of irreducible factors of a polynomial, *Acta Arith.* **34** (1979), 391–401.

[55] _____, On a question of Lehmer. Abelian functions and transcendental numbers (Colloq., École Polytech., Palaiseau, 1979), *Mém. Soc. Math. France (N.S.)* **1980/81**, 35–39.

[56] _____, Mahler's measure of a polynomial in function of the number of its coefficients, *Canad. Math. Bull.* **34** (1991), 186–195.

[57] _____, Mahler's measure of a polynomial in terms of the number of its monomials, *Acta Arith.* **123** (2006), 201-231.

[58] _____, A note on integer symmetric matrices and Mahler's measure, *Canad. Math. Bull.* (to appear).

[59] E. Dobrowolski, W. Lawton, A. Schinzel, On a problem of Lehmer, *Studies in pure mathematics*, 135–144, Birkhäuser, Basel, 1983.

[60] C. Doche, On the spectrum of the Zhang-Zagier height, *Math. Comp.* **70** (2001), 419–430.

[61] _____, Zhang-Zagier heights of perturbed polynomials, 21st Journées Arithmétiques (Rome, 2001). *J. Théor. Nombres Bordeaux* **13** (2001), 103–110.

[62] A. Dubickas, On a conjecture of A. Schinzel and H. Zassenhaus, *Acta Arith.* **63** (1993), 15–20.

[63] _____, On algebraic numbers of small measure, *Liet. Mat. Rink.* **35** (1995), 421–431; translation in *Lithuanian Math. J.* **35** (1995), 333–342 (1996).

[64] _____, The maximal conjugate of a non-reciprocal algebraic integer, *Liet. Mat. Rink.* **37** (1997), 168–174; translation in *Lithuanian Math. J.* **37** (1997), 129–133 (1998).

[65] _____, On algebraic numbers close to 1, *Bull. Austral. Math. Soc.* **58** (1998), no. 3, 423–434.

[66] _____, On the measure of a nonreciprocal algebraic number, *Ramanujan J.* **4** (2000), 291–298.

[67] _____, Mahler measures close to an integer, *Canad. Math. Bull.* **45** (2002), 196–203.

[68] _____, Nonreciprocal algebraic numbers of small measure, *Comment. Math. Univ. Carolin.* **45** (2004), 693–697.

[69] _____, Mahler measures generate the largest possible groups, *Math. Res. Lett.* **11** (2004), 279–283.

[70] _____, On numbers which are Mahler measures, *Monatsh. Math.* **141** (2004), 119–126.

[71] _____, Algebraic, arithmetic and geometric properties of Mahler measures, *Proc. Inst. Math. (Nat. Acad. Sc. Belarus)* **13(1)** (2005), 70–74.

[72] A. Dubickas, S.V. Konyagin, On the number of polynomials of bounded measure, *Acta Arith.* **86** (1998), 325–342.

[73] A. Dubickas, C.J. Smyth, The Lehmer constants of an annulus, *J. Théor. Nombres Bordeaux* **13** (2001), 413–420.

[74] _____, _____, On the metric Mahler measure, *J. Number Theory* **86** (2001), 368–387.

[75] A. Durand, On Mahler's measure of a polynomial, *Proc. Amer. Math. Soc.* **83** (1981), 75–76.

[76] G. Everest, Measuring the height of a polynomial, *Math. Intelligencer* **20** (1998), no. 3, 9–16.

[77] G. Everest, B. ní Fhlathúin, The elliptic Mahler measure, *Math. Proc. Cambridge Philos. Soc.* **120** (1996), 13–25.

[78] G. Everest, C. Pinner, Bounding the elliptic Mahler measure, II. *J. London Math. Soc.* (2) **58** (1998), 1–8. Corrigendum **62** (2000), 640.

[79] G. Everest, T. Ward, *Heights of polynomials and entropy in algebraic dynamics*, Universitext. Springer-Verlag London Ltd., London, 1999.

[80] V. Flammang, Two new points in the spectrum of the absolute Mahler measure of totally positive algebraic integers, *Math. Comp.* **65** (1996), 307–311.

[81] _____, Inégalités sur la mesure de Mahler d'un polynôme, *J. Théor. Nombres Bordeaux* **9** (1997), 69–74.

[82] V. Flammang, M. Grandcolas, G. Rhin, Small Salem numbers, *Number theory in progress*, 1 (Zakopane-Kościelisko, 1997), 165–168, de Gruyter, Berlin, 1999.

[83] V. Flammang, G. Rhin, J.-M. Sac-Épée, Integer transfinite diameter and polynomials with small Mahler measure, *Math. Comp.* **75** (2006), 1527–1540.

[84] M. Hindry, J.H. Silverman, On Lehmer's conjecture for elliptic curves, *Séminaire de Théorie des Nombres, Paris 1988–1989*, 103–116, Progr. Math. **91**, Birkhäuser Boston, Boston, MA, 1990.

[85] E. Hironaka, The Lehmer polynomial and pretzel links, *Canad. Math. Bull.* **44** (2001), 440–451. Correction: **45** (2002), 231.

[86] G. Höhn, N.-P. Skoruppa, Un résultat de Schinzel, *J. Théor. Nombres Bordeaux* **5** (1993), 185.

[87] J. Hunter, Algebraic integers on the unit circle, *Math. Chronicle* **11** (1982), 37–47.

[88] J.L.V.W. Jensen, Sur un nouvel et important théorème de la théorie des fonctions, *Acta Math.* **22** (1899), 359–364.

[89] V. Kessler, On the minimum of the unit lattice, *Sém. Théor. Nombres Bordeaux* (2) **3** (1991), 377–380.

[90] L. Kronecker, Zwei Sätze über Gleichungen mit ganzzahligen Coefficienten, *J. Reine Angew. Math.* **53** (1857), 173–175.

[91] N. Kurokawa, A q-Mahler measure, *Proc. Japan Acad. Ser. A Math. Sci.* **80** (2004), no. 5, 70–73.

[92] M. Langevin, Calculs explicites de constantes de Lehmer, *Groupe de travail en théorie analytique et élémentaire des nombres, 1986–1987*, 52–68, Publ. Math. Orsay, 88-01, Univ. Paris XI, Orsay, 1988.

[93] M. Laurent, Minoration de la hauteur de Néron-Tate, *Seminar on number theory, Paris 1981–82* (Paris, 1981/1982), 137–151, Progr. Math. **38**, Birkhäuser Boston, Boston, MA, 1983.

[94] D.H. Lehmer, Factorization of certain cyclotomic functions, *Ann. of Math.* (2) **34** (1933), 461–479.

[95] ———, The vanishing of Ramanujan's function $\tau(n)$, *Duke Math. J.* **14** (1947), 429–433.

[96] D.A. Lind, Ergodic automorphisms of the infinite torus are Bernoulli, *Isr. J. Math.* **17** (1974), 162–168.

[97] C.W. Lloyd-Smith, Algebraic numbers near the unit circle, *Acta Arith.* **45** (1985), 43–57.

[98] R. Louboutin, Sur la mesure de Mahler d'un nombre algébrique, *C. R. Acad. Sci. Paris Sér. I Math.* **296** (1983), 707–708.

[99] J. McKee, C. Smyth, Salem numbers, Pisot numbers, Mahler measure, and graphs, *Experiment. Math.* **14** (2005), 211–229.

[100] ———, ———, Integer symmetric matrices of small index and small Mahler measure, Preprint.

[101] K. Mahler, On some inequalities for polynomials in several variables, *J. London Math. Soc.* **37** (1962), 341–344.

[102] ———, A remark on a paper of mine on polynomials, *Illinois J. Math.* **8** (1964), 1–4.

[103] ———, An inequality for the discriminant of a polynomial, *Michigan Math. J.* **11** (1964), 257–262.

[104] D.W. Masser, Counting points of small height on elliptic curves, *Bull. Soc. Math. France* **117** (1989), 247–265.

[105] E.M. Matveev, On the cardinality of algebraic integers, (Russian) *Mat. Zametki* **49** (1991), no. 4, 152–154, translation in *Math. Notes* **49** (1991), no. 3–4, 437–438.

[106] ———, On a connection between the Mahler measure and the discriminant of algebraic numbers, (Russian) *Mat. Zametki* **59** (1996), no. 3, 415–420, 480, translation in *Math. Notes* **59** (1996), no. 3–4, 293–297

[107] ———, On the successive minima of the extended logarithmic height of algebraic numbers, (Russian) *Mat. Sb.* **190** (1999), no. 3, 89–108, translation in *Sb. Math.* **190** (1999), no. 3–4, 407–425.

[108] M. Meyer, Le problème de Lehmer: méthode de Dobrowolski et lemme de Siegel "à la Bombieri-Vaaler", *Publ. Math. Univ. P. et M. Curie (Paris VI), Problèmes Diophantiens* , **90** (1988-89), no. 6, 15 pp.

[109] M. Mignotte, Entiers algébriques dont les conjugués sont proches du cercle unité, *Séminaire Delange-Pisot-Poitou, 19e année: 1977/78, Théorie des nombres*, Fasc. 2, Exp. No. 39, 6 pp., Secrétariat Math., Paris, 1978.

[110] ———, Approximation des nombres algébriques par des nombres algébriques › grand degré, *Ann. Fac. Sci. Toulouse Math.* (5) **1** (1979), 165–170.

[111] ———, Sur les nombres algébriques de petite mesure, *Mathematics*, pp. 65–80 CTHS: Bull. Sec. Sci., III, Bib. Nat., Paris, 1981.

[112] ———, Sur un théorème de M. Langevin, *Acta Arith.* **5** (1989), 81–86.

[113] M. Mignotte, M. Waldschmidt, On algebraic numbers of small height: linear forms in one logarithm, *J. Number Theory* **4** (1994), 43–62.

[114] M.J. Mossinghoff, Polynomials with small Mahler measure, *Math. Comp.* **67** (1998), 1697–1705, S11–S14.

[115] ———, *Tables of polynomials of small Mahler measure.*
http://www.cecm.sfu.ca/~mjm/Lehmer/lists/SalemList.html

[116] C. Notari, Sur le produit des conjugués à l'extérieur du cercle unité d'un nombre algébrique, *C. R. Acad. Sci. Paris Sér. A–B* **286** (1978), A313–A315.

[117] A.M. Ostrowski, On an inequality of J. Vicente Gonçalves, *Univ. Lisboa Revista Fac. Ci. A* (2) **8** (1960), 115–119.

[118] H. Oyanagi, q-analogues of Mahler measures, *J. Ramanujan Math. Soc.* **19** (2004), 203–212.

[119] L. Panaitopol, Minorations pour les mesures de Mahler de certains polynômes particuliers, *J. Théor. Nombres Bordeaux* **12** (2000), 127–132.

[120] T.A. Pierce, The numerical factors of the arithmetical functions $\prod_{i=1}^{n}(1 - \alpha_i^m)$, *Ann. of Math.* **18** (1916-17), 53–64.

[121] C.G. Pinner, J.D. Vaaler, The number of irreducible factors of a polynomial. I, *Trans. Amer. Math. Soc.* **339** (1993), 809–834; II, *Acta Arith.* **78** (1996), 125–142; III, *Number theory in progress*, Vol. 1 (Zakopane-Kościelisko, 1997), 395–405, de Gruyter, Berlin, 1999.

[122] I.E. Pritsker, An areal analog of Mahler's measure, *Illinois J. Math.* (to appear).

[123] I.E. Pritsker, Polynomial inequalities, Mahler's measure, and multipliers, *this volume*, 255–76.

[124] Q.I. Rahman, G. Schmeisser, *Analytic theory of polynomials*, London Mathematical Society Monographs. New Series **26**. The Clarendon Press, Oxford University Press, Oxford, 2002.

[125] U. Rausch, On a theorem of Dobrowolski about the product of conjugate numbers, *Colloq. Math.* **50** (1985), 137–142.

[126] K. Reidemeister, *Knot theory*, Transl. from the German and ed. by Leo F. Boron, Charles O. Christenson, and Bryan A. Smith. (English) Moscow, Idaho, USA: BCS Associates. XV, 143 p. (1983), German: *Knotentheorie*, Reprint of 1932 original, Berlin-Heidelberg-New York: Springer-Verlag. VI, 74 S. (1974).

[127] G. Rhin, A generalization of a theorem of Schinzel, *Colloq. Math.* **101** (2004), 155–159.

[128] G. Rhin, J.-M. Sac-Épée, New methods providing high degree polynomials with small Mahler measure, *Experiment. Math.* **12** (2003), 457-461.

[129] G. Rhin, C. Smyth, On the absolute Mahler measure of polynomials having all zeros in a sector, *Math. Comp.* **64** (1995), 295–304.

[130] ———, ———, On the Mahler measure of the composition of two polynomials, *Acta Arith.* **79** (1997), 239–247.

[131] G. Rhin, Q. Wu, On the absolute Mahler measure of polynomials having all zeros in a sector. II, *Math. Comp.* **74** (2005), 383–388.

[132] ———, ———, On the smallest value of the maximal modulus of an algebraic integer, *Math. Comp.* **76** (2007), 1025–1038.

[133] R. Salem, A remarkable class of algebraic integers. Proof of a conjecture of Vijayaraghavan, *Duke Math. J.* **11** (1944), 103–108.

Algebraic numbers and Fourier analysis, D.C. Heath and Co., Boston, Mass.

Samuels, Lower bounds on the projective heights of algebraic points, *Acta Arith.* (2006), 41-50.

Schinzel, Reducibility of lacunary polynomials. I, *Acta Arith.* **16** (1969/1970), 123-159.

———, On the product of the conjugates outside the unit circle of an algebraic number, *Acta Arith.* **24** (1973), 385-399; Addendum: **26** (1974/75), 329-331.

[138] ———, On the number of irreducible factors of a polynomial, *Topics in number theory (Proc. Colloq., Debrecen, 1974)*, pp. 305-314. Colloq. Math. Soc. Janos Bolyai, Vol. 13, North-Holland, Amsterdam, 1976.

[139] ———, *Selected topics on polynomials*, University of Michigan Press, Ann Arbor, Mich., 1982.

[140] ———, On the number of irreducible factors of a polynomial. II, *Ann. Polon. Math.* **42** (1983), 309-320.

[141] ———, The Mahler measure of polynomials, *Number theory and its applications* (Ankara, 1996), 171-183, Lecture Notes in Pure and Appl. Math. **204**, Dekker, New York, 1999.

[142] ———, *Polynomials with special regard to reducibility*, Encyclopedia of Mathematics and its Applications **77**. Cambridge University Press, Cambridge, 2000.

[143] ———, On values of the Mahler measure in a quadratic field (solution of a problem of Dixon and Dubickas), *Acta Arith.* **113** (2004), 401-408.

[144] A. Schinzel, H. Zassenhaus, A refinement of two theorems of Kronecker, *Michigan Math. J.* **12** (1965), 81-85.

[145] W.M. Schmidt, Heights of points on subvarieties of G_m^n, *Number theory* (Paris, 1993-1994), 157-187, London Math. Soc. Lecture Note Ser. **235**, Cambridge Univ. Press, Cambridge, 1996.

[146] I. Schur, Über Potenzreihen, die im Innern des Einheitskreises beschränkt sind. I, II, *J. Reine Angew. Math.* **147** (1917), 205-232; **148** (1918), 122-145; English translation: On power series which are bounded in the interior of the unit circle. I, II. *Methods in operator theory and signal processing: Oper. Theory, Adv. Appl.* **18** (1986), 31-59, 68-88.

[147] C.L. Siegel, Algebraic integers whose conjugates lie in the unit circle, *Duke Math. J.* **11**, (1944), 597-602.

[148] D.S. Silver, S.G. Williams, Mahler measure, links and homology growth, *Topology* **41** (2002), 979-991.

[149] ———, ———, Mahler measure of Alexander polynomials, *J. London Math. Soc.* (2) **69** (2004), 767-782.

[150] J.H. Silverman, Small Salem numbers, exceptional units, and Lehmer's conjecture, Symposium on Diophantine Problems (Boulder, CO, 1994). *Rocky Mountain J. Math.* **26** (1996), 1099-1114.

[151] ———, A lower bound for the canonical height on elliptic curves over abelian extensions, *J. Number Theory* 104 (2004), 353-372.

[152] ———, *The arithmetic of dynamical systems*, Springer Verlag (to appear).

[153] C. Sinclair, The range of multiplicative functions on $\mathbb{R}[x]$, $\mathbb{C}[x]$ and $\mathbb{Z}[x]$. Preprint 2007.

[154] H.-P. Skoruppa, *Heights*, Graduate lecture course, Bordeaux 1999, http://wotan.algebra.math.uni-siegen.de/~countnumber/D/

[155] C.J. Smyth, On the product of the conjugates outside the unit circle of an algebraic integer, *Bull. London Math. Soc.* **3** (1971), 169-175.

[156] ———, *Topics in the theory of numbers*, PhD thesis, Univ. of Cambridge 1972.

[157] _____, On the measure of totally real algebraic integers. I, *J. Austral. Math. S Ser. A* **30** (1980/81), 137–149; II, *Math. Comp.* **37** (1981), 205–208.

[158] C.L. Stewart, Algebraic integers whose conjugates lie near the unit circle, *Bull. So Math. France* **106** (1978), 169–176.

[159] _____, On a theorem of Kronecker and a related question of Lehmer, *Séminaire de Théorie des Nombres 1977–1978*, Exp. No. 7, 11 pp., CNRS, Talence, 1978.

[160] J.D. Vaaler, Mahler's measure, *Pacific Institute for the Mathematical Sciences Magazine* **7** (2003), no. 2, 30–34.

[161] A.J. Van der Poorten, J.H. Loxton, Multiplicative relations in number fields, *Bull. Austral. Math. Soc.* **16** (1977), 83–98.

[162] J. Vicente Gonçalves, L'inégalité de W. Specht, *Univ. Lisboa. Revista Fac. Ci. A. Ci. Mat.* (2) **1** (1950), 167–171.

[163] F. Rodriguez Villegas, Modular Mahler measures. I, *Topics in number theory (University Park, PA, 1997)*, 17–48, Math. Appl. **467**, Kluwer Acad. Publ., Dordrecht, 1999.

[164] P. Voutier, An effective lower bound for the height of algebraic numbers, *Acta Arith.* **74** (1996), 81–95.

[165] M. Waldschmidt, Nombres transcendants et groupes algébriques, *Astérisque* **69–70**. Société Mathématique de France, Paris, 1979. 218 pp.

[166] _____, A lower bound for linear forms in logarithms, *Acta Arith.* **37** (1980), 257–283.

[167] _____, Sur le produit des conjugués extérieurs au cercle unité d'un entier algébrique, *Enseign. Math.* (2) **26** (1980), 201–209 (1981).

[168] _____, Minorations de combinaisons linéaires de logarithmes de nombres algébriques, *Canad. J. Math.* **45** (1993), 176–224.

[169] _____, Diophantine approximation on linear algebraic groups. Transcendence properties of the exponential function in several variables, Grundlehren der Mathematischen Wissenschaften **326**. Springer-Verlag, Berlin, 2000.

[170] E. Weiss, *Algebraic number theory*, McGraw-Hill Book Co., Inc., New York-San Francisco-Toronto-London 1963.

[171] D. Zagier, Algebraic numbers close to both 0 and 1, *Math. Comp.* **61** (1993), 485–491.

[172] T. Zaïmi, Minoration d'un produit pondéré des conjugués d'un entier algébrique totalement réel, *C. R. Acad. Sci. Paris Sér. I Math.* **318** (1994), 1–4.

[173] S. Zhang, Positive line bundles on arithmetic surfaces, *Ann. of Math.* (2) **136** (1992), 569–587.

Printed in the United States
by Baker & Taylor Publisher Services

Printed in the United States
by Baker & Taylor Publisher Services